Inhaltsverzeichnis.
Contents. — Table des matières.

FORTSCHRITTE DER CHEMIE ORGANISCHER NATURSTOFFE

PROGRESS IN THE CHEMISTRY OF ORGANIC NATURAL PRODUCTS

PROGRÈS DANS LA CHIMIE DES SUBSTANCES ORGANIQUES NATURELLES

HERAUSGEGEBEN VON   EDITED BY   RÉDIGÉ PAR

L. ZECHMEISTER
CALIFORNIA INSTITUTE OF TECHNOLOGY, PASADENA

DREIUNDZWANZIGSTER BAND
TWENTY-THIRD VOLUME   VINGT-TROISIÈME VOLUME

VERFASSER   AUTHORS   AUTEURS
J. ENGEL · W. GRASSMANN · K. HANNIG · H. HÖRMANN · L. M. JACK-
MAN · R. T. JONES · K. KÜHN · A. NORDWIG · S. PEAT · F. SCHLENK ·
H. H. SCHLUBACH · W. A. SCHROEDER · J. R. TURVEY

MIT 58 ABBILDUNGEN   WITH 58 FIGURES   AVEC 58 ILLUSTRATIONS

1965

WIEN · SPRINGER-VERLAG · NEW YORK

© 1965 BY SPRINGER-VERLAG / WIEN
SOFTCOVER REPRINT OF THE HARDCOVER 1ST EDITION 1965

LIBRARY OF CONGRESS CATALOG CARD NUMBER AC 39-1015

ISBN 978-3-7091-7141-7   ISBN 978-3-7091-7139-4 (eBook)
DOI 10.1007/978-3-7091-7139-4

Titel Nr. 8230

Some Applications of Nuclear Magnetic Resonance Spectroscopy in Natural Product Chemistry. By L. M. JACKMAN, Department of Chemistry, University of Melbourne, Australia

Polysaccharides of Marine Algae.

By STANLEY PEAT and J. R. TURVEY, Bangor, North Wales.

Contents.

I. Introduction.

The seaweeds, that is, the macroscopic marine algae, consist of a group of photosynthetic plants, in which it is believed that evolution has not proceeded as far as in land plants. In general, the morphology of algae is much simpler than that of land plants. There is a more limited diversification of cells according to function and, indeed, many unicellular algae are known which are related botanically to the seaweeds. Insofar as the metabolism of the algae has been studied, it appears to resemble in many respects that of land plants, although the metabolic end-products may differ (44).

We are concerned in this article primarily with the metabolism of carbohydrates, and one important way in which algae appear to differ from land plants is in the type and variety of seaweed polysaccharides, some of which have no counterpart on land.

For example, although one class of algae (Chlorophyceae) contains polysaccharides which, in chemical structure and in location in the plant are closely allied to the characteristic polysaccharides of land plants such as starch and cellulose, yet the Chlorophyceae contain, in addition, polysaccharides which appear to have no representative in land plants. Thus, the major polysaccharide constituents of the red and brown marine algae are not found in land plants, and these are perhaps most strikingly exemplified by the sulphated polysaccharides which occur in the three algal classes constituting the bulk of the seaweeds.

The functions in seaweeds of the sulphated polysaccharides are, as yet, incompletely defined but two obvious functions are related to their ion-exchange properties on the one hand, and to their ability to form gels on the other. Seaweeds live in a saline environment, with sodium as the principal cation present. Clearly, the algal cells must possess a mechanism for ion-exchange with their environment in order that essential cations, such as potassium and calcium, may be selectively absorbed. That some sulphated polysaccharides of algae can function in this way has been demonstrated (40). The gelling and water-retaining properties of these polysaccharides must also contribute to the physical security of the alga and help to avoid desiccation of the algal cells, when the seaweed is exposed at low tide.

The most extensively studied of the algal polysaccharides are those that occur in the macroscopic seaweeds, which generally belong to one of the three algal classes, Chlorophyceae (green algae), Rhodophyceae (red algae) and Phaeophyceae (brown algae). Other classes, such as Myxophyceae (blue-green algae), Xanthophyceae (yellow-green algae), and the many, unicellular species of yet other classes, have been less extensively studied and will be omitted from this article, largely for the sake of brevity. In general, each of the three major algal classes has its characteristic polysaccharides although in certain species one or more of these polysaccharides may be absent or be replaced by others.

A few of the algal polysaccharides are of commercial importance and the structural chemistry of these has received closer study. Alginic acid and the galactans of red algae, such as agar and carrageenan, are examples of such commercial products. In other cases, little or nothing is yet known of the detailed chemical structures. Indeed, many of the polysaccharides studied in the past are, almost certainly, mixtures of two or more chemical species. This aspect of the study of polysaccharides presents one of the most challenging problems in the field of naturally-occurring polymers. By what criteria can one ensure that a given preparation of a polysaccharide contains only one chemical species? It is generally assumed that when repeated attempts at fractionation fail to yield fractions which differ significantly in physical properties or in chemical constitution, then the polysaccharide is chemically homogeneous.

Certain aspects of this assumption deserve comment. In the first place, new methods for fractionating polymers are constantly being developed and these methods may often lead to the separation of fractions which are structurally different although earlier methods had appeared to show homogeneity. Secondly, most naturally-occurring polymers consist of molecules covering a range of molecular weights and even having small variations in the content of a given structural feature. A sufficiently precise method of fractionation may well pick out variations of this sort and yet, from the chemists viewpoint, it is convenient to regard such a molecular spectrum as constituting a single polysaccharide species. On the other hand, the industrial user may be profoundly interested in minor variations of this sort, since they may affect the physical properties of the polysaccharide.

Since this account is concerned primarily with the chemical structure of certain of these polysaccharides, but little reference will be made to their physico-chemical properties or to their industrial utilisation. For more detailed treatments of these aspects, several excellent texts are available (*126*, *133*). The authors do, however, feel that the relationship of algal polysaccharides to one another and to the polysaccharides of land plants may be of some interest from the evolutionary point of view. Where there are obvious similarities, either in chemical structure or in function, these will be discussed. It is therefore relevant to consider the location within the seaweed of the various types of polysaccharides. Broadly speaking, polysaccharides occur at three main sites, and fulfil three main functional roles: (a) as relatively insoluble, cell-wall material, often fibrous in character, (b) as hot-water soluble mucilages, which act as a cement for the cell-wall materials and which may also be present as a protective, intercellular material, and (c) as intracellular food reserves. As with all classifications, this may be an over-simplification, and some polysaccharides may belong to more than one category, or may not fit readily into any.

II. Skeletal Polysaccharides of the Algal Cell-Wall.

The principal skeletal material of most land plants is cellulose, which occurs in a fibrous form in the cell-wall. The fibres consist of aggregates of partially crystalline bundles (fibrils), which in turn are constituted of largely parallel chains of glucose residues, the latter being glycosidically joined by β-1,4-links which confer a rectilinear structure on the macromolecule.

It is unfortunate but inevitable that the term "cellulose" has different connotations in different contexts. This is mainly due to the emphasis on physical characteristics, such as microscopic appearance, degree of crystallinity, X-ray diffraction pattern, average length of fibrils, and even solubility. In certain circumstances, it has been found convenient (by paper manufacturers, in particular) to define cellulose in terms of a purely arbitrary procedure for isolation from plant material. For example, "α-cellulose" is the residue obtained after plant material has been submitted to a rigidly defined extraction procedure. Other products of this method of extraction are named β- and γ-cellulose, and are generally held to be hydrolytically degraded α-cellulose. As might be expected this residue, α-cellulose, can, and often does, contain other polysaccharides in admixture with the β-1,4-glucan (of high molecular weight) which is the chemist's conception of "cellulose". Polysaccharides which may be intimately associated with the glucan in the microfibrillar structure of the plant cell-wall are termed "hemicelluloses", or are designated according to the monosaccharide units of which they are composed, e. g. xylans, glucomannans.

The examination of cellular structures by X-ray diffraction has yielded information of the first importance in regard to cell-wall morphology, and the elegant studies of Preston et al. (35, 87, 88, 94) in particular, have demonstrated the role of cellulose and cellulose-like constituents. Here again it must be clearly recognised that "cellulose" refers to an entity which displays a characteristic X-ray diffraction pattern, and a specialised nomenclature has grown up which classifies the physical forms of cellulose known to the X-ray crystallographer. Cellulose I, in these terms, gives a characteristic and well-defined X-ray diffraction pattern, has a high degree (70%) of crystallinity, shows typical staining reactions, and is hydrolysed by acid to glucose alone. Cellulose I is believed to be the physical form normally found in the higher plants. Cellulose II gives a different, but also well-defined, diffraction pattern. It appears during the mercerisation of cotton and is thus to be regarded as an alkali-modified cellulose. The modification finds expression also in the microfibrillar structure and this is explicable on the view that cellulose II has a lower degree of crystallinity than has cellulose I. Cellulose IV, prepared by the action of boiling glycerol on cellulose II, also shows a characteristic diffraction pattern but these modifications of cellulose I are of importance here only because their formation draws attention to the fact that chemical treatment of a native cellulose may alter its X-ray diffraction pattern and, hence, that the apparently-characterising pattern observed with a chemically-cleaned specimen of cell-wall is not necessarily proof of the existence in the original cell-wall of the cellulose so identified (87).

Furthermore, the observation of any one of these diffraction patterns cannot, per se, be taken as definitive evidence of the presence of a cellulose. Thus, it has been shown that a residue prepared from the red seaweed, *Porphyra umbilicalis*, under the specific experimental conditions prescribed for the isolation of "α-cellulose", gives a pattern closely resembling that of cellulose II, and yet this preparation consists largely of a mannan (88).

1. Cellulose in Algae.

Many early investigators were concerned with establishing whether the algal cell-wall contains cellulose and, usually, its typical staining reactions and its solubility in cuprammonium solution were used as criteria for its presence. In general, the examination was applied to the so-called cellulose isolated from algae by the "α-cellulose" procedure already mentioned. In 1930, SPONSLER (127) examined such a fraction from the green alga, *Valonia*, and showed that it gave an X-ray diffraction pattern typical of cellulose I. Later investigators (92) examined the crude fibre from a number of algal cell-walls (both red and green algae) and, on the basis of staining reactions, solubility in cuprammonium solution, and formation of soluble acetyl derivatives, concluded that many red and brown algae contained cellulose in amounts ranging from 2 to 15%.

The first significant chemical evidence of the existence of cellulose in algal cell-walls came from studies of cell-wall preparations from two *Laminaria* species and *Fucus vesiculosus* by PERCIVAL and ROSS (114). Standard procedures were used to isolate the fibrous cell-wall, which was then purified by precipitation from its cuprammonium solution. The products gave typical staining reactions and also furnished glucose (in 80% yield) on hydrolysis. Acetolysis gave cellobiose octaacetate, and on periodate oxidation, the fractions were oxidised to the same extent as the celluloses of land plants. From the periodate oxidation data it was also concluded that the average length of the cellulose chains was 160 glucose units. The X-ray diffraction pattern of one of these preparations was described as being characteristic of normal cellulose (presumably cellulose I). Certain fresh-water algae have also been shown to contain cellulosic material of chain length greater than 100 glucose units and which, on methylation and hydrolysis, gives predominantly 2,3,6-tri-O-methyl-D-glucose, thus establishing the presence of a β-1,4-linked glucan (59).

PRESTON and his co-workers have used the techniques of X-ray diffraction and, later, of electron microscopy, to study the nature of the algal cell-wall. On the basis of X-ray diffraction patterns, NICOLAI and PRESTON (94) distinguished three groups in the filamentous green algae. The first group (Group 1), consisting of *Valonia, Chaetomorpha melagonium* and all the Cladophorales except the *Spongomorpha* group, contain in their cell-walls a highly crystalline glucan of the cellulose I type. The cellulose is present in a crossed fibrillar structure and on being hydrolysed it yields glucose as the sole monosaccharide unit. Group 2, typified by *Halicystis*, contains a cellulose with a lower degree of crystallinity and no marked orientation in the fibrillar structure. The principle spacings in the X-ray diffraction pattern approximate to those of cellulose II and, furthermore, treatment of the cellulose with glycerol converts it into a form showing the cellulose IV pattern. In Group 3, which includes among other orders the Chlorococcales and the Ulotrichales, the skeletal material is well orientated and highly crystalline but gives a pattern unlike that of either cellulose I or cellulose II. It is concluded that this may not be a β-1,4-glucan.

The majority of the red algae show crystalline, microfibrillar structures in the cell-wall, which all give X-ray diffraction patterns similar to that observed with Group 2 of the green algae (88). On being hydrolysed, the cell-wall material of some of these red seaweeds gives mixtures of monosaccharides in which nevertheless glucose usually predominates. For example, the microfibrils of *Rhodymenia palmata* yield, when hydrolysed, approximately equal quantities of xylose and glucose (87). On the other hand, the polysaccharides of *Porphyra umbilicalis* microfibrils are composed of mannose units only. These microfibrils are readily eroded by treatment with alkali. In the brown algae, the cell-wall microfibrils show some degree of crystallinity but the X-ray diffraction pattern is not that of cellulose I (35). Chemical treatment of the microfibrils, as for instance the standard method for the preparation of an α-cellulose fraction, may result in some modification of the physical form and the appearance of a cellulose I diffraction pattern.

2. Mannans.

As previously mentioned, the cell-wall microfibrils of certain algae appear to consist of mannans, or of complex polysaccharides containing xylose and other sugars as well as glucose. In many of these cases, the cell-wall material is not so insoluble in mild reagents as is cellulose and the polysaccharides may be extracted by dilute alkali or even by hot water. There would appear in such cell material to be a range of polysaccharides differing mainly in molecular weight. The longer chains act as fibres but the smaller chains are present as an amorphous matrix which is partly soluble in dilute alkali, and the shortest chains may even be extracted by hot water. In *Porphyra umbilicalis*, for example, a small quantity of mannan is extractable by cold water, more by dilute alkali, and the major portion is only solubilised by hot, concentrated alkali. Whether there are minor structural differences between these mannan fractions has not yet been established, but it is believed that they are all structurally similar and differ only in molecular size. For this reason it is convenient to consider the mannans and xylans of algae as skeletal polysaccharides although a portion of the polysaccharide may be present as a water-soluble mucilage rather than as characteristic fibres.

In red algae, the most completely studied mannan is that from *Porphyra umbilicalis*. When the alga was extracted exhaustively with water to remove soluble polysaccharides, the residue contained, among other polysaccharides, a mannan which was extracted with dilute alkali and purified by formation of the insoluble copper complex (*64*). Acidification of the complex regenerated the mannan as a fibrous material, which was then insoluble even in strong alkali. Methylation of the mannan, followed by hydrolysis with formic acid, afforded 2,3,4,6-tetra-O-methyl- and 2,3,6-tri-O-methyl-*D*-mannose in proportions suggesting one non-reducing chain end for each 12–13 mannose units, and a structure containing only 1,4-linkages. On periodate oxidation, all the mannose units were destroyed and hence no 1,3-linkages could be present. A low optical rotation for the mannan suggested a predominance of linkages having the β-configuration. No evidence for branching in the molecule was obtained other than the fact that the polysaccharide is insoluble, which suggests a degree of polymerisation much greater than the value of 12–13 units suggested by the evidence from methylation.

In the green alga, *Codium fragile*, over 20% of the dry weight of the alga consists of a mannan, much of which can be solubilised by 20% sodium hydroxide (*61, 70*). After removal of this fraction, the residual polysaccharide in the weed contained only mannose units, suggesting that, in this alga, mannan replaces cellulose as the principal skeletal material. The alkali-extracted mannan had a low optical rotation and, on oxidation with periodate, consumed 0.8 mol. of periodate per mannose unit, thus suggesting an essentially β-1,4-linked mannan, which would consume 1.0 mol. of periodate. Confirmation came from a study of the oligosaccharides produced by the action of a fungal hemicellulase preparation on the mannan. In addition to mannose, β-1,4-linked manno-biose and mannotriose were identified (70).

In contrast with the algae, land plants rarely synthesize a mannan containing solely mannose units, but it is perhaps significant that, where mannose does occur, it is combined by β-1,4-linkages. The extremely acid-stable mannan in the seeds of vegetable ivory has chains of mannose units linked in this way, but it also has galactose units attached as side chains. In the glucomannans of various trees, units of glucose and mannose occur joined by β-1,4-linkages.

3. Xylans.

Polysaccharides consisting predominantly of xylose units are wide-spread throughout the algae. Nevertheless, in only a few cases have the structures been investigated. For instance, the green algae have for long been known to contain cell-wall polysaccharides, which, on being hydrolysed, liberate xylose, among other sugars (35, 120, 122), but only from *Caulerpa filiformis* has a pure xylan been obtained (74). After removal of water-soluble polysaccharides from this alga, a mild chlorite treatment of the residue was followed by extraction with sodium hydroxide. This extract contained a crude xylan which, after purification, gave xylose only on hydrolysis and had $[\alpha]_D — 35°$, suggesting a predominance of β-linkages. Periodate effected but little oxidation of the xylan and, when the xylan which had been so treated was hydrolysed, most of the xylose was recovered unchanged. This resistance to oxidation would be expected if 1,3-linkages predominate. Methylation, followed by hydrolysis of the methylated xylan gave the expected 2,4-di-O-methyl-D-xylose in a molar percentage of 95.2 and an amount of trimethyl xylose which corresponded to one non-reducing end group per 34 xylose residues. From this evidence it is concluded that the *Caulerpa* xylan consists of an essentially unbranched chain of 1,3-linked β-D-xylopyranose units. Other green algae are known to contain xylans in the cell-wall fibres (62). Extraction of these fibres with dilute alkali yielded poly-saccharides containing xylose as the principal sugar unit, and the resistance of these products to oxidation with periodate again suggests a pre-ponderance of 1,3-linkages.

In red algae xylose is found frequently in hydrolysates of the poly-saccharides, the proportions in which it is present varying from mere traces to its being, as xylan, the major polysaccharide constituent, as in the species *Rhodymenia palmata* ("dulse") (15, 113). Whether a polysaccharide, which is extracted as readily with water as is the xylan

of dulse, should be regarded as a skeletal polysaccharide, is open to
question. Certainly, hydrolysis of the cell-wall microfibrils of dulse yields
xylose as well as glucose (in equal quantities), but whether the xylose
thus shown to be present in the cell-wall is the building unit of a polysaccharide identical with the xylan extractable from the weed by cold
water, or whether in fact the two xylans are different has not yet been
settled. For convenience, therefore, the water-soluble xylan from dulse
will be discussed under this heading although it may be more correctly
described as a mucilage or even a food-reserve polysaccharide.

Methylation of the xylan from dulse, followed by hydrolysis, afforded
2,3,4-tri-O-methyl-, 2,4-di- and 2,3-di-O-methyl-, and 2-O-methyl-
D-xylose. This evidence suggests that the xylan is branched and contains
both 1,3- and 1,4-linkages. On being oxidised by periodate, 80% of the
xylose units of the polysaccharide were attacked by the reagent and
must, therefore, be solely 1,4-linked (*113*). The molecular size of the
xylan was placed at about 40 xylose units with, on average, one branch
point in each molecule. That the two types of linkage (1,3- and 1,4-)
are present in the same molecule and that the xylan of dulse is not a
mixture of structurally different xylans, was demonstrated by HOWARD (*60*)
Fragmentation of the xylan by a bacterium isolated from sheep rumen
furnished a trisaccharide and a tetrasaccharide having the structures (**1**)
and (**2**), respectively *(Chart 1)*.

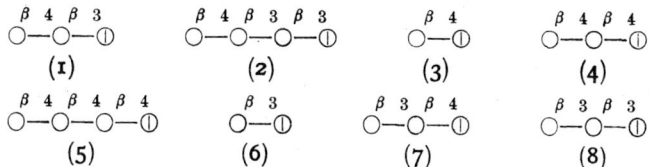

KEY: —○— xylopyranosyl unit. —① reducing end unit.

Chart 1. Oligosaccharides from the xylan of dulse.

Further evidence on the structure of this polysaccharide has come
from a study of its periodate oxidation, which has indicated that a
small proportion of the 1,3-links in the molecule are adjacent to one
another, the majority, however, being flanked by 1,4-links. Confirmation
comes from the isolation of oligosaccharides (**3** to **8**) some of which contain
contiguous 1,4-linkages (**4** and **5**) but only one (**8**) containing contiguous
1,3-linkages (and this was obtained in trace quantity only) when the
xylan was fragmented with an enzyme from *Myrothecium* (*81*).

A polysaccharide fraction from the red alga *Nemalion multifidum*
has been reported to contain xylose, mannose and ester sulphate although
it has not been established whether these units are present in one and
the same polysaccharide. From a study of the hydrolysis products of

the methylated polysaccharide it was concluded (96) that both 1,3- and 1,4-linked xylose units are present, the sulphate groups being attached to mannose units. Many other red algae contain xylans which are probably skeletal polysaccharides but the only available evidence on their structures suggests that 1,4-linkages predominate.

4. Conclusions.

In considering the skeletal polysaccharides of seaweeds and their relationship to those of land plants, it must be borne in mind that the environment of a plant is of particular importance. In seaweeds, it is imperative that the structure be pliant and flexible if the weed is not to be broken by wave action. Land plants, on the other hand, require some rigidity of structure and one would expect, therefore, that, on land, the main structures of the plant will have considerable deposits of cellulose. In seaweeds, it is only in the individual cell-walls that rigidity is desirable since an unbroken envelope for the cell contents must be maintained. In keeping with this conception, it is found that seaweeds contain a much smaller proportion of cellulose than do land plants. Furthermore, the characteristic properties of cellulose I, such as a high degree of crystallinity and very long chains, would contribute a rigidity which would be detrimental to a sea plant. Indeed, any relatively insoluble and fibrous material would probably suffice.

Although cellulose I is found in algae, it has hitherto been detected only in some of the Chlorophyceae. In the rest of the green seaweeds, and in the red and brown seaweeds, where cellulose does occur it is less well orientated and, probably, less crystalline than cellulose I. In some species we see that cellulose can be replaced by a mannan or xylan provided these are sufficiently insoluble. Where mannans occur, they are predominantly β-1,4-linked and in consequence linear, thus forming a relatively insoluble and fibrous structure. The only known analogies with mannans of land plants are with (a) the mannan of vegetable ivory; this mannan, however, contains a small proportion of sugar residues other than β-1,4-linked mannose; with (b) the glucomannans of certain trees and tubers, in which both β-1,4-linked mannose and glucose units occur, and with (c) the galactomannans of legume seeds, which have the β-1,4-linked mannose backbone but have side branches of galactose residues. The skeletal xylans in algae differ from those of land plants in respect of the modes of polymerisation. For example, esparto xylan, and most other land plant xylans have mainly β-1,4-links, whereas in *Caulerpa* xylan the linkage is β-1,3. In red algae, both linkage types occur, but the 1,4-linkage usually predominates. It will be interesting to see if the brown algae synthesise xylans similar to either the xylan of *Caulerpa* or that of the red algae.

References, pp. 39—45.

III. Food Reserve Polysaccharides.

Land plants usually store carbohydrates in the form of starch, which occurs as characteristically shaped granules in tubers and seeds. In addition, certain species of maize lay down large deposits of a polysaccharide, phytoglycogen, which is allied in structure to the glycogen of animal tissue. The Compositae, however, store carbohydrate in the form of a fructosan, inulin, rather than as starch. Certain algae synthesize starch-type polysaccharides but, so far, no fructose polymers have been recognised in algae. The starch of land plants generally consists of a mixture of the two fractions, amylose and amylopectin. Amylose is an essentially linear polymer of α-1,4-linked glucopyranose units with a chain length of 200–300 units. Amylopectin, on the other hand, is constituted of congeries of short chains (average 20–25 units in length) of α-1,4-linked glucopyranose residues, these chains being mutually joined by α-1,6-linkages to form a randomly branched structure with a degree of polymerisation of several thousand glucose units. Phytoglycogen resembles amylopectin in general structure but is even more highly ramified, and it does not occur as granules.

1. Starch (Green Algae).

The early literature contains many reports of the occurrence of starch-like polysaccharides in green algae but only recently has a study been made of purified starches from this source. MACKIE and PERCIVAL (75) obtained a mixture of polysaccharides from *Caulerpa filiformis* and succeeded in separating from it a pure glucan. The glucan contained 99% of *D*-glucose, had $[\alpha]_D + 154°$ and stained purple with iodine. It was also hydrolysed by α- and β-amylases and by yeast isoamylase, an enzyme which debranches amylopectins and glycogens. Attempts to separate the starch into an amylose and an amylopectin fraction were unsuccessful. Evidence obtained by methylating and then hydrolysing the polysaccharide confirmed that it was an amylopectin-type molecule with 1,4-linked glucose units as main structure and some branching at position 6. The ratio of branch points to total glucose units corresponded to a branch-length* of 23. Evidence from periodate oxidation also favoured a structure in which the chain-forming linkages were 1,4-glucosidic, the branch linkages 1,6-glucosidic, and the branch-length averaged 21 units. The polysaccharide thus closely resembles the amylo-

---

* The "branch-length" of a polysaccharide is defined as the ratio of the total sugar units in the molecule to those units which constitute non-reducing chain ends. In the case of amylopectin-type polysaccharides it is a measure of the average length of each branch. An alternative term is "chain-length", and MANNERS (80) suggests the use of the symbol $\overline{CL}$ for this quantity.

pectin component of the starch of land plants. It has been suggested (69) that during the isolation and purification of this polysaccharide, some degradation may have occurred and that an amylose fraction may have been lost in this way.

In a recent study (69), starches were extracted from five species of green algae, were purified by precipitation as the iodine complexes, and then separated into amylose and amylopectin components. A comparison was made between these components and those obtained from potato starch. The starches from *Cladophora rupestris, Enteromorpha compressa*, and *Codium fragile* contained amylose and amylopectin in ratios similar to those of land-plant starches, and these components were closely related in most of their properties and enzymic degradation patterns to those from potato starch. Starch from *Ulva lactuca* had a higher proportions (37%) of amylose than the other starches (16–22%) but was otherwise similar. The polysaccharide obtained from *Chaetomorpha capillaris* had all the properties typical of an amylopectin, although it is possible that the conditions used in its extraction could have eliminated any amylose fraction originally present. A starch-type polysaccharide was obtained in an impure form from *Acrosiphonia centralis* but detailed examination of it has not been made (97).

The X-ray diffraction patterns given by the grains of a number of algal starches have been studied (83). In certain cases, the patterns obtained resembled those of cereal starches, while others showed the pattern characteristic of tuber starches. Certain algal starches gave a third pattern, which differed from these in that it showed only one well-defined diffraction line. The new pattern was interpreted as meaning that these starches have a low degree of molecular orientation, crystallisation being largely absent. In the first two starch types, some degree of crystallinity may exist and therefore the patterns characteristic of land-plant starches are observed. Investigation of the breakdown of granular starches from both red and green algae by amylolytic enzymes shows that, in general, algal starch grains are more readily attacked than are cereal or tuber starches, this being particularly evident with *Ulva* starch (84). Here again is an indication that algal starches have a lower degree of crystallinity than cereal or tuber starches so that penetration of the enzyme occurs more readily.

It is worthy of note that the fresh-water alga, *Nitella translucens* (Order, Characeae) contains a starch with a smaller proportion (12%) of amylose than is usually found; the amylopectin fraction resembles land-plant amylopectins (1). Many unicellular green algae are also known to contain starches and, in the few cases where a detailed comparison has been made, a close resemblance to land-plant starches was noted (24, 39).

2. Floridean Starch (Red Algae).

The characteristic reserve polysaccharide of the red algae (Rhodo-phyceae) is a type of starch which, because it was first recognised in members of the class Florideophyceae, was termed floridean starch. This occurs as small, characteristically-shaped granules in certain cells. The granules stain red-violet with iodine but the stain is less intense than that given by land-plant starches and is more akin to that shown by animal glycogens (65). KYLIN first isolated the grains of floridean starch from *Furcellaria fastigiata* and showed that it was a glucan, susceptible to hydrolysis by malt diastase, indicating a predominance of α-1,4-glucosidic linkages and hence a structural relationship to starch (65).

An examination of the starch from *Dilsea edulis* by the method of periodate oxidation appeared to indicate the presence of 1,3-linkages, since the starch was largely resistant to this reagent (17). The starch of *D. edulis* was re-examined by O'COLLA (95), who concluded that at least 50% of the linkages were α-1,4-glucosidic inasmuch as (i) maltose was obtained in this yield by β-amylolysis and (ii) methylation of the polysaccharide, followed by hydrolysis, gave 2,3,6-tri-O-methyl-*D*-glucose in 42% yield. That floridean starch resembled the amylopectin of land-plant starches in being a highly-branched molecule was convincingly shown by PEAT and his co-workers (103, 104), who isolated the starch from *D. edulis*. This preparation had an average branch-length of 15 glucose units, and was attacked by β-amylase, α-amylase, as well as by R-enzyme, which is known to attack amylopectins but not glycogens. It is possible that floridean starch contains a small proportion of α-1,3-linkages, in addition to the 1,4- and 1,6-linkages, since nigerose (3-O-α-*D*-gluco-pyranosyl-*D*-glucose) has been detected in partial acid and in enzymic hydrolysates of the starch (103, 104). A sample of floridean starch from *D. edulis*, which contained traces of a galactan impurity, was examined by HIRST and his colleagues (43) and they reported a branch-length of 9 glucose units. In a later study, during which this starch was further purified by ultracentrifugal sedimentation, the branch-length was reported as 19 units and a resemblance to amylopectin in certain physical properties was noted (48). It is not yet clear why these discrepancies occur in the values found for the branch-length.

A survey of other red algae as sources of granular floridean starch has been made (82) and a number of these starches have been examined for grain size, susceptibility to enzymic attack and for branch length (values found, 10–14 glucose units). In general, the grain size is smaller than that of land-plant starches and the grains are irregular in shape. The grains are also hydrolysed by amylolytic enzymes at a faster rate than are cereal or tuber starches, indicating a lower degree of crystallinity within the granule. Samples of floridean starch from four red algae have been compared by Peat and his co-workers (*106*). All the samples display similar colours with iodine, the absorption maxima being between 525 and 535 mμ, and show no wide divergencies of branch-lengths (10–13 glucose units).

An assessment of the available evidence indicates, therefore, that floridean starch is structurally similar to the amylopectin component of land-plant starches but has a somewhat shorter branch-length and gives a redder stain with iodine (λ_{max} at 530 mμ, compared with 540 mμ for amylopectins). Perhaps the biggest difference between floridean starch and land-plant starches lies in the absence of an amylose component in the former. In this it is perhaps more closely related to the starches of the waxy cereals.

3. Laminarin (Brown Algae).

Brown algae, unlike the red and green algae, do not synthesize starch-type polysaccharides; instead they store carbohydrate in the form of a glucan, laminarin. The fronds of most brown algae, in particular of the *Laminaria* group, may contain up to 50% of the dry-weight as laminarin. That this polysaccharide is a food reserve is shown by its large seasonal variation, this being at a maximum in autumn (*21*).

Some confusion arose in the past because laminarin exists in two forms, water-insoluble laminarin, which occurs in *L. cloustoni* and to a lesser extent in *L. saccharina*, and soluble laminarin, which is the main form in *L. digitata*. The chemical structures of the two forms appear to be very similar and they will be dealt with as one polysaccharide except where specific differences have been noted.

Insoluble laminarin is readily extracted from the macerated fronds by cold, dilute acid, from which it spontaneously precipitates on being kept (*13*), and the soluble form is obtained by precipitation with ethanol or basic lead acetate from a similar extract of the appropriate algal species (*115*). That both forms are glucans was early shown by the isolation from their acid hydrolysates of *D*-glucose in 99% yield. Partial hydrolysis by enzymes from snail juice or by oxalic acid gave, among other products, the disaccharide, laminaribiose (3-O-β-*D*-glucopyranosyl-*D*-glucose (9 in *Chart 2*), indicating the presence of β-1,3-linked glucose units (*14*). Early methylation studies (*13*) confirmed that a high proportion of 1,3-linkages was present, since in these experiments 2,4,6-tri-O-methyl-*D*-glucose was isolated on hydrolysis of methylated laminarin. Furthermore, the optical rotation of laminarin, $[\alpha]_D - 13°$, suggested that the glycosidic linkages had the β-configuration. Later, tetra-O-methyl-*D*-glucopyranose was isolated from hydrolysates of methylated laminarin in amounts which indicated the presence of one non-reducing end group per 20 glucose units, i. e. a unit chain length of 20 for both the soluble and insoluble forms (*31*). Further confirmation of this basic structure came from the observation that laminarin is largely resistant to periodate oxidation, a characteristic of 1,3-linked hexosans.

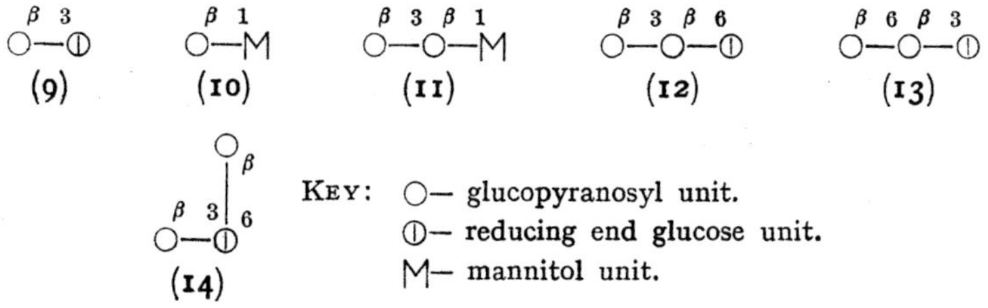

Chart 2. Oligosaccharides from Laminarin.

A more detailed structure for laminarin came from a study of the products obtained by partial acid hydrolysis of the polysaccharide (*107, 108*). A curious observation was then made, namely that in addition to glucose, the monosaccharide fraction contained *D*-mannitol, which was shown to be an integral part of the laminarin molecule by the isolation from the oligosaccharide fractions of 1-O-β-*D*-glucopyranosyl-*D*-mannitol (**10**), 1-O-β-laminaribiosyl-*D*-mannitol (**11**), and higher oligosaccharides, which were also terminated at the reducing end by mannitol. The disaccharide fraction of the laminarin hydrolysate contained, in addition to laminaribiose, small quantities of gentiobiose, suggesting the presence of β-1,6-linkages as well as the predominant β-1,3-linkages. Confirmation came from the isolation (from the trisaccharide fraction) of the two trisaccharides, 6-O-β-laminaribiosyl-*D*-glucose (**12**), and 3-O-β-gentiobiosyl-*D*-glucose (**13**), as well as the expected laminaritriose. Significantly, no gentiotriose was isolated, suggesting that the 1,6-linkages do not occur in sequences. From these results, the picture of laminarin that emerged was of a chain of glucose units, in which the predominant glycosidic linkage was β-1,3 but that a few, perhaps randomly distributed, β-1,6-linkages also occurred. Estimates of the proportion of chains terminated by mannitol made by periodate oxidation (*2, 131*) and by treatment of the polysaccharide with alkali, when only those chains which are not terminated by mannitol will be degraded to *D*-glucometasaccharinic acid (*33*), give values ranging from 30 to 60%. It has been suggested that a mannitol unit may terminate more than one laminarin chain (*47*) but more recent evidence favours the view that only one chain is attached to each mannitol unit (*3*).

A question obviously calling for decision is whether the few 1,6-linkages constitute links in an essentially linear chain or whether they are present as points of branching in the molecule. The former view is favoured by PEAT and his colleagues (*107, 108*) on the basis of results from periodate oxidation and because the trisaccharide, 3,6-di-O-β-D-glucopyranosyl-D-glucose (14) although painstakingly sought, could not be detected in the products of partial hydrolysis of laminarin. Estimation by isothermal distillation of the molecular weight of a methylated laminarin has indicated an average degree of polymerisation (D. P.) of 58 units, and for the mannitol-terminated constituent a D. P. of 65. Since the average branch-length found by end-group assay was about 23 units, it is suggested that some branching occurs, the 1,6-linkages constituting the points of branching (2).

The two forms of laminarin appear to differ very little in physical properties, reducing power or molecular weight. Chemically, they differ only in the content of mannitol and of 1,6-linkages, the soluble form having a slightly higher proportion of each (107, 108). Whether this difference is responsible for the difference in water-solubility of the two forms, or whether the solubility is a function of either the ease of aggregation or of the presence of other protective colloids, is not yet known. Commercially, laminarin as such is not important although chemically-sulphated laminarin is of medical interest as a possible blood anticoagulant.

4. Conclusions.

In surveying the relationship of algae to land plants, the nature of the food reserve polysaccharides is interesting. The green algae, which are perhaps the most closely related in a physiological sense to land plants, synthesize a granular starch, which is almost indistinguishable from the starch of land plants in that it contains the amylose as well as the amylopectin component. The only significant difference, which does not appear to be related to difference of chemical structure, is that the starch grains from some green algae have a low degree of crystallinity and are therefore less resistant to amylolytic attack than are tuber or cereal starches. In red algae, a starch-type polysaccharide is synthesized, but this is not a typical starch as it contains no amylose component and it differs also in branch-length from amylopectin. The grains are smaller, less regular in shape and have a lower degree of crystallinity than starch grains from land plants. The brown algae are, in respect of carbohydrate storage, the furthest removed from land plants in that they do not synthesize starch but instead store food in the form of mannitol and laminarin. Laminarin is more closely related to the β-glucans of cereal gums than to starches. What is perhaps significant is that laminarin-type materials occur in a number of unicellular algae (other than Phaeophyceae) and here the polysaccharides are present as semicrystalline "globules", easily distinguished under the microscope and originally mistaken for starch granules. These polysaccharides, leucosin and chrysolaminarin (11, 20), probably represent an evolutionary development from the amorphous laminarin of brown algae towards an ordered granular structure, paralleling that of the starch grain.

That laminarin-type polysaccharides are not entirely absent from green algae has recently been demonstrated for *Ulva*, *Acrosiphonia* and *Cladophora* species, in some of which a β-1,3-linked glucan is present, and latterly it has been shown that alkaline extracts of *Caulerpa filiformis* contain appreciable quantities of a laminarin-type polysaccharide (74). In each of these green algae the main carbohydrate storage type is starch, as has already been stated.

IV. Polysaccharides Containing Sulphate Esters.

The continuous matrix in the cell-walls of algae is partly soluble in water, the constituents of such extracts, the mucilages, being mainly polysaccharide in nature. Furthermore, the cells of macroscopic algae are usually embedded in an amorphous, gel-like material which is extractable by water once the fronds have been disintegrated by chemical or mechanical means. This soluble material may be similar in composition to the cell-wall mucilage and these two materials will, therefore, be considered together. It should be emphasized that there is no sharp line of demarcation between the relatively-insoluble skeletal material and the mucilages also present in the cell-wall. As has been discussed previously, some constituents of the mucilage may also be present in a modified form as skeletal material (e. g., xylans, mannans) and in other cases, such as alginic acid, the mucilage should perhaps be regarded as skeletal. Salts of alginic acid occur as major constituents in the middle lamella and in the primary cell-wall of most brown algae but, since the alginates are readily soluble in dilute sodium carbonate, it is perhaps more convenient on technical grounds to regard alginic acid as a mucilage.

The most important features of the majority of these mucilages is that they contain either uronic acid units (i. e. they are polyuronides) or sulphate hemi-ester groups attached to sugar units. In certain cases, both uronic acids and sulphate groups occur in the same polysaccharide. The functions of these acidic polysaccharides have been discussed briefly in the Introduction, particularly as being related to their ion-exchange and water retaining properties. Although sulphated polysaccharides are absent from land plants (with the possible exception of certain ferns), polyuronides are frequently encountered as gum exudates, pectins, seed gums, and in the hemicellulose fractions of corn and some trees. For convenience the sulphated polysaccharides will be considered first.

1. Fucoidin (Brown Algae).

The intercellular tissues of brown algae, particularly the Fucaceae, contain a polysaccharide sulphate, fucoidin. The fucoidin content of these algae varies with the season and with the depth of immersion of the seaweed, being at its highest in autumn and in those species (e. g. *Pelvetia canaliculata*) which occur high in the intertidal zone (*22*). Fucoidin is obtained by extraction of the alga with dilute acid and fractional precipitation with alcohol of the extracted polysaccharides. Further purification is necessary to give a preparation free from other poly-saccharides, such as laminarin and alginic acid (*23*). A calcium salt of fucoidin, prepared in this way, contained 38.3% sulphate and 56.7% fucose, but small amounts of galactose and xylose were also present in this sample of the polysaccharide. Fractionation of a fucoidin preparation from *Ascophyllum nodosum* on diethylaminoethyl cellulose gave a xylan together with several fractions which contained both fucose and galactose residues. These fractions differed from each other in sulphate content (*68*) and, since they contained no xylose, it would seem that this pentose is not an integral part of the fucoidin molecule but that galactose may well be a minor constituent. Similar evidence has been adduced (*121*) regarding the fucoidin from *Macrocystis pyrifera*, which it is suggested contains residues of *L*-fucose and *D*-galactose in the molar ratio 18:1.

The structure of fucoidin is not yet established completely but certain of the structural features have been elucidated. Methylation of fucoidin from *Fucus vesiculosus* gave a product which contained much of the original combined sulphate (*30*). From the complete acid hydrolysate of this product, 2,3-di-O-methyl-*L*-fucose (one part), 3-O-methyl-*L*-fucose (three parts) and unmethylated *L*-fucose (one part) were isolated. The high proportion of the monomethyl sugar suggests a predominance of either fucose 4-sulphate units glycosidically linked through position 2, or alternatively fucose 2-sulphate units linked through position 4. Treatment of a sulphuric ester of a carbohydrate with alkali has been used to assign a position to the sulphate group, since elimination of this group takes place readily only if an adjacent hydroxyl group in a *trans* configuration is available for anhydro-ring formation, or, in the case of a hexose 6-sulphate, if position 3 carries a free hydroxyl group and so allows of the formation of a 3,6-anhydro-ring (*112*). When fucoidin was treated with alkali (*30*), only 10% of the sulphate was eliminated, suggesting that the predominant structural unit is *L*-fucose 4-sulphate linked through position 2 (*15*). The presence of fucose in the hydrolysate of completely

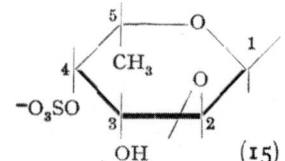

Predominant structural units in fucoidin.

methylated fucoidin suggests either that the molecule may be branched with position 3 of fucose units involved in the branch points, or that some of the fucose units carry two sulphate groups.

Confirmation of the nature of the glycosidic linkage came from the study of the products obtained by partial acetolysis of the fucoidin from *F. vesiculosus*, followed by reduction of the deacetylated fragments with sodium borohydride, when 2-O-α-L-fucopyranosyl-L-fucitol was identified in the products (*99*). More recently, 2-, 3- and 4-O-α-L-fucopyranosyl-L-fucose have been isolated (*34*) from an acetolysate of a commercial sample of fucoidin (from *F. vesiculosus*). The inference is that this sample of fucoidin contains 3- and 4-linked fucose units in addition to the predominant 2-linkage. It is possible that different samples of fucoidin have slightly different structures, and that the term fucoidin refers to a mixture of polyfucoses, in which the predominant linkage is to position 2 of *L*-fucose 4-sulphate units. Some members of the mixture may well have a few 1,3- and 1,4-linkages, possibly constituting branch points.

In addition to fucoidin, the brown alga *Ascophyllum nodosum* contains a small proportion of a second sulphated polysaccharide, ascophyllan (*66*, *53*) which occurs with alginic acid. This polysaccharide contains fucose, xylose, glucuronic acid and ester sulphate but no details on its structure have yet been published. It is interesting that it contains both sugar sulphate ester groups and uronic acid residues, and electrophoretic evidence makes it probable that these are parts of the same molecule.

2. Galactan Sulphates of Red Algae.

The galactans of red algae constitute a group of polysaccharides characteristic of this algal class and not yet found extensively in other classes. They usually occur in the cell-wall and in intercellular elements and are mucilaginous in nature. Many of them have a high content of ester-sulphate groups although a few have only traces of this constituent. In spite of a material resemblance in physical properties, they exhibit several different structural features and some classification is called for. O'COLLA (*96*) distinguishes three types of galactan mucilages; (a) agarose type, (b) ϰ-carrageenan type, and (c) miscellaneous (e. g., the mucilage of *Dilsea edulis*). This classification is convenient and will be used here.

(a) Agarose Type.

(i) *Agar (formerly Agar-agar)*. This name is given to a commercial product obtained by extraction of several species of red algae. As such, it usually contains a mixture of polysaccharides but the predominant component is a galactan, with which is combined a very small proportion of ester sulphate. Many of the early structural studies lose much of their value because commercial agar was used and this is usually a complex mixture of polysaccharides. The principle constituents of technically purified agar are D-galactose (50–65%), L-galactose (1%), 3,6-anhydro-L-galactose (30–50%) and ester sulphate (0.1–5%).

The constitution varies somewhat according to the species from which the agar is obtained, that from *Gelidium amansii* having twice as much D-galactose as 3,6-anhydro-L-galactose, whereas that from *Gracilaria confervoides* contains these two sugars in equal amounts. As such it is obvious that the term, agar, covers a wide range of similar molecular species and that structural investigations should be carried out only on an agar obtained from a single algal species. The reported presence of small amounts of other sugars (pentoses, uronic acids) as constituents of agar may be due to contamination although persistent reports of the presence of D-glucuronic acid in divers agar specimens points to its being a possible structural constituent.

Early structural studies on agar, i.e. on a commercial agar, were based on the methylation technique. Methanolysis of methylated agar gave the methyl glycosides of 2,4,6-tri-O-methyl-D-galactose, and an oligosaccharide residue, which on further methylation, followed by hydrolysis, gave 3,6-anhydro-2,4-di-O-methyl-L-galactose (50). These two sugars were also isolated by a similar procedure from hydrolysates of methylated agar (45). These studies established the presence in agar of units of D-galactose linked through position 3, and of 3,6-anhydro-L-galactose. The mode of linkage of the anhydrosugar in agar from *Gelidium amansii* was established by Araki (5, 6), who isolated 3,6-anhydro-2-O-methyl-L-galactose, as its dimethyl acetal, from a hydrolysate of the methylated agar. The glycosidic linkage to the 3,6-anhydrogalactopyranose units thus involves position 4.

It has been suggested that the anhydrosugar might be an artefact introduced either during the extraction of agar from the seaweed with dilute sodium carbonate, or during the methylation process, since it is known that L-galactose 6-sulphate is readily converted into 3,6-anhydro-L-galactose by treatment with dilute alkali. Nevertheless, extraction of agar from the seaweed under very mild conditions and at neutral pH still gives a product containing the anhydrosugar rather than the 6-sulphate. It may well be that the anhydrosugar units arise by an enzymic desulphation of L-galactose 6-sulphate units during the growth of the alga and, indeed, an enzyme of this type has been isolated from one of the red algae (117). Since most specimens of agar contain, in addition to the anhydrosugar, small proportions of L-galactose and of ester sulphate, it is possible that these represent units of L-galactose 6-sulphate, which have not been converted into the anhydrosugar.

Further information on the structure of the agar from *G. amansii* has come from studies of the oligosaccharides produced by fragmentation of the polysaccharide in various ways. Partial hydrolysis with N-sulphuric acid at 100° gave a mixture from which a crystalline disaccharide, agarobiose, was isolated (*6, 7*). The structure of this sugar was established by the methylation method as 3,6-anhydro-4-O-(β-D-galactopyranosyl)-L-galactose (**16**), the β-linkage being inferred from the low optical rotation ([α]$_D$-5.8°). A variety of organisms from sea-water, and certain gastropods are known to produce enzymes which decompose agar. By the action of such an enzyme from the marine bacterium *Pseudomonas kyotoensis*, ARAKI and ARAI (*9, 10*) isolated a crystalline disaccharide and an amorphous tetrasaccharide from agar. The disaccharide, *neo*agarobiose, was readily hydrolysed by dilute acid to give 3,6-anhydro-L-galactose and D-galactose, the ease of hydrolysis suggesting a 3,6-anhydrogalactosidic linkage. Methylation and identification of the products of hydrolysis established its structure to be 3-O-(3,6-anhydro-α-L-galactopyranosyl)-D-galactose (**17**). The isolation of agarobiose on the one hand and of *neo*agarobiose on the other prompted the suggestion that the principal polysaccharide component of agar is

(**16**)
Agarobiose.

(**17**)
*Neo*agarobiose.

(**18**)
Tetrasaccharide from agar.

a linear chain of alternating 3-linked *D*-galactose and 4-linked 3,6-anhydro-*L*-galactose residues (*9*, *10*). Some support for this structure came from a study of the tetrasaccharide mentioned above. This was shown to consist of two *neo*agarobiose residues joined as in (*18*).

The proposed structure for agar accords with the evidence for some agar specimens but not for those specimens in which the molar ratio, *D*-galactose to 3,6-anhydro-*L*-galactose is greater than unity. A possible reason for this discrepancy is provided by the observation, made by Araki, that acetylated agar could be separated into two components, one soluble and the other insoluble in chloroform (*4*). The soluble fraction was termed agarose acetate and the insoluble fraction agaropectin acetate. Many of the structural studies carried out by Araki and his collaborators (*8*) refer to the agarose component of agar, and it is this component which is believed to have the structure postulated above. The agaropectin fraction contains some of the structural features of agarose, but undoubtedly possesses others, which have yet to be elucidated. The small sulphate content, the uronic acids and the *L*-galactose are believed to occur in the agaropectin fraction.

Certain specimens of commercial agar have been reported as containing pyruvic acid as a minor constituent (1%) (*54*, *55*). Methanolysis of one of these agar specimens gave, among other products, an acidic substance which, on hydrolysis, furnished 3,6-anhydro-*L*-galactose, *D*-galactose and pyruvic acid. The structure of the acidic substance was established as the dimethylacetal of (*19*). This is basically pyruvic acid linked as a ketal to positions 4 and 6 of the *D*-galactose residue in agarobiose.

Since not all anhydrogalactose-containing specimens give the disaccharide (*19*) it would appear that the pyruvic acid is not an artefact produced by acidic degradation of 3,6-anhydrogalactose during the methanolysis of agar.

(ii) *Funori*. The mucilage of the red alga *Gloiopeltis furcata*, funori, is used as an adhesive and sizing-agent in parts of Asia. It contains residues of *D*-galactose, *L*-galactose and 3,6-anhydro-*L*-galactose in the molar ratios 12 : 1 : 8 (*56*), but it differs from agar in having a sulphate content of 18% (as SO_4). That the mucilage has structural similarities to agar was shown by the isolation in good yield of agarobiose dimethyl-acetal on methanolysis of funori. Little is yet known of the distribution of the sulphate within the molecule.

(iii) *Porphyran.* *Porphyra umbilicalis* (laverbread) contains a high proportion of a mucilage, porphyran, which is composed of residues of *D*- and *L*-galactose, 3,6-anhydro-*L*-galactose, 6-O-methyl-*D*-galactose, and ester sulphate (*105*). There appears to be some seasonal variation in the proportions of these sugars in porphyran, and other factors, such as the location of the alga on the shore, also affect the proportions (*119*). On cautious hydrolysis with hydrochloric acid, *L*-galactose 6-sulphate was identified among the products and it has been suggested that most of the sulphate present in the alga occurs in this form (*129*). Treatment of porphyran with alkali, or with an enzyme extracted from the alga, results in the removal of most of the sulphate and, in the process, the *L*-galactose constituents are converted into 3,6-anhydro-*L*-galactose residues (*116, 117*). On this basis it has been suggested that in this alga *L*-galactose 6-sulphate is probably a biological precursor of 3,6-anhydro-*L*-galactose in the sense that, when the polysaccharide is first formed, it contains no 3,6-anhydrogalactose but subsequent enzymic desulphation leads to the appearance of the anhydrosugar as an integral part of the polysaccharide molecule.

Disaccharides from porphyran.

Further information on the structure of porphyran came from a study of oligosaccharides produced by partial acid hydrolysis (*130*). These included four disaccharide monosulphates, *L*-galactose 6-sulphate being a common constituent unit. The structure of each was established by periodate oxidation and by treatment with alkali, and these structures are shown in (20) to (23). As is to be expected, it is found that desulphation with alkali of (20) furnishes agarobiose and of (22), *neo*agarobiose. Disaccharides (21) and (23) differ from (20) and (22), respectively, only in having a methyl ether grouping on position 6 of the *D*-galactose units. A repeating sequence has been suggested for porphyran, in which 3-linked *D*-galactose residues alternate with 4-linked *L*-galactose 6-sulphate residues, with the variations that some of the *D*-galactose carries an O-methyl group at position 6 and that some of the sulphated units have been desulphated to form 3,6-anhydro-*L*-galactose (*130*).

(b) *κ-Carrageenan Type*.

(i) *Carrageenan*. The water-soluble mucilages of many members of the family Gigartinaceae contain a high proportion of a galactan sulphate, carrageenan. This mucilage is of commercial importance in the food, textile and pharmaceutical industries. The principle commercial sources are the *Chondrus* and *Gigartina* species, mixtures of these often being used.

An early study (*49*) of extracts from these species indicated the presence of two fractions, one extracted by cold and the other by hot water, the products obtained in this way having different physical properties. It was later suggested that these differences might be accounted for by differences in the nature of the cations associated with the sulphate-ester anions (*26*). The polysaccharide contains 29–35% sulphate (as SO_4) and 33–44% galactose, according to the species from which it is extracted. The galactose is predominantly the *D*-isomer but the presence of small amounts of the *L*-isomer has been established for the carrageenan from some *Chondrus* species (*63, 86, 110*). The existence of an acid-labile constituent of carrageenan had been suspected for many years and it was finally established that this was 3,6-anhydro-*D*-galactose by the isolation of the diethyl dithioacetal of this sugar from the products of mercaptolysis of carrageenan (*100, 110*).

The mode of linkage of the *D*-galactose units was established by the methylation technique; hydrolysis of the methylated carrageenan from *Chondrus crispus* furnished 2,6-di- and 2-O-methyl-*D*-galactose (*26*). Since the sulphate ester links in the original carrageenan were largely resistant to the action of alkali, it was concluded that position 4 was the site of attachment of the sulphate group, and hence that the glycosidic linkage engaged position 3. The alternative unit, 4-linked galactose 3-sulphate, has a hydroxyl group on position 2, adjacent and *trans* to the sulphate group and hence this should allow elimination of the sulphate with ethylene-oxide ring formation, when treated with alkali. These conclusions were supported when methylation and hydrolysis of a carrageenan sample, previously degraded with oxalic acid (during which treatment 25% of the sulphate was removed) gave 2,4,6-tri-O-methyl-*D*-galactose, in addition to a di-O-methyl- and tetra-O-methyl-*D*-galactopyranose (see also *38*). It was further suggested that the 2-O-methyl-D-galactose isolated in the first instance might arise from points of branching in the molecule, position 6 being the point of attachment of these branches (*63*).

In 1953 it was shown that an aqueous solution of carrageenan could be separated into two polysaccharide components by addition of potassium chloride (*123*, *124*) thus confirming earlier indications. The component (40%), which gelled at a concentration of 0.15 M-potassium chloride, was termed $\varkappa$-carrageenan and the component (45%), soluble in this reagent, λ-carrageenan. The fractions differed in sulphate content, in optical rotation, and in their content of 3,6-anhydro-D-galactose, the $\varkappa$-component containing 24% of this sugar, the λ-fraction considerably less (*100*, *101*). Furthermore, the L-galactose seemed to be present predominantly in the λ-fraction. Subsequently it was shown that the λ-component could be further separated by ethanol precipitation into a fraction containing no L-galactose units and a second enriched in this isomer (*125*).

(24)

(25)
Carrabiose.

The structure of $\varkappa$-carrageenan has been established by O'NEILL (*101*), using the technique of mercaptolysis. Treatment of the polysaccharide with cold, concentrated hydrochloric acid and ethanthiol causes scission of glycosidic linkages and stabilisation of the resultant reducing sugars by formation of their diethyl dithioacetals (mercaptals). Destruction of 3,6-anhydrogalactose units by the acid is thus prevented. From the fragments obtained by such a treatment, a disaccharide mercaptal was obtained, reductive desulphurisation (with Raney nickel) of which gave a non-reducing 'disaccharide'. The structure of this disaccharide was established as 3,6-anhydro-1-deoxy-4-O-(β-D-galactopyranosyl)-D-galactitol (**24**), and hence the original disaccharide, carrabiose, must have the structure shown in (25). The ratio of galactose to 3,6-anhydrogalactose in the *ϰ*-fraction is near unity, and it has been suggested that this component consists of alternating units of *D*-galactose 4-sulphate linked to 3,6-anhydro-*D*-galactose as shown in (26) (*101*). The sulphate ester group is shown as located on the 4-position of the *D*-galactose units, in agreement with the earlier methylation studies and with the infrared spectrum (*67*). This is obviously only a picture of the gross structure of *ϰ*-carrageenan and it may need modification and amplification in the light of future work. For example, there is an indication that some degree of branching occurs and that sulphate groups may be found at position 6 of some of the galactose units.

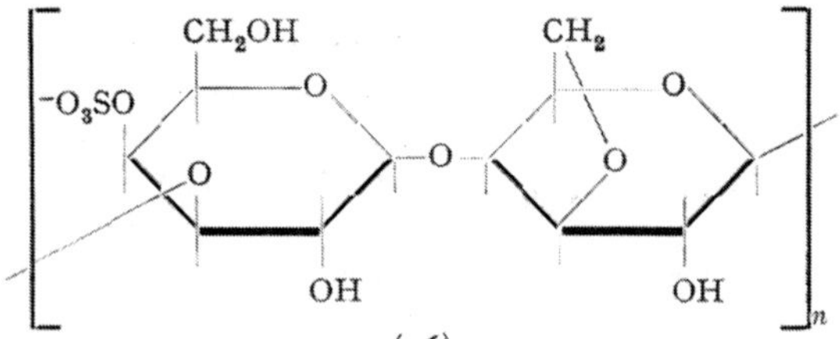

(26)

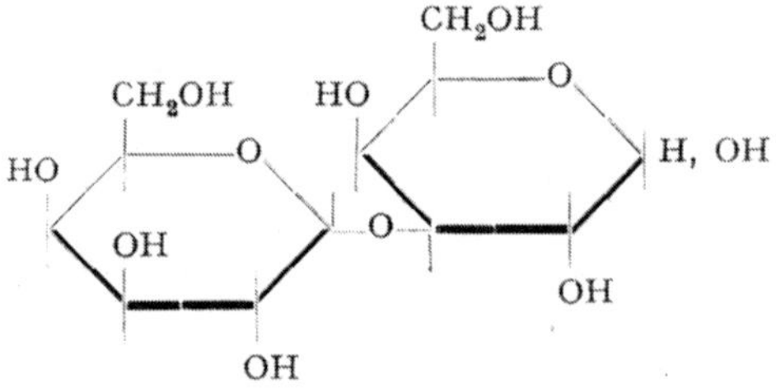

(27)
Disaccharide from *λ*-carrageenan.

The *λ*-component of carrageenan has received some attention but its structure is more complex. When *λ*-carrageenan was fractionally precipitated with ethanol, it gave a major fraction, which was investigated by the acetolysis method (*85*). The deacetylated fragments consisted of a polymer-homologous series of oligosaccharides constituted of *D*-galactose units only. The disaccharide member was shown to be 3-O-α-*D*-galactopyranosyl-*D*-galactose (27), and the trisaccharide, the next higher member. This work indicates that *λ*-carrageenan contains sequences of 1,3-linked *D*-galactose units. Other linkage types are almost certainly present. For instance, the polysaccharide was oxidised by periodate, albeit to a small extent only (*85*). More recently, Rees has shown that *λ*-carrageenan from *Chondrus crispus* contains one third of its sulphate ester groups so situated on the galactose units that alkali (in the presence of borohydride, see *116*) removed them with concomitant formation of 3,6-anhydrogalactose (*118*). Mild acid hydrolysis of the alkali-treated fraction and further reduction of the fragments with borohydride, gave as one of the products a reduced sugar sulphate, tentatively identified as 3,6-anhydro-*D*-galactitol 2-sulphate. The alkali-treated fraction also furnished carrabiose diethyl dithioacetal on being mercaptolysed, thus displaying a resemblance to the *x*-fraction. There is therefore some indication that one of the building units in *λ*-carrageenan is 4-linked *D*-galactose 2,6-disulphate. A study of the rates of sulphate ester removal from alkali-treated *λ*-carrageenan by acid hydrolysis tends to confirm that two types of sulphate linkage are present, one type (40%) involving position 4 of galactose units (as in *x*-carrageenan) and the other type (60%) most probably at position 2 of 3,6-anhydrogalactose units (*118*). REES believes that the *λ*-carrageenan examined by him contains a major component, which is the biological precursor of *x*-carrageenan (26) and, has the structure shown in (28).

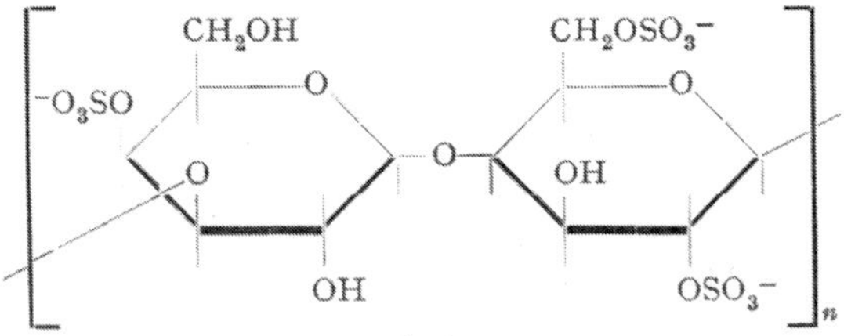

(28)

Component of *λ*-carrageenan.

(ii) *Furcellaran (Danish Agar)*. The red alga, *Furcellaria fastigiata*, contains a galactan sulphate which is extracted on a commercial scale and termed Danish Agar. This polysaccharide contains residues of *D*-galactose, 3,6-anhydro-*D*-galactose and ester sulphate. Acetylative desulphation (which is accompanied by considerable depolymerisation), followed by methylation of the sulphur-free product and then hydrolysis, furnished tetra-, 2,4,6-tri- and 2,4-di-O-methyl-*D*-galactose (*28*). This suggests the presence of predominantly 3-linked *D*-galactose units and some branching involving position 6. The resemblance of furcellaran to *x*-carrageenan was noted by PAINTER (*102*) who showed, for example, that the polysaccharide gelled in the presence of potassium chloride. Mercaptolysis also gave the mercaptals of *D*-galactose, 3,6-anhydro-*D*-galactose and carrabiose (25), again suggesting a structure similar to *x*-carrageenan (*102*). The main distinguishing feature of furcellaran is a somewhat lower content of sulphate ester (19% as SO_4) than occurs in carrageenan (26%).

(iii) *Other Carrageenan Types.* The mucilage of *Iridophycus flaccidum* consists of a galactan sulphate, iridophycin, in which each *D*-galactose unit bears a sulphate ester group (*51*). By the methylation technique it was established that the polysaccharide contains a high proportion of galactose residues linked through position 3. Using an enzymatic method, YAPHE (*135*) has detected 36% of a *x*-carrageenan-type polysaccharide in iridophycin. At present there is no evidence as to whether the polysaccharide contains 3,6-anhydrogalactose residues.

Extraction of *Hypnea spicifera* with hot water yields a solution of a polysaccharide, which is completely precipitated (as a gel) by addition of potassium chloride (*29*). The constituent units of this polysaccharide are galactose, 3,6-anhydrogalactose and ester sulphate. Evidence from methylation of the polysaccharide and examination of the hydrolysis products suggests that the galactose units are glycosidically linked through position 3 and have sulphate ester groups at position 4. Species of *Eucheuma* also contain polysaccharides with properties similar to *x*-carrageenan (*89, 90, 91*).

(c) Miscellaneous.

(i) *The Mucilage of Dilsea edulis.* Extraction of this red alga with water or dilute acid yields an extremely viscous solution of a galactan sulphate which contains *D*-galactose (70%), *D*-xylose (7%), *D*-glucuronic acid (10%), ester sulphate (9.7%) and a trace of 3,6-anhydrogalactose (*16, 18, 36*). Acetylative desulphation, followed by methylation and methanolysis, gave the methyl glycosides of tetra- and 2,4,6-tri-O-methyl-*D*-galactose, together with traces of the 2,3,6-isomer, and a di-O-methyl galactose. The predominant linkage is thus 1,3-glycosidic but a few 1,4-linkages are also present and perhaps some degree of branching occurs (*16*). Application of the BARRY degradation method (*19*) to this polysaccharide has given further information on the structure. In this method, the polysaccharide is oxidised with periodate and any oxidised fragments are eliminated from the polysaccharide by subsequent treatment with phenylhydrazine in aqueous acetic acid, leaving only the unoxidised fragments in their original form. Application of this technique to the *Dilsea* galactan showed that 31.5% of the sugar residues were oxidised initially and that the unoxidised product was a polymer containing only galactose and ester sulphate. Repetition of the degradation procedure resulted in only small losses of galactose units, presumably from the chain ends. Thus, there appears to be a resistant core to the molecule consisting of sulphated 1,3-linked galactose units. A detailed examination of the low-molecular weight products from the degradation procedure led to the postulation of a structure in which a repeating unit of nine galactose and four galactose 6-sulphate units, all 1,3-linked, had four side-chains containing both 1,3- and 1,4-linked galactose units, 1,3-linked xylose units, with 3,6-anhydrogalactose and glucuronic acid as end groups (*18, 36*).

REES (*116*) has recently suggested some modification to this structure. The infrared spectrum of the galactan indicates that much of the sulphate occurs on position 4 of the galactose units, but that small amounts of the sulphate are present as 4-linked galactose 6-sulphate units, which are converted into 3,6-anhydrogalactose units on treatment of the poly-saccharide with alkali. REES also considers that xylose is not an integral part of the molecule but is present in a contaminating polysaccharide. The main core of the molecule is a chain of 1,3-linked galactose and galactose 4-sulphate units as in λ-carrageenan (**29**). Another portion of the molecule contains an alternating sequence of 3-linked galactose (or galactose 4-sulphate) units and of 4-linked galactose (or galactose 6-sulphate) units (**30**).

—◯— galactopyranosyl unit.

Extraction of *Dumontia incrassata* with dilute acid gives a water-soluble galactan sulphate structurally related to the mucilage of *Dilsea edulis* (*37*). This polysaccharide is constituted predominantly of 3-linked galactose sulphate units but, in addition, there may be present a few 1,4-links. It differs from the *Dilsea* polysaccharide in having about double the content of ester sulphate.

(ii) *Polysiphonia fastigiata.* Extraction of this red alga with very dilute acid affords a mucilage, the hydrolysate of which contains galactose, 6-O-methylgalactose, 3,6-anhydrogalactose, ester sulphate and xylose (*73*). The remarkable feature of this polysaccharide (or mixture of polysacchari-des) is the presence of both *D*- and *L*-isomers of galactose (ratio, 2 : 1), of 6-O-methylgalactose (*D* : *L* ratio, 9 : 7), and of the anhydro sugar (*D* : *L* ratio, 1 : 1). Furthermore, hydrolysis of the polysaccharide with dilute acid gave, among other products, a sugar sulphate which was tentatively identified as a mixture of *D*- and *L*-galactose 6-sulphate. Complete methylation of the polysaccharide, followed by hydrolysis, gave many fragments, among which 2,4-di-O-methylgalactose, containing both *D* and *L* forms, predominated. The dimethylgalactose probably arose from 3-linked galactose 6-sulphate units. The polysaccharide is obviously complex and needs much further investigation before a structure can be proposed.

3. Polysaccharide Sulphates of Green Algae.

The water-soluble polysaccharides of green algae are complex mixtures, sometimes including starch, and frequently containing heteropoly-saccharides. The heteropolysaccharides usually contain a number of different sugar residues, including uronic acids, and are sulphated. From the studies carried out to date, certain structural similarities have emerged and, broadly speaking, two main types of heteropolysaccharide have emerged. In the first type, the main constituent sugars are L-arabinose and D-galactose together with ester sulphate, and in the second type the characteristic building units are D-glucuronic acid, L-rhamnose, D-xylose and ester sulphate. A provisional classification into these two types is used here for convenience, although it should be emphasized that, as more knowledge accumulates, other types will probably be encountered.

(a) Heteropolysaccharides Containing Sulphated Arabogalactans.

(i) *Cladophora rupestris*. An aqueous extract of this alga contains proteins, a glucan and a complex polysaccharide, cladophoran (*42*). Cladophoran migrated as a single component on electrophoresis and contained residues of L-arabinose, D-galactose, L-rhamnose, and D-xylose in the approximate molar ratios, $9.3 : 7.0 : 2.5 : 1.0$, together with ester sulphate. The polysaccharide was difficult to methylate but a methyl derivative (methoxyl content, 25.1%) gave on methanolysis and hydrolysis of the products 2,3,4,6-tetra-, 2,3,5-tri-, 2,4-di-, and 2-O-methyl-D-galactose; unmethylated D-galactose; 2- and 3-O-methyl-L-arabinose and unmethylated L-arabinose; 2,3,4-tri-, and 2,3-di-O-methyl-D-xylose; 2,4-di-, 3.4-di-, and 4-O-methyl-L-rhamnose. From this it would appear that the polysaccharide is highly branched and contains 1,3-linked L-arabinose, D-galactose and L-rhamnose residues, together with 1,4-linked xylose residues.

Partial acid hydrolysis gave some oligosaccharides which contained only arabinose residues, and later studies (*109*) have failed to yield oligo-saccharides containing more than one sugar residue. This suggests either that cladophoran is a mixture of sulphated homopolysaccharides, or that it contains chains, each a polymer of one particular sugar only, the chains being glycosidically linked to each other. That some of the chains could be highly branched was suggested by the results of periodate oxidation (*42*). Oxidation destroyed all the xylose units indicating that this sugar was either linked through position 4, or occupied non-reducing chain-ends; two thirds of the galactose was oxidised and again it is probable that these units are non-reducing end groups. The oxypolysaccharide recovered after the periodate treatment was constituted mainly of arabinose, galactose and ester sulphate units. In an extension of this work using the Barry degradation technique (98), it was confirmed that all of the xylose and much of the galactose was destroyed by one treatment with periodate. After three Barry degradations, the oxypolysaccharide remaining contained units of arabinose, galactose and rhamnose in the approximate molar ratios 1.8:1.8:1.0, together with one sulphate group per 3 or 4 monosaccharide units. Methylation of this oxypolysaccharide (26.9% methoxyl), followed by hydrolysis gave 2,4-di- and 2-O-methyl-L-arabinose; 2,4,6-tri- and 6-O-methyl-D-galactose with some free galactose; 2,4-di-O-methylrhamnose and rhamnose. This study, therefore, lends further support to the idea that the polysaccharide is highly branched and that the oxidation-resistant core of the molecule has residues linked as shown below.

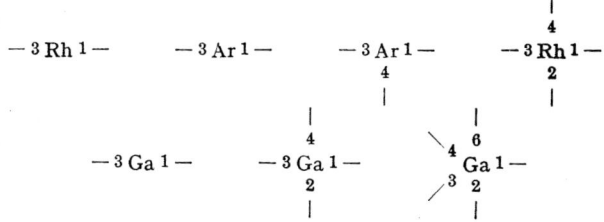

KEY: Rh = L-rhamnose; Ar = L-arabinose; Ga = D-galactose.

The sites occupied by some of the sulphate groups are the 6-position of D-galactose residues and the 3-position of L-arabinose units. This follows from the isolation of D-galactose 6-sulphate and L-arabinose 3-sulphate from a partial acid hydrolysate of the polysaccharide (77).

(ii) *Caulerpa filiformis*. An aqueous extract of this green alga contains a complex polysaccharide, which has 17.6% of ester sulphate. On hydrolysis, L-arabinose, D-galactose, D-xylose and D-mannose were obtained in the molar proportions 1:5:2:2. Small amounts of L-rhamnose, L-fucose, ribose, D-glucose, and 3,6-anhydrogalactose were also isolated (76). A partial fractionation of the polysaccharide was obtained by treatment of an aqueous solution of the free acid form with a saturated solution of barium hydroxide. The three fractions so obtained differed significantly in their contents of the various sugar residues, suggesting that the original polysaccharide was a mixture of several molecular species. Galactose and arabinose residues were relatively more abundant in one fraction, and xylose and mannose more abundant in another. It is not yet possible to say whether the fractions are mixtures of an arabogalactan with a xylan and a mannan or a xylomannan, or whether there is a range of polysaccharides with varying proportions of sugar residues. The isolation of a xylan from this particular algal species (74) suggests that the xylose may arise from a separate xylan. Some of the sulphate is linked to galactose residues since a galactose monosulphate was isolated from an acid hydrolysate of the polysaccharide. On periodate oxidation of the polysaccharide, little oxidant was consumed and the derived oxypoly-saccharide still contained residues of all the sugars. At present no other structural information is available.

Polysaccharides with similar compositions also occur in *C. racemosa* and *C. sertularioides* (76).

(iii) *Codium fragile.* Aqueous extraction of this alga furnished a polysaccharide mixture, from which a sulphated arabogalactan was obtained by chromatography on diethylaminoethyl cellulose (70). Upon hydrolysis, galactose and arabinose in the molar ratio 3 : 1 were the only sugars obtained. The sulphur content was reduced by approximately 4% (from 12.7%) on treatment with alkali and 3,6-anhydrogalactose was formed as a result of this treatment. This suggests that some of the ester sulphate is located on position 6 of the galactose residues, but the infrared spectrum suggested, in addition, the presence of some sulphate in an axial position, presumably the 4-position of a galactose unit.

(b) Heteropolysaccharides Containing Glucuronic Acid.

(i) *Ulva lactuca.* Extraction of this green alga with water or dilute sodium carbonate yielded a water-soluble polysaccharide with the percentage composition, D-xylose, 9.4; L-rhamnose, 31; D-glucose, 7.7; D-glucuronic acid, 19.2; and sulphate, 15.9 (as SO_4) (25). The sulphate groups were resistant to hydrolysis by alkali and it was suggested that they were located on xylose or rhamnose residues. Complete methylation of the polysaccharide was difficult but a partly methylated product, when hydrolysed, gave 2,3-di- and 2,3,4-tri-O-methyl-L-rhamnose, 2,3-di- and 2,3,4-tri-O-methyl-D-xylose, among other products. Methyl-D-glucoses, present in the hydrolysate, probably arose from a contaminating starch. A relatively high yield of trimethyl-L-rhamnose and of trimethyl-D-xylose suggested that these units probably arose from the non-reducing end groups of short side-chains, but little further information could be obtained on the other fragments. It was significant, however, that periodate effected but little oxidation of the polysaccharide, suggesting the presence of both sulphate-ester and glycosidic linkages in positions such that oxidisable diglycol groups were essentially absent.

In a later study (78) of a similar, and indeed probably identical, polysaccharide from *U. lactuca*, partial acid hydrolysis and separation of the acidic oligosaccharides gave an aldobiuronic acid (as Ba salt) in a 17% yield. This acid had $[\alpha]_D — 22°$ and on esterification, followed by reduction with potassium borohydride, it gave a neutral disaccharide.

Methylation assay on the one hand and periodate oxidation on the other established that this disaccharide was 4-O-β-D-glucopyranosyl-L-rhamnose. The aldobiuronic acid was thus 4-O-β-D-glucopyranuronosyl-L-rhamnose (31), and the fact that this constituted 75–78% of all the acidic oligosaccharides in the partial hydrolysate suggests that most of the glucuronic acid is linked in this way.

(31)
Aldobiuronic acid.

Since then it has been shown that the polysaccharide obtained by cold-water extraction of this alga displays heterogeneity in the ultracentrifuge. Fractionation of the polysaccharide on diethylaminoethyl cellulose (chloride form) gave three distinct fractions, one of which was shown to be homogeneous in the ultracentrifuge (III). Furthermore, since each fraction contained the same sugars as well as sulphate in approximately the same proportions, it was concluded that the fractions differed mainly in molecular weight. Periodate oxidation of the polysaccharide before and after partial desulphation with methanolic hydrogen chloride established that desulphation liberated α-glycol groups in rhamnose, and, to a small extent, in xylose, indicating that some of these sugar residues carried sulphate groups. The site of the sulphate groups on rhamnose units was established by infrared spectroscopy, an absorption peak at 850 cm^{-1} being characteristic of ester sulphate on a secondary hydroxyl group in an axial position, i. e. on $C_{(2)}$ of a rhamnose

(32) 2-O-Methylxylose.

unit. The sulphate groups on xylose were assigned to the 2-position since alkaline desulphation of the polysaccharide, followed by opening with sodium methoxide of the anhydro-ring so formed and then hydrolysis of the polysaccharide, gave 2-O-methylxylose (**32**, p. 33), in agreement with the reaction sequence shown above. Beyond these observations, no other structural information is available on this polysaccharide.

(ii) *Acrosiphonia centralis.* A water-soluble polysaccharide, obtained by extraction of this alga with 1% ammonium oxalate solution, was constituted of sugar units in the following molar proportions: D-galactose, 0.1; D-xylose, 1.6; L-rhamnose, 1.4; D-mannose, 0.2; D-glucose, 1.0. In addition, both sulphate (7.8%) and uronic acid (19.3%) were present (97). Acetylation of the polysaccharide and extraction of the acetate with chloroform gave a soluble, glucose-rich polysaccharide acetate, and an insoluble acetate fraction ([α]$_D$ — 35°). The latter was simultaneously deacetylated and methylated, and the product on being hydrolysed gave the following: 2,3-di- (3 parts) and 2,3,4-tri-O-methyl-D-xylose (1 part); 2,3-di- (3 parts) and 2,3,4-tri-O-methyl-L-rhamnose (5 parts). In addition, a series of methylated oligouronic acids was obtained. The simplest of the methylated uronic acids, i. e. the aldobiuronic acid, was converted into the methyl ester-methyl glycoside, reduced with borohydride (—COOCH$_3$ → —CH$_2$OH) and then acid hydrolysed. The products were 2,3,4-tri-O-methyl-glucose and 2-O-methyl-rhamnose. The suggestion is that the original aldobiuronic acid might be 4-O-β-D-glucopyranuronosyl-L-rhamnose and that in the original polysaccharide this was linked at position 3 of the rhamnose unit. This view received support when a partial hydrolysate of the polysaccharide gave, among other products, aldobi-, aldotri-, and aldotetra-uronic acids. A detailed structural analysis of the aldobiuronic acid showed it to be 4-O-β-D-

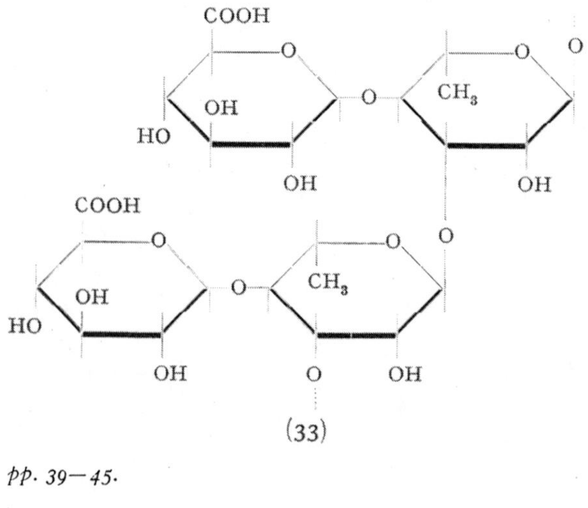

(33)

glucopyranuronosyl-*L*-rhamnose or, perhaps, its derived lactone. The acidic trisaccharide contained two glucuronic acid and one rhamnose units, while a higher oligosaccharide fraction from the partial hydrolysate contained the two sugars in a 1 : 1 ratio. This evidence is taken to mean that the polysaccharide is constituted in part of chains of 1,3-linked *L*-rhamnose units with *D*-glucuronic acid addenda at position 4 of each rhamnose unit (33).

(iii) *Enteromorpha compressa*. A hot-water extract of this alga contains starch and a heteropolysaccharide (79). After removal of the starch, the residual polysaccharide had $[\alpha]_D - 87°$ and contained uronic acid (18.3%) and sulphate (16%). A hydrolysate of the polysaccharide contained rhamnose (45%), xylose (15%), small amounts of glucose, and an aldobiuronic acid, which was believed to be 4-O-β-*D*-glucuronosyl-*L*-rhamnose. Desulphation without degradation of the polysaccharide was accomplished by treatment with methanolic hydrogen chloride and the polysaccharide was subjected to periodate oxidation both before and after desulphation. The desulphated material consumed more oxidant than the original and the oxypolysaccharide derived from the desulphated polysaccharide contained a lower proportion of both xylose and rhamnose than did the oxidised, original polysaccharide. This finding suggests that both xylose and rhamnose units carry sulphate ester groups, in such positions as to render them resistant to periodate oxidation. Since the rhamnose appears to be glycosidically-linked through position 4, the sulphate ester must be on either positions 2 or 3. The 2-position is favoured inasmuch as the infrared spectrum of the polysaccharide indicates the presence of sulphate in an axial position. In the stable Cl-chair conformation, rhamnose has an axial hydroxyl group at position 2.

4. Conclusions.

In seeking for a relationship between algae of different classes on the one hand and land plants on the other, the group of sulphated polysaccharides appear, at first sight, to be of little significance. Some general observations may, however, be relevant to this issue.

The brown seaweeds synthesise a fucosan ester-sulphate, fucoidin, as the principal sulphated polysaccharide which appears to be found only in this class of algae. A polysaccharide, ascophyllan, constituted of fucose, glucuronic acid and ester-sulphate, which is found in a brown alga may be related, superficially at least, to those polysaccharides of green algae which contain units of rhamnose, glucuronic acid and ester sulphate. Again, various reports of the occurrence in polysaccharides from red algae (e. g., agar and the mucilage of *Dilsea edulis*) of glucuronic acid residues may point to a similar type of polysaccharide in this algal class.

The galactan sulphates of red algae appear to be characteristic of this class. Variations in the constitution of these polysaccharides can be a useful means of relating, in a botanical connotation, the species within the Rhodophyceae, as has been previously pointed out by Stoloff and Silva (*128*) and by Yaphe (*135*), and briefly mentioned in this article. Little relationship with polysaccharides of other algal classes is apparent, except for the sulphated arabogalactans of green algae in which sulphate groups are attached to the galactose members of the chain. Polysaccharides of *Caulerpa* species have units of this kind and that from *Codium fragile* contains both galactose 4- and 6-sulphate units. In the sulphated polysaccharides of green algae, nevertheless, sugar units other than galactose may also carry sulphate groups (e. g., arabinose in *Cladophora rupestris* mucilage).

Possible relationships of the sulphated polysaccharides of green algae to ascophyllan (brown algae) and to certain galactan sulphates of red algae have already been mentioned. Another relationship appears to subsist between those polysaccharides of green algae which contain units of glucuronic acid linked to rhamnose, and certain plant gums (e. g., that of *Sterculia caudata*) which also contain glucuronic acid linked to rhamnose, although the linkages in the two types of polysaccharide appear to be different (*126, 133*).

V. Alginic acid.

Polysaccharides which are constituted solely of uronic acid units, polyuronides, are comparatively rare in nature. One example is pectic acid, which occurs as a partial methyl ester, pectin, in the primary cell-wall and intercellular layers of all land plants, particularly in fruits and young tissues, and is a polymer of *D*-galacturonic acid. Pectin, so far as the authors are aware, has not been found in any of the seaweeds, but there occurs in large abundance in the brown algae a polyuronide, algin, which is exceptional in that (i) it appears to be confined to the Phaeophyceae and (ii) it is mainly composed of polymerised *D*-mannuronic acid. By virtue of its unusual constitution, it is difficult to place algin in any of the preceding categories of marine polysaccharides. Algin is the name given to the mixed salts of alginic acid, which is primarily a structural component of the plant, as is evidenced by its occurrence in both stipes and fronds, in the middle lamella and primary cell-walls, constituting between 20 and 40% of the dry weight of the alga. Considerable seasonal variation occurs, and the depth of immersion of the alga also influences the content (*21*).

The algin is usually extracted by dissolution of the algae in dilute alkali, e. g., in sodium carbonate, and alginic acid is precipitated either as the insoluble calcium salt, or as the free acid by acidification of the extract. The free acid has $[\alpha]_D$ — 113° to — 148°. Commercially, the acid, or a soluble salt such as the sodium salt, is used extensively in the food, pharmaceutical, cosmetic, textile and paper industries (126, 133). Certain esters of alginic acid have also found industrial uses.

Chemically, alginic acid is extremely resistant to acid hydrolysis, the vigorous conditions necessary to hydrolyse it leading to extensive degradation of the constituent sugars, and this has prevented a full study of its chemical constitution. NELSON and CRETCHER (93) were the first to isolate and identify a salt of D-mannuronic acid from a hydrolysate of alginic acid. Later, on the basis of X-ray diffraction patterns, a formal resemblance to cellulose was noted and the structure postulated as being a chain of β-1,4-linked mannopyranuronic acid units (12, 72). Methylation of a degraded alginic acid, followed by hydrolysis, gave a 50% yield of 2,3-di-O-methyl-D-mannuronic acid (57) in agreement with the structure suggested above, and further confirmation of this structure came from periodate oxidation studies (71). In a more detailed examination, HIRST et al. (27) prepared a fully methylated alginic acid and reduced the methyl ester with lithium aluminium hydride, thus converting the units of mannuronic acid methyl ester into mannose units. Subsequent hydrolysis was thereby facilitated and an 88% yield of 2,3-di-O-methyl-D-mannose was obtained, with only traces of a monomethyl mannose and an unidentified dimethyl sugar. This suggests that the molecule contains sequences of mannuronic acid units and that it is not a highly branched structure (34, p. 38).

That this picture of alginic acid is oversimplified was demonstrated by FISCHER and DÖRFEL (41), who observed that when alginic acid was hydrolysed and the derived uronic acids were lactonised, L-guluronolactone was obtained in addition to the expected D-mannuronolactone. That L-guluronic acid is a constituent of alginic acid has since been amply confirmed. Analyses of alginic acids from several species of brown algae has indicated a variation from 20 to 65% in the content of guluronic acid units (52, 134).

Do the two uronic acids occur as units in one polymer, or is alginic acid a mixture of two polymers, one a polymeric D-mannuronic acid, the other derived from L-guluronic acid? Several attempts at fractionation have been made and subfractions differing in the proportions of the two uronic acids have been obtained (46, 52, 134). So far, however, no fraction has been obtained which contains only one of these uronic acids. FREI and PRESTON (46) examined the X-ray powder diagrams given by alginic acids extracted by various procedures from *Himanthalia* and suggest that two different polysaccharides are present; (a) a poly-mannuronic acid, which is present mainly in the intercellular mucilage, and (b) a polyguluronic acid found primarily in the cell wall. These authors also suggest that the X-ray diffraction pattern, previously attributed to polymannuronic acid, may in fact be due to the polyguluronic acid. Nevertheless strong evidence that the two uronic acids can be integral parts of the same molecule is presented in the recent studies of Hirst et al. (58). These authors converted alginic acid (from *Laminaria digitata*) into the 2,3-di-O-propionyl derivative, which is soluble in "diglyme", and then reduced the carboxyl groups with diborane to the corresponding primary alcohol groups, the propionyl groups also being removed at the same time. There was thus produced a neutral polysaccharide, which yielded on partial hydrolysis, 4-O-β-D-mannosyl-D-mannose and, significantly, a disaccharide constituted of D-mannose and L-gulose. The structure of this "mixed" disaccharide was established as 4-O-β-D-mannosyl-L-gulose thus proving that the two uronic acids can occur in the same molecule, the amount of disaccharide isolated presumably eliminating the possibility of its production by acid reversion.

It is seen that the fine structure of alginic acid has not yet been elucidated and it may well be that the term alginic acid refers to a family of polysaccharides ranging from poly-D-mannuronic acid to poly-L-guluronic acid with intermediate polymers containing both uronic acids as shown by (34), (35) and (36).

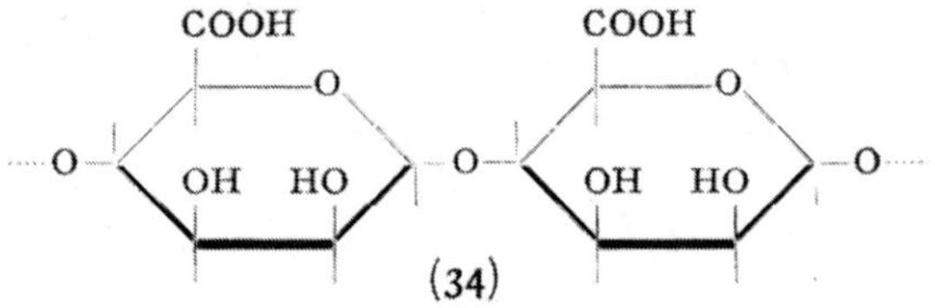

(34)

β-D-Mannuronic acid units in alginic acid.

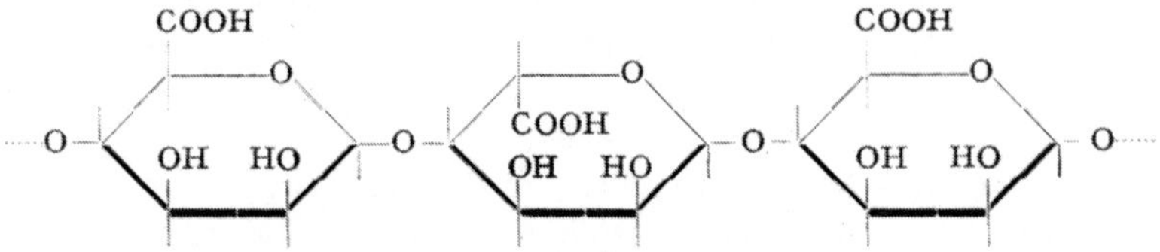

(35)

D-Mannuronic acid and L-guluronic acid units in alginic acid.

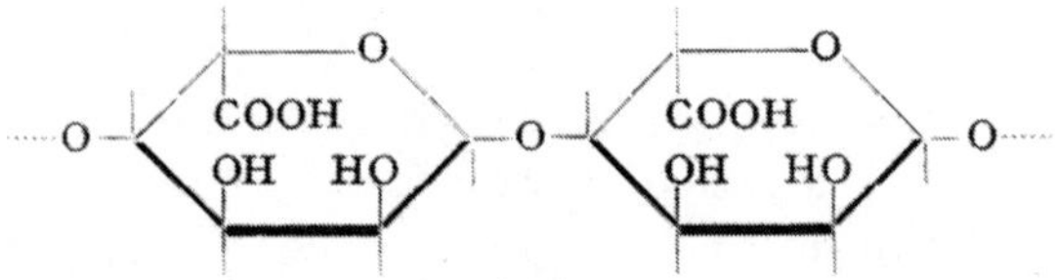

(36)

Poly-L-guluronic acid.

The configuration of the chains in alginic acid is of interest in view of the formal resemblance to cellulose. Evidence from X-ray diffraction measurements suggests a periodicity along the chain axis of 8.7 Å compared with 10.3 Å for cellulose. This has been taken to mean that the units in alginic acid are in the 1 C-chair conformation and, consequently, are subtended at an angle of nearly 90° to one another, whereas in cellulose they are in the C 1-chair conformation and the chains are extended. Molecular weight measurements, carried out by various methods on commercial sodium alginate samples, indicate values ranging from 32,000 to 200,000 (*32, 132*).

It is reasonable to suggest that alginic acid fulfils a somewhat similar rôle in the brown algae to that of pectin in land plants, being part of the amorphous matrix of the cell-wall. Just why it should be found in brown algae and not in the other classes, is a matter for conjecture. Although *D*-glucuronic acid is a constituent of certain polysaccharides of both green algae and, to a smaller extent, of red algae, neither of these algal classes appears to synthesise a polysaccharide akin to alginic acid.

It will be apparent from what has been said in this article that generalisations on seaweed polysaccharides must be made with the greatest caution and then only as hypotheses which will point the way to further research. For instance, we have said that pectin has not been found in marine algae and this is true today, but it may not be true tomorrow. A recent discovery which is significant is that a fresh-water alga, *Nitella translucens*, belonging to the Characeae class and with a close botanical relationship to the green algae, synthesises a poly-galacturonic acid akin to pectic acid (*1 a*).

References.

1. ANDERSON, D. M. W. and N. J. KING: Polysaccharides of the Characeae. Part I. Preliminary Examination of a Starch-type Polysaccharide from *Nitella translucens*. J. Chem. Soc. (London) **1961**, 2914.

1 a. — — Polysaccharides of the Characeae. Part IV. A Non-esterified Pectic Acid from *Nitella translucens*. J. Chem. Soc. (London) **1961**, 5333.

2. ANDERSON, F. B., E. L. HIRST, D. J. MANNERS and A. G. ROSS: The Constitution of Laminarin. Part III. The Fine Structure of Insoluble Laminarin. J. Chem. Soc. (London) **1958**, 3233.

3. ANNAN, W. D., E. L. HIRST and D. J. MANNERS: The Position of Mannitol in Laminarin. Chem. and Ind. **1962**, 984.

4. ARAKI, C.: Agar-agar. III. Acetylation of the Agar-like Substance of *Gelidium amansii* L. J. Chem. Soc. Japan **58**, 1338 (1937).

5 — Chemical Studies of Agar-agar. X. Isolation of 2-Methyl-3,6-anhydro-L-galactose Dimethyl Acetal. J. Chem. Soc. Japan **61**, 775 (1940).

6. — Chemical Studies on Agar-agar. XIII. 1. Separation of Agarobiose from the Agar-agar-like Substance of *Gelidium amansii* by Partial Hydrolysis. J. Chem. Soc. Japan **65**, 533 (1944).

7. Araki, C.: Chemical Studies on Agar-agar. XIII. 2. Chemical Constitution of Agarobiose. J. Chem. Soc. Japan 65, 627 (1944).

8. — Seaweed Polysaccharides. Proc. Intern. Congr. Biochem. 4th Congr. Symposium No. 1, 1 (1958).

9. Araki, C. and K. Arai: The Chemical Constitution of Agar-agar. XVIII. Isolation of a New Crystalline Disaccharide by Enzymic Hydrolysis of Agar-agar. Bull. Chem. Soc. Japan 29, 339 (1956).

10. — — Studies on the Chemical Constitution of Agar-agar. XX. Isolation of a Tetrasaccharide by Enzymatic Hydrolysis of Agar-agar. Bull. Chem. Soc. Japan 30, 287 (1957).

11. Archibald, A. R., D. J. Manners and J. F. Riley: Structure of a Reserve Polysaccharide (Leucosin) from Ochromonas malhamensis. Chem. and Ind. 1958, 1516.

12. Astbury, W. T.: Structure of Alginic Acid. Nature 155, 667 (1945).

13. Barry, V. C.: Preparation, Properties and Mode of Occurrence of Laminarin. Sci. Proc. Roy. Dublin Soc. 21, 615 (1938).

14. — Hydrolysis of Laminarin. Isolation of a New Glucose Disaccharide. Sci. Proc. Roy. Dublin Soc. 22, 423 (1941).

15. Barry, V. C. and T. Dillon: The Xylan of Rhodymenia palmata. Nature 146, 620 (1940).

16. — — A Galactan Sulphuric Ester from Dilsea edulis. Proc. Roy. Irish Acad. 50 B, 349 (1945).

17. Barry, V. C., T. G. Halsall, E. L. Hirst and J. K. N. Jones: The Polysaccharides of the Florideae. Floridean Starch. J. Chem. Soc. (London) 1949, 1468.

18. Barry, V. C. and J. McCormick: Properties of Periodate-oxidised Polysaccharides. VI. The Mucilage from Dilsea edulis. J. Chem. Soc. (London) 1957, 2777.

19. Barry, V. C. and P. W. D. Mitchell: Properties of Periodate-oxidised Polysaccharides. Part IV. The Products Obtained on Reaction with Phenylhydrazine. J. Chem. Soc. (London) 1954, 4020.

20. Beattie, A., E. L. Hirst and E. Percival: Studies on the Metabolism of the Chrysophyceae. Biochem. J. 79, 531 (1961).

21. Black, W. A. P.: The Seasonal Variation in Chemical Constitution of the Sub-littoral Seaweeds Common to Scotland. J. Soc. Chem. Ind. 67, 165 (1948).

22. — The Seasonal Variation in the Combined L-Fucose Content of the Common British Laminariaceae and Fucaceae. J. Sci. Food Agric. 5, 445 (1954).

23. Black, W. A. P., E. T. Dewar and F. N. Woodward: Manufacture of Algal Chemicals. IV. Laboratory-scale Isolation of Fucoidin from Brown Marine Algae. J. Sci. Food Agric. 3, 122 (1952).

24. Bourne, E. J., M. Stacey and I. A. Wilkinson: Composition of the Polysaccharide Synthesized by Polytomella caeca. J. Chem. Soc. (London) 1950, 2694.

25. Brading, J. W. E., M. M. T. Georg-Plant and D. M. Hardy: The Polysaccharide from the Alga, Ulva lactuca. Purification, Hydrolysis, and Methylation of the Polysaccharide. J. Chem. Soc. (London) 1954, 319.

26. Buchanan, J., E. E. Percival and E. G. V. Percival: The Polysaccharides of Carragheen Moss (Chondrus crispus). Part I. The Linkage of the D-Galactose Residues and the Ethereal Sulphate. J. Chem. Soc. (London) 1943, 51.

27. Chanda, S. K., E. L. Hirst, E. G. V. Percival and A. G. Ross: The Structure of Alginic Acid. II. J. Chem. Soc. (London) 1952, 1833.

28. CLANCY, M. J., K. WALSH, T. DILLON and P. S. O'COLLA: The Gelatinous Polysaccharide of *Furcellaria fastigiata*. Sci. Proc. Roy. Dublin Soc. A 1, No. 5, 197 (1960).

29. CLINGMAN, A. L. and J. R. NUNN: Red-seaweed Polysaccharides. III. Polysaccharide from *Hypnea spicifera*. J. Chem. Soc. (London) **1959**, 493.

30. CONCHIE, J. and E. G. V. PERCIVAL: Fucoidin. Part II. The Hydrolysis of a Methylated Fucoidin Prepared from *Fucus vesiculosus*. J. Chem. Soc. (London) **1950**, 827.

31. CONNELL, J. J., E. L. HIRST and E. G. V. PERCIVAL: The Constitution of Laminarin. Part I. An Investigation on Laminarin isolated from *Laminaria cloustoni*. J. Chem. Soc. (London) **1950**, 3494.

32. COOK, W. H. and D. B. SMITH: Molecular Weight and Hydrodynamic Properties of Sodium Alginate. Canad. J. Biochem. and Physiol. **32**, 227 (1954).

33. CORBETT, W. M., J. KENNER and G. N. RICHARDS: Carbonyl Oxycellulose. Chem. and Ind. **1953**, 462.

34. CÔTÉ, R. H.: Disaccharides from Fucoidin. J. Chem. Soc. (London) **1959**, 2248.

35. CRONSHAW, J., A. MYERS and R. D. PRESTON: A Chemical and Physical Investigation of the Cell-walls of some Marine Algae. Biochim. Biophys. Acta **27**, 89 (1958).

36. DILLON, T. and J. McKENNA: The Mucilage of *Dilsea edulis*. Proc. Roy. Irish Acad. **53 B**, 45 (1950).

37. — — The Mucilage of *Dumontia incrassata*. Nature **165**, 318 (1950).

38. DILLON, T. and P. O'COLLA: Acetolysis of Carrageen Mucilage. Nature **145**, 749 (1940).

39. EDDY, B. P., I. D. FLEMING and D. J. MANNERS: α-1 : 4-Glucosans. Part IX. The Molecular Structure of a Starch-type Polysaccharide from *Dunaliella bioculata*. J. Chem. Soc. (London) **1958**, 2827.

40. EPPLEY, R. W.: Sodium Exclusion and Potassium Retention by the Red Marine Alga, *Porphyra perforata*. J. Gen. Physiol. **41**, 901 (1958).

41. FISCHER, F. G. und H. DÖRFEL: Die Polyuronsäuren der Braunalgen. Z. physiol. Chem. **302**, 186 (1955).

42. FISHER, I. S. and E. PERCIVAL: The Water-soluble Polysaccharides of *Cladophora rupestris*. J. Chem. Soc. (London) **1957**, 2666.

43. FLEMING, I. D., E. L. HIRST and D. J. MANNERS: α-1 : 4-Glucosans. Part IV. A Re-examination of the Molecular Structure of Floridean Starch. J. Chem. Soc (London) **1956**, 2831.

44. FOGG, G. E.: The Metabolism of Algae. London: Methuen and Co. 1953.

45. FORBES, I. A. and E. G. V. PERCIVAL: Studies on Agar-agar. Part II. The Isolation of Derivatives of 3 : 6-Anhydro-*L*-galactose from Agar and the Synthesis of their Enantiomorphs. J. Chem. Soc. (London) **1939**, 1844.

46. FREI, E. and R. D. PRESTON: Configuration of Alginic Acid in Marine Brown Algae. Nature **196**, 130 (1962).

47. GOLDSTEIN, I. J., F. SMITH and A. M. UNRAU: Constitution of Laminarin. Chem. and Ind. **1959**, 124.

48. GREENWOOD, C. T. and J. THOMSON: Physicochemical Studies on Starches. Part XXIII. Some Physical Properties of Floridean Starch and the Characterization of Structure-type of Branched α-1,4-Glucans. J. Chem. Soc. (London) **1961**, 1534.

49. HAAS, P. and B. RUSSELL-WELLS: Carrageen *(Chondrus crispus)*. IV. The Hydrolysis of Carrageen Mucilage. Biochem. J. **23**, 425 (1929).

50. HANDS, S. and S. PEAT: Anhydro-*L*-galactose in Agar-agar. Chem. and Ind. **1938**, 937.

51. Hassid, W. Z.: The Structure of Sodium Sulfuric Acid Ester of Galactan from *Irideae laminarioides* (Rhodophyceae). J. Amer. Chem. Soc. **57**, 2046 (1935).
52. Haug, A.: Ion Exchange Properties of Alginate Fractions. Acta Chem. Scand. **13**, 1250 (1959).
53. Haug, A. and B. Larsen: The Solubility of Alginate at Low pH. Acta Chem. Scand. **17**, 1653 (1963).
54. Hirase, S.: Studies on the Chemical Constitution of Agar-agar. XIX. Pyruvic Acid as a Constituent of Agar-agar (Part 2). Isolation of a Pyruvic Acid-linking Disaccharide Derivative from the Methanolysis Products of Agar. Bull. Chem. Soc. Japan **30**, 70 (1957).
55. — Studies on the Chemical Constitution of Agar-agar. XIX. Pyruvic Acid as a Constituent of Agar-agar (Part 3). Structure of the Pyruvic Acid-linking Disaccharide Derivative Isolated from the Methanolysis Products of Agar. Bull. Chem. Soc. Japan **30**, 75 (1957).
56. Hirase, S., C. Araki and T. Itô: Isolation of Agarobiose Derivative from the Mucilage of *Gloiopeltis furcata*. Bull. Chem. Soc. Japan **31**, 428 (1958).
57. Hirst, E. L., J. K. N. Jones and W. O. Jones: The Structure of Alginic Acid. Part I. J. Chem. Soc. (London) **1939**, 1880.
58. Hirst, E. L., E. Percival and J. K. Wold: The Structure of Alginic Acid. Part IV. Partial Hydrolysis of the Reduced Disaccharide. J. Chem. Soc. (London) **1964**, 1493.
59. Hough, L., J. K. N. Jones and W. H. Wadman: An Investigation of the Polysaccharide Components of Certain Fresh-water Algae. J. Chem. Soc. (London) **1952**, 3393.
60. Howard, B. H.: Hydrolysis of the Soluble Pentosans of Wheat Flour and *Rhodymenia palmata* by Ruminal Micro-organisms. Biochem. J. **67**, 643 (1957).
61. Iriki, Y. and T. Miwa: Chemical Nature of the Cell Wall of the Green Algae, *Codium*, *Acetabularia*, and *Halicoryne*. Nature **185**, 178 (1960).
62. Iriki, Y., T. Suzuki, K. Nishizawa and T. Miwa: Xylan of Siphonaceous Green Algae. Nature **187**, 82 (1960).
63. Johnston, R. and E. G. V. Percival: The Polysaccharides of Carragheen. Part III. Confirmation of the 1 : 3-Linkage in Carragheenin, and the Isolation of *L*-Galactose Derivatives from a Resistant Fragment. J. Chem. Soc. (London) **1950**, 1994.
64. Jones, J. K. N.: The Structure of the Mannan present in *Porphyra umbilicalis*. J. Chem. Soc. (London) **1950**, 3292.
65. Kylin, Z.: Zur Biochemie der Meeresalgen. Z. physiol. Chem. **83**, 171 (1913).
66. Larsen, B. and A. Haug: Free-boundary Electrophorisis of Acidic Poly-saccharides from the Marine Alga *Ascophyllum nodosum* (L) Le Jol. Acta Chem. Scand. **17**, 1646 (1963).
67. Lloyd, A. G., K. S. Dodgson, R. G. Price and F. A. Rose: Infrared Studies on Sulphate Esters. I. Polysaccharide Sulphates. Biochim. Biophys. Acta **46**, 108 (1961).
68. Lloyd, K. O.: Ph. D. Thesis, Wales (1960).
69. Love, J., W. Mackie, J. W. McKinnell and E. Percival: Starch-type Polysaccharides Isolated from the Green Seaweeds, *Enteromorpha compressa*, *Ulva lactuca*, *Cladophora rupestris*, *Codium fragile* and *Chaetomorpha capillaris*. J. Chem. Soc. (London) **1963**, 4177.
70. Love, J. and E. Percival: Polysaccharides of the Green Seaweed *Codium fragile*. Biochem. J. **84**, 29 P (1962).
71. Lucas, H. J. and W. T. Stewart: Oxidation of Alginic Acid by Periodic Acid. J. Amer. Chem. Soc. **62**, 1792 (1940).

72. LUNDE, G., E. HEEN und E. ÖY: Untersuchungen über Alginsäure. 1. Über die Konstitution der Alginsäure. Kolloid-Z. **83**, 196 (1938).

73. MCKENZIE, M. E.: Ph. D. Thesis, Edinburgh (1953).

74. MACKIE, I. M. and E. PERCIVAL: The Constitution of Xylan from the Green Seaweed *Caulerpa filiformis*. J. Chem. Soc. (London) **1959**, 1151.

75. — — Polysaccharides from the Green Seaweed *Caulerpa filiformis*. Part II. A Glucan of the Amylopectin Type. J. Chem. Soc. (London) **1960**, 2381.

76. — — Polysaccharides from the Green Seaweeds of *Caulerpa* spp. Part III. Detailed Study of the Water-soluble Polysaccharides of *C. filiformis*: Comparison with the Polysaccharides Synthesised by *C. racemosa* and *C. sertularioides*. J. Chem. Soc. (London) **1961**, 3010.

77. MACKIE, W. and E. PERCIVAL: The Site of Ester Sulphate Groups in the Polysaccharide from the Green Seaweed *Cladophora rupestris*. Biochem. J. **91**, 5 P (1964).

78. MCKINNELL, J. P. and E. PERCIVAL: The Acid Polysaccharide from the Green Seaweed *Ulva lactuca*. J. Chem. Soc. (London) **1962**, 2082.

79. — — Structural Investigations on the Water-soluble Polysaccharide of the Green Seaweed *Enteromorpha compressa*. J. Chem. Soc. (London) **1962**, 3141.

80. MANNERS, D. J. and A. R. ARCHIBALD: α-1 : 4-Glucosans. Part V. End Group Assay of Glycogens by Periodate Oxidation, and the Oxidation of Maltose by Sodium Metaperiodate. J. Chem. Soc. (London) **1957**, 2205.

81. MANNERS, D. J. and J. P. MITCHELL: The Fine-structure of *Rhodymenia palmata* Xylan. Biochem. J. **89**, 92 P (1963).

82. MEEUSE, B. J. D., M. ANDRIES and J. A. WOOD: Floridean Starch. J. Exp. Botany **11**, 129 (1960).

83. MEEUSE, B. J. D. and D. R. KREGER: X-Ray Diffraction of Algal Starches. Biochim. Biophys. Acta **35**, 26 (1959).

84. MEEUSE, B. J. D. and B. N. SMITH: A Note on the Amylolytic Breakdown of some Raw Algal Starches. Planta **57**, 624 (1962).

85. MORGAN, K. and A. N. O'NEILL: Degradative Studies on λ-Carrageenin. Canad. J. Chem. **37**, 1201 (1959).

86. MORI, T.: Seaweed Polysaccharides. Adv. Carbohydrate Chem. **8**, 315 (1953).

87. MYERS, A. and R. D. PRESTON: Fine Structure in the Red Algae. II. The Structure of the Cell Wall of *Rhodymenia palmata*. Proc. Roy. Soc. (London) B **150**, 447 (1959).

88. — — Fine Structure in the Red Algae. III. A General Survey of Cell-wall Structure in the Red Algae. Proc. Roy. Soc. (London) B **150**, 456 (1959).

89. NAKAMURA, T.: Mucilage from the Red Seaweed *Eucheuma spinosum*. Changes in Chemical Composition Due to Various Treatments. Bull. Japan Soc. Sci. Fisheries **20**, 501 (1954).

90. — Mucilage from the Seaweed *Eucheuma muricatum*. IV. Bull. Japan Soc. Sci. Fisheries **23**, 647 (1958).

91. — Mucilage from the Seaweed *Eucheuma muricatum*. VI. Bull. Japan Soc. Sci. Fisheries **24**, 285 (1958).

92. NAYLOR, G. L. and B. RUSSELL-WELLS: On the Presence of Cellulose and its Distribution in the Cell Walls of Brown and Red Algae. Ann. Botany **48**, 635 (1934).

93. NELSON, W. L. and L. H. CRETCHER: The Alginic Acid from *Macrocystis pyrifera*. J. Amer. Chem. Soc. **51**, 1914 (1929).

94. NICOLAI, E. and R. D. PRESTON: Cell Wall Studies in the Chlorophyceae. 1. A General Survey of Submicroscopic Structure in Filamentous Species. Proc. Roy. Soc. (London) B **140**, 244 (1952).

95. O'Colla, P.: Floridean Starch. Proc. Roy. Irish Acad. **55 B**, 321 (1953).

96. — Mucilages. In: R. A. Lewin, Physiology and Biochemistry of Algae, p. 344. New York: Academic Press. 1962.

97. O'Donnell, J. J. and E. Percival: Structural Investigations on the Water-soluble Polysaccharides from the Green Seaweed *Acrosiphonia centralis (Spongomorpha arcta)*. J. Chem. Soc. (London) **1959**, 2168.

98. — — The Water-soluble Polysaccharides of *Cladophora rupestris*. Part II. Barry Degradation and Methylation of the Degraded Polysaccharide. J. Chem. Soc. (London) **1959**, 1739.

99. O'Neill, A. N.: Degradative Studies on Fucoidin. J. Amer. Chem. Soc. **76**, 5074 (1954).

100. — 3,6-Anhydro-*D*-galactose as a Constituent of ϰ-Carrageenin. J. Amer. Chem. Soc. **77**, 2837 (1955).

101. — Derivatives of 4-O-β-*D*-Galactopyranosyl-3,6-anhydro-*D*-galactose from ϰ-Carrageenin. J. Amer. Chem. Soc. **77**, 6324 (1955).

102. Painter, T. J.: The Polysaccharides of *Furcellaria fastigiata*. 1. Isolation and Partial Mercaptolysis of a Gel-fraction. Canad. J. Chem. **38**, 112 (1960).

103. Peat, S., J. R. Turvey and J. M. Evans: The Structure of Floridean Starch. Part I. Linkage Analysis by Partial Acid Hydrolysis. J. Chem. Soc. (London) **1959**, 3223.

104. — — — The Structure of Floridean Starch. Part II. Enzymic Hydrolysis and Other Studies. J. Chem. Soc. (London) **1959**, 3341.

105. Peat, S., J. R. Turvey and D. A. Rees: Carbohydrates of the Red Alga, *Phorphyra umbilicalis*. J. Chem. Soc. (London) **1961**, 1590.

106. Peat, S., J. R. Turvey and P. Wootton-Davies: Unpublished work.

107. Peat, S., W. J. Whelan and H. G. Lawley: Structure of Laminarin. Part I. Main Polymeric Linkage. J. Chem. Soc. (London) **1958**, 724.

108. — — — Structure of Laminarin. Part II. Minor Structural Features. J. Chem. Soc. (London) **1958**, 729.

109. Percival, E. E.: Personal communication.

110. — The Mercaptolysis of the Polysaccharide from *Chondrus crispus*. Chem. and Ind. **1954**, 1487.

111. Percival, E. and J. K. Wold: The Acid Polysaccharide from the Green Seaweed *Ulva lactuca*. Part II. The Site of the Ester Sulphate. J. Chem. Soc. (London) **1963**, 5459.

112. Percival, E. G. V.: Carbohydrate Sulphates. Quart. Rev. (Chem. Soc. London) **3**, 369 (1949).

113. Percival, E. G. V. and S. K. Chanda: The Xylan of *Rhodymenia palmata*. Nature **166**, 787 (1950).

114. Percival, E. G. V. and A. G. Ross: Marine Algal Cellulose. J. Chem. Soc. (London) **1949**, 3041.

115. — — Fucoidin. I. The Isolation and Purification of Fucoidin from Brown Seaweeds. J. Chem. Soc. (London) **1950**, 717.

116. Rees, D. A.: Estimation of the Relative Amounts of Isomeric Sulphate Esters in Some Sulphated Polysaccharides. J. Chem. Soc. (London) **1961**, 5168.

117. — Enzymic Synthesis of 3:6-Anhydro-*L*-galactose within Porphyran from *L*-Galactose 6-Sulphate Units. Biochem. J. **81**, 347 (1961).

118. — The Carrageenan System of Polysaccharides. Part I. The Relation between the ϰ- and λ-Components. J. Chem. Soc. (London) **1963**, 1821.

119. Rees, D. A. and E. Conway: The Structure and Biosynthesis of Porphyran: A Comparison of some Samples. Biochem. J. **84**, 411 (1962).

120. ROELOFSEN, P. A., V. C. DALITZ and C. F. WIJNMAN: Constitution, Submicroscopic Structure and Degree of Crystallinity of the Wall of *Halicystis osterhoutii*. Biochim. Biophys. Acta **11**, 344 (1953).

121. SCHWEIGER, R. G.: Methanolysis of Fucoidan. II. The Presence of Sugars Other than *L*-Fucose. J. Organ. Chem. (USA) **27**, 4270 (1962).

122. SISSON, W. A.: Some X-ray Observations Regarding the Membrane Structure of *Halicystis*. Contribs. Boyce Thompson Inst. **12**, 31 (1941).

123. SMITH, D. B. and W. H. COOK: Fractionation of Carrageenin. Arch. Biochem. Biophys. **45**, 232 (1953).

124. SMITH, D. B., W. H. COOK and J. L. NEAL: Physical Studies on Carrageenin and Carrageenin Fractions. Arch. Biochem. Biophys. **53**, 192 (1954).

125. SMITH, D. B., A. N. O'NEILL and A. S. PERLIN: Studies on the Heterogeneity of Carrageenin. Canad. J. Chem. **33**, 1352 (1955).

126. SMITH, F. and R. MONTGOMERY: The Chemistry of Plant Gums and Mucilages. New York: Reinhold Publ. Corp. 1959.

127. SPONSLER, O. L.: New Data on Cellulose Space Lattice. Nature **125**, 633 (1930).

128. STOLOFF, L. and P. SILVA: An Attempt to Determine Possible Taxonomic Significance of the Properties of Water Extractable Polysaccharides in Red Algae. Econ. Botany **11**, 327 (1957).

129. TURVEY, J. R. and D. A. REES: Isolation of *L*-Galactose-6-sulphate from a Seaweed Polysaccharide. Nature **189**, 831 (1961).

130. TURVEY, J. R. and T. P. WILLIAMS: Sugar Sulphates from the Mucilage of *Porphyra umbilicalis*. Proc. 4th Intern. Seaweed Sympos., p. 370. Paris: Pergamon Press. 1964.

131. UNRAU, A. M. and F. SMITH: A Chemical Method for the Determination of the Molecular Weight of Certain Polysaccharides. Chem. and Ind. **1957**, 330.

132. VINCENT, D. L., D. A. I. GORING and E. G. YOUNG: A Comparison of the Properties of Various Preparations of Sodium Alginate. J. Appl. Chem. (London) **5**, 374 (1955).

133. WHISTLER, R. L. and J. N. BEMILLER: Industrial Gums. Polysaccharides and Their Derivatives. New York: Academic Press. 1959.

134. WHISTLER, R. L. und K. W. KIRBY: Notiz über die Zusammensetzung der Alginsäure von *Macrocystis pyrifera*. Z. physiol. Chem. **314**, 46 (1959).

135. YAPHE, W.: The Determination of x-Carrageenin as a Factor in the Classification of the Rhodophyceae. Canad. J. Bot. **37**, 751 (1959).

(Received, October 21, 1964.)

Der Kohlenhydratstoffwechsel in Gerste, Hafer und Rispenhirse.

Von H. H. SCHLUBACH, München.

Mit 1 Abbildung.

Inhaltsübersicht.

Der Verfasser hat früher über die Kohlenhydrate der Gräser, des Roggens und des Weizens in diesen „Fortschritten" berichtet (25, 26).

I. Die löslichen Kohlenhydrate in der Gerste und ihre Konstitution.

Als erster beobachtete KÜHNEMANN (*17*) außer kristallisiertem Zucker in ungekeimter Gerste eine in Wasser schwerer lösliche Substanz, die links drehte und deshalb „Sinistrin" benannt wurde. Von ihr verschieden, beobachtete er eine zweite Substanz, die in Wasser leichter löslich war und aus dieser Lösung durch Alkohol ausgefällt werden konnte. Dextrine ließen sich aber weder in der ungekeimten noch in der gekeimten Gerste nachweisen.

Im Unterschied zum Roggen konnte MÜNTZ (*20*) in der Gerste keine „Synanthrose" feststellen.

Dagegen gelang es aber TANRET (*45*), aus Gerste eine Verbindung zu gewinnen, die er mit dem aus Roggen erhaltenen *„livosine"* für identisch hielt. Während sie in den grünen Körnern noch verhältnismäßig reichlich vorgefunden wurde, verschwand sie in den reifen Körnern vollständig.

COLIN (*9*) und BELVAL (*7*) beobachteten Fructoseanhydride in den Achsen und Blattscheiden der Gerste.

Literaturverzeichnis: SS. 57—60.

Archbold und Barter (*4*) konnten aus getrockneten Gerstenblättern, die sie geschnitten hatten, als die Ähren aus ihren Schäften hervorkamen, durch Auskochen mit 75—80% Äthanol und fraktionierte Fällungen (zuletzt der Barytverbindungen) zwei Verbindungen mit den Drehungen $[\alpha]_D^{15} = -27°$ und $-37°$ isolieren, von denen die erstere sich als leichter löslich in Wasser und rascher hydrolysierbar durch Hefeinvertin erwies als die letztere. Bei der Hydrolyse wurden 94% Fructose und als Rest ein anderer Zucker erhalten.

Haworth, Hirst und Lyne (*16*) haben diese Verbindungen näher analysiert. Die bei $[\alpha]_D^{20} = -35°$ (Wasser) drehende Verbindung, welche noch 6% Wasser enthielt, ließ sich zu 99% zu einem Gemisch von 93% Fructose und etwa 6% Aldose hydrolysieren. Bei der Acetylierung mit Essigsäureanhydrid in Pyridin wurde, neben einer in Wasser schwer löslichen Acetylverbindung von $[\alpha]_D^{20} = -27°$ (Chloroform), eine solche von $[\alpha]_D^{20} = +11°$ erhalten. Der Methyläther von $[\alpha]_D^{20} = -50°$ (Chloroform) ergab bei der Hydrolyse die 1,3,4-Trimethylfructose.

Eine gute Übersicht über den Stand unserer damaligen Kenntnisse über die Polyfructosane hat Archbold gegeben (*3*).

In weiteren eingehenden Untersuchungen haben Archbold und Mitarbeiter (*4, 5*) die Rolle der Fructosane beim Kohlenhydratstoffwechsel in der Gerste und die Änderungen im Kohlenhydratgehalt in den verschiedenen Organen der Gerstenpflanze während einer Vegetationsperiode verfolgt. Vor allen Dingen haben sich Archbold und Mukerjee (*5*) mit der Frage beschäftigt, wieweit die in den Blättern, Blattscheiden, Achsen und unreifen Ähren gebildeten oder angesammelten Kohlenhydrate dem Aufbau der Stärke in der reifen Ähre dienen. Hierzu hat Müntz (*20*) bereits 1878 bemerkt, daß sich in den unreifen Ähren das Verhältnis von Polyfructosan (Synanthrose) zu Stärke bis zur Reife von 45,0% : 24,55% am 25. Mai auf 6,85% : 68,75% am 12. Juli verschiebt. Und er hat daher geschlossen: [das Fructosan] „est remplacé par l'amidon qui se forme, sans aucun doute, à ses dépens".

Zu einem ähnlichen Ergebnis ist 1924 Belval (*7*) beim Weizen gelangt. Da jedoch die Fructosankonzentration von 6,07% auf 0,76% fiel, während diejenige der Stärke von 4,94% auf 61,84% anstieg, hat er geschlossen, daß nur ein Teil der Stärke aus dem in den Ähren enthaltenen Fructosan aufgebaut werden konnte. Bei der Gerste begegnete Archbold schon 1938 (*2*) ähnlichen Verhältnissen und schloß daraus: „At present all that can be said is that fructosan may be in part the precursor of starch." In der oben angeführten Arbeit von Archbold und Mukerjee (*5*) ist die Frage der Herkunft der zum Aufbau der Stärke dienenden Kohlenhydrate eingehend untersucht. Die Forscherinnen sind dabei zu dem Ergebnis gelangt, daß diese nicht nur durch die in den Blättern erfolgende Assimilation geliefert werden, sondern auch durch die in den Blattscheiden, den Achsen und den Ähren sich vollziehende, und daß aus diesen Ursprüngen mindestens 80% des zum Aufbau der Stärke erforderlichen Materials geliefert werden und nur etwa 20% aus den angesammelten Polyfructosanen. Damit wurde eine schon 1901 von Dehérain und Dupont (13) geäußerte Ansicht bestätigt, daß beim Weizen die in den Halmen erfolgende Assimilation einen wesentlichen Faktor für den Aufbau der Stärke bildet.

Russell (24) hat die Wirkung von Mineralsalzmangel auf den Fructosanstoffwechsel der Gerstenpflanze untersucht. Danach senkt Kaliummangel, steigert aber Phosphormangel den Fructosangehalt. Auf gleiche Zuckergehalte bezogen, enthält der Halm mehr Fructose als die Blüte.

Richards und Shih (23) haben die Wirkung des Kaliums auf Wassergehalt und Zusammensetzung der Blätter verfolgt.

In Gerstenwurzeln konnte Hassid (15) ein wasserlösliches Glucosan nachweisen.

Täufel und Müller (46) bestätigten, daß auch in der reifen Gerste noch geringe Mengen einer Verbindung enthalten sind, die bei der Hydrolyse Fructose liefert.

Schlubach und Rathje (41) gewannen durch Alkoholextraktion aus unreifen Gerstenähren ein Polyfructosan, das sie „Kritesin" genannt haben. Seine Bausteinanalyse nach der Methylierungsmethode ergab jedoch, daß die Verbindung nicht einheitlich war, in der Hauptsache aber in ihr die Fructosereste in der 1,2-Stellung miteinander verbunden sind; sie ist mithin nach dem Inulintyp* gebaut und insofern von der von Haworth (16) beschriebenen verschieden.

In einer eingehenden Untersuchung erhielt MacLeod (18) aus ungekeimter Gerste durch Wasserextraktion ein Kohlenhydratgemisch, das durch Trennung an einer Kohlesäule nach Whistler und Durso (47) und Papierchromatographie in eine Reihe von Oligosacchariden zerlegt werden konnte. Ein Trisaccharid wurde so als eine Gluco-difructose charakterisiert. Unter der Annahme, daß nur jeweils eine endständige Glucose im Molekül enthalten ist, konnte auch die Anwesenheit einer Gluco-tetra- und Gluco-pentafructose wahrscheinlich gemacht werden. Bei der höchstmolekularen Komponente wurde aus dem Verhältnis von Glucose : Fructose 1 : 10 auf einen Durchschnitts-Polymerisationsgrad von 11 Hexoseeinheiten geschlossen.

Porter und Edelman (21) konnten in Extrakten aus Gerstenhalmen eine Reihe von Oligo- und Polysacchariden nachweisen, bei der ein ab-

---

* Polyfructosane mit 2—1 Bindungen zwischen den Fructofuranose-Bausteinen, wie im Inulin, werden, als nach dem „Inulintyp" gebaut, von denjenigen mit 2—6 Bindungen, wie im *Phleum pratense*, als „Phleintyp" unterschieden.

Literaturverzeichnis: SS. 57—60.

nehmender R_f-Wert mit geringer werdendem Glucosegehalt und verlangsamter Inversionsgeschwindigkeit durch Hefeninvertin parallel liefen.

SCHLUBACH und HABERLAND (32) konnten das Kritesin rein erhalten, indem sie von Gerstenhalmen ausgingen, die zwei Wochen vor der Blüte geschnitten und von den Ähren befreit waren. Durch fraktionierte Fällung der Chloroformlösungen der Acetylverbindungen mit Petroläther und Verseifung wurde ein Polysaccharid von $[\alpha]_D^{20} = -39,3°$ (Wasser, $c = 1$) erhalten, das einen Aldosewert von 4,6% zeigte. Bei der Methylierungsanalyse fielen außer einem Gemisch der 2,3,4,6-Tetramethylglucose mit der 1,3,4,6-Tetramethylfructose, die 3,4,6-Trimethylfructose und die 3,4-Dimethylfructose im Verhältnis von 7 : 10 : 5, bezogen auf Fructosen, an. Der Befund von SCHLUBACH und RATHJE (41), daß das Kritesin nach dem Inulintyp gebaut ist, konnte also bestätigt werden. Und auf Grund der analytischen Daten wurde das folgende Formelbild vorgeschlagen:

$$\text{Gl 1} - \begin{bmatrix} \text{2 Fr 1} - \text{2 Fr 1} - \text{2 Fr 1} \\ \text{6}| \\ \text{2 Fr} \qquad \text{Kritesin.} \end{bmatrix}_5 - \text{2 Fr}$$

Aus den Mutterlaugen konnte ein anderes, Hordeacin benanntes Polysaccharid isoliert werden, das eine Drehung von $[\alpha]_D^{20} = -33,5°$ (Wasser, $c = 1,2$) zeigte und nach Hydrolyse einen Aldosewert von 9,4%. Bei der Hydrolyse des Methyläthers wurde die Tetramethylfraktion, die aus einem Gemisch der 2,3,4,6-Tetramethylglucose mit der 1,3,4,6-Tetramethylfructose im Verhältnis 1 : 3 bestand, die Trimethylfraktion, die aus der 1,3,4-Trimethylfructose, und die Dimethylfraktion, die aus der 3,4-Dimethylfructose bestanden, im Verhältnis 2 : 5 : 4 gefunden. Hieraus, wie aus dem Aldosewert von 9,4%, wurde auf ein Polyfructosan vom Phleintyp mit 2—6 Bindungen und einer Teilchengröße von 11 Hexoseeinheiten geschlossen und das folgende Formelbild entworfen:

$$\text{Gl 1} - \begin{bmatrix} \text{2 Fr 6} - \text{2 Fr 6} - \text{2 Fr 6} \\ \text{1}| \\ \text{2 Fr} \qquad \text{Hordeacin.} \end{bmatrix}_2 - \text{2 Fr 6} - \text{2 Fr}$$

Nach diesen Werten waren dem Hordeacin von HAWORTH und Mitarb. (16) noch etwa 20% Kritesin beigemengt gewesen, dem Kritesin von SCHLUBACH und RATHJE (41) noch etwas Hordeacin.

Da hier aus dem gleichen Organ, den Gerstenhalmen, ein Polyfructosan vom Inulintyp (Kritesin) neben einem solchen vom Phleintyp (Hordeacin) angetroffen wurde, mußte die frühere Annahme, daß die ersteren in den Ähren, die letzteren in den Halmen gebildet werden, aufgegeben werden.

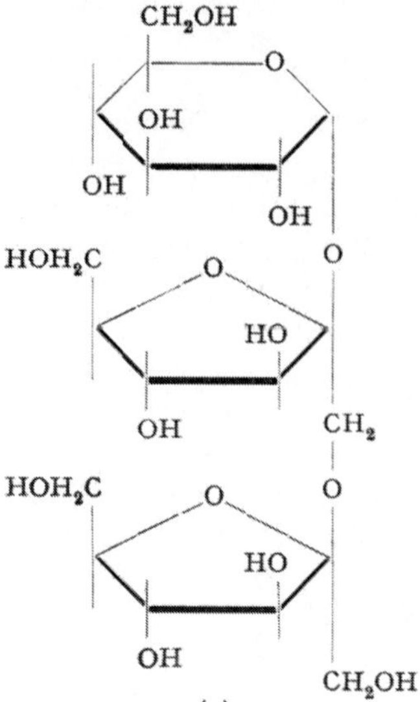

CH$_2$OH

OH

OH

OH

HOH$_2$C

O

O

HO

OH   CH$_2$

HOH$_2$C

O

O

HO

OH   CH$_2$OH

(1)
Isokestose.

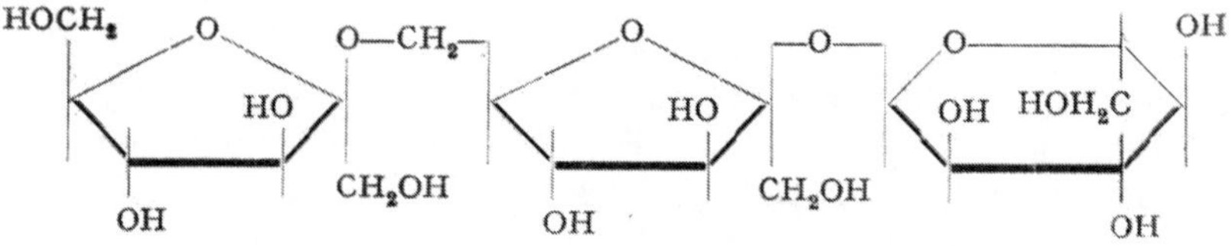

HOCH$_2$ O   O—CH$_2$—   O   —O—   O   OH

HO   HO   OH HOH$_2$C

OH   CH$_2$OH   OH   CH$_2$OH   OH

(2)
Kestose.

CH$_2$OH

O

OH

OH

OH

HOH$_2$C   O   —O—H$_2$C   O   O

HO   HO

OH   OH   CH$_2$

HOH$_2$C   O   O

HO

OH   CH$_2$OH

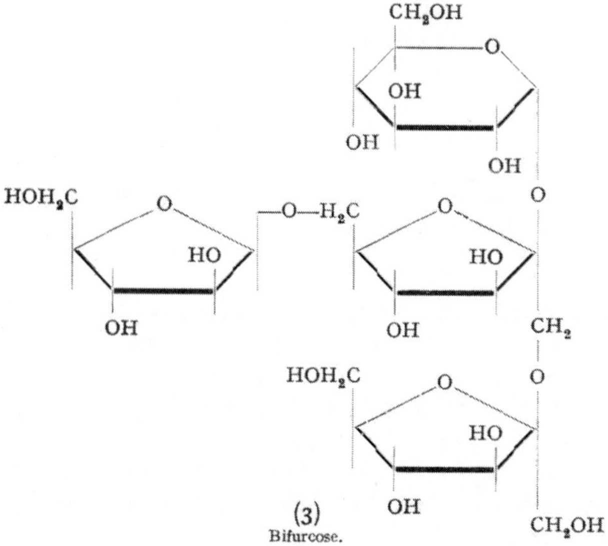

(3)
Bifurcose.

Literaturverzeichnis: SS. 57—60.

Da sich die beiden Polyfructosane der Gerste in die beim Roggen gefundene Reihe nach ihren Drehungen einfügen ließen *(Abb. 1)*, lag die Annahme nahe, daß sich auch ihr Aufbau in analoger Weise vollzieht, wie dies beim Roggen nachgewiesen ist. Von SCHLUBACH, BERNDT und

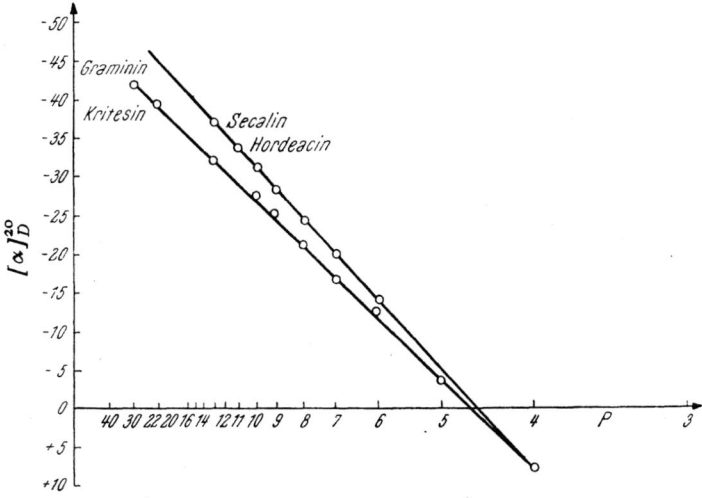

Abb. 1. Die Beziehungen zwischen den Drehungen und Polymerisationsstufen (P) der beiden Reihen von Oligo- und Polysacchariden in den Roggenhalmen und der Gerste. [Aus: Liebigs Ann. Chem. 614, 126 (1958); s. auch 665, 192 (1963).]

CHIEMPRASERT (*29*) wurden deshalb die begleitenden Oligosaccharide herausgearbeitet. Neben einer Bestätigung des Kritesins und Hordeacins konnten die Isokestose (**1**), die Kestose (**2**), die Bifurcose (**3**) und ein Pentasaccharid vom Phleintyp nachgewiesen werden, also mit Ausnahme des letzteren die gleichen Oligosaccharide wie beim Roggen. Der Aufbau der Polysaccharide in der Gerste erfolgt also in der gleichen Weise wie im Roggen.

II. Die löslichen Kohlenhydrate im Hafer und ihre Konstitution.

MÜNTZ (*20*) konnte, im Unterschied zum Roggen, im Hafer *(Avena sativa)* keine „Synanthrose" finden. Ebensowenig konnte TANRET (*45*) weder im grünen noch im reifen Hafer „lévosine" beobachten.

Erst SCHULZE und FRANKFURT (*44*) gelang es, aus Hafersamen eine kohlenhydrat-artige Substanz zu isolieren, welche in konzentriertem Äthanol schwerer löslich war als die Saccharose. In einer weiteren Arbeit hat SCHULZE (*43*) die Reinigung der „Secalin" genannten Verbindung beschrieben. Über die Strontiumverbindung und fraktionierte Fällung der wässerigen Lösungen mit Äthanol erhielt er ein Polysaccharid mit

4*

der Drehung $[\alpha]_D = -31,7°$ (Wasser, $c = 7,0$), das er als identisch mit der aus grünen Roggenpflanzen gewonnenen „Secalose" ansah.

Rawton (22) konnte aus dem Pericarp des Haferkornes drei Verbindungen kristallisiert erhalten. Von diesen wurde die eine als Vanillinglucosid angesprochen und ihr die erregende Wirkung des Hafers auf Pferde zugeschrieben. Colin und Belval (10) bestätigten das Vorkommen von „lévulosane" in den unreifen Halmen und den reifenden Körnern. Im Unterschied zum Roggen, dem Weizen und der Gerste, verschwindet aber nach Belval (7) die linksdrehende Verbindung bei der Reife vollständig, und im Hafermehl ist infolgedessen von ihr nichts mehr enthalten.

Morris (19) erhielt aus Haferkörnern zwei Polysaccharide, welche als Lichenin und ein Araban angesprochen wurden. Aus jungen Haferhalmen wurde von Schlubach und Müller (39) durch Alkoholextraktion ein Polysaccharid von $[\alpha]_D^{20} = -38,2°$ (Wasser, $c = 1$) isoliert, das bei der Hydrolyse nur Fructose ergab. Bei der Hydrolyse des Methyläthers wurden die 1,3,4,6-Tetramethylfructose, die 1,3,4-Trimethylfructose und die 3,4-Dimethylfructose im Verhältnis $1 : 5 : 1$ erhalten. Danach ist dieses „Avenarin" benannte Polyfructosan nach dem Phleintyp gebaut. In Übereinstimmung hiermit drehte die Acetylverbindung positiv: $[\alpha]_D^{20} = +10,1°$ (Chloroform, $c = 1$). Die Teilchengröße sollte 7 Fructoseeinheiten oder einem Vielfachen dieser Zahl entsprechen.

Aus unreifen Haferähren konnten dann Schlubach und Hauschildt (33) ein ähnliches Polysaccharid gewinnen. Es drehte $[\alpha]_D^{20} = -30,4°$ (Wasser, $c = 1$) und ergab bei der Säurehydrolyse außer Fructose einen Aldosewert von 2,8%. Bei der Hydrolyse des Methyläthers wurden die 1,3,4,6-Tetramethylfructose, die 1,3,4-Trimethylfructose und eine als 1,3-Dimethylfructose angesprochene Verbindung im Verhältnis von $1 : 3 : 1$ erhalten. Abweichend von den im Roggen, dem Weizen und der Gerste in den Ähren angetroffenen Polyfructosanen vom Inulintyp wurden also in den Haferähren, ebenso wie in den Halmen, Polyfructosane vom Phleintyp gefunden. Da bei der Invertinspaltung außer unangegriffenem Polyfructosan nur Fructose, aber keine Oligosaccharide zu beobachten waren, wurde angenommen, daß bei ersterer zunächst nur Seitenketten abgetrennt werden.

(4)

Neokestose.

Literaturverzeichnis: SS. 57—60.

Um den Weg des Aufbaues der Polyfructosane im Hafer kennen-zulernen, wurden von SCHLUBACH und BERNDT (27) Haferhalme vor und nach der Blüte untersucht. Aus den Extrakten aus etwa drei Wochen vor dem Erscheinen der Ähren geschnittenem Material wurden außer 72,5% an Mono- und Disacchariden 22,9% eines Gemisches von Oligo- und Polysacchariden erhalten. Durch seine Zerlegung an einer Kohle-säule und Feintrennung durch Papierchromatographie konnten außer Saccharose die Trisaccharide Neokestose (4) und die Kestose (2) sowie das Tetrasaccharid Bifurcose (3) nachgewiesen werden. Außerdem wurde ein der letzteren isomeres Tetrasaccharid entdeckt. Es ist durch Trans-fructosidierung auf dem Glucoseteil der Neokestose entstanden und wurde deshalb „Neobifurcose" benannt.

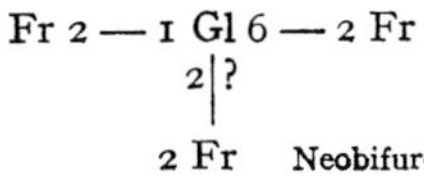

$$\text{Fr } 2 - 1 \text{ Gl } 6 - 2 \text{ Fr}$$
$$2 | ?$$
$$2 \text{ Fr} \quad \text{Neobifurcose.}$$

Nach diesem Befund wurde in dem nach der Blüte geschnittenen Material ein noch komplexeres, durch weitere Transfructosidierung in verschiedenen Richtungen aufgebautes Gemisch von Polysacchariden erwartet. Es wurden Oligo- und Polysaccharide aller Stufen von den Polymerisationsgraden 4 bis 20 gefunden. Bei der Analyse nach der Methylierungsmethode wurde festgestellt, daß, wie bei den Gräsern, die Transfructosidierung im allgemeinen bei der sechsten Hydroxylgruppe der Fructosehälfte der Saccharose ansetzt, sie also zum Phleintyp gehören. Und daß insofern die Befunde von SCHLUBACH und MÜLLER (39) sowie von SCHLUBACH und HAUSCHILDT (33) bestätigt wurden; daß sie sich aber in geringem Maße wie bei der Neokestose und der Neobifurcose auch auf die Glucosehälfte erstreckt hat. Es wird daher nach dem folgenden Reaktionsschema ein Aufbau in zwei parallelen Reihen angenommen:

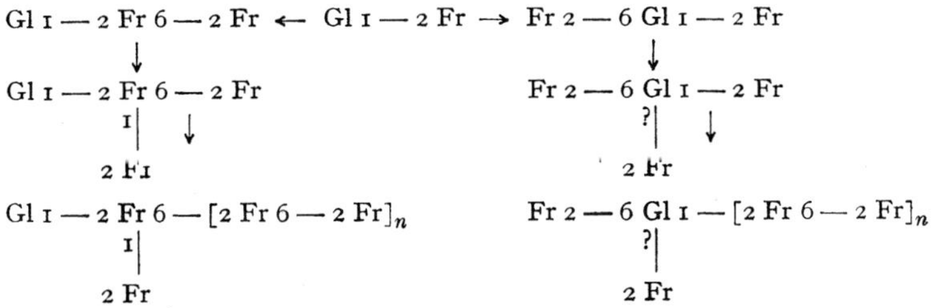

$$\text{Gl } 1 - 2 \text{ Fr } 6 - 2 \text{ Fr} \leftarrow \text{Gl } 1 - 2 \text{ Fr} \rightarrow \text{Fr } 2 - 6 \text{ Gl } 1 - 2 \text{ Fr}$$
$$\downarrow \qquad\qquad\qquad \downarrow$$
$$\text{Gl } 1 - 2 \text{ Fr } 6 - 2 \text{ Fr} \qquad\qquad \text{Fr } 2 - 6 \text{ Gl } 1 - 2 \text{ Fr}$$
$$1 | \quad \downarrow \qquad\qquad\qquad ? | \quad \downarrow$$
$$2 \text{ Fr} \qquad\qquad\qquad\qquad 2 \text{ Fr}$$
$$\text{Gl } 1 - 2 \text{ Fr } 6 - [2 \text{ Fr } 6 - 2 \text{ Fr}]_n \qquad \text{Fr } 2 - 6 \text{ Gl } 1 - [2 \text{ Fr } 6 - 2 \text{ Fr}]_n$$
$$1 | \qquad\qquad\qquad\qquad\qquad ? |$$
$$2 \text{ Fr} \qquad\qquad\qquad\qquad\qquad 2 \text{ Fr}$$

Schema 1. Aufbau der Oligo- und Polysaccharide im Hafer in zwei parallelen Reihen (27).

Während noch die Oligosaccharide des Hafers manche Ähnlichkeit mit denjenigen des Roggens und Weizens aufweisen, nähern sich die Polysaccharide mit zunehmendem Polymerisationsgrad denjenigen der

Gräser. Von den letzteren unterscheiden sie sich durch eine geringere Verzweigung. Die Mittelstellung des Hafers zwischen den Gräsern und den Getreidearten gelangt also auch in diesen Verhältnissen zum Ausdruck.

III. Die löslichen Kohlenhydrate in der Rispenhirse und ihre Konstitution.

Unter den zahlreichen Hirsearten hat die Rispenhirse *(Panicum miliaceum)* schon frühzeitig eine große Bedeutung gewonnen. Nach HAHN (*14*) soll sie die älteste der landwirtschaftlich genutzten Kulturpflanzen gewesen sein. Während aber in der Folgezeit ihr Anbau in Westeuropa seit dem Mittelalter in dem Maße zurückgegangen ist, in dem die Breinahrung durch das Brot verdrängt wurde, bildet sie heute noch für die Bevölkerung einer vom nordwestlichen China bis zum Kaukasus reichenden Zone, außerdem in Teilen Afrikas einen Hauptbestandteil der Nahrung. Ihr Anteil an den Anbauflächen der Getreidearten der Welt übertrifft daher denjenigen des Roggens und erreicht nahezu zwei Drittel derjenigen von Mais und Reis.

Ungeachtet dieser Bedeutung lagen bisher keine Untersuchungen über den Kohlenhydratstoffwechsel in der Rispenhirse vor. Lediglich in der Zuckerhirse *(Sorghum saccharatum)* wurde von verschiedenen Forschern die Anwesenheit von Saccharose, Glucose und Fructose festgestellt. Diese Lücke in unserer Kenntnis des Kohlenhydratstoffwechsels wichtiger Getreidearten ist jetzt durch eine Untersuchung von SCHLUBACH und ZIEGLER (*42*) geschlossen.

Nach den bei anderen Getreidearten gemachten Erfahrungen, daß die stärkste Ansammlung von löslichen Oligo- und Polysacchariden in ihnen vor dem Ährenschieben angetroffen wird, wurde dieser Zeitpunkt für den Schnitt der Rispenhirse gewählt. Das Gemisch von löslichen Kohlenhydraten wurde in der üblichen Weise durch Alkoholextraktion und nachfolgende Enteiweißung gewonnen und durch eine kombinierte Anwendung der Trennungsmethoden an einer Kohlesäule, einer Sephadexsäule und der Papierchromatographie in seine Komponenten zerlegt. Es wurden so die folgenden Kohlenhydrate isoliert und in ihren Konstitutionen bestimmt:

a) Ein Polyfructosan vom Phleintyp mit einer Kette von 16 Hexoseeinheiten, die nur zwei Verzweigungen enthielt. Nach den analytischen Daten wurde dafür das folgende Formelbild vorgeschlagen:

$$\text{Gl } 1 - 2\text{ Fr } 6 - 2\text{ Fr } 6 - 2\text{ Fr } 6 - [2\text{ Fr } 6]_9 - 2\text{ Fr}$$
$$\overset{1}{\big|} \qquad\qquad \overset{1}{\big|}$$
$$2\text{ Fr} \qquad\qquad 6\text{ Fr}$$

Hexadecasaccharid aus der Rispenhirse.

Literaturverzeichnis: SS. 57—60.

In seinen Eigenschaften kommt es dem aus Weizen (*37*) isolierten Dodekasaccharid nahe.

   b) Vier Oligosaccharide:

   Die Kestose Gl 1 — 2 Fr 6 — 2 Fr.
   Die Isokestose Gl 1 — 2 Fr 1 — 2 Fr.
   Die Neokestose Fr 2 — 1 Gl 6 — 2 Fr.
   Die Bifurcose Gl 1 — 2 Fr 6 — 2 Fr.
$$\left| 1 \right.$$
   2 Fr

   c) Glucose mit 4,2 und 6,1%.
   Fructose mit 1,1 und 1,0%.
   Saccharose mit 54,4 und 54,0%.

Grundsätzlich werden also in der Rispenhirse analoge Polyfructosane vom Phleintyp gebildet wie in den Gräsern und den anderen Getreidearten. In dem Grade ihrer Verzweigung stehen sie zwischen den unverzweigten Polysachariden der Gräser und den stark verzweigten der anderen Getreidearten. Auch der Aufbau durch Transfructosidierung auf der sechsten Hydroxylgruppe der Fructosehälfte der Saccharose erfolgt in der gleichen Weise. Die Neokestose (4) läßt erkennen, daß daneben in geringem Ausmaße, ähnlich wie beim Hafer (*27*), auch der Aufbau auf der Glucosehälfte der Saccharose eintreten kann. Aber es besteht insofern ein großer Unterschied gegenüber den Gräsern und anderen Getreidearten, als der Gesamtanteil der Oligo- und Polysaccharide an den löslichen Kohlenhydraten nur ein geringer ist. Während dieser z. B. beim *Lolium perenne* 92,2% (*35*), beim Weizen 29,0% (*37*) und beim Hafer 22,9% (*27*) betragen hatte, erreichte er bei der Rispenhirse nur 3,3%. Andererseits, während er bei den Gräsern und anderen Getreidearten immer nur einige wenige Prozente an Saccharose betragen hatte, stieg er bei der Rispenhirse auf 54%. Es ist also bei ihr die Polymerisation der Saccharose nur in ganz geringem Maße eingetreten. Und sie nähert sich in ihrem Kohlenhydratstoffwechsel der Zuckerhirse, in der nur Saccharose angetroffen wurde.

IV. Die enzymatische Steuerung des Kohlenhydratstoffwechsels in den Getreidearten.

Für die enzymatische Steuerung der Transfructosidierung sind außer der Saccharase und der Phleinase auch die Inulase, wenn auch in geringer Konzentration, wirksam, wie aus der Auffindung der Isokestose (1) und der Bifurcose (3) hervorgeht. Ob für die Transfructosidierung auf der Glucosehälfte der Saccharose wie bei der Bildung der Neokestose (4) ein besonderes Enzym anzunehmen ist, bleibt noch eine offene Frage.

Grundlage für das Verständnis des Aufbaues der Polyfructosane in den Getreidearten sind die Vorstellungen über die enzymatische Gruppenübertragung der Fructofuranose-Hälfte der Saccharose, die Transfructosidierung, wie sie von Dedonder (*11, 12*) sowie von Bacon und Edelman (*6*) entwickelt worden sind.

Zur Deutung der außer in der Gerste schon vorher beim Roggen festgestellten gleichzeitigen Bildung von Polyfructosanen vom Inulin- *und* Phleintyp wurde angenommen, daß in den Getreidearten *zwei* verschiedene Polyasen wirksam sind (*36*). Bei der einen, der Inulase, erfolgt die Transfructosidierung auf der ersten, bei der anderen, der Phleinase, auf der sechsten Hydroxylgruppe der Fructofuranose-Hälfte der Saccharose. Bei den Inulin führenden Pflanzen ist nur die Inulase, bei den Gräsern nur die Phleinase wirksam. Bei den Getreidearten sind *beide* Enzymarten am Aufbau beteiligt. Als erste Stufe dieser simultanen Wirkung auf die erste *und* die sechste Hydroxylgruppe wurde beim Roggen das Tetrasaccharid Bifurcose (*3*) entdeckt (*36*).

Der erste Nachweis der so geforderten Verschiedenheit der Inulase und der Phleinase konnte erbracht werden, indem in dem aus *Aspergillus niger* und *A. oryzae* erhaltenen Enzymgemisch die Inulase durch Streptomycin gehemmt wurde, die Phleinase nicht (*34*). Weiter konnte gezeigt werden, daß auch in einem im Hefeninvertin enthaltenen Enzymgemisch die Inulase sowohl durch Streptomycin wie auch durch Isonicotinsäurehydrazid und p-Aminosalicylsäure gehemmt wird (*31*). Es konnte auch gezeigt werden, daß auch das die Saccharose spaltende Enzym, die Saccharase oder β-Fructofuranosidase, von der Inulase verschieden ist, wie dies schon früher angenommen war (*1*), da sie durch Streptomycin nicht gehemmt wird. Weiter konnte wahrscheinlich gemacht werden, daß auch die Phleinase von der Saccharase verschieden ist, indem durch partielle Schwermetallvergiftung der Quotient Saccharase-Phleinasewirkung verschoben werden konnte (*40*).

Auf Grund dieser Befunde sowie der Wirkungsweise der aus *Bacillus mesentericus* erhaltenen Lävanpolyase, welche Saccharose quantitativ in Glucose und Lävan zu verwandeln vermag (*28*), wurde jetzt für die Wirkung der Polyfructosanasen das folgende Schema vorgeschlagen:

Die Saccharose wird durch die Saccharase, die β-Fructofuranosidase, in Glucose und den Fructofuranose-Enzymkomplex zerlegt. Aus dem letzteren löst als Acceptor die Polyase den Fructofuranose-Rest ab. Und dieser baut unter Übertragung des Restes auf unveränderte Saccharose oder Oligosaccharide die Polyfructosane auf. Dieser letztere Vorgang konnte auch in vitro durchgeführt werden, indem es gelang, ein Phlein[20] in zwei Stufen in einer konzentrierten Saccharoselösung bei Gegenwart von Hefeinvertin zu einem Phlein[41] aufzubauen (*38*), entsprechend der in der Natur beobachteten, bei *Lolium perenne* (*35*) und *Phleum pratense* (*30*)

Literaturverzeichnis: SS. 57—60.

im Laufe der Vegetationsperiode eintretenden Erhöhung des Polymerisationsgrades (P). Dieser Aufbauweg gilt für den Fall, daß die Transfructosidierung auf der Fructose-Hälfte der Saccharose erfolgt. Es werden so Polyfructosane erzeugt, in denen Glucose stets das eine Ende der Kette bildet und ihr Anteil 1/P beträgt. Wenn dagegen die primäre Transfructosidierung auf der Glucose-Hälfte der Saccharose erfolgt, wie dies durch die Auffindung des Trisaccharids Neokestose im Weizen (*37*), im Hafer (*27*) und in der Rispenhirse (*42*) und in der nächsten Stufe der Entdeckung des Tetrasaccharids Neobifurcose zu folgern ist, so bildet die Glucose nicht mehr das Endglied der Kette.

Wenn endlich, wie bei den Gräsern gefunden wurde, die Transfructosidierung primär auf der Fructose zu einer Difructose (*25*) geschieht, werden so glucose-freie Polyfructosane aufgebaut.

V. Zur Geschichte der Kultur der Getreidearten.

Vergleicht man die Zunahme der Anbauflächen, welche in historischer Zeit die vier Getreidearten Rispenhirse, Gerste, Weizen und Roggen erfahren haben, mit den Gehalten an Oligo- und Polyfructosanen in ihnen, so ergibt es sich, daß die ersteren in dem Maße erweitert wurden, in dem die letzteren zunehmen. Der Mensch hat also den Anbau derjenigen Getreidearten bevorzugt, in denen die Ansammlung von Polyfructosanen am stärksten erfolgt ist. Das ist verständlich, wenn man bedenkt, daß das Ziel dieser Entwicklung die Gewinnung von möglichst stärkereichen Getreidekörnern gewesen ist, die letzteren aber, wie oben ausgeführt, wenigstens zum Teil bei der Reife, aus den angesammelten Polyfructosanen aufgebaut werden. In Fortsetzung dieser Tendenz kann daher erwartet werden, daß es möglich ist, durch Züchtung von Getreidearten, welche noch stärker als bisher Polyfructosane ansammeln, auch noch höhere Körnererträge zu erreichen.

Literaturverzeichnis.

1. ADAMS, M., N. K. RICHTMYER and C. S. HUDSON: Some Enzymes Present in Highly Purified Invertase Preparations; a Contribution to the Study of Fructofuranosidases, Galactosidases, Glucosidases and Mannosidases. J. Amer. Chem. Soc. **65**, 1369 (1943).
2. ARCHBOLD, H. K.: Physiological Studies in Plant Nutrition. VII. The Role of Fructosans in the Carbohydrate Metabolism of the Barley Plant. 1. Materials Used and Methods of Sugar Analysis Employed. — 2. Seasonal Changes in the Carbohydrates, with a Note on the Effect of Nitrogen Deficiency. Ann. Botany [N. S.] **2**, 183, 403 (1938).
3. — Fructosans in the Monocotyledons. New Phytologist **39**, 185 (1940).
4. ARCHBOLD, H. K. and A. M. BARTER: A Fructose Anhydride from the Leaves of the Barley Plant. Biochem. J. **29**, 2689 (1935).

5. Archbold, H. K. and B. N. Mukerjee: Physiological Studies in Plant Nutrition. XII. Carbohydrate Change in the Several Organs of the Barley Plant during Growth, with especial Reference to the Development and Ripening of the Ear. Ann. Botany [N. S.] 6, 1 (1942).
6. Bacon, J. S. D. and J. Edelman: The Action of Invertase Preparations. Arch. Biochem. 28, 467 (1950).
7. Belval, H.: La genèse de l'amidon dans les céréales. Rev. gén. bot. 36, 308, 337, 395 (1924).
8. Blanchard, P. H. and N. Albon: The Inversion of Sucrose; a Complication. Arch. Biochem. 29, 220 (1950).
9. Colin, H.: La genèse des lévulosanes chez les végétaux. Bull. soc. chim. biol. (Paris) 7, 173 (1925).
10. Colin, H. et H. Belval: Les lévulosanes dans les céréales. C. R. hebd. séances Acad. Sci. 177, 973 (1923).
11. Dedonder, R.: Les glucides du Tompinambour. IV. Isolement, analyse et structure des premiers termes de la série des polyosides. C. R. hebd. séances Acad. Sci. 232, 1134 (1950).
12. Dedonder, R. et Cl. Buvry: Les glucides du Topinambour. III. Données sur leur synthèse biologique. C. R. hebd. séances Acad. Sci. 231, 790 (1950).
13. Dehérain, P. P. et C. Dupont: Sur l'origine de l'amidon du grain du blé. C. R. hebd. séances Acad. Sci. 133, 774 (1901).
14. Hahn, E.: Die Hirse, seine geographische Verbreitung und seine Bedeutung für die älteste Kultur. Z. Ethnologie (1894).
15. Hassid, W. Z.: A Water-Soluble Glucosan from Barley Roots. J. Amer. Chem. Soc. 61, 1223 (1939).
16. Haworth, W. N., E. L. Hirst and R. R. Lyne: A Water-soluble Polysaccharide from Barley Leaves. Biochem. J. 31, 786 (1937).
17. Kühnemann, G.: Untersuchung der ungekeimten Gerste auf Zucker und Dextrin. Ber. dtsch. chem. Ges. 8, 387 (1875).
18. MacLeod, A. M.: Untersuchungen über freie Zucker im Gerstenkorn. 2. Mitt.: Verteilung einzelner Zuckerfraktionen. J. Inst. Brewing 58 [N. S. 49], 363 (1952) [Chem. Zbl. 1953, 4389].
19. Morris, D. L.: Lichenin and Araban in Oats (Avena sativa). J. Biol. Chem. 142, 881 (1942).
20. Müntz, A.: Sur la maturation de la graine du seigle. C. R. hebd. séances Acad. Sci. 87, 679 (1878).
21. Porter, H. K. and J. Edelman: Fructosans of Barley. Biochem. J. 50, xxxiii (1952).
22. Rawton, O. de: Sur la composition de l'avoine. C. R. hebd. séances Acad. Sci. 125, 797 (1897).
23. Richards, F. J. and S.-H. Shih: Physiological Studies in Plant Nutrition. X. Water Content of Barley Leaves as Determined by the Interaction of Potassium with certain other Nutrient Elements. Part II. The Relationship between Water Content and Composition of the Leaves. Ann. Botany [N. S.] 4, 403 (1940).
24. Russell, R. S.: Physiological Studies in Plant Nutrition. IX. The Effect of Mineral Deficiency on the Fructosan Metabolism of the Barley Plant. Ann. Botany [N. S.] 2, 865 (1938).
25. Schlubach, H. H.: Der Kohlenhydratstoffwechsel der Gräser. Fortschr. Chem. organ. Naturstoffe 15, 1 (1958).
26. — Der Kohlenhydratstoffwechsel im Roggen und Weizen. Fortschr. Chem. organ. Naturstoffe 19, 291 (1961).

27. SCHLUBACH, H. H. und J. BERNDT: Untersuchungen über Polyfructosane. LIX: Der Kohlenhydratstoffwechsel im Hafer. Liebigs Ann. Chem. **647**, 41 (1961).

28. — — Untersuchungen über Polyfructosane. LXIV: Über das Lävane aufbauende Enzym aus *Bacillus mesentericus*. Liebigs Ann. Chem. **665**, 200 (1963).

29. SCHLUBACH, H. H., J. BERNDT und T. CHIEMPRASERT: Untersuchungen über Polyfructosane. LXII: Der Aufbau der Polyfructosane in der Gerste. Liebigs Ann. Chem. **665**, 191 (1963).

30. SCHLUBACH, H. H. und L. GASSMANN: Untersuchungen über Polyfructosane. XLI: Über den Kohlenhydratstoffwechsel im *Phleum pratense*. Liebigs Ann. Chem. **594**, 33 (1955).

31. SCHLUBACH, H. H. und M. GREHN: Untersuchungen über Polyfructosane. LX: Zwei Polyfructosanasen im Hefeninvertin. Liebigs Ann. Chem. **647**, 51 (1961).

32. SCHLUBACH, H. H. und E. HABERLAND: Untersuchungen über Polyfructosane. LIII: Über das Kritesin und Hordeacin der Gerste. Liebigs Ann. Chem. **614**, 119 (1958).

33. SCHLUBACH, H. H. und P. HAUSCHILDT: Untersuchungen über Polyfructosane. XXXI: Über das Aigilopsin. Liebigs Ann. Chem. **578**, 201 (1952).

34. SCHLUBACH, H. H., G. HOHN und K. REPENNING: Untersuchungen über Polyfructosane. LVI: Zwei Polyfructosanasen aus *Aspergillus niger*. Liebigs Ann. Chem. **627**, 123 (1959).

35. SCHLUBACH, H. H. und K. HOLZER: Untersuchungen über Polyfructosane. XXXVIII: Über den Kohlenhydratstoffwechsel in *Lolium perenne*. Liebigs Ann. Chem. **587**, 111 (1954).

36. SCHLUBACH, H. H. und H. O. A. KOEHN: Untersuchungen über Polyfructosane. LV: Die Bildung der verzweigten Polyfructosane in den Roggenhalmen. Liebigs Ann. Chem. **614**, 126 (1958).

37. SCHLUBACH, H. H. und F. LEDERER: Untersuchungen über Polyfructosane. LVIII: Der Kohlenhydratstoffwechsel im Weizen. Liebigs Ann. Chem. **635**, 154 (1960).

38. SCHLUBACH, H. H. und H. LÜBBERS: Untersuchungen über Polyfructosane. XLII: Über die Phleine und ihren enzymatischen Aufbau. Liebigs Ann. Chem. **594**, 41, und zwar 52 (1955).

39. SCHLUBACH, H. H. und H. MÜLLER: Untersuchungen über Polyfructosane. XXVII: Über das Avenarin. Liebigs Ann. Chem. **572**, 106 (1951).

40. SCHLUBACH, H. H. und G. NEURATH: Untersuchungen über Polyfructosane. LII: Die Polyfructosanase-Aktivität des Hefeinvertins. Liebigs Ann. Chem. **606**, 134 (1957).

41. SCHLUBACH, H. H. und E. RATHJE: Untersuchungen über Fructoseanhydride. XXVI: Über das Kritesin. Liebigs Ann. Chem. **561**, 180 (1949).

42. SCHLUBACH, H. H. und R. ZIEGLER: Untersuchungen über Polyfructosane. LXVI: Der Kohlenhydratstoffwechsel in der Rispenhirse *(Panicum miliaceum)*. Liebigs Ann. Chem. **677**, 165 (1964).

43. SCHULZE, E.: Über die Verbreitung des Rohrzuckers in den Pflanzen, über seine physiologische Rolle und über lösliche Kohlenhydrate, die ihn begleiten. II. Z. physiol. Chem. **27**, 267 (1899).

44. SCHULZE, E. und S. FRANKFURT: Über die Verbreitung des Rohrzuckers in den Pflanzen, über seine physiologische Rolle und über lösliche Kohlenhydrate, die ihn begleiten. Z. physiol. Chem. **20**, 511 (1895).

45. TANRET, CH.: Sur la lévosine, nouveau principe immédiat de quelques céréales. C. R. hebd. séances Acad. Sci. **112**, 293 (1891).

46. Täufel, K. und K. Müller: Über den Gehalt der Gerste und von Malz an Mono- und Oligosacchariden. Z. Unters. Lebensmittel **83**, 49 (1942).

47. Whistler, R. L. and D. F. Durso: Chromatographic Separation of Sugars on Charcoal. J. Amer. Chem. Soc. **72**, 677 (1950).

(Eingelaufen am 21. September 1964.)

The Chemistry of Biological Sulfonium Compounds.

By **Fritz Schlenk**, Argonne, Illinois.

With 2 Figures.

Acknowledgement. The author's research has been sponsored by the United States Atomic Energy Commission.

I. Introduction.

Sulfonium compounds have been known to the organic chemist for almost 90 years. During most of this period they remained curiosities, well suited for didactic purposes to illustrate certain similarities to ether adducts and to organic ammonium compounds. Their exploration remained largely in the academic realm, because no unusual practical applications were found in spite of a steady increase in the attention paid to them. An event that occasionally puts a class of compounds into the center of interest is the recognition of its occurrence in nature and of special biological significance. The first natural sulfonium compound was discovered less than 20 years ago, somewhat by chance rather than as the climax of some coveted biochemical problem. Gradually, however, the unique role of sulfonium compounds, especially of S-adenosyl-methionine, in enzymatic group transfer reactions such as trans-methylation was recognized, and the recent upsurge of interest has been exceptional.

The present review is directed toward an interpretation of the properties and functions of biological sulfonium compounds with frequent reference to their structural chemistry.

The data presented here may suggest novel or improved analytical techniques and new interpretations of the mechanism of group transfer. Furthermore, a stimulus may be derived for the design of sulfonium compound analogues. The availability of competitive analogues of biochemical key compounds has aided in the understanding of the function of virtually all metabolic intermediates, prosthetic groups of enzymes, vitamins, amino acids, and other compounds. The pattern will be similar in sulfonium biochemistry.

Most reviews of this Series develop a rather complete picture of results in advanced fields. The present paper deals with the early phases of an emerging section of biochemistry. It will become apparent that the unsolved problems are numerous, and an attempt to attract the interest and participation of as yet unpledged investigators appears timely.

II. The Discovery of Sulfonium Compounds.

The first sulfonium compound, $(CH_3)_2\overset{\oplus}{S}CH_2\overset{\ominus}{COO}$, was obtained by Brown and Letts in 1878 (21). Several homologues and a variety of other relatively simple organic sulfonium compounds were prepared by Letts (87). A few years later, the studies were extended by Carrara (32) who prepared, among other compounds, dimethyl-β-propiothetin, $(CH_3)_2\overset{\oplus}{S}CH_2CH_2\overset{\ominus}{COO}$. Brown and Letts stated that the first investigation was undertaken to prepare a sulfur compound corresponding to betaine in the nitrogen series: "We have given the substance the name *thetine* to recall its relation to betaine and the fact that it contains sulfur" (*21*).

Subsequent chemical studies are summarized in the reviews of CHALLENGER (*34, 37*), INGOLD (*69*), and GOERDELER (*57*); not until about twenty-five years ago was the occurrence and biological significance of sulfonium compounds in nature suspected.

In 1935, HAAS investigated the strong odor given off by crushed seaweeds of the genus *Polysiphonia*. Dimethylsulfide was identified as the product of an enzymatic process (*65*). The parent substance was identified in 1948 by CHALLENGER and SIMPSON (*41*) as dimethyl-β-propiothetin. Meanwhile, the stage was set for biochemical investigations on sulfonium compounds by the discovery of methionine by MUELLER (*108*), the concept of transmethylation and discovery of homocysteine by DU VIGNEAUD and his school (*174–176*), and by the studies of CHALLENGER and his associates on biological methylations (*34*). The resemblance of sulfonium compounds to the quaternary ammonium compounds choline, $(CH_3)_3\overset{\oplus}{N}CH_2CH_2OH$, and betaine, $(CH_3)_3\overset{\oplus}{N}CH_2CO\overset{\ominus}{O}$, may have been of great importance in suggesting experiments concerning their methyl donor capacity. Thus, the first experiments with aceto- and propiothetin (*12, 46, 174*) as well as S-methylmethionine antedate the discovery of sulfonium compounds in any natural source.

S-Methylmethionine [(3-amino-3-carboxypropyl)dimethylsulfonium salt] (1) has found particular attention after its discovery in higher plants by SHIVE and co-workers (*91, 92*) and by CHALLENGER and HAYWARD (*36*).

$$\begin{array}{c} CH_3 \\ \diagdown \overset{\oplus}{S}-CH_2-CH_2-\underset{\underset{NH_2}{|}}{CH}-CO\overset{\ominus}{O} \\ \diagup \\ CH_3 \end{array}$$

(1)
S-Methylmethionine.

The significant steps leading to the discovery of the most important biological sulfonium compound, S-adenosylmethionine, were as follows: LIPMANN (*89*) suggested the involvement of the long known nucleoside 5'-methylthioadenosine in transmethylation. BORSOOK and DUBNOFF (*19*) recognized that methionine as such is not a methyl donor in vitro; the interaction of adenosine triphosphate was demonstrated and various intermediates containing both methionine and adenosine triphosphate were suggested. The success in the transmethylating experiments with ammonium compounds and thetins suggested that methionine has to be converted to an "onium" compound to become active for group transfer (*46*). This "active methionine" was eventually isolated by Cantoni (*24, 25*) and identified as S-adenosylmethionine (**2**).

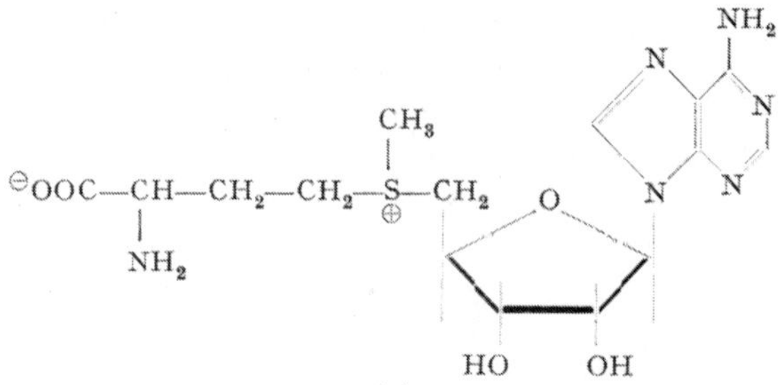

(**2**)
S-Adenosylmethionine.

Baddiley (*8*) has corroborated this structure by synthesis. Extensive experimentation with S-adenosylmethionine became possible when practical procedures for biosynthesis by yeast (*137*) and simple methods for isolation had been elaborated (*135*).

Other aspects of these developments have been summarized elsewhere (*29, 121 a, 133, 176*).

III. The Chemistry of Sulfonium Compounds.

1. Formation and Structure.

A brief survey of some chemical features of sulfonium compounds will help in the understanding of problems encountered in their synthesis, stability, the ways of decomposition, and intramolecular effects of the sulfonium group. Moreover, it will be seen that some of the chemical reactions of the sulfonium compounds have an enzymatic analogy.

The formation of sulfonium compounds is based on the presence of two pairs of unshared electrons in the thioethers. Thus, they are capable of adding a proton. The analogy to the formation of oxonium compounds (**3**) from ethers is obvious:

$$R_1 : \overset{..}{\underset{..}{O}} : R_2 + H^{\oplus} \rightarrow \left[R_1 : \overset{H}{\underset{..}{\overset{..}{O}}} : R_2 \right]^{\oplus}$$

(**3**)

The tendency of the proton to stay attached to the sulfur as well as to the oxygen, however, is rather limited. Stable sulfonium compounds are obtained by the attachment of organic radicals, especially by reaction with alkyl halogenides (**4**).

References, pp. 104—112.

$$R_1 : \overset{..}{\underset{..}{S}} : R_2 + R_3 X \rightarrow \left[R_1 : \overset{..}{\underset{R_3}{S}} : R_2 \right]^{\oplus} + X^{\ominus}$$

(4)

The formation of sulfonium compounds resembles the reaction of tertiary amines with alkyl halogenides (5).

$$R_1 : \overset{..}{\underset{R_3}{N}} : R_2 + R_4 X \rightarrow \left[R_1 : \overset{R_4}{\underset{R_3}{\overset{..}{N}}} : R_2 \right]^{\oplus} + X^{\ominus}$$

(5)

In contrast to the ammonium compounds, however, the sulfonium ion has one residual pair of unshared electrons which explains several differences in chemical behavior.

For the synthesis, it is usually convenient to start with the thioether that has the two largest of the three substituents of the contemplated sulfonium compound. The reactivity of alkylhalogenides decreases with the chain length. For example, S-ethylmethionine (6) is more readily

$$CH_3 - \overset{\oplus}{\underset{C_2H_5}{S}} - CH_2CH_2\overset{}{\underset{NH_2}{CH}}COOH$$

(6)

S-Ethylmethionine.

obtained from ethionine and methyl iodide than from methionine and ethyliodide. The reaction of ethionine with ethyliodide is very slow (91). In a similar fashion, β-bromopropionic acid reacts with dimethylsulfide, but not with diethylsulfide.

Further details of preparative methods have been reviewed by GOERDELER (57) and by SCHÖBERL and LANGE (140, 141).

2. Mechanisms of Decomposition and Rearrangement.

Of special importance are the intramolecular effects of the sulfonium configuration on adjoining groups.

An example is dimethylsulfoniopyruvic acid (7) which, according to BLAU and STUCKWISCH (14), in titration behaves as a dibasic acid (Chart 1). Two end points of titration are reached with methyl orange and phenolphthalein, respectively. The corresponding methyl and ethyl esters were found to be monobasic acids.

Comparison with the corresponding trimethylammonium derivative of pyruvic acid, $(CH_3)_3\overset{\oplus}{N}CH_2COCOOH$, showed that it behaves like a monobasic acid. In contrast to the sulfonium compound the nitrogen valence shell cannot be expanded; the electron withdrawing effect and resonance stabilization are characteristic of the sulfonium group only.

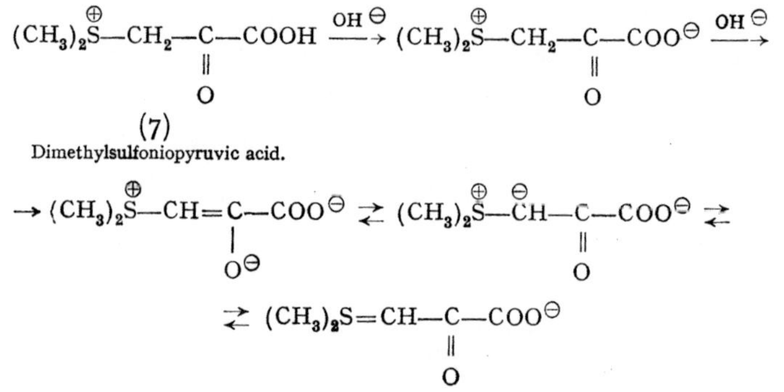

$$(CH_3)_2\overset{\oplus}{S}-CH_2-\underset{\underset{O}{\|}}{C}-COOH \xrightarrow{\text{OH}^\ominus} (CH_3)_2\overset{\oplus}{S}-CH_2-\underset{\underset{O}{\|}}{C}-COO^\ominus \xrightarrow{\text{OH}^\ominus}$$

(7)

Dimethylsulfoniopyruvic acid.

$$\rightarrow (CH_3)_2\overset{\oplus}{S}-CH=\underset{\underset{O^\ominus}{|}}{C}-COO^\ominus \rightleftarrows (CH_3)_2\overset{\oplus}{S}-\overset{\ominus}{C}H-\underset{\underset{O}{\|}}{C}-COO^\ominus \rightleftarrows$$

$$\rightleftarrows (CH_3)_2S=CH-\underset{\underset{O}{\|}}{C}-COO^\ominus$$

Chart 1. Dimethylsulfoniopyruvic Acid as a Dibasic Acid.

Interesting examples of group elimination attributable to a sulfonium structure in the molecule have been furnished by CRANE and RYDON (43) as well as MAMALIS and RYDON (93).

Dimethylethylsulfonium iodides, substituted in position 2 of the ethyl group by aromatic radicals, were used. Dimethyl-2-benzoyloxyethyl-sulfonium iodide (8) was studied in detail. This compound is decomposed

$$[C_6H_5-COO-CH_2CH_2-\overset{\oplus}{S}(CH_3)_2]\,I^\ominus$$

(8)

Dimethyl-2-benzoyloxyethyl-sulfonium iodide.

in alkaline solution with remarkable ease. In 0.01 M solution in 60% alcohol at 25° and pH 11, one equivalent of acid was liberated in four minutes. An involvement of the sulfonium group is demonstrated by the fact that the corresponding methylthioether is stable (43). The fission proceeds as follows:

$$[C_6H_5-COO-CH_2-CH_2-\overset{\oplus}{S}(CH_3)_2]\,I^\ominus + \text{NaOH} \rightleftarrows$$

$$\rightleftarrows C_6H_5-COONa + [CH_2=CH-\overset{\oplus}{S}(CH_3)_2]\,I^\ominus + H_2O$$

(9)

The reversibility of this step is explained by activation of the double bond in the dimethylvinylsulfonium iodide (9) caused by the adjacent sulfonium group. Facile reaction with anions is observed.

$$[CH_2=CH-\overset{\oplus}{S}(CH_3)_2]\,I^\ominus + \text{NaOH} \rightarrow CH\equiv CH + S(CH_3)_2 + \text{NaI} + H_2O$$

(9)

Dimethylvinylsulfonium iodide.

References, pp. 104—112.

Further elimination requires elevated temperature. As a by-product, a small quantity of 2-hydroxyethyldimethylsulfonium iodide (10) is found:

$$[HOCH_2—CH_2—\overset{\oplus}{S}(CH_3)_2]\,I^{\ominus}$$
$$(10)$$
2-Hydroxyethyldimethylsulfonium iodide.

The reaction of (8) with NaOH resembles the much studied analogous elimination of hydrogen halides from 2-halogenoethyltrimethylammonium halides:

$$X—CH_2—\overset{\overset{\displaystyle HO^{\ominus}}{\underset{|}{H}}}{CH}—\overset{\oplus}{N}(CH_3)_3 \rightarrow X^{\ominus} + CH_2{=}CH—\overset{\oplus}{N}(CH_3)_3 + H_2O$$

Under the influence of the ammonium pole electrons are withdrawn which facilitates the removal of the α-hydrogen. Far more rigorous conditions are required, however, than for the formation of the vinyl-sulfonium compound. MAMALIS and RYDON (93) interpret this as indicative of an unsaturated intermediate (Chart 2).

$$C_6H_5—COO—CH_2—\overset{\overset{\displaystyle HO^{\ominus}}{\underset{|}{H}}}{CH}—\overset{\oplus}{S}(CH_3)_2 \rightarrow H_2O + C_6H_5—COO—CH_2—CH{=}S(CH_3)_2 \rightarrow$$
$$\rightarrow C_6H_5—COO^{\ominus} + CH_2{=}CH—\overset{\oplus}{S}(CH_3)_2$$

Chart 2. The Action of Alkali on Dimethyl-2-benzoyloxyethylsulfonium iodide.

Such a rearrangement is not possible in the ammonium series because nitrogen, in contrast to sulfur, cannot expand its outer shell to ten electrons.

Similar examples of the great reactivity of sulfonium compounds have been reported by BÖHME and his co-workers (15–17). Especially γ-keto-sulfonium salts and β-sulfonyl-sulfonium salts are decomposed rapidly:

$$R—CO—CH_2—CH_2—\overset{\oplus}{S}{=}(CH_3)_2 + OH^{\ominus} \rightarrow R—CO—CH{=}CH_2 + S(CH_3)_2 + H_2O$$
$$C_6H_5—SO_2—CH_2—CH_2—\overset{\oplus}{S}(CH_3)_2 + OH^{\ominus} \rightarrow$$
$$\rightarrow C_6H_5—SO_2—CH{=}CH_2 + S(CH_3)_2 + H_2O$$

Chart 3. The Decomposition of γ-Ketosulfonium Salts and of β-Sulfonylsulfonium Salts.

Desaturation is observed but, in contrast to the compounds discussed on p. 66, the sulfonium structure is abolished in both instances.

Other sulfonium compounds lend themselves to special rearrangements resulting in C-alkylations. Phenacylsulfonium salts, for example, can form substituted thioethers:

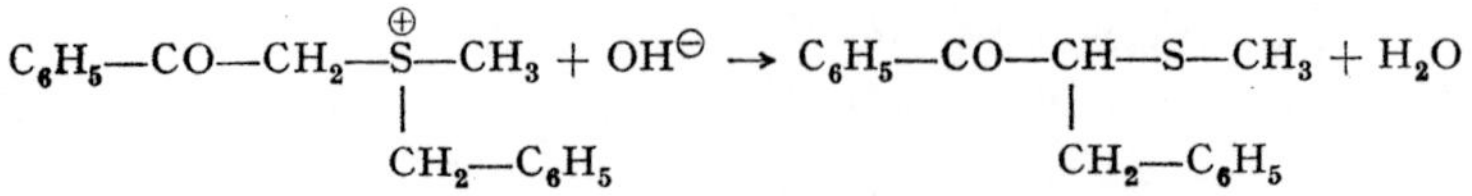

$$C_6H_5—CO—CH_2—\overset{\oplus}{\underset{|}{S}}—CH_3 + OH^\ominus \rightarrow C_6H_5—CO—\underset{|}{CH}—S—CH_3 + H_2O$$
$$CH_2—C_6H_5 \qquad\qquad\qquad CH_2—C_6H_5$$

More drastic conditions are required for the following reaction (67):

$$[C_6H_5—CH_2—\overset{\oplus}{S}(CH_3)_2] Br^\ominus + NaNH_2 \rightarrow$$
$$\rightarrow o\text{-}CH_3—C_6H_5—CH_2—S—CH_3 + NaBr + NH_3$$

The substituents of the sulfonium group, the nature of the anion, solvent, pH, and the temperature have great influence on the rate and course of the decomposition of this class of compounds (69).

The closest resemblance of chemical methyl transfer to enzymatic transmethylation has been observed by CHALLENGER and his associates (35). They found that dimethylacetothetin (11) under anhydrous conditions on heating transfers one of its methyl groups to aromatic amines such as aniline, p-toluidine or β-naphthylamine:

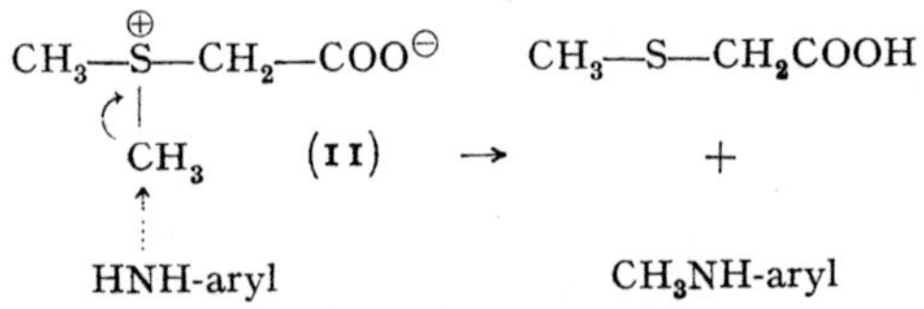

$$CH_3—\overset{\oplus}{S}—CH_2—COO^\ominus \qquad CH_3—S—CH_2COOH$$
$$\underset{\uparrow}{\overset{|}{CH_3}} \quad (11) \quad \rightarrow \qquad\qquad +$$

HNH-aryl CH₃NH-aryl

Chart 4. Non-enzymatic Transmethylation with Dimethylacetothetin.

The authors (35) assume that the positive charge on the sulfonium pole, $(CH_3)_2—\overset{\oplus}{S}—$, attracts the electrons of the $\overset{\delta\oplus}{CH_3}—S$ link. The methyl group becomes slightly electropositive, and nucleophilic attack by the unshared electrons of the —NH₂ group of the arylamine probably gives a transition compound which breaks down to N-methylarylamine and methylthioacetic acid.

Such nonenzymatic transmethylation under rigorous conditions was found possible also with phenolic acceptors (35).

3. Stereoisomerism of the Sulfonium Group.

Of particular interest is the ability of sulfonium compounds to form stereoisomers if the substituents on the sulfur are different from each other:

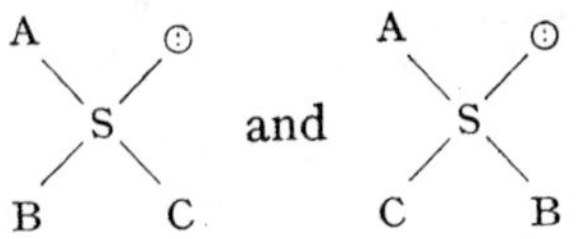

In these compounds the sulfur is believed to be in the center of a tetrahedron with three corners occupied by the organic substituents. The fourth corner is the site of the unshared electrons. The separation of such stereoisomers was attempted repeatedly, but no unequivocal success was achieved until, in 1900, POPE and PEACHEY (*121*) and SMILES (*154*) reported the resolution of the stereoisomers of several compounds including, for example (12) and (13).

$$C_2H_5(CH_3)\overset{\oplus}{S}—CH_2COOH \qquad C_2H_5(CH_3)\overset{\oplus}{S}—CH_2COC_6H_5$$

(12)   (13)

Methylethylthetin.   Methylethylphenacylsulfonium salt.

The ease of racemization apparently has discouraged subsequent investigators from extensive studies. Since none of the compounds was of natural origin, the incentive of singling out one of the stereoisomers as the biological form did not offer itself. The discrimination of enzymes against one of the isomers of S-adenosylmethionine was observed much later (see p. 81).

IV. In vitro Synthesis and Properties of Biological Sulfonium Compounds.

1. Dimethyl-β-propiothetin.

Dimethyl-β-propiothetin (15) can be synthesized by the methylation of β-methylthiopropionic acid (14) (*13, 52*). An alternative is the reaction of dimethylsulfide with β-chloropropionic acid (*57*). The compound crystallizes from alcohol and forms a Reinecke salt of limited solubility (*30*).

$$CH_3S—CH_2CH_2COOH + CH_3I \rightarrow (CH_3)_2\overset{\oplus}{S}—CH_2CH_2COOH + I^{\ominus}$$

(14)   (15)

β-Methylthiopropionic acid.   Dimethyl-β-propiothetin.

$$(CH_3)_2S + ClCH_2CH_2COOH \qquad\qquad —Cl^{\ominus}$$

Heating of dimethyl-β-propiothetin in water or dilute acid gives β-methylthiopropionic acid. Dimethylsulfide and acrylic acid are by-products in this process. In alkaline medium, however, the latter compounds are the main products of hydrolysis:

$$(15) \rightarrow (CH_3)_2S + CH_2{=}CHCOOH + H^{\oplus}$$

CHALLENGER and his associates have elaborated extensive analytical procedures for the detailed exploration of the chemistry of thetins (*34*). Especially the separation of mercaptans and dialkylsulfides by mercuric salts has furthered progress in the study of sulfonium compounds. Mercuric acetate and cyanide usually give no precipitate with dialkylsulfides, while mercuric chloride leads to precipitates (*42*). The reaction of mercuric ion with alkylthiols is so striking that it has led to the name "mercaptan".

2. S-Methylsulfonium Salt of Methionine (S-Methylmethionine).

The thetin-like analogue of methionine, S-methylmethionine (16), (3-amino-3-carboxypropyl)dimethylsulfonium salt, $(CH_3)_2\overset{\oplus}{S}$—$CH_2CH_2CH$-$(NH_2)\overset{\ominus}{C}OO$, has been prepared for biochemical experimentation by Toennies and Kolb (170).

Methionine and methyl iodide, dissolved in a mixture of glacial acetic acid and formic acid, were permitted to react at room temperature for about 24 hours. After removal of the excess of methyl iodide and the solvents under reduced pressure, the sirup-like product crystallized readily from alcohol as the iodide. The bromide and chloride have less tendency to crystallize, but they can be readily prepared from the iodide by ion exchange. Methylation of dimethylsulfate is an alternative way of preparation (3).

S-Methylmethionine forms a Reinecke salt and a phosphotungstate of low solubility. The chemical properties of the compound found renewed interest when its occurrence in plant material was discovered (92).

S-Methylmethionine is completely decomposed by boiling in presence of two equivalents of alkali for one hour (92). Dimethylsulfide and homoserine (18) are formed; the latter is a secondary product arising from homoserine lactone (17) (115):

$$(CH_3)_2\overset{\oplus}{S}\text{—}CH_2CH_2CH(NH_2)COO^{\ominus} \rightarrow (CH_3)_2S +$$

(16)

S-Methylmethionine.

$$+ \overset{\overline{\qquad O \qquad}}{CH_2CH_2CH(NH_2)C{=}O} \rightarrow HOCH_2CH_2CH(NH_2)COOH$$

(17) **(18)**

Homoserine lactone. Homoserine.

Unpublished experiments of the author have shown that hydrolysis by alkali in tritiated water leads to homoserine containing firmly bound tritium. Apparently, an equilibrium exists between the sulfonium compound and a desaturated derivative:

$$(CH_3)_2\overset{\oplus}{S}\text{—}CH_2CH_2CH(NH_2)COO^{\ominus} \rightleftarrows$$
$$\rightleftarrows (CH_3)_2S{=}CHCH_2CH(NH_2)COO^{\ominus} + H^{\oplus};$$
$$H^{\oplus} + TOT \rightleftarrows T^{\oplus} + HOT|$$
$$T^{\oplus} + (CH_3)_2S{=}CHCH_2CH(NH_2)COO^{\ominus} \rightleftarrows (CH_3)_2\overset{\oplus}{S}\text{—}CHTCH_2CH(NH_2)COO^{\ominus}$$

Chart 5. The Incorporation of Tritium from Tritiated Water into S-Methylmethionine.

The formation and decomposition of methionine sulfonium salts in acid solution has been studied in detail by Lavine and his associates (81). In strong acid, displacement of one of the groups contributing to the sulfonium function occurs. The relative electron release effects of the three substituents determine which group is removed.

The conversion of S-methylmethionine, according to the reaction,

$$(CH_3)_2\overset{\oplus}{S}-CH_2CH_2CH(NH_2)COOH \rightleftarrows$$
$$\rightleftarrows CH_3SCH_2CH_2CH(NH_2)COOH + CH_3OH + H^{\oplus}$$

depends on the nature and concentration of the acid (*54, 81, 82*). When boiled with HI, both methyl groups are eventually converted to methyl iodide. On the other hand, boiling with 16 to 18 N sulfuric acid decomposes little of the compound. When methionine was boiled under these conditions for 30 minutes with an excess of methanol, a quantitative conversion to S-methylmethionine was achieved. With methyl-labeled methionine, there was almost no dilution of the label by the excess of methanol which shows that the reaction is nearly irreversible and is going from methionine to S-methylmethionine. Because of the ease of the isolation of the latter as the phosphotungstate, the method has been used for the isolation of methionine from protein hydrolyzates by way of S-methylmethionine (*179*).

Various alcohols other than methanol can react with methionine to form sulfonium derivatives. The ethyl, propyl, isopropyl, *tert*-butyl, benzyl, and allyl sulfonium derivatives of methionine have been prepared

$$HOOC-\underset{\underset{NH_2}{|}}{CH}-CH_2-\underset{\underset{CH_3}{|}}{\overset{\oplus}{S}}-CH_2CH_2\underset{\underset{NH_2}{|}}{CH}-COOH$$

(19)
S-(α-Aminopropionyl)-methionine.

in this way (*82*). It should be emphasized that serine failed to react despite extensive variation of experimental conditions. The expected product, S-(α-aminopropionyl)-methionine (**19**), the methyl sulfonium salt of cystathionine, is a coveted hypothetical intermediate in sulfur amino acid metabolism (*123, 169*); so far, it has not been isolated or synthesized (*38, 82*).

Similarly unsuccessful has been a search for the methylsulfonium derivative of S-methylcysteine (**20**). When S-methylcysteine was treated

$$(CH_3)_2\overset{\oplus}{S}-CH_2CH(NH_2)COOH$$
(20)
S-Dimethylcysteine.

with methanol in 18 N sulfuric acid, no S-dimethylcysteine was obtained. Instead, much dimethylsulfide was formed. It appears that S-dimethylcysteine, if formed at all, is very unstable. Attempts to obtain this compound by methylation of S-methylcysteine with methyl iodide were unsuccessful under conditions that lead readily from methionine to S-methylmethionine. It appears that the proximity of the amino group blocks methylation (*82*).

3. S-Adenosylmethionine and Related Compounds.

a. *In vitro Synthesis of S-Adenosylmethionine.*

5′-Methylthioadenosine (5′-deoxy-5′-methylthioadenosine; **21**) has figured prominently in the synthesis of S-adenosylmethionine. Its structure (*162*) has been established conclusively by degradation and

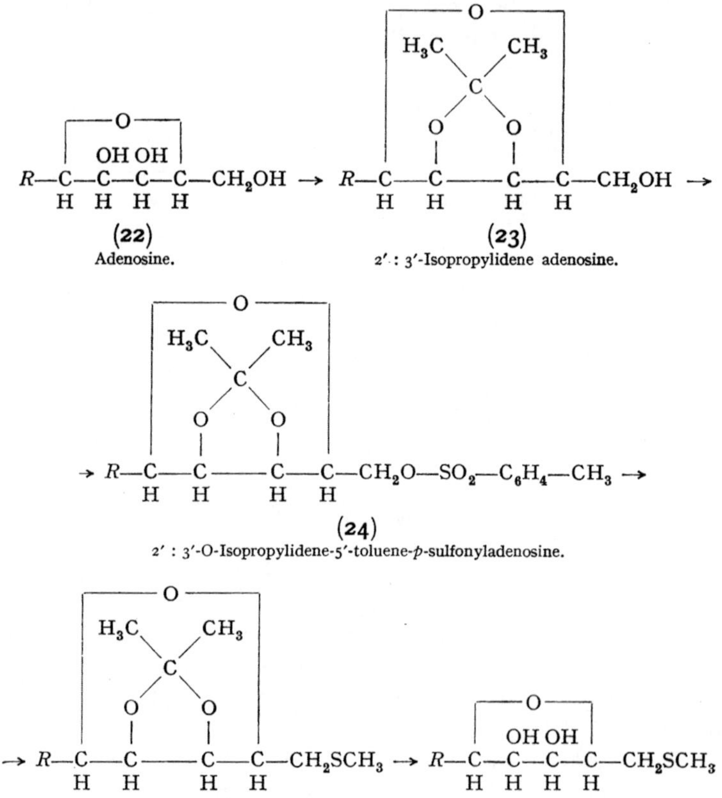

Chart 6. The Synthesis of 5′-Methylthioadenosine and 5′-Methylthioinosine (R = adenine or hypoxanthine).

synthesis; the latter was accomplished independently in several laboratories (*5, 130, 180, 182, 183*). BADDILEY (*5*) started with adenosine (9-β-D-ribofuranosyl adenine) and substituted the primary hydroxyl group at carbon atom 5 of the ribose by a methylthiol group. The steps were as follows *(Chart 6)*: Conversion of adenosine (**22**) to 2′ : 3′-isopropylidene adenosine (**23**) and introduction of a toluene-*p*-sulfonyl group in the 5′-position of this compound (**24**). The latter was replaced by a methylthio group using the method of RAYMOND (*125*), with improvements (**25**). The final step was the removal of the isopropylidene group, leading to the methylthionucleoside (**21**). Despite modification

(**21**)
5'-Methylthioadenosine.

of various steps, very small yields of 5'-methylthioadenosine were obtained. The procedure worked very well, however, with inosine (9-β-D-ribofuranosyl hypoxanthine) as the starting material, and the resulting 5'-deoxy-5'-methylthioinosine could be identified with the product that had been obtained by deamination of 5'-methylthioadenosine (*80*).

The principal difficulty in the synthesis of S-adenosylmethionine lies in securing sufficient quantities of the key intermediate, 5'-methylthioadenosine. The procedure of WEYGAND (*182*) offers little advantage over the method of BADDILEY (*5*). Fortunately, this compound can be isolated in quantity from yeast (*155, 181*).

A synthesis of S-adenosylmethionine was achieved by reaction of 2-amino-4-bromobutyric acid with 5'-methylthioadenosine in a mixture of formic and acetic acid. The yield remained very small (*8*).

$$\text{Adenine-ribose-S---CH}_3 + \text{BrCH}_2\text{CH}_2\text{CH(NH}_2)\text{COOH} \rightarrow$$

$$\rightarrow \overset{\oplus}{\underset{\underset{\text{CH}_3}{|}}{\text{Adenine-ribose-S}}}\text{---CH}_2\text{CH}_2\text{CH(NH}_2)\text{COOH} + \text{Br}^{\ominus}$$

An alternative route would be the condensation of methionine and 5'-deoxy-5'-halogenoadenosine. Compounds of the latter type are exceedingly unstable, and their use as a reaction component has as yet not been realized.

More successful has been the methylation of S-adenosylhomocysteine (**26**) with methyliodide:

$$\text{Adenine-ribose-S---CH}_2\text{CH}_2\text{CH(NH}_2)\text{COOH} + \text{CH}_3\text{I} \rightarrow$$

(**26**)
S-Adenosylhomocysteine.

$$\rightarrow \overset{\oplus}{\underset{\underset{\text{CH}_3}{|}}{\text{Adenine-ribose-S}}}\text{---CH}_2\text{CH}_2\text{CH(NH}_2)\text{COOH} + \text{I}^{\ominus}$$

S-Adenosylhomocysteine [S-(5'-deoxyadenosine-5')-L-homocysteine] (26) could be synthesized from 2' : 3'-O-isopropylidene-5'-toluene-p-sulfonyl-adenosine (24) and the disodium derivative of homocysteine, either in liquid ammonia at low temperature, or in dimethylformamide at 100° (9). The isopropylidene derivative of S-adenosylhomocysteine is obtained, and the masking group is removed by hydrolysis. Improvements of the procedure have been reported by Sakami (128). For methylation, the product was treated in acetic acid with an excess of methyliodide. The reaction remains incomplete even after prolonged periods, but more drastic conditions cannot be used because of the instability of the product, S-adenosylmethionine. The latter is racemic with respect to the sulfonium group.

While none of these total syntheses of S-adenosylmethionine is of practical value in competition with biosynthesis and isolation from natural sources, especially yeast, their value lies in the confirmation of the structure suggested by degradation of the biological material (6). Especially important is the establishment of the β-glycosidic bond which had been inferred only from the participation of adenosine triphosphate in the biosynthesis.

b. S-Adenosyl-3-methylmercaptopropylamine.

A related compound, decarboxylated S-adenosylmethionine, 5'-deoxy-5'-S-(3-methylthiopropylamino) sulfonium adenosine, also termed S-adenosyl-(5')-3-methylmercaptopropylamine (27), has been synthesized

$$
\overset{\oplus}{\underset{\overset{|}{CH_3}}{\text{Adenine-ribose-S}}}-CH_2CH_2CH_2NH_2
$$

(27)
S-Adenosyl-(5')-3-methylmercaptopropylamine.

by Jamieson (71). For this, 2',3'-isopropylidene-5'-toluene-p-sulfonyl-adenosine (24) was condensed with 3-thiopropylamine. The acetonide group was removed, the thioether was methylated and isolated as the flavianate. This compound acts as propylamino group donor in the enzymatic synthesis of spermidine and spermine (see p. 96).

c. Hydrolysis of S-Adenosylmethionine.

The chemical properties of S-adenosylmethionine comprise many of the features that can be predicted from the behavior of sulfonium compounds in general. However, special properties are provided by the fact that the nucleoside and the amino acid substituent on the sulfonium group are of a type not included in the earlier chemical studies of sulfonium compounds.

References, pp. 104—112.

Hydrolysis of S-adenosylmethionine (2) leads to a variety of fragments whose nature depends on the experimental conditions. In all instances the degradation involves the bonds between the characteristic structural units of the molecule as illustrated.

(2)
S-Adenosylmethionine (and hydrolysis).

The glycosidic bond A shows stability toward acid similar to that observed in other adenine ribosides. Heating at 100° for 5 hours in 0.1 N HCl completes the hydrolysis (116). Adenine, 5-methylthioribose, and homoserine (split D) are the products under these conditions. Heating at pH 5 to 7 at 100° for 20 to 30 minutes leaves the ribosidic bond (A) almost completely intact, and scission D is the predominant reaction; 5′-methylthioadenosine and homoserine are isolated (25, 136). The latter is formed from homoserine lactone (17) which is the primary product (115). Reaction D may be interpreted, therefore, as an intramolecular nucleophilic attack of the carboxyl group on carbon atom 4 of the amino acid part of the molecule, resulting in a displacement of the thionucleoside moiety.

Reaction B hat not been observed under acid conditions. The linkage C between the sulfur atom and the methyl group shows particular resistance. No conditions of hydrolysis have been found that lead from S-adenosylmethionine to S-adenosylhomocysteine or homocysteine. This firm attachment of the methyl group is in striking contrast to the ease of enzymatic resolution of bond C. Boiling of S-adenosylmethionine with HI has apparently not yet been explored.

This analytical method of determination of sulfur-bound methyl groups has been applied successfully to methionine (10). Under the rigorous conditions of the procedure one of the first products formed would be 5′-methylthioadenosine. The release of methyliodide from the latter by treatment with HI has been investigated by KUHN et al. (79), but the results were discouraging because of the concomitant formation of methyl mercaptan. LAVINE et al. (83) have resumed these studies. By incorporation of β-iodopropionic acid as an alkylating agent into the reaction mixture, a sulfonium salt is formed that decomposes almost quantitatively with liberation of methyl iodide:

$$\text{Adenine-ribose-S—CH}_3 + \text{ICH}_2\text{CH}_2\text{COOH} \rightarrow$$

$$\rightarrow \text{Adenine-ribose-}\overset{\oplus}{\underset{\underset{\text{CH}_3}{|}}{\text{S}}}\text{—CH}_2\text{CH}_2\text{COOH} + \text{I}^{\ominus} \rightarrow$$

$$\rightarrow \text{Adenine-ribose-S—CH}_2\text{CH}_2\text{COOH} + \text{CH}_3\text{I}$$

(28)

Chart 7. The Demethylation of 5'-Methylthioadenosine by HI in Presence of β-Iodo-propionic Acid.

Strict adherence to the conditions specified by these authors is required for a satisfactory yield of methyl iodide, and a number of complications have been pointed out including rearrangement of methylthioribose to a mercaptal. The adenosylthiopropionic acid (28) formed in the process is decomposed under the rigorous conditions.

d. Effect of Alkali on S-Adenosylmethionine and Related Compounds.

The effect of alkali on S-adenosylmethionine has found much attention in recent years. Methionine can be obtained (reactions A and B of (2), p. 75) by heating in 0.1 N NaOH at 100° for 10 minutes; adenosine has not been isolated under these conditions nor under any other circumstances of chemical hydrolysis (136). A study of the alkaline degradation under more lenient conditions has shown that severance of the glycosidic bond (split A of 2) is always the primary step (116). In 0.1 N NaOH at room temperature the formation of adenine is complete after 10 minutes. The process can be observed by spectrophotometry because of the shift of the absorption maximum from 260 mμ to 268 mμ (Fig. 1).

Adenosine, 5'-methylthioadenosine and adenosylhomocysteine suffer no destruction under these circumstances. It became probable, therefore, that the labilization of the glycosidic bond is a special effect of the sulfonium group in the molecule. Support for this concept was provided by studying a related, but somewhat simpler, sulfonium compound, 5'-dimethylthioadenosine sulfonium salt (29). It was prepared by methylation of 5'-methylthioadenosine (21) (116).

$$\underset{\text{(21)}}{\text{Adenine-ribose-S—CH}_3} + \text{CH}_3\text{I} \rightarrow \underset{\text{(29)}}{\text{Adenine-ribose-}\overset{\oplus}{\text{S}}(\text{CH}_3)_2} + \text{I}^{\ominus}$$

In contrast to the stability of 5'-methylthioadenosine, the resulting dimethylthioadenosine sulfonium compound showed the same lability

References, pp. 104—112.

toward alkali as does S-adenosylmethionine. Thus, the sulfonium group
is responsible for the labilization of the glycosidic bond (*116*). Effects
of the sulfonium group extending to distant bonds in the molecule were

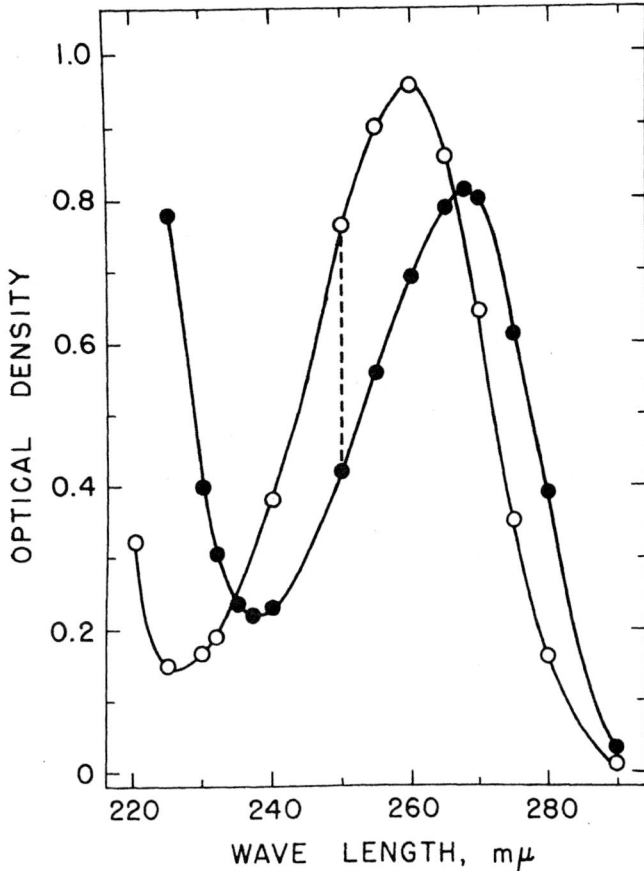

Fig. 1. Absorption spectra of S-adenosylmethionine (○) and adenine (●). The measurement of S-adenosylmethionine was carried out at pH 9.7 and 7.0 with identical result; measurement in 0.1 N NaOH is precluded
because of the instability of the compound. Adenine was split from S-adenosylmethionine by exposure to
0.1 N NaOH at 25° for 10 minutes, it was measured in this solution. The concentration of each compound
was 0.062 μmole per ml. Maximum decrease in optical density during the reaction, S-adenosylmethionine →
→ adenine + S-pentosylmethionine, occurs at 250 mμ. According to PARKS and SCHLENK (*116*). [From:
J. Biol. Chem. **230**, 295 (1958).]

known previously (Chapter III, p. 64), but earlier experiences did not
include labilization of a glycosidic bond. The sensitivity of the ribosidic
bond in S-adenosylmethionine closely resembles that of the analogous
linkage between ribose and the pyridinium base in the nicotinamide
nucleotides (*75, 139*) and in nicotinamide riboside (*132*):

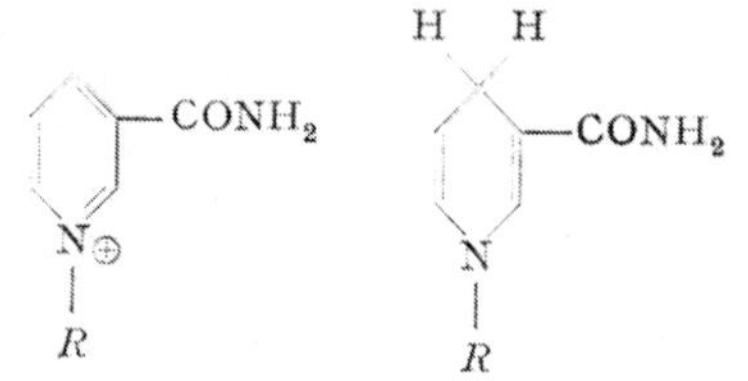

H H

Chart 8. The Linkage between Nicotinamide and Ribose in the Nicotinamide Nucleotides. (Left: oxidized state; right: reduced state; R = ribose.)

In the reduced form with trivalent nitrogen the ribosidic bond is stable, but it is very labile in the quaternary pyridinium compounds. In the adenosine sulfonium compounds a glycosidic structure similar to that of pyridinium ribosides would be attained by an intramolecular transmethylation leading to a quaternary nitrogen atom in position 9 of the adenine. An alternative to intramolecular methylation at position 9 would be position 7 of the adenine with subsequent rearrangement of the double bond to confer the quaternary character to nitrogen atom 9.

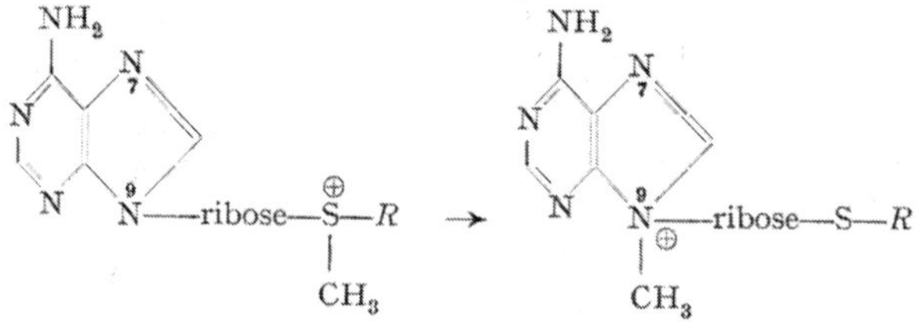

Chart 9. Hypothetical Intramolecular Transmethylation in S-Adenosylsulfonium Compounds.

The surmised lability of purine ribosides containing a quaternary nitrogen atom has found incidental support by recent observations of LAWLEY and BROOKES (84). These authors were concerned with the non-enzymatic methylation of nucleic acid components, particularly guanine derivatives which were readily methylated in position 7 of the purine ring. Rearrangement and facile splitting of the glycosidic bond was observed. In alkali, further destruction of the methyl purine compound to a methylformamido derivative occurred as shown in Chart 10.

If these mechanisms would apply to the adenosine sulfonium compounds, the aglycon obtained by alkali should be 9-methyladenine or a methylformamido compound similar to that in Chart 10. However, only adenine was isolated, and the methyl group remained attached to the sulfur in accord with scission A of Formula (2), p. 75. Thus, the hypothesis of labilization by intramolecular transmethylation had to be abandoned. It remains probable, however, that quaternization of nitrogen 9 of the adenine occurs in some other way under the influence of the sulfonium group.

References, pp. 104—112.

A more successful interpretation of the alkaline hydrolysis of S-adenosylmethionine has been given resently by BADDILEY and his associates (7, 55). The original observations (116) concerning the hydrolysis

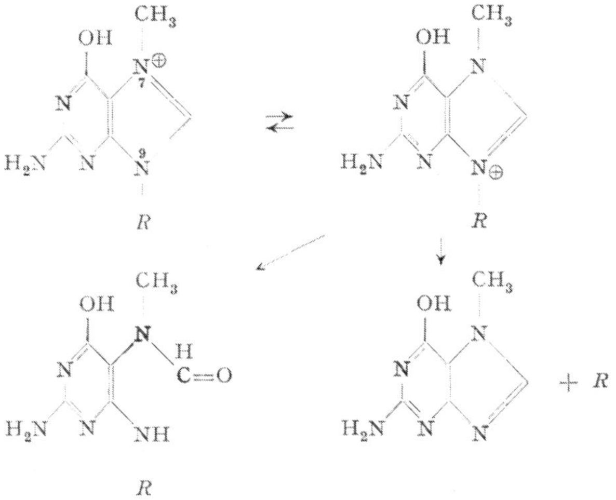

Chart 10. Methylation, Rearrangement and Splitting of Guanosine Derivatives.

to adenine and a carbohydrate sulfonium compound were confirmed and extended by similar experiments with synthetic uracil and hypoxanthine dimethylsulfonium nucleosides as well as with the *n*-butyl sulfonium riboside (30). It was shown in all instances that the sensitivity toward alkali is independent of the aglycone and is an effect of the sulfonium group. The adenine dimethylsulfonium nucleoside was used for exploratory experiments, and a reaction sequence was proposed *(Chart 11)* which indicates formation of a 4 : 5-olefinic intermediate (31). However, in the reversal of the elimination process the addition of the nucleophilic ion occurred at position 1 of the sugar. This was shown in experiments in which sodium methoxide was used. Had the addition of methoxide taken place at position 4, an acyclic sulfonium sugar would have been formed.

The formation of the ethylenic bond at position 4 is attended by loss of asymmetry, and formation of derivatives of *D*-ribose (32) and *L*-lyxose (33) is expected. No separation of these isomeric sugars was achieved by paper chromatography or electrophoresis. Partial inversion at carbon atom 4 was examined, therefore, by a degradation process separating the center of asymmetry at carbon atom 4 from the other asymmetric carbons of the pentose. Since 5'-dimethylthioadenosine was not available in sufficient quantity, the experiments were done with the corresponding methylglycoside. The steps are represented by *Chart 12.*

Methyl-(5-deoxy-5-dimethylsulfonium)-β-D-ribofuranoside (34) was converted by periodic acid to the dialdehyde (35). The latter was reduced to the dialcohol (36) which was split by acid. 1-Deoxy-1-dimethylsulfonium-L-glycerol (37) was obtained (acetate, $[\alpha]_D^{20} = 38°$; iodide, $[\alpha]_D^{20} = 27°$). The compound was pure as judged by comparison with the synthetic D-isomer. Treatment of the methyl glycoside with sodium methoxide in methanol prior to the degradation led to the racemic glycerol sulfonium derivative (acetate, $[\alpha]_D^{21} = 0.5°$; iodide, $[\alpha]_D^{20} = 0.2°$).

It is highly probable that the mechanism demonstrated by BADDILEY and his associates (7) with methyl(5-deoxy-5-dimethylsulfonium)-β-D-

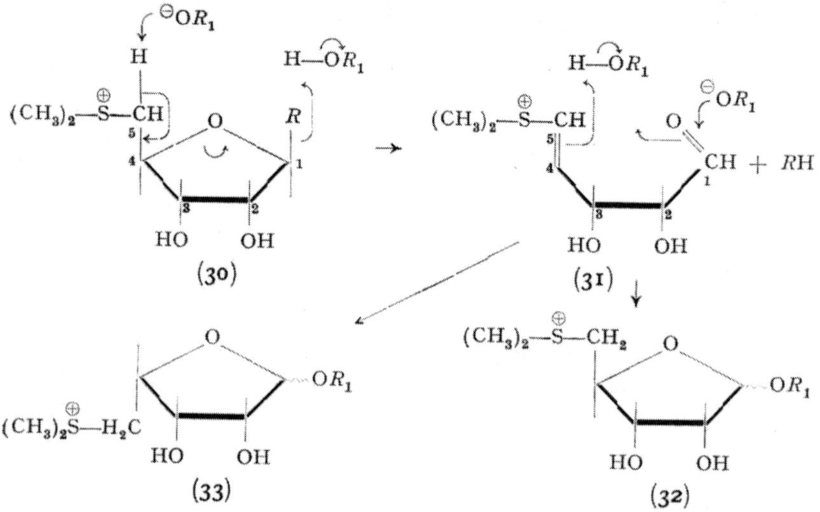

Chart 11. The Action of Alkali on Ribose Sulfonium Nucleosides.

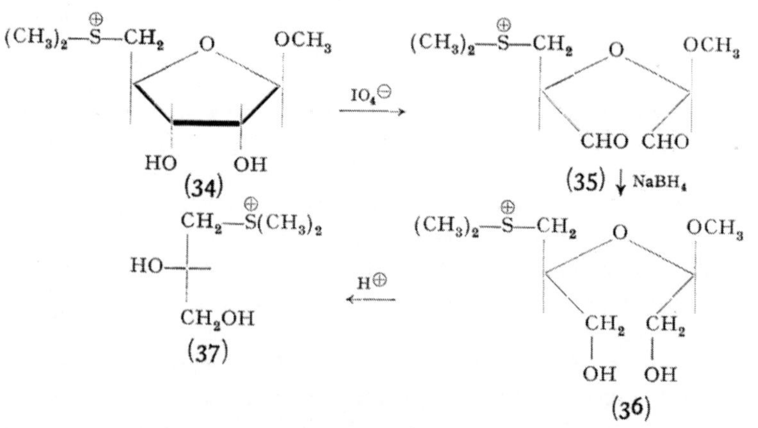

Chart 12. The Degradation of Methyl-(5-deoxy-5-dimethylsulfonium)-β-D-ribofuranoside (34).

ribofuranoside in sodium methoxide applies also to the adenine sulfonium ribosides. Supporting evidence has been obtained by the alkaline hydrolysis of S-adenosylmethionine in tritiated water (*134*). The expected quantity of one tritium atom per molecule was incorporated into the S-pentosylmethionine. This is consistent with the intermediary formation of an ethylenic bond. It was observed, however, that on termination of the exposure to alkali prior to completion of the hydrolysis much tritium was incorporated into the residual S-adenosylmethionine. Therefore, the desaturation of the ribose precedes the severance of adenine, and the process represented by Chart II, p. 80 occurs in two discrete steps.

The tritiated S-adenosylmethionine obtained in this way had retained its full activity as methyl donor in the enzymatic methylation of homocysteine to methionine (*133*) which shows that either no inversion at carbon 4 had occurred or that the *L*-lyxose sulfonium compound is as active in this enzymatic system as the *D*-ribose derivative. A detailed exploration with the biological methyl donor, S-adenosylmethionine, patterned after the experiment with the dimethyl sulfonium methyl riboside (Chart 12) will clarify details.

e. Diastereoisomers of S-Adenosylmethionine.

Of particular interest is the diastereoisomerism of S-adenosylmethionine which is caused by the disparity of the substituents of the sulfonium center and of the α-carbon atom of the amino acid moiety. Although stereoisomerism of sulfonium compounds has been familiar to the organic chemist for a long time (see p. 68) no biological evaluation has been possible prior to the discovery of S-adenosylmethionine. The biochemical effects of such diastereoisomerism have been explored by DE LA HABA et al. (*66*). The studies of these authors have included also the effect of the isomerism of the amino acid moiety. The following S-adenosylmethionine preparations were available:

I. (—)-*S-Adenosyl-L-methionine*, synthesized by yeast or liver enzymes from *L*-methionine and adenosine triphosphate. $[\alpha]_D^{25} = 47.2$ to $48.5°$.

II. (±)-*S-Adenosyl-L-methionine*, obtained from enzymatically synthesized S-adenosyl-*L*-homocysteine by methylation with CH_3I. $[\alpha]_D^{25} = 52.2°$.

III. (+)-*S-Adenosyl-L-methionine*, obtained by enzymatic resolution of preparation II. $[\alpha]_D^{25} = 57°$.

IV. (±)-*S-Adenosyl-D-methionine*, obtained by methylation of synthetic S-adenosyl-*D*-homocysteine with CH_3I. $[\alpha]_D^{25} = 16°$.

Since the absolute configuration at the sulfonium center is not known, the lower value of optical rotation of the biological Compound I compared with Compound III has been used as a basis to designate I as (—)-S-adenosyl-*L*-methionine.

The preparations were tested with guanidinoacetate methylpherase:
Guanidinoacetate + S-adenosylmethionine → creatine + S-adenosylhomocysteine.

The methyl group transfer approached 100% with preparation I, and 50% with preparation II. Compound III showed very little activity, the small effect perhaps being caused by incomplete enzymatic resolution. Compound IV showed very low but measurable activity, which indicates that the stereospecificity for the amino acid configuration is not absolute.

The enzyme from yeast which cleaves the sulfonium compound to 5'-methylthioadenosine and homoserine lactone (p. 96) showed during the first 30 minute period of incubation twice the effect with Compound I as compared with Compound II. Finally, the catechol O-methyltransferase system was used in testing these isomers. While Compound I showed full effect, at most 50% of Compound II was used, and virtually none of Compound III.

All these data suggest that (—)-S-adenosyl-L-methionine is the biological methyl donor.

The yeast *Candida utilis* incorporates D-methionine almost as well as L-methionine from the culture medium into S-adenosylmethionine. By alkaline hydrolysis of the sulfonium compound to methionine and analysis of the latter with D- and with L-amino acid oxidase it was ascertained that the D-form of methionine predominates in this isomer which presumably is (—)-S-adenosyl-D-methionine (*136*). The compound was found to be almost inactive with guanidinoacetate methyltransferase and with S-adenosylmethionine-homocysteine methyltransferase (*145*); the enzymes for these tests had been obtained from liver and from bacteria, respectively. It would be interesting to test enzymes from *C. utilis* from which the anomalous sulfonium compound was isolated.

The high stereospecificity of enzymes operating with S-adenosylmethionine was not observed in experiments with methylethylacetothetin (*66*). In the enzymatic transfer of the alkyl group to L-homocysteine 90% utilization of the racemic mixture was observed. It was considered, therefore, that the enzyme does not exhibit stereospecificity toward the sulfonium group, or that the compound is racemized during the experiment.

The ease with which racemization of sulfonium stereoisomers occurs may cause concern about the uniformity of S-adenosylmethionine preparations available for experimentation. It is possible that at least partial racemization occurs during the isolation procedure. This would be noticed only in experiments in which the concentration of the sulfonium donor is the limiting factor. In most experiments recorded in the literature the transmethylating capacity of the donor is not exhausted beyond 50% and, therefore, the presence of inactive isomers would not be noted.

References, pp. 104—112.

V. Origin and Metabolism of Biological Sulfonium Compounds.

1. Dimethyl-β-propiothetin.

A plausible route of biosynthesis of dimethyl-β-propiothetin has been reported by GREENE (59). The marine alga *Ulva lactuca* was incubated with S^{35}-, $C^{14}H_3$-, and α-C^{14}-labeled methionine. Dimethyl-β-propiothetin (38)

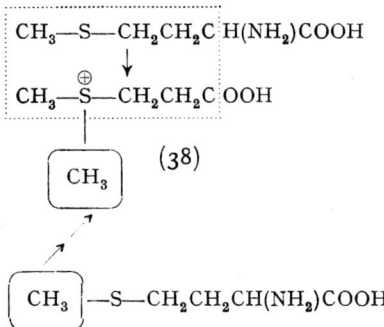

Chart 13. The Biosynthesis of Dimethyl-β-propiothetin from Methionine.

was isolated from the cells as the Reinecke salt. The recovery of all labels according to *Chart 13* in the thetin shows that the core of methionine is used after decarboxylation and deamination. Oxidation of the original α-carbon leads to the carboxyl group, and a second methyl group is added to the sulfur to yield the sulfonium configuration. The latter was found to be derived from another molecule of methionine, and the ratio of labeling achieved was 1.5 to 1.8 as compared with a theoretical maximum of 2.0. One may surmise that dilution from an endogenous methyl source accounts for the difference. S-Adenosylmethionine is perhaps the immediate methyl donor, although no experiments with it were included. Extraneous labeled S-methylmethionine appears to be excluded as a precursor, because very little of its radioactivity was recovered in dimethyl-β-propiothetin. However, as GREENE (59) has pointed out, the algal cell membrane may act as a permeability barrier. No intermediates of the biosynthesis could be identified by paper chromatography of ethanol-water extracts of the algae.

The concentration of dimethyl-β-propiothetin in some algal species is very high. In *Polysiphonia lanosa* as much as 40 to 60 μmoles per gram of moist cells has been found (30). This indicates an unusual expenditure of methionine, and the cells must have a very effective system for the biosynthesis of this amino acid.

Unfortunately, nothing is known about transmethylating enzymes in algal species. It would be interesting to test whether dimethyl-β-propiothetin constitutes the principal methyl donor in these cells. Considerable quantities of this compound find their way into marine organisms that ingest algae. Motohiro (99) has reported that the chum salmon, a Northern Pacific species, contains dimethyl-β-propiothetin which it derives from the alga *Limacina helicina*. In red salmon only small quantities of the sulfonium compound were found which may indicate differences in storage ability or in feeding habits.

It has been speculated (59) that the methylester of methylthiopropionic acid (39) may be derived from dimethyl-β-propiothetin (38):

$$(CH_3)_2\overset{\oplus}{S}-CH_2CH_2COO^\ominus \rightarrow CH_3S-CH_2CH_2COOCH_3$$
$$(38) \qquad\qquad\qquad (39)$$

The ester has been found in pineapple *(Ananas sativus)* (64). An intra- or intermolecular transmethylation of one of the sulfur-bound methyl groups of the thetin appears possible, but no enzymatic process has been established so far. There is also the disparity of the sources of the compounds: Neither has the methylthiopropionic ester been found in algae, nor has dimethyl-β-propiothetin been demonstrated in pineapple. Indeed, the interest in the thetin would increase greatly if it were found in plants other than algae.

The chemical decomposition of dimethyl-β-propiothetin to dimethyl-sulfide and acrylic acid has an enzymatic counterpart. Cantoni and Anderson (30) have found that the marine algae in which the compound occurs, notably *Polysiphonia lanosa*, contain an enzyme which is bound to protoplasmic particles. It begins to act on dimethyl-β-propiothetin as soon as the cellular integrity is destroyed. Thus, on grinding the algae the odor of dimethylsulfide is noticed (36, 65). By extraction with phosphate buffer the enzyme is separated from debris, but it remains bound to cellular fragments that are sedimentable by high speed centrifugation. It acts in presence of reduced glutathione or other sulf-hydryl compounds:

$$(CH_3)_2\overset{\oplus}{S}-CH_2CH_2COOH \longrightarrow (CH_3)_2S + CH_2=CHCOOH + H^\oplus$$

The enzyme is rather specific: in addition to dimethyl-β-propiothetin only dimethylacetothetin was decomposed, but slowly; betaine and choline were not attacked. The identification of dimethylsulfide was achieved by aspiration into palladium chloride solution. Steam distillation and chromatography permitted the identification of acrylic acid (30). The role of this enzyme and the high concentration of its substrate in algae need further explanation.

Decomposition of dimethyl-β-propiothetin has been described in a bacterial system by Wagner and Stadtman (178). The organism was a *Clostridium*, isolated from soil by enrichment culture. It resembled

Cl. propionicum in its metabolic pattern, but morphological differences preclude identity. Because of its obligate anaerobic character, the utilization of an energy-rich compound, dimethyl-β-propiothetin, as the major source of carbon was of special interest. Energetic coupling of the cleavage of the sulfonium compound with cellular metabolism seemed possible, perhaps through the synthesis of adenosine triphosphate. This, however, has not yet been verified. The decomposition proceeds according to the equation,

$$3\,(CH_3)_2\overset{\oplus}{S}-CH_2CH_2COOH + 2\,H_2O \rightarrow$$
$$\rightarrow 2\,CH_3CH_2COOH + CH_3COOH + CO_2 + 3\,(CH_3)_2S + 3\,H^{\oplus}$$

Acrylic acid is believed to be an intermediary product (*178*).

2. S-Methylsulfonium Salt of Methionine (S-Methylmethionine).

There is so far only one report about the biosynthesis of S-methyl-methionine. GREENE (*60*) has observed a transmethylation from S-adenosylmethionine to methionine, resulting in S-methylmethionine:

S-Adenosyl-*L*-methionine + *L*-methionine →
→ S-Adenosyl-*L*-homocysteine + S-methyl-*L*-methionine.

An enzyme from Jack bean plants has been used. The reaction is apparently irreversible. This is one of the infrequent instances of formation of an "-onium compound" at the expense of S-adenosyl-methionine, other examples being N-methylnicotinamide (*23*) and choline (*2, 73*).

The formation of S-methylmethionine would be far more intelligible if the cell would acquire by this step greater versatility in trans-methylations or other group transfer reactions. However, research to date has not borne out such a role; the only transmethylation with this donor recognized with certainty is the formation of methionine from homocysteine (*143, 144*). This is not unique for S-methylmethionine, because S-adenosylmethionine also can serve as a methyl donor in the formation of methionine from homocysteine (*146*). A function as a methyl reservoir has been considered. Because of the irreversibility of the reaction, however, the methyl supply could be mobilized only by producing methionine. The latter would serve for protein synthesis, or it could be converted again to S-adenosylmethionine by reaction with adenosine triphosphate. The latter would be an energetically costly cycle leading to the initial methyl donor, S-adenosylmethionine.

It is probable that future research will uncover a more meaningful metabolic role of S-methylmethionine.

Very little is known about the catabolism of S-methylmethionine other than by transmethylation. Experiments of CHALLENGER (*40*) indicate an enzyme in molds that liberates dimethylsulfide, presumably according to the cleavage of (16), p. 70. The occurrence of this reaction in bacteria, higher plants, or mammalian cells has not yet been demonstrated. It is tempting to speculate, however, that dimethylsulfone, which has been found occasionally in higher plants and in mammalian tissue (76, 119, 127) stems from S-methylmethionine by an enzymatic sequence:

$$\text{S-Methylmethionine} \rightarrow (CH_3)_2S \rightarrow (CH_3)_2SO_2$$

Experiments of Snow (156) and of Leaver and Challenger (86) lend support to this concept.

3. S-Adenosylmethionine.

The events leading to the isolation and identification of S-adenosylmethionine have been reviewed on p. 63. Several reactions have been recognized by which this substance is synthesized in cells. However, only one of them can be formulated in a satisfactory way:

$$L\text{-Methionine} + \text{adenosine triphosphate} \rightarrow$$
$$\rightarrow \text{S-adenosylmethionine} + \text{pyrophosphate} + \text{phosphate}.$$

This reaction was discovered by Cantoni and his associates (31, 106), and its details have been reviewed recently by Mudd (104). The orthophosphate is derived from the terminal phosphate group of adenosine triphosphate. Enzymes catalyzing this reaction have been obtained from mammalian liver (24), yeast (27), Escherichia coli (165), and barley extract (103). The mode of action seems to be the same in all instances. The quantities of the sulfonium compound found in cells and in tissues rarely exceed 0.1 μmole per gram (11, 112, 133). Because of this low concentration it would be cumbersome to isolate the material from tissues. Instead, incubation of a mixture of adenosine triphosphate and methionine with liver enzyme in vitro has been recommended as a preliminary to isolation (26). Most convenient, however, is the incubation of intact, metabolizing yeast cells with methionine (135, 137). The cells furnish the requisite adenosine triphosphate, and the sulfonium compound is accumulated in the vacuoles (163) in quantities amounting to several per cent of the dry weight of the yeast. Facile isolation from the yeast cells by acid extraction and chromatography has made this material generally available. Advantage is taken of the great retentiveness of cation exchange resins for S-adenosylmethionine (see Fig. 2).

In yeast cells, two more reactions have been observed that lead to S-adenosylmethionine, but in both instances the details of the mechanism are not yet understood (133). The first of these occurs with S-adenosylhomocysteine as the precursor (47):

S-Adenosylhomocysteine + unidentified methyl source → S-adenosylmethionine.

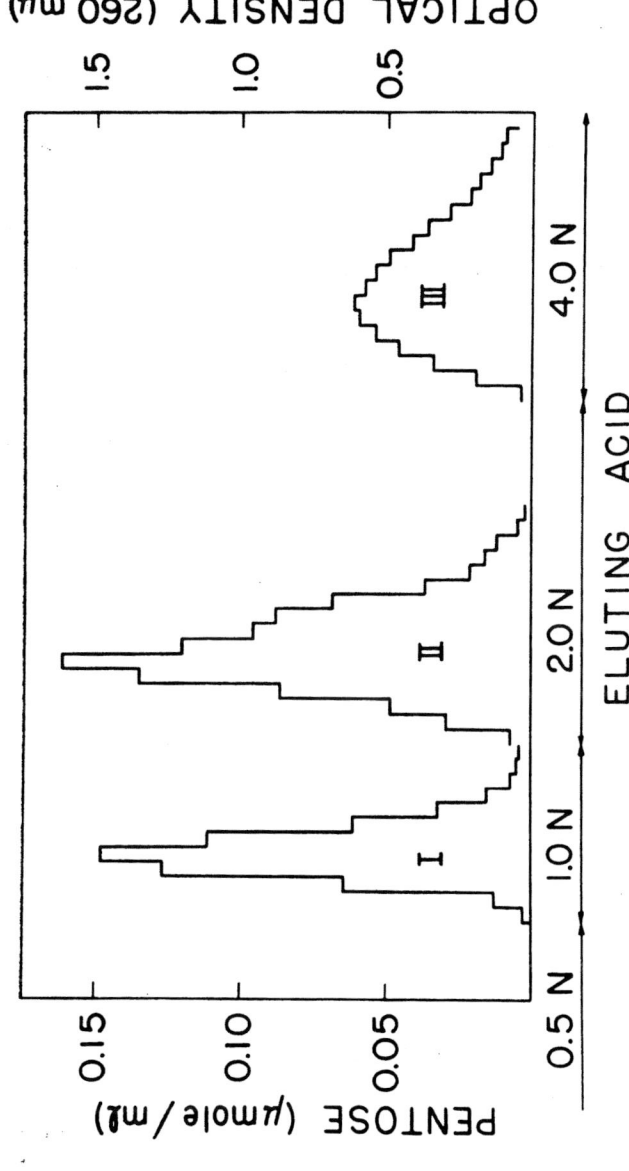

Fig. 2. Separation of S-adenosylmethionine (III) from its degradation products, S-pentosylmethionine (I), and adenine (II), by ion exchange chromatography. A column of Dowex 50 H⊕, 8.0 × 0.7 cm, was used with hydrochloric acid as the eluting agent. (I) was determined by the orcinol test, (II) and (III) by spectrophotometry. S-Methylmethionine and 5'-methylthioadenosine (not shown here) coincide approximately with peak (II). [From (136) and unpublished data.]

By using suitably labeled S-adenosylhomocysteine it has been ascertained that the starting material is not decomposed to serve merely as a nonspecific source of precursor units. Attempts to reconstruct the process in vitro with isolated enzymes have failed, so far. The importance of this reaction, if its more general occurrence can be established, lies in the fact that it constitutes a mechanism of regeneration of S-adenosyl-methionine after its donor function in transmethylation. Such a mechanism has to be invoked to explain the discrepancy between the high rate of methyl transfer in cells and tissues and the low concentration of the methyl donor (*133*).

The second reaction leads from 5'-methylthioadenosine and an as yet unidentified four-carbon amino acid precursor to S-adenosyl-methionine (*138*):

$$\text{Adenine-ribose-S—CH}_3 + \text{amino acid precursor} \rightarrow$$

$$\rightarrow \text{adenine-ribose—}\overset{\oplus}{\underset{|}{\text{S}}}\text{—CH}_2\text{CH}_2\text{CH(NH}_2)\text{COO}^{\ominus}$$
$$\text{CH}_3$$

Here too we seem to deal which a regeneration process by which the cell recovers 5'-methylthioadenosine; the latter is formed when S-adenosylmethionine acts as an enzymatic donor of its amino acid moiety (see p. 96).

In a similar category is the observation that an S-adenosylmethionine requiring mutant of *Aerobacter aerogenes* can use a mixture of adenine and S-ribosylmethionine with equal efficiency as a substitute (*146*). An enzymatic conversion,

$$\text{S-Ribosylmethionine} + \text{adenine} \rightarrow \text{S-adenosylmethionine,}$$

may be surmised. Details of the process are still unknown. One difference from the biosynthetic processes starting with S-adenosylhomocysteine or with 5'-methylthioadenosine is that these compounds are part of metabolic schemes while S-ribosylmethionine so far has to be considered an artificial degradation product of S-adenosylmethionine. None of the known nucleosidases splits the latter compound.

VI. Metabolic Functions of Biological Sulfonium Compounds.

1. Transmethylations.

a. General Considerations.

Among the biochemical functions of natural sulfonium compounds, only their methyl donor capacity in enzymatic transmethylation has been studied extensively. Unequivocal proof of such processes (*175*), especially of the in toto transfer of methyl groups was possible only after

References, pp. 104—112.

tracer techniques became available. Earlier experiments, extending throughout the past century (*88*), should be termed more appropriately biological methylation. The concept of organic radicals and their introduction into, and removal from, molecules by the methods of the organic chemist were tacitly assumed to have their replica in biochemistry. However, the unsatisfactory state of the scientific foundation of this conclusion was recognized, and it is best summarized in a paper of HIS (*68*) who, in 1887, had studied the biological methylation of pyridine in dogs:

"For the understanding of chemical reactions occurring in the organism, the mere fact that a methyl group is added to the pyridine molecule resulting in its change into an ammonium base does not constitute a gain in our knowledge, unless and until it is demonstrated where this methyl group originates and in what way it is added. Thus, further investigations should first answer the question whether the methyl is added as such, perhaps by means of a carrier analogous to the iodine in methyl iodide, or whether the methyl group is to be considered merely a residue of a complex molecule from which it has remained after secondary oxidations and splittings." (Translation.)

This statement has validity to this day and applies to many investigations of the present time.

Accepting the criteria of HIS (*68*), such as a clearly identified methyl donor as well as acceptor, and an in vitro enzymatic process, we may term the experiments of BORSOOK and DUBNOFF (*19*, *46*) the first adequately documented transmethylation. The following reactions were established:

$$(CH_3)_2\overset{\oplus}{S}-CH_2COOH + H-S-CH_2CH_2CH(NH_2)COOH \rightarrow$$
$$(40)$$
$$\rightarrow CH_3S-CH_2COOH + CH_3S-CH_2CH_2CH(NH_2)COOH + H^{\oplus}$$

$$(CH_3)_2\overset{\oplus}{S}CH_2CH_2COOH + HSCH_2CH_2CH(NH_2)COOH \rightarrow$$
$$(38)$$
$$\rightarrow CH_3SCH_2CH_2COOH + CH_3SCH_2CH_2CH(NH_2)COOH + H^{\oplus}$$

The satisfaction with this success and with subsequent investigations on this system is somewhat dampened by the fact that acetothetin (40), which to this date has not been found in nature, is a better methyl donor than propiothetin (38). The latter has been found only in aquatic plants (*34*, *35*) while the enzyme used in most instances has been derived from mammalian liver. This enzyme has now been highly purified (*78*). With the liver enzyme only the simple thetins serve as methyl donors. S Adenosylmethionine and S-methylmethionine require another liver enzyme for the methylation of homocysteine (*146*). An enzyme from Jack bean seeds *(Canavalia ensiformis)* uses all donors except dimethylacetothetin (*1*).

With respect to transmethylation, S-methylmethionine has many features in common with the thetins. It was synthesized by Toennies (*170*) prior to its isolation from plant material by Shive and his co-workers (*92*) and by Challenger and Hayward (*36*). The first, and so far only, in vitro transmethylation with the compound was achieved by Shapiro (*144, 145*):

$$(CH_3)_2\overset{\oplus}{S}\!-\!CH_2CH_2CH(NH_2)COOH + H\!-\!S\!-\!CH_2CH_2CH(NH_2)COOH \rightarrow$$
$$\rightarrow 2\ CH_3S\!-\!CH_2CH_2CH(NH_2)COOH + H^{\oplus}$$

The enzyme has been purified to a high degree and seems to have a rather wide distribution in microorganisms, higher plant material, and animal tissue (*146*). S-Methylmethionine, the donor, has been found in a variety of higher plants such as spinach (*92*), asparagus (*36*), green tea leaves (*77*), mint and pelargonium (*118*), but not in animal tissue. It may be a useful methyl donor toward homocysteine in animals when ingested with the diet. The availability of the synthetic material has led to frequent inclusion in specificity tests in the study of enzyme systems that operate with S-adenosylmethionine as the methyl donor. Only in the methylation of homocysteine was S-methylmethionine found to be active; in this system it sometimes surpasses S-adenosylmethionine in reactivity (*151, 152*).

It appeared possible that S-methylmethionine, because of its wide distribution in plants, would turn out to be the typical methyl donor in higher plants. In the few instances, however, in which cell-free enzyme and sulfonium donors have been used, S-adenosylmethionine rather than S-methylmethionine was found to be the specific methyl donor. These studies include hordenine (*102*), trigonelline (*72*), and the O-methylation of norbelladine (*94*).

Compared with S-methylmethionine and the thetins, S-adenosylmethionine appears to be by far the most versatile methyl donor (*28, 107*). The number and variety of methyl compounds in nature is well known, and S-adenosylmethionine has been established as the source of the methyl group in nearly all instances in which the details of biogenesis have been established in vitro. Exceptions are the terminal methyl groups of aliphatic chains, certain angular methyl groups of aromatic or hydroaromatic compounds that stem from the methyl group of acetic or propionic acid, and a few instances in which a direct transfer of the methyl group of methyltetrahydrofolic acid is indicated (*107*). Over-all results with C^{14}-methyl-labeled methionine are often proffered as evidence of transmethylation. Such experiments, however, should be classified as suggestive rather than conclusive. They do not ascertain the in toto transfer of the methyl group, nor do they settle the question whether methionine donates the methyl group after conversion to S-adenosylmethionine, S-methylmethionine, or some other derivative. It appears desirable that this type of transmethylation experiment be rendered conclusive by using sulfonium donors, which are readily available. However, it has to be admitted that this may not always be possible, because some of the transmethylating enzymes are difficult to separate from intact cells.

Experiments with an extraneous supply of sulfonium compounds for living organisms pose the problem that cell membranes are often an impediment to the uptake and distribution of sulfonium compounds. For example, yeast cells have an exceptional ability to produce and store S-adenosylmethionine, but an external supply is not readily assimilated from the culture medium (*163*). In a similar fashion, the intraperitonal injection of the sulfonium compound into the rat is virtually without effect on its level in the liver. Experiments with *Lupinus* and *Convallaria* (*133*) showed little if any uptake of S-adenosylmethionine through the root system. On the other hand, certain bacterial species can take up S-adenosylmethionine from the medium(*146*), and a mutant of *Aerobacter aerogenes* has been described which depends on the sulfonium compound for growth (*147*).

b. The Diversity of Methyl Acceptors.

The diversity of transmethylations has been stressed repeatedly (*29, 149*), and no attempt will be made here to compile a complete list of the numerous reactions. A summary of the various types of acceptors will illustrate some of the salient points.

Transfer of the methyl group to nitrogen includes as acceptors primary, secondary, and tertiary amines, or the nitrogen of heterocyclic compounds. Examples are, the formation of creatine, choline, anserine, sarcosine, carnitine, adrenaline, 5-methylhistamine, N-methylnicotinamide, N-methyl containing alkaloids, and methylated nucleic acid bases. The latter have found particular interest in recent years because of the possible alteration of the genetic message by methylation (*18*).

Sulfur as an acceptor is exemplified by homocysteine which reacts with any of the biological sulfonium methyl donors in presence of the requisite enzyme. In one instance methionine itself has been found to act as a methyl acceptor by reaction with S-adenosylmethionine (p. 85). In this way, a sulfonium compound, S-methylmethionine, is formed at the expense of another sulfonium group. A similar mechanism may account for the biosynthesis of dimethyl-β-propiothetin (see p. 83). Thiopurines and their ribosides act as methyl acceptors (*126*).

Transfer of a methyl group to carbon has been established in several instances, but the enzymes involved have not been purified to any great extent. The transfer of the methyl group of S-adenosylmethionine to position 28 of ergosterol is the first example of establishment of a carbon to carbon linkage by transmethylation (*113, 117*). In numerous instances, the evidence of such transfer is only suggestive in that merely a C^{14}-labeled methyl group from methionine has been recovered in the reaction product. A circuitous route is not excluded in these cases. In many organisms a considerable fraction of the methyl groups of extraneously supplied methionine is oxidized and may enter into one carbon unit metabolism by routes other than transmethylation. For example, the methyl group of thymine deoxyriboside does not originate from methionine (*70*).

A special process akin to transmethylation leads to cyclopropane fatty acids such as lactobacillic acid (*cis*-11,12-methylene-octadecanoic acid). The methylene group of the cyclopropane ring is derived from S-adenosylmethionine (*110, 111*).

Transfer of the methyl group from S-adenosylmethionine to oxygen may involve esterification or attachment to phenolic groups. An example of ester formation is the enzymatic production of magnesium protoporphyrin monomethyl ester in *Rhodopseudomonas spheroides*; S-adenosylmethionine is the donor (*167*). Only suggestive data are available concerning the methyl ester groups in pectins.

The evidence of transmethylation is much better concerning the formation of methylated phenolic compounds. Metanephrine is an example of this, and an exceptional variety of methoxyl groups has since been shown to originate by transmethylation (*4*). The specificity of the enzymes is often rather wide, and numerous non-biological compounds are methylated as well. An unusual enzymatic process of restricted occurrence is the formation of methanol from S-adenosylmethionine; it may be visualized as a transmethylation with the $OH^{\ominus}$ group of water as the acceptor (*4*).

Presumably the largest store of methyl groups in nature is bound to the phenolic groups of lignin. Their origin and the stage at which they are attached deserve exploration.

Selenium, tellurium, and arsenic can serve as methyl acceptors. The history and circumstances of the isolation of methylated products of these elements constitutes one of the fascinating chapters of early biochemistry and toxicology (*33, 34, 39, 45*). There are as yet no experiments on record in which cell-free enzymes, well-defined donors, and specific acceptors were used.

A recent development of particular importance is the participation of cobalamine enzymes in methyl transfer. Apparently, the cobalt atom can accept methyl groups from S-adenosylmethionine (*185*).

From an ever increasing number of transmethylations being reported, one gains the impression that S-adenosylmethionine outranks S-methylmethionine and dimethyl-β-propiothetin overwhelmingly as the methyl donor in biological systems. The diversity of acceptors, many of them non-biological molecules, makes it clear that the acceptor specificity is far less stringent than the donor specificity. Indeed, one is sometimes at a loss to understand the significance of certain transmethylations, and some of them may be entirely fortuitous.

c. Analytical Techniques.

For the quantitative study of transmethylation, any component of the general system,

$$R_1—\overset{\oplus}{\underset{\underset{CH_3}{|}}{S}}—R_2 + \text{methyl acceptor} \rightarrow R_1—S—R_2 + \text{methylated product} + H^{\oplus}$$

can be used for assay provided the identity of all reaction components is established and side reactions are excluded. In most cases a proton is released in the process, and in sufficiently purified systems this can be measured with bicarbonate in a manometric apparatus (95). Tracer techniques have figured prominently in the study of transmethylations, especially in combination with chromatographic procedures. The strongly cationic properties of the methyl sulfonium donors permit chromatographic separation from the reaction products. Moreover, S-adenosylmethionine and S-adenosylhomocysteine can be measured readily by ultraviolet spectrophotometry, but first they have to be separated from each other, because their absorption spectra are virtually identical. The identity of S-adenosylmethionine can be ascertained by its unique sensitivity to alkali as explained on p. 76.

The combination of small scale ion-exchange chromatography with tracer assay has particular advantages for routine assays such as needed in the course of enzyme purification.

The procedure is exemplified by the method of SHAPIRO and YPHANTIS (151): Columns of Dowex 50 Li$^{\oplus}$ (20 × 3 mm) are used to retain methyl labeled S-adenosylmethionine from 0.2 ml reaction mixture. The reaction product, in this case methyl-labeled methionine, passes the column and is washed quantitatively by a few small charges of water into a vial for scintillation counting. Several modifications of this technique have appeared including the use of acid resin in a system for the assay of guanidinoacetate methyltransferase (129).

Unfortunately, no specific reagent for the detection of sulfonium compounds on paper chromatograms has been found. Various sprays recommended for the detection of sulfur compounds, including the platinum reagent (172), are rather insensitive toward sulfonium compounds. Ninhydrin is useful for the detection of S-methylmethionine, and the coincidence of a ninhydrin spot with the ultraviolet quenching zone reveals S-adenosylmethionine or S-adenosylhomocysteine. The availability of integrating paper strip counters facilitates transmethylation studies with labeled reaction components.

d. Attempts to Explain the Mechanism of Transmethylation.

In all enzymatic group transfer reactions the characterization of the proteins and their binding sites, the orientation of the substrates on the proteins, the isolation of enzyme-substrate-acceptor complexes, and, ultimately, a chemical formulation of the intermediary processes in the action of the enzyme are the desired goal.

The enzymology of transmethylation is still far from this state. Details of the mechanism beyond the donor and acceptor specificity and a few of the characteristics of some of the enzymes are not yet known. The number of methyltransferases that have been purified significantly is still very small. The methyl transfer reactions with sulfonium donors are strongly exergonic; the free energy of hydrolysis of the sulfonium compounds at physiological pH values resembles that of the terminal phosphate of adenosine triphosphate (29, 104). All enzymatic transmethylations observed to this date are irreversible which complicates the conventional approaches of the enzymologist.

In a few instances, metal ions such as Mg^{++}, Zn^{++}, Mn^{++}, or Cd^{++} activate the process (4, 150). In the case of catechol-O-methyltransferase it has been suggested that the metal serves as a chelating agent to bring the substrate, donor, and enzyme together (4).

The great diversity of acceptors of the methyl group complicates a unified concept of transmethylation. In the case of purines and of phenolic compounds, for example, the methyl group can be transferred to various positions of the ring system of the acceptor. This makes it difficult to visualize a rigid arrangement of donor and substrate on the enzyme protein. A stepwise reaction including a methylated enzyme protein as an intermediate would facilitate the understanding of transmethylation. Durell and Cantoni (49) have tested this possibility with thetin-homocysteine methylpherase. The reaction might be visualized as occurring in two steps:

$$
\begin{array}{cc}
\underset{\underset{R}{\overset{|}{\underset{|}{S^{\oplus}}}}}{CH_3 \diagdown \diagup C^{14}H_3} + E \rightleftharpoons E^{\oplus}{-}C^{14}H_3 + \underset{\underset{R}{\overset{|}{\underset{|}{S}}}}{\overset{CH_3}{\overset{|}{S}}} ;
\end{array}
$$

$$E^{\oplus}{-}C^{14}H_3 + HSR \rightleftharpoons E + C^{14}H_3SR + H^{\oplus}$$

Chart 14. Hypothetical Steps in Enzymatic Transmethylation.

After exposure to methyl-labeled substrate the enzyme protein was isolated and examined for labeling. The protein was precipitated at low temperature with alcohol or with ammonium sulfate, followed by repeated reprecipitation. No evidence was found to suggest that a methylated enzyme functions as an intermediate in transmethylation.

References, pp. 104—112.

FROMM and NORDLIE (56) have carried out kinetic measurements with the corresponding enzyme from rat liver. They observed competitive inhibition by homoserine. The latter may be considered a homologue of homocysteine. The inhibition was non-competitive with respect to dimethyl-β-propiothetin. It appears, therefore, that the two substrates have separate sites of attachment on the enzyme. Obviously, for inter-action one has to assume proximity of these sites.

Transmethylation may be considered a nucleophilic substitution in which the acceptor is the attacking agent. The result is displacement of a thioether (S-adenosylhomocysteine, methionine, methylthiopropionic acid) by formation of a new bond between the carbon of the methyl group and the acceptor. The mode of action of the enzyme protein is still obscure.

Even less understood in its detail is the transfer of the methyl group of S-adenosylmethionine to yield the methylene group of cyclopropane fatty acids (110, 111). Branched methyl intermediates (a), branched methylene (b), cyclopropene (c), and methyl olefinic intermediates (d) have been considered as precursors:

$$
\begin{array}{cc}
\text{CH}_3 & \text{CH}_2 \\
\overset{\text{H}}{\underset{\text{H}}{|}}\ \overset{|}{\underset{\text{H}}{}}\ \overset{\text{H}}{\underset{\text{H}}{}}\ \overset{\text{H}}{\underset{\text{H}}{}} & \overset{\text{H}}{}\ \overset{\|}{}\ \overset{\text{H}}{}\ \overset{\text{H}}{} \\
-\text{C}-\text{C}-\text{C}-\text{C}-\quad(a) & -\text{C}-\text{C}-\text{C}-\text{C}-\quad(b)
\end{array}
$$

An alternative is the formation of a methylene group on the sulfonium sulfur of S-adenosylmethionine (120). For this, one would have to postulate a rather unique enzyme to convert the methyl group to a methylene group in an irreversible fashion (A = adenine, R = ribose):

$$
\text{A}-R-\text{S}-(\text{C}_4\text{H}_8\text{NO}_2) \underset{\text{CH}_2}{\overset{\|}{}} + \ \underset{\text{H H H H}}{-\text{C}-\text{C}=\text{C}-\text{C}-} \ \rightarrow
$$

$$
\rightarrow \text{A}-R-\text{S}-(\text{C}_4\text{H}_8\text{NO}_2) + \ \underset{\text{H H}}{-\text{C}-\text{C}}\overset{\text{CH}_2}{\underset{\text{H H}}{-\text{C}-\text{C}-}}
$$

If the methylene and methyl form of the sulfonium compound would be in equilibrium, deuterium or tritium-labeled methyl groups would exchange the hydrogen isotope with protons from the solvent water. So far, such a process has never been observed in transmethylation.

Of fundamental importance is the observation of DU VIGNEAUD and his school (173, 176, 177), especially RACHELE (122), that the methyl group is transferred as a unit. In all instances that have been studied with tracer techniques, no exchange of the hydrogen of the methyl group with protons from the solvent water was observed.

2. Biosynthesis of Spermidine and Spermine.

Tracer experiments with labeled methionine in cultures of *Neurospora crassa* gave the first indication that the propylamino part of spermine and spermidine is derived from methionine (*58*). Many details were clarified by Tabor and Rosenthal (*164, 165*) by using cell-free purified enzymes from *Escherichia coli*. Adenosylsulfonium compounds are key intermediates in the process: The first step is the formation of S-adenosyl-methionine, followed by decarboxylation:

$$\text{Methionine} + \text{ATP} \rightarrow \text{Adenine-ribose}\overset{\oplus}{\underset{\overset{|}{CH_3}}{S}}\text{—CH}_2\text{CH}_2\text{CH(NH}_2)\text{COOH}$$

$$\downarrow$$

$$\text{Adenine-ribose}\overset{\oplus}{\underset{\overset{|}{CH_3}}{S}}\text{—CH}_2\text{CH}_2\text{CH}_2\text{NH}_2 + CO_2$$

The decarboxylated S-adenosylmethionine [S-adenosyl-(5′)-3-methyl-mercapto-propylamine] reacts with putrescine to yield spermidine in the following way:

$$\text{Adenine-ribose}\overset{\oplus}{\underset{\overset{|}{CH_3}}{S}}\text{—CH}_2\text{CH}_2\text{CH}_2\text{NH}_2 + H_2\text{N(CH}_2)_4\text{NH}_2 \rightarrow$$

$$\rightarrow \text{Adenine-ribose—S—CH}_3 + H_2\text{N(CH}_2)_4\text{NH(CH}_2)_3\text{NH}_2 + H^{\oplus}$$

In the latter reaction, 5′-methylthioadenosine is formed which, in some fashion, may re-enter the cycle by being converted to S-adenosylmethionine (*138*). The formation of spermine from spermidine presumably involves repetition of the propylamino group transfer. However, a separate enzyme is necessary (*166*).

It would be very interesting to know why the enzyme effecting the transfer of the propylamino group is specific for removing this group rather than the methyl group. The absence of the carboxyl group from the donor apparently is a prerequisite for entering into the last reaction formulated. The enzyme catalyzing this neither splits the α-aminobutyryl group nor transfers it to putrescine.

3. Formation of Homoserine Lactone and Homoserine from S-Adenosylmethionine.

The enzymatic formation of homoserine lactone from S-adenosyl-methionine was discovered by Shapiro and Mather (*148*) in extracts of *Aerobacter aerogenes* (*Chart 15*). The conversion of homoserine lactone to homoserine appears to be non-enzymatic. MUDD (*100*) has found the same enzyme in yeast. He tested the interesting possibility that the reaction leading to homoserine might be more complex than represented above. It was surmised that 2-amino-3-butenoic acid (**41**), a hitherto unknown compound, might be a transitory intermediate. This was tested by carrying out the reaction in tritiated water (*101*) *(Chart 16)*.

$$\text{Adenine-ribose}\overset{\oplus}{\underset{\underset{(2)}{\overset{|}{\text{CH}_3}}}{\text{S}}}\text{—CH}_2\text{CH}_2\text{CH(NH}_2)\text{COO}^{\ominus}$$

$$\underset{(21)}{\text{Adenine-ribose—S—CH}_3} + \underset{\downarrow\ (17)}{\text{CH}_2\text{CH}_2\text{CH(NH}_2)\text{C}=\text{O}}$$

$$\underset{(18)}{\text{HOCH}_2\text{CH}_2\text{CH(NH}_2)\text{COOH}}$$

Chart 15. The Enzymatic Decomposition of S-Adenosylmethionine (**2**) to 5'-Methyl-thioadenosine (**21**), Homoserine Lactone (**17**) and Homoserine (**18**).

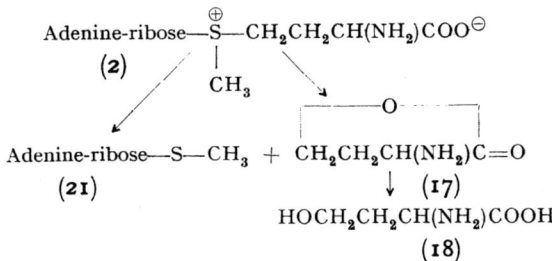

$$\text{Adenine-ribose}\overset{\oplus}{\underset{\underset{\text{CH}_3}{|}}{\text{S}}}\text{—CH}_2\text{—CH}_2\text{—CH(NH}_2)\text{COO}^{\ominus}$$

$$\underset{(41)}{\text{Adenine-ribose—S—CH}_3} + [\text{CH}_2{=}\text{CH—CH(NH}_2)\text{COO}^{\ominus}] + \text{H}^{\oplus}$$

$$\underset{\text{T}}{\text{CH}_2\text{—CH—CH(NH}_2)\text{C}=\text{O}}$$

Chart 16. Hypothetical Mechanism of Enzymatic Degradation of S-Adenosyl-methionine. Search for 2-Amino-3-butenoic Acid (**41**) as an Intermediate by Using Tritiated Water.

Virtually no tritium was found in the resulting homoserine lactone, and this mechanism, if operative at all, is not the principal route of enzymatic decomposition of S-adenosylmethionine. In non-enzymatic hydrolysis by heating near neutrality in tritiated water solution, the incorporation was somewhat larger, but it still did not exceed a few per cent of the value expected if 2-amino-3-butenoic acid were an obligatory intermediate. Thus, the mechanism represented in Chart 15 is the principal route in enzymatic and non-enzymatic formation of homoserine lactone. This reaction constitutes an interesting example of an intramolecular nucleophilic attack of the carboxyl group, by which 5'-methylthioadenosine is displaced.

VII. Derivatives and Analogues of Biological Sulfonium Compounds.

1. Analogues for Metabolic Studies.

Derivatives of the simple thetins and of S-methylmethionine include many variations that can be obtained by organic synthesis. The enzymatic synthesis of these substances has not yet been studied sufficiently to attempt the production of analogues by this route. In contrast, an increasing number of studies on modifications of S-adenosylmethionine is appearing. Its analogues usually have been obtained by biosynthesis with precursors that contain the modification intended for the final product. The utilization or exclusion of these precursors reflects on the specificity of the enzymes forming the sulfonium compound.

Sulfonium compound analogues comprise the following groups:

a. Compounds in which the methyl group is replaced by ethyl and higher homologues, or with substitution of the methyl hydrogen atoms, such as in the trifluoromethyl derivative.

b. Selenonium analogues.

c. Compounds containing modifications in the amino acid component.

d. Sulfonium derivatives that are modified in the adenosine region of S-adenosylmethionine.

The ethyl group as a substitute for the methyl group has found the greatest attention, and all experiments pertaining to this have their root in the synthesis of ethionine by DYER (50). The literature concerning this amino acid analogue has grown in recent years to exceptional proportions (51, 158). Most investigators, however, have been concerned with over-all nutritional observations, or with the competitive effects of ethionine with methionine in protein synthesis. The concept of trans-ethylation is foreshadowed in the work of DU VIGNEAUD and his school (175). In dietary experiments with rats, homocystine and dimethylaceto-thetin or dimethylpropiothetin proved nutritionally adequate (98). Methylethylthetin proved less effective than dimethylthetin, and diethyl-thetin was ineffective in promoting growth. Incidental to these experiments it was found that sulfocholine, $(CH_3)_2\overset{\oplus}{S}—CH_2CH_2OH$, cannot replace choline in nutritional experiments; it is highly toxic (97).

Subsequent in vitro investigations on thetin homocysteine methyltransferase from mammalian liver have shown that the enzyme is rather non-specific toward donor compounds (Table 1).

A bicarbonate medium was used, and the carbon dioxide liberated by the proton resulting from transmethylation was measured (p. 93). Similar data for the enzyme from horse liver have been reported by DURELL et al. (48). By transethylation with homocysteine, ethionine is formed which accounts for the insufficiency or deleterious effects of the ethyl group containing sulfonium compounds in animal nutrition.

Table 1. Activity of Various Sulfonium Halides as Substrates of Thetin — Homocysteine Transmethylase from Rat Liver.

Compound	Initial relative reaction rate
Dimethylacetothetin chloride......................	100.0
Ethylmethylacetothetin chloride...................	85.0
Diethylacetothetin chloride	0
Dimethyl-α-propiothetin bromide..................	51.9
Dimethyl-β-propiothetin bromide..................	24.9
Ethylmethyl-β-propiothetin bromide	33.0
Dimethyl-γ-butyrothetin bromide..................	23.6
Methionine methylsulfonium chloride	17.5
Sulfocholine iodide...............................	0
Trimethylsulfonium chloride	16.5
Triethylsulfonium chloride	0
Ethyldimethylsulfonium iodide	6.6
Butyldimethylsulfonium iodide	4.0

The results are expressed as percentages of the initial rate obtained with dimethylacetothetin chloride. Adapted from Maw (96).

Other products of transethylation may also contribute to the harmful effects of ethionine (158, 161). Already, there is evidence of the formation of ethyl analogues of phospholipid cholines, creatine, N-ethylhistidine, and N-ethylcarnosine, and transfer of the ethyl group in vivo to various nucleic acids has been reported (160).

A systematic exploration of transethylation with isolated enzymes is now facilitated by the availability of S-adenosylethionine by the biosynthetic method of Parks (114). In his procedure L-ethionine is incorporated into the culture medium of vigorously metabolizing yeast. In the cells, the amino acid analogue interacts with adenosine triphosphate and forms the sulfonium compound. Labeled modifications are obtained by employing the requisite precursors (135, 159, 160). In liver, the ethyl sulfonium analogue appears to be metabolized less rapidly than S adenosylmethionine as judged by the accumulation of the ethyl compound in this tissue (160). Transethylations are either less numerous or less effective than transmethylations. A systematic exploration of this field portends important information. Recently, this problem has received special interest by the discovery of ethionine as a natural product in some microorganisms (53, 90).

Attempts to isolate higher alkyl homologues of S-adenosylmethionine and S-adenosylethionine have failed, so far. The following compounds were tested as precursors in the biosynthesis by yeast: S-n-propylhomocysteine, S-*n*-butylhomocysteine, and S-*iso*-amylhomocysteine. Very small quantities, if any, of the corresponding adenosine sulfonium compounds are formed (*160*).

Other modifications of S-adenosylmethionine include the substitution of S-trifluoromethylhomocysteine (*44*) for methionine. Like ethionine, this amino acid analogue reacts with adenosine triphosphate to give S-adenosyltrifluoromethionine. STEKOL (*160*) attributes the high toxicity of this compound to competitive inhibition of the metabolism of S-adenosylmethionine. Enzymatic transfer of the trifluoromethyl group is probable, but details are not yet known.

Of particular interest is the selenium analogue of S-adenosyl-methionine (*42*). Selenomethionine can substitute for methionine in the enzymatic formation (*105*):

$$\text{Adenine-ribose—} \overset{\oplus}{\underset{|}{\text{Se}}} \text{—CH}_2\text{CH}_2\text{CH(NH}_2\text{)COOH}$$
$$\text{CH}_3$$

$$(42)$$

In the biochemical literature, this is usually designated as Se-adenosyl-selenomethionine. The ethyl selenonium derivative has also been prepared (*160*). Se-adenosylselenomethionine participates in trans-methylations (*105*). Both the methyl- and ethylselenonium analogues are in some instances more efficient alkyl donors than the natural sulfonium compounds (*160*). The problem has great practical importance because selenomethionine is found in plants from selenium-containing soils (*153*). Its conversion to protein-bound selenomethionine may account for part of its toxicity; the selenonium derivative contributes to this by altering the rate of methyl distribution (*20, 126a, 160*).

Modifications of the nucleoside part of S-adenosylmethionine have been explored very little, because such compounds are rather inaccessible. The enzyme synthesizing S-adenosylmethionine is very specific in its requirement for adenosine triphosphate; no other nucleoside triphosphate, not even inosine triphosphate, reacts measurably with methionine.

The enzymatic exchange of the adenine of S-adenosylmethionine for other bases has not yet been observed. It will be remembered that from the nicotinamide nucleotides a large variety of analogues have been obtained in this way (*74*).

Deamination of S-adenosylmethionine with nitrous acid leads to the hypoxanthine derivative in which, however, the amino group of the methionine part is also removed (*25*). This compound has not yet been fully characterized. One may surmise that the 6-amino group of the purine part is as essential as in many nucleotide coenzymes.

2. Sulfonium Derivatives of Protein-bound Methionine.

The sulfur atom of protein-bound methionine can react in the same fashion as methionine to yield sulfonium derivatives, provided the site is accessible to the reagent. No enzymatic processes of this type have been found so far.

Alkylation of protein-bound methionine was described by WIND-MUELLER et al. (*184*) in a study of the effects of ethylene oxide on proteins. The practical importance lies in the extensive use of this reagent as a fumigating agent. Its high reactivity alters a number of compounds including methionine. In proteins the nutritional value is lowered, and toxic effects are possible. The reaction is exemplified by N-acetylmethionine from which S-(2-hydroxyethyl)-N-acetylmethionine (**43**) is formed:

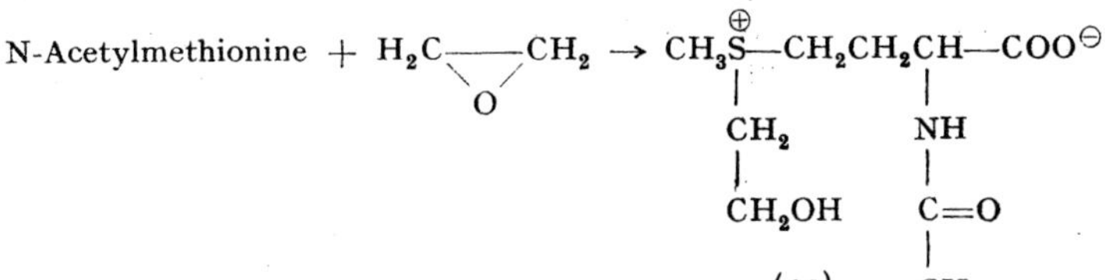

$$N\text{-Acetylmethionine} + H_2C \underset{O}{\overset{}{\diagdown\!/}} CH_2 \rightarrow CH_3\overset{\oplus}{S}-CH_2CH_2CH-COO^{\ominus}$$

$$
\begin{array}{cc}
 & \\
CH_2 & NH \\
| & | \\
CH_2OH & C{=}O \\
& | \\
(43) & CH_3
\end{array}
$$

S-(2-Hydroxyethyl)-N-acetylmethionine.

More specific are the studies of GUNDLACH et al. (*62, 63*) on the effects of iodoacetic acid on the methionine in ribonuclease. The inactivation of this enzyme formerly had been interpreted as an interaction with the sulfhydryl groups of cysteine. When the exploration of the amino acid composition of ribonuclease showed that no sulfhydryl group is present in the molecule, the interpretation had to be revised. It was found that the methionine in ribonuclease reacts readily with iodoacetic acid leading to the carboxymethylated enzyme:

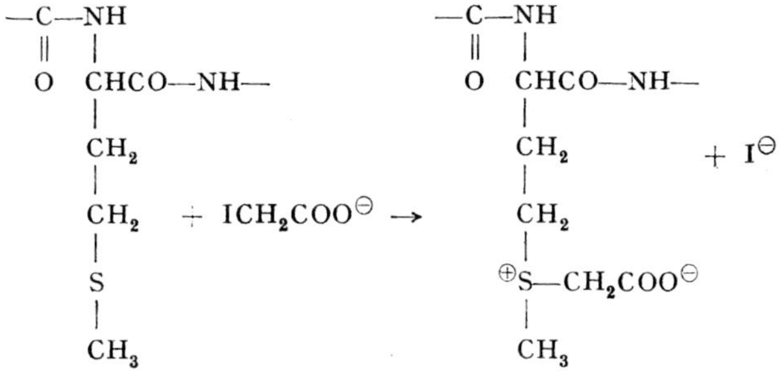

$$
\begin{array}{lll}
-C-NH & & -C-NH \\
\;\|\;\;\; | & & \;\|\;\;\; | \\
O \;\; CHCO-NH- & & O \;\; CHCO-NH- \\
\quad\; | & & \quad\; | \\
\quad CH_2 & & \quad CH_2 \qquad\quad + I^{\ominus} \\
\quad\; | & & \quad\; | \\
\quad CH_2 & +\; ICH_2COO^{\ominus} \rightarrow & \quad CH_2 \\
\quad\; | & & \quad\; | \\
\quad\; S & & \;\;\overset{\oplus}{S}-CH_2COO^{\ominus} \\
\quad\; | & & \quad\; | \\
\quad CH_3 & & \quad CH_3
\end{array}
$$

Chart 17. S-Carboxymethylation of Enzyme-bound Methionine.

The reaction occurs in the range of pH 4 to 8 under lenient conditions, and the presence of carboxymethylated methionine in proteins explains the methionine deficiency in amino acid analyses of such material, because the hydrolysis occurs in various ways, and methionine is regenerated only in part. With free carboxymethylmethionine the following concurrent processes *(Chart 18)* were observed *(62, 171)*.

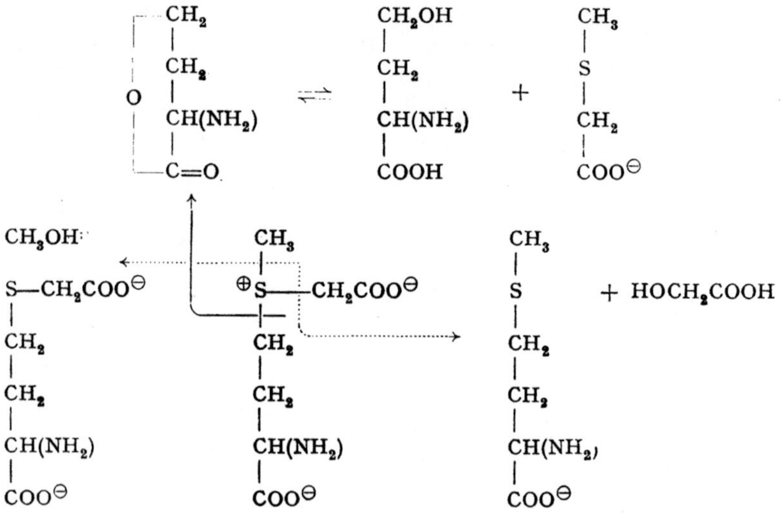

$$
\begin{array}{ccc}
\text{CH}_2 & \text{CH}_2\text{OH} & \text{CH}_3 \\
| & | & | \\
\text{CH}_2 & \text{CH}_2 & \text{S} \\
\text{O} \quad | & | \quad + & | \\
\text{CH(NH}_2) & \text{CH(NH}_2) & \text{CH}_2 \\
| & | & | \\
\text{C=O} & \text{COOH} & \text{COO}^{\ominus}
\end{array}
$$

$$
\begin{array}{ccc}
\text{CH}_3\text{OH} & \text{CH}_3 & \text{CH}_3 \\
& | & | \\
\text{S—CH}_2\text{COO}^{\ominus} & {}^{\oplus}\text{S——CH}_2\text{COO}^{\ominus} & \text{S} \quad + \text{ HOCH}_2\text{COOH} \\
| & | & | \\
\text{CH}_2 & \text{CH}_2 & \text{CH}_2 \\
| & | & | \\
\text{CH}_2 & \text{CH}_2 & \text{CH}_2 \\
| & | & | \\
\text{CH(NH}_2) & \text{CH(NH}_2) & \text{CH(NH}_2) \\
| & | & | \\
\text{COO}^{\ominus} & \text{COO}^{\ominus} & \text{COO}^{\ominus}
\end{array}
$$

Chart 18. Hydrolytic Cleavages of S-Carboxymethylmethionine.

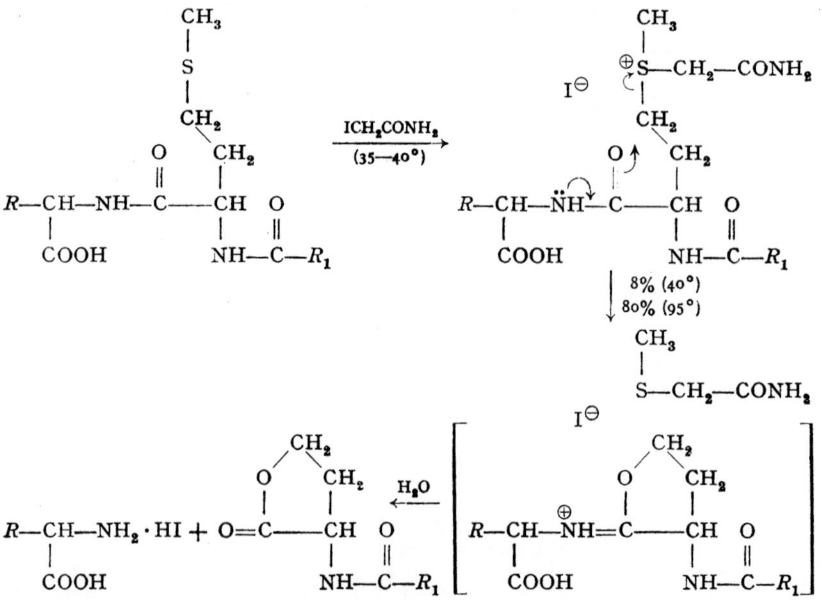

Chart 19. Selective Splitting of a Peptide Chain by Introduction of a Sulfonium Group.

References, pp. 104—112.

An interesting application of the change in stability resulting from a sulfonium group has been reported by GROSS and WITKOP (61) and by LAWSON (85): Protein-bound methionine is converted by iodoacetic acid or by iodoacetamide into the corresponding sulfonium compound. This is followed by hydrolysis which leads to scission of the peptide bond that adjoins the sulfonium group. The yield is not quantitative (Chart 19).

The alteration of protein-bound methionine to a sulfonium derivative has been used in recent years by several investigators (109, 142, 168) for exploration of the accessibility of the methionine in enzyme proteins and of the effect that the sulfonium group exerts on the enzymatic activity of these derivatives (124, 131, 157). An elaborate application of such protein modification has been conceived by BURR and KOSHLAND (22). A special residue, termed "reporter group" is introduced into the enzyme near the site of reaction with the substrate. Depending on the distance of the reporter group from the site of binding of the substrate, the orientation of the latter, conformational changes of the protein, and other properties are revealed. The method by which one observes the effects depends on the type of the changes and on the nature of the reporter group. Alteration in the absorbancy, fluorescence, electron spin resonance, and many other parameters may be used. The first examples of reporter groups are sulfonium derivatives of the methionine molecule in chymotrypsin near the serine which constitutes the active center:

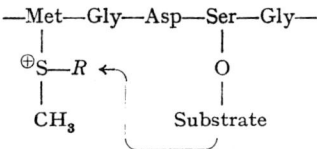

Chart 20. Introduction of a Sulfonium "Reporter Group" near Active Site of Chymotrypsin.

An ideal reporter group would leave the enzyme activity intact, yet it would be sensitive to changes imposed by the substrate. This situation was very nearly realized by using 2-bromoacetamido-4-nitropheno l for production of the reporter group in chymotrypsin. Its chromophoric group responds readily to the environment and has an absorption maximum near 400 mμ. In this range the enzyme and its substrates do not absorb; hence, observation of the reporter group is possible by spectrophotometry. For example, tests with benzoyl-L-phenylalanine and benzoyl-D-phenyl-alanine as substrates of the chymotrypsin with the sulfonium reporter group showed marked differences in the absorption spectra. Many possibilities and some limitations of this technique have been pointed out by BURR and KOSHLAND (22).

The introduction of sulfonium reporter groups into enzymes is particularly promising because of the sensitivity of this group to intramolecular changes and to various influences in the vicinity of the sulfonium pole.

References.

1. ABRAHAMSON, L. and S. K. SHAPIRO: Biosynthesis of Methionine; Partial Purification and Properties of Homocysteine Methyltransferase of Jack Bean Meal. Arch. Biochem. Biophys. 109, 376 (1965).
2. ARTOM, C. and N. H. EUDY: Methylation of Phosphatidyl Monomethylethanolamine in Liver Preparations. Biochem. Biophys. Res. Comm. 15, 201 (1964).
3. ATKINSON, R. O. and F. POPPELSDORF: Sulfonium Derivatives of DL-Methionine and 5-(2-Methyl-thioethyl)hydantoin. J. Chem. Soc. (London) 1951, 1378.
4. AXELROD, J.: The Formation and Metabolism of Physiologically Active Compounds by N- and O-Methyltransferases. In: S. K. Shapiro and F. Schlenk (Edit.), Transmethylation and Methionine Biosynthesis, p. 71. Chicago and London: Univ. of Chicago Press. 1965.
5. BADDILEY, J.: Adenine 5'-Deoxy-5'-methylthiopentoside (Adenine Thiomethyl Pentoside): A Proof of Structure and Synthesis. J. Chem. Soc. (London) 1951, 1348.
6. BADDILEY, J., G. L. CANTONI and G. A. JAMIESON: Structural Observations on "Active Methionine". J. Chem. Soc. (London) 1953, 2662.
7. BADDILEY, J., W. FRANK, N. A. HUGHES and J. WIECZORKOWSKI: The Hydrolysis of Sulphonium Nucleosides and Glycosides by Alkali. J. Chem. Soc. (London) 1962, 1999.
8. BADDILEY, J. and G. A. JAMIESON: Synthesis of Active Methionine. J. Chem. Soc. (London) 1954, 4280.
9. — — Synthesis of S-(5'-Deoxyadenosine-5')-homocysteine, a Product from Enzymic Methylations Involving "Active Methionine". J. Chem. Soc. (London) 1955, 1085.
10. BAERNSTEIN, H. D.: The Sulfur Distribution in Proteins. II. The Combined Methods for the Determination of Cystine, Methionine, and Sulfates in Hydriodic Acid Digests. J. Biol. Chem. 115, 33 (1936).
11. BALDESSARINI, R. J. and I. J. KOPIN: Assay of Tissue Levels of S-Adenosylmethionine. Analyt. Biochem. 6, 289 (1963).
12. BENNETT, M. A.: The Replaceability of DL-Methionine in the Diet of the Albino Rat with DL-Methionine Sulfone and DL-Methionine Methylsulfonium Chloride. J. Biol. Chem. 141, 573 (1941).
13. BLAU, N. F. and C. G. STUCKWISCH: A New Synthesis of Dimethyl-β-propiothetin Hydrochloride. J. Amer. Chem. Soc. 73, 2355 (1951).
14. — — The Conjugative Effect of the Dimethylsulfonium Group in an Aliphatic System. J. Organ. Chem. (USA) 22, 82 (1957).
15. BÖHME, H. und P. HELLER: Über β- und γ-Ketosulfonium-Salze und ihren Zerfall in wäßriger Lösung. Chem. Ber. 86, 443 (1953).
16. — — Über α- und β-Sulfonylsulfonium-Salze. Chem. Ber. 86, 785 (1953).
17. BÖHME, H. und W. KRAUSE: Über Dialkyl-phenacyl-sulfoniumsalze und ihre Spaltung mit wäßrigen Laugen. 1. Mitt. über Sulfoniumsalze. Chem. Ber. 82, 426 (1949).
18. BOREK, E. and P. R. SRINIVASAN: Alteration of the Macromolecular Structure of Nucleic Acids by Transmethylation. In: S. K. Shapiro and F. Schlenk (Edit.), Transmethylation and Methionine Biosynthesis, p. 115. Chicago and London: Univ. of Chicago Press. 1965.

19. BORSOOK, H. and J. W. DUBNOFF: On the Role of Oxidation in the Methylation of Guanidoacetic Acid. J. Biol. Chem. **171**, 363 (1947).

20. BREMER, J. and Y. NATORI: Behavior of some Selenium Compounds in Trans methylation. Biochim. Biophys. Acta **44**, 367 (1960).

21. BROWN, C. and E. A. LETTS: On Dimethylthetine and its Derivatives. Trans. Roy. Soc. Edinburgh **28**, 571 (1878).

22. BURR, M. and D. E. KOSHLAND, Jr.: Use of "Reporter Groups" in Structure— Function Studies of Proteins. Proc. Nat. Acad. Sci. (USA) **52**, 1017 (1964).

23. CANTONI, G. L.: Methylation of Nicotinamide with a Soluble Enzyme System from Rat Liver. J. Biol. Chem. **189**, 203 (1951).

24. — Synthesis and Transfer of the Labile Methyl Group. In: W. D. McElroy and B. Glass (Edit.), Phosphorus Metabolism, Vol. II, p. 129. Baltimore: Johns Hopkins Press. 1952.

25. — S-Adenosylmethionine; a New Intermediate Formed Enzymatically from *L*-Methionine and Adenosinetriphosphate. J. Biol. Chem. **204**, 403 (1953).

26. — S-Adenosylmethionine (AMe). In: D. Shemin (Edit.), Biochemical Preparations, Vol. V, p. 58. New York: Wiley. 1957.

27. — Preparation of S-Adenosylmethionine. In: S. P. Colowick and N. O. Kaplan (Edit.), Methods in Enzymology, Vol. III, p. 600. New York: Academic Press. 1957.

28. — Onium Compounds and their Biological Significance. In: M. Florkin and H. S. Mason (Edit.), Comparative Biochemistry, Vol. I, p. 181. Amsterdam: Elsevier. 1960.

29. — S-Adenosylmethionine—Revisited. In: S. K. Shapiro and F. Schlenk (Edit.), Transmethylation and Methionine Biosynthesis, p. 21. Chicago and London: Univ. of Chicago Press. 1965.

30. CANTONI, G. L. and D. G. ANDERSON: Enzymatic Cleavage of Dimethylpropiothetin by *Polysiphonia lanosa*. J. Biol. Chem. **222**, 171 (1956).

31. CANTONI, G. L. and J. DURELL: Activation of Methionine for Transmethylation. II. The Methionine-activating Enzyme; the Mechanism of the Reaction. J. Biol. Chem. **225**, 1033 (1957).

32. CARRARA, G.: Sopra alcune tetine isomere. Gazz. chim. ital. **23**, 493 (1893).

33. CERWENKA, E. A., Jr. and W. C. COOPER: Toxicology of Selenium and Tellurium and their Compounds. Arch. Environmental Health **3**, 189 (1961).

34. CHALLENGER, F.: Aspects of the Organic Chemistry of Sulphur. New York: Academic Press. 1959.

35. CHALLENGER, F., R. BYWOOD, P. THOMAS and B. J. HAYWARD: Studies on Biological Methylation. XVII. The Natural Occurrence and Chemical Reactions of some Thetins. Arch. Biochem. Biophys. **69**, 514 (1957).

36. CHALLENGER, F. and B. J. HAYWARD: Occurrence of a Methylsulfonium Derivative of Methionine-(α-aminodimethyl-γ-butyrothetin) in Asparagus. Chem. and Ind. **1954**, 729.

37. CHALLENGER, F. and C. HIGGINBOTTOM: The Production of Trimethylarsine by *Penicillium brevicaule*. Biochem. J. **29**, 1757 (1935).

38. CHALLENGER, F. and H. D. HOLLINGWORTH: The Abnormal Hydrolysis of Methyl β-Methylthio- and β-Ethylthio-propionate. J. Chem. Soc. (London) **1959**, 61.

39. CHALLENGER, F., D. B. LISLE and P. B. DRANSFIELD: Studies on Biological Methylation. XIV. The Formation of Trimethylarsine and Dimethyl Selenide in Mould Cultures from Methyl Sources Containing ^{14}C. J. Chem. Soc. (London) **1954**, 1760.

40. CHALLENGER, F. and Y. C. LIU: Elimination of Methanethiol and Methyl Sulfide from Methylmercapto and Dimethylsulfonium Compounds by Molds. Rec. trav. chim. Pays-Bas **69**, 334 (1950).

41. CHALLENGER, F. and M. I. SIMPSON: Biological Methylation. XII. Precursor of the Dimethylsulfide Evolved by *Polysiphonia fastigiata.* J. Chem. Soc. (London) **1948**, 1591.

42. CHALLENGER, F. and J. M. WALSHE: Methyl Mercaptan in Relation to Foetor Hepaticus. Biochem. J. **59**, 372 (1955).

43. CRANE, W. C. and H. N. RYDON: The Alkaline Fission of some 2-Substituted Dimethylethylsulphonium Iodides. J. Chem. Soc. (London) **1947**, 766.

44. DANNLEY, R. L. and R. G. TABORSKY: Synthesis of *DL*-S-Trifluoro-methyl-homocysteine (Trifluoromethylmethionine). J. Organ. Chem. (USA) **22**, 1275 (1957).

45. DRANSFIELD, P. B. and F. CHALLENGER: Formation of Dimethylselenide in Mold Cultures in Presence of *D*- and *L*-Methionine. J. Chem. Soc. (London) **1955**, 1153.

46. DUBNOFF, J. W. and H. BORSOOK: Dimethylthetin and Dimethyl-β-propio-thetin in Methionine Synthesis. J. Biol. Chem. **176**, 789 (1948).

47. DUERRE, J. A. and F. SCHLENK: Formation and Metabolism of S-Adenosyl-*L*-homocysteine in Yeast. Arch. Biochem. Biophys. **96**, 575 (1962).

48. DURELL, J., D. G. ANDERSON and G. L. CANTONI: The Synthesis of Methionine by Enzymic Transmethylation. I. Purification and Properties of Thetin Homocysteine Methylpherase. Biochim. Biophys. Acta **26**, 270 (1957).

49. DURELL, J. and G. L. CANTONI: The Synthesis of Methionine by Enzymic Transmethylation. III. Mechanism of the Reversible Polymerization of Thetin-homocysteine Methylpherase and its Relation to the Mechanism of Methionine Synthesis. Biochim. Biophys. Acta **35**, 515 (1959).

50. DYER, H. M.: Evidence of the Physiological Specificity of Methionine in Regard to the Methylthiol Group: The Synthesis of S-Ethylhomocysteine (Ethionine) and a Study of its Availability for Growth. J. Biol. Chem. **124**, 519 (1938).

51. FARBER, E., K. H. SHULL, S. VILLA-TREVINO, B. LOMBARDI and M. THOMAS: Biochemical Pathology of Acute Hepatic Adenosine Triphosphate Deficiency. Nature **203**, 34 (1964).

52. FERGER, M. F. and V. DU VIGNEAUD: Oxidation *in vivo* of the Methyl Groups of Choline, Betaine, Dimethylthetin, and Dimethyl-β-propiothetin. J. Biol. Chem. **185**, 53 (1950).

53. FISHER, J. F. and M. F. MALLETTE: The Natural Occurrence of Ethionine in Bacteria. J. Gen. Physiol. **45**, 1 (1961).

54. FLOYD, N. F. and T. F. LAVINE: Isolation of Methionine from Protein Hydrolysates as the Methylsulfonium Salt. J. Biol. Chem. **207**, 119 (1954).

55. FRANK, W., J. WIECZORKOWSKI, N. A. HUGHES and J. BADDILEY: The Alkali Hydrolysis of Sulphonium Nucleosides. Proc. Chem. Soc. (London) **1961**, 449.

56. FROMM, H. J. and R. C. NORDLIE: On the Purification and Kinetics of Rat Liver Thetin—Homocysteine Transmethylase. Arch. Biochem. Biophys. **81**, 363 (1959).

57. GOERDELER, J.: Methoden zur Herstellung und Umwandlung von Sulfonium-verbindungen. In: E. Müller (Edit.), Houben-Weyl, Methoden der organischen Chemie, Bd. 9, S. 171. Stuttgart: Thieme. 1955.

58. GREENE, R. C.: Incorporation of the Carbon Chain of Methionine into Spermidine. J. Amer. Chem. Soc. **79**, 3929 (1957).

59. — Biosynthesis of Dimethyl-β-propiothetin. J. Biol. Chem. **237**, 2251 (1962).

60. GREENE, R. C. and N. B. DAVIS: Biosynthesis of S Methylmethionine in the Jack Bean. Biochim. Biophys. Acta **43**, 360 (1960).
61. GROSS, E. and B. WITKOP: Selective Cleavage of the Methionyl Peptide Bonds in Ribonuclease with Cyanogen Bromide. J. Amer. Chem. Soc. **83**, 1510 (1961).
62. GUNDLACH, H. G., S. MOORE and W. H. STEIN: The Reaction of Iodoacetate with Methionine. J. Biol. Chem. **234**, 1761 (1959).
63. GUNDLACH, H. G., W. H. STEIN and S. MOORE: The Nature of the Amino Acid Residues Involved in the Inactivation of Ribonuclease by Iodoacetate. J. Biol. Chem. **234**, 1754 (1959).
64. HAAGEN-SMIT, A. J., J. G. KIRCHNER, C. L. DEASY and A. N. PRATER: Chemical Studies of Pineapple (*Ananas sativus*). II. Isolation and Identification of a Sulfur Containing Ester in Pineapple. J. Amer. Chem. Soc. **67**, 1651 (1945).
65. HAAS, P.: The Liberation of Methyl Sulphide by Seaweed. Biochem. J. **29**, 1297 (1935).
66. HABA, G. DE LA, G. A. JAMIESON, S. H. MUDD and H. H. RICHARDS: S-Adenosylmethionine: The Relation of Configuration at the Sulfonium Center to Enzymatic Reactivity. J. Amer. Chem. Soc. **81**, 3975 (1959).
67. HAUSER, C. R., S. W. KANTOR and W. R. BRASEN: Rearrangement of Benzyl Sulfides to Mercaptans and of Sulfonium Ions to Sulfides Involving the Aromatic Ring by Alkali Amides. J. Amer. Chem. Soc. **75**, 2660 (1953).
68. HIS, W.: Über das Stoffwechselprodukt des Pyridins. Arch. exp. Pathol. Pharmakol. **22**, 253 (1887).
69. INGOLD, G. K.: Structure and Mechanism in Organic Chemistry. Ithaca, N. Y.: Cornell Univ. Press. 1953.
70. JAENICKE, L. und C. KUTZBACH: Folsäure und Folat-Enzyme. Fortschr. Chem. organ. Naturstoffe **21**, 183 (1963).
71. JAMIESON, G. A.: The Synthesis of 5'-Deoxy-5'-S-(3-methylthiopropylamine) sulfoniumadenosine ("Decarboxylated S-Adenosylmethionine"). J. Organ. Chem. (USA) **28**, 2397 (1963).
72. JOSHI, J. G. and P. HANDLER: Biosynthesis of Trigonelline. J. Biol. Chem. **235**, 2981 (1960).
73. KANESHIRO, T. and J. H. LAW: Phosphatidylcholine Synthesis in *Agrobacterium tumefaciens*. J. Biol. Chem. **239**, 1705 (1964).
74. KAPLAN, N. O.: The Pyridine Coenzymes. In: P. D. Boyer, H. Lardy and K. Myrbäck (Edit.), The Enzymes, Vol. 3 b, p. 105. New York: Academic Press. 1960.
75. KAPLAN, N. O., S. P. COLOWICK and C. C. BARNES: Effect of Alkali on Diphosphopyridine Nucleotide. J. Biol. Chem. **191**, 461 (1951).
76. KARRER, P., C. H. EUGSTER und D. K. PATEL: Über Inhaltsstoffe einiger Equisetum Arten. Helv. Chim. Acta **32**, 2397 (1949).
77. KIRIBUCHI, T. and T. YAMANISHI: Studies on the Flavor of Green Tea. IV. Dimethylsulfide and its Precursor. Agr. Biol. Chem. (Tokyo) **27**, 56 (1963).
78. KLEE, W. A.: Thetinhomocysteine Methylpherase: A Study in Molecular Organization. In: S. K. Shapiro and F. Schlenk (Edit.), Transmethylation and Methionine Biosynthesis, p. 220. Chicago and London: Univ. of Chicago Press. 1965.
79. KUHN, R., L. BIRKOFER und F. W. QUACKENBUSH: Jodometrische Titration von SH-Gruppen; Mikromethode zur Bestimmung von Cystein und Methionin in Proteinen. Ber. dtsch. chem. Ges. **72**, 407 (1939).
80. KUHN, R. und K. HENKEL: Über die Senkung der Körpertemperatur durch Adenylthiomethylpentose. Z. physiol. Chem. **269**, 41 (1941).

81. Lavine, T. F. and N. F. Floyd: The Decomposition of Methionine in Sulfuric Acid. J. Biol. Chem. **207**, 97 (1954).

82. Lavine, T. F., N. F. Floyd and M. S. Cammaroti: The Formation of Sulfonium Salts from Alcohols and Methionine in Sulfuric Acid. J. Biol. Chem. **207**, 107 (1954).

83. — — — Decomposition of 5'-Methylthioadenosine. Biochem. Biophys. Res. Comm. **1**, 156 (1959).

84. Lawley, P. D. and P. Brookes: Further Studies on the Alkylation of Nucleic Acids and their Constituent Nucleotides. Biochem. J. **89**, 127 (1963).

85. Lawson, W. B., E. Gross, C. M. Foltz and B. Witkop: Specific Cleavage of Methionyl Peptides. J. Amer. Chem. Soc. **83**, 1509 (1961).

86. Leaver, D. and F. Challenger: Biological Methylation. XVI. Alkyl Methyl Sulfides Evolved from the Urine of Dogs by Boiling Alkali. J. Chem. Soc. (London) **1957**, 39.

87. Letts, E. A.: Action of Iodoacetic and Bromoacetic Ethyl Ether on Sulphide of Methyl. Trans. Roy. Soc. Edinburgh **28**, 618 (1878).

88. Lieben, F.: Geschichte der physiologischen Chemie. Wien: Deuticke. 1935.

89. Lipmann, F.: Metabolic Generation and Utilization of Phosphate Bond Energy. Adv. Enzymology **1**, 99 (1941).

90. Loerch, J. D. and M. F. Mallette: Ethionine Biosynthesis in *Escherichia coli.* Arch. Biochem. Biophys. **103**, 272 (1963).

91. McRorie, R. A., M. R. Glazener, C. G. Skinner and W. Shive: Microbiological Activity of the Methylsulfonium Derivative of Methionine. J. Biol. Chem. **211**, 489 (1954).

92. McRorie, R. A., G. L. Sutherland, M. S. Lewis, A. D. Barton, M. R. Glazener and W. Shive: Isolation and Identification of a Naturally Occurring Analog of Methionine. J. Amer. Chem. Soc. **76**, 115 (1954).

93. Mamalis, P. and H. N. Rydon: The Alkaline Fission of 2-Aroyloxyethyl-dimethylsulphonium Iodides: The Evaluation of Hammett's Substituent Constants for some *ortho*-Substituents. J. Chem. Soc. (London) **1955**, 1049.

94. Mann, J. D., H. M. Fales and S. H. Mudd: Alkaloids and Plant Metabolism. VI. O-Methylation *in vitro* of Norbelladine, a Precursor of Amaryllidaceae Alkaloids. J. Biol. Chem. **238**, 3820 (1963).

95. Maw, G. A.: Thetin—Homocysteine Transmethylase. Biochem. J. **63**, 116 (1956).

96. — Thetin—Homocysteine Transmethylase. Some Further Characteristics of the Enzyme. Biochem. J. **70**, 168 (1958).

97. Maw, G. A. and V. du Vigneaud: An Investigation of the Biological Behavior of the Sulfur Analogue of Choline. J. Biol. Chem. **176**, 1029 (1948).

98. — — Compounds Related to Dimethylthetin as Sources of Labile Methyl Groups. J. Biol. Chem. **176**, 1037 (1948).

99. Motohiro, T.: Petroleum Odor in Canned Chum Salmon. Mem. Fac. Fisheries Hokkaido Univ. **10**, no. 1 (1962) [Chem. Abstr. **59**, 12090 (1963)].

100. Mudd, S. H.: Enzymatic Cleavage of S-Adenosylmethionine. J. Biol. Chem. **234**, 87 (1959).

101. — The Mechanism of the Enzymatic Cleavage of S-Adenosylmethionine to α-Amino-γ-butyrolactone. J. Biol. Chem. **234**, 1784 (1959).

102. — S-Adenosylmethionine Requirement for Plant Transmethylations. Biochim. Biophys. Acta **37**, 164 (1960).

103. — S-Adenosylmethionine Formation by Barley Extracts. Biochim. Biophys. Acta **38**, 354 (1960).

104. — The Mechanism of the Enzymatic Synthesis of S-Adenosylmethionine. In: S. K. Shapiro and F. Schlenk (Edit.), Transmethylation and Methionine Biosynthesis, p. 33. Chicago and London: Univ. of Chicago Press. 1965.

105. Mudd, S. H. and G. L. Cantoni: Selenomethionine in Enzymatic Trans methylations. Nature **180**, 1052 (1957).

106. — Activation of Methionine for Transmethylation. III. The Methionine-activating Enzyme of Baker's Yeast. J. Biol. Chem. **231**, 481 (1958).

107. — — Biological Transmethylation, Methyl Group Neogenesis and other "One-carbon" Metabolic Reactions Dependent upon Tetrahydrofolic Acid. In: M. Florkin and E. H. Stotz (Edit.), Comprehensive Biochemistry, Vol. 15, p. 1. Amsterdam: Elsevier Publ. Co. 1964.

108. Mueller, J. H.: A New Sulfur-containing Amino Acid Isolated from the Hydrolytic Products of Protein. J. Biol. Chem. **56**, 157 (1923).

109. Neumann, N. P., S. Moore and W. H. Stein: Modification of the Methionine Residues in Ribonuclease. Biochemistry **1**, 68 (1962).

110. O'Leary, W. M.: S-Adenosylmethionine in the Biosynthesis of Bacterial Fatty Acids. J. Bacteriol. **84**, 967 (1962).

111. — Transmethylation in the Biosynthesis of Cyclopropane Fatty Acids. In: S. K. Shapiro and F. Schlenk (Edit.), Transmethylation and Methionine Biosynthesis, p. 94. Chicago and London: Univ. of Chicago Press. 1965.

112. Pansuwana, P.: S-Adenosylmethionine in Rat Liver. Summary Report, Biol. and Med. Research Div., Argonne National Laboratory, ANL-6368, p. 100 (1963).

113. Parks, L. W.: S-Adenosylmethionine and Ergosterol Synthesis. J. Amer. Chem. Soc. **80**, 2023 (1958).

114. — S-Adenosylethionine and Ethionine Inhibition. J. Biol. Chem. **232**, 169 (1958).

115. Parks, L. W. and F. Schlenk: Formation of α-Amino-γ-butyrolactone from S-Adenosylmethionine. Arch. Biochem. Biophys. **75**, 291 (1958).

116. — — The Stability and Hydrolysis of S-Adenosylmethionine; Isolation of S-Ribosylmethionine. J. Biol. Chem. **230**, 295 (1958).

117. Parks, L. W., J. R. Turner and R. L. Larson: Transmethylation in Yeast Sterol Synthesis. In: S. K. Shapiro and F. Schlenk (Edit.), Transmethylation and Methionine Biosynthesis, p. 85. Chicago and London: Univ. of Chicago Press. 1965.

118. Peyron, L.: Precursors of Certain Sulfur Compounds Encountered with Essential Oils. Bull. Soc. Franc. Physiol. Végétale **7**, 46 (1961) [Chem. Abstr. **56**, 12014 (1962)].

119. Pfiffner, J. J. and H. B. North: Dimethylsulfone, a Constituent of the Adrenal Gland. J. Biol. Chem. **134**, 781 (1940).

120. Pohl, S., J. H. Law and R. Ryhage: The Path of Hydrogen in the Formation of Cyclopropane Fatty Acids. Biochim. Biophys. Acta **70**, 583 (1963).

121. Pope, W. J. and S. J. Peachey: Asymmetric Optically Active Sulphur Compounds. D-Methylethylthetine Platinichloride. J. Chem. Soc. (London) **77**, 1072 (1900).

121a. Price, C. C. and S. Oae: Sulfur Bonding. New York: Ronald Press. 1962.

122. Rachele, J. R., E. J. Kuchinskas, F. H. Kratzer and V. du Vigneaud: Hydrogen Isotope Effect in the Oxidation *in vivo* of Methionine Labeled in the Methyl Group. J. Biol. Chem. **215**, 593 (1955).

123. Ragland, J. B. and J. L. Liverman: S-Methyl-L-cysteine as a Naturally Occurring Metabolite in *Neurospora crassa*. Arch. Biochem. Biophys. **65**, 574 (1956).

124. Ray, W. J., Jr., H. G. Latham, Jr., M. Katsoulis and D. E. Koshland, Jr.: Evidence for Involvement of a Methionine Residue in the Enzymatic Action of Phosphoglucomutase and Chymotrypsin. J. Amer. Chem. Soc. **82,** 4743 (1960).

125. Raymond, A. L.: Thiosugars. J. Biol. Chem. **107,** 85 (1934).

126. Remy, C. N.: Methylation of Synthetic and Normal Purines, Pyrimidines, and Ribonucleosides. In: S. K. Shapiro and F. Schlenk (Edit.), Transmethylation and Methionine Biosynthesis, p. 107. Chicago and London: Univ. of Chicago Press. 1965.

126a. Rosenfeld, I. and O. A. Beath: Selenium: Geobotany, Biochemistry, Toxicity and Nutrition. New York: Academic Press. 1964.

127. Ruzicka, L., M. W. Goldberg und H. Meister: Inhaltsstoffe des Blutes. 1. Mitt. Isolierung von Dimethylsulfon aus Rinderblut. Helv. Chim. Acta **23,** 559 (1940).

128. Sakami, W.: S-Adenosyl-*L*-homocysteine. In: A. Meister (Edit.), Biochemical Preparations, Vol. 8, p. 8. New York: Wiley and Sons. 1961.

129. Salvatore, F. and F. Schlenk: A New Assay of Guanidinoacetate Methyltransferase. Biochim. Biophys. Acta **59,** 700 (1962).

130. Satoh, K.: The Structure of Adenylthiomethylpentose. I, II, and III. J. Biochemistry (Tokyo) **40,** 485, 557, 563 (1953).

131. Schachter, H. and G. H. Dixon: Identification of the Methionine Involved in the Active Center of Chymotrypsin. Biochem. Biophys. Res. Comm. **9,** 132 (1962).

132. Schlenk, F.: Nicotinamide Riboside. Arch. Biochem. **3,** 93 (1943).

133. — Biochemical and Cytological Studies with Sulfonium Compounds. In: S. K. Shapiro and F. Schlenk (Edit.), Transmethylation and Methionine Biosynthesis, p. 48. Chicago and London: Univ. of Chicago Press. 1965.

134. Schlenk, F. and J. L. Dainko: The Alkaline Hydrolysis of S-Adenosylmethionine in Tritiated Water. Biochem. Biophys. Res. Comm. **8,** 24 (1962).

135. Schlenk, F., J. L. Dainko and S. M. Stanford: Improved Procedure for the Isolation of S-Adenosylmethionine and S-Adenosylethionine. Arch. Biochem. Biophys. **83,** 28 (1959).

136. Schlenk, F. and R. E. de Palma: The Formation of S-Adenosylmethionine in Yeast. J. Biol. Chem. **229,** 1037 (1957).

137. — — The Preparation of S-Adenosylmethionine. J. Biol. Chem. **229,** 1051 (1957).

138. Schlenk, F. and D. J. Ehninger: Observations on the Metabolism of 5'-Methylthioadenosine. Arch. Biochem. Biophys. **106,** 95 (1964).

139. Schlenk, F., H. v. Euler, H. Heiwinkel, W. Gleim und H. Nyström: Die Einwirkung von Alkali auf Cozymase. Z. physiol. Chem. **247,** 23 (1937).

140. Schöberl, A. und G. Lange: Über eine Sulfoniumsalzbildung bei der Addition von Mercaptocarbonsäuren an ungesättigte Säuren. Angew. Chem. **64,** 224 (1952).

141. — — Über die Addition von Mercaptocarbonsäuren an ungesättigten Säuren und eine neue Darstellungsweise von Sulfoniumsalzen. Liebigs Ann. Chem. **599,** 140 (1956).

142. Schramm, H. J. und W. B. Lawson: Modifizierung eines Methioninrestes in Chymotrypsin durch einfache Benzolderivate. Z. physiol. Chem. **332,** 97 (1963).

143. Shapiro, S. K.: The Biosynthesis of Methionine from Homocysteine and Methylmethionine Sulfonium Salt. Biochim. Biophys. Acta **18,** 134 (1955).

144. — Biosynthesis of Methionine from Homocysteine and S-Methylmethionine in Bacteria. J. Bacteriol. **72,** 730 (1956).

145. SHAPIRO, S. K.: Adenosylmethionine-Homocysteine Transmethylase. Biochim. Biophys. Acta **29**, 405 (1958).

146. — The Function of S-Adenosylmethionine in Methionine Biosynthesis. In: S. K. Shapiro and F. Schlenk (Edit.), Transmethylation and Methionine Biosynthesis, p. 200. Chicago and London: Univ. of Chicago Press. 1965.

147. SHAPIRO, S. K., P. LOHMAR and M. HERTENSTEIN: Utilization of S-Adenosyl-methionine for the Biosynthesis of Methionine. Arch. Biochem. Biophys. **100**, 74 (1963).

148. SHAPIRO, S. K. and A. N. MATHER: The Enzymatic Decomposition of S-Adenosyl-L-methionine. J. Biol. Chem. **233**, 631 (1958).

149. SHAPIRO, S. K. and F. SCHLENK (Edit.): Transmethylation and Methionine Biosynthesis. Chicago and London: Univ. of Chicago Press. 1965.

150. SHAPIRO, S. K. and D. A. YPHANTIS: Evidence for the Participation of Metals in a Transmethylation. Arch. Biochem. Biophys. **82**, 477 (1959).

151. — — Assay of S-Methylmethionine and S-Adenosylmethionine Homocysteine Transmethylases. Biochim. Biophys. Acta **36**, 241 (1959).

152. SHAPIRO, S. K., D. A. YPHANTIS and A. ALMENAS: Partial Purification and Properties of S-Adenosylmethionine—Homocysteine Methyltransferase. J. Biol. Chem. **239**, 1551 (1964).

153. SHRIFT, A.: Biological Activities of Selenium Compounds. Bot. Rev. **24**, 550 (1958).

154. SMILES, S.: A Contribution to the Stereochemistry of Sulphur. An Optically Active Sulphine Base. J. Chem. Soc. (London) **77**, 1174 (1900).

155. SMITH, R. L., E. E. ANDERSON, Jr., R. N. OVERLAND and F. SCHLENK: The Occurrence, Formation, and Isolation of Thiomethyladenosine. Arch. Biochem. Biophys. **42**, 72 (1953).

156. SNOW, G. A.: The Metabolism of Compounds Related to Ethanethiol. Biochem. J. **65**, 77 (1957).

157. STARK, G. R. and W. H. STEIN: Alkylation of the Methionine Residues of Ribonuclease in 8 M Urea. J. Biol. Chem. **239**, 3755 (1964).

158. STEKOL, J. A.: Biochemical Basis for Ethionine Effects on Tissues. Adv. Enzymology **25**, 369 (1963).

159. — Newer Methods for Preparation of S-Adenosylmethionine and Derivatives. In: S. P. Colowick and N. O. Kaplan (Edit.), Methods in Enzymology, Vol. 6, p. 566. New York: Academic Press. 1963.

160. — Formation and Metabolism of S-Adenosyl Derivatives of S-Alkylhomo-cysteines in the Rat and Mouse. In: S. K. Shapiro and F. Schlenk (Edit.), Transmethylation and Methionine Metabolism, p. 231. Chicago and London: Univ. of Chicago Press. 1965.

161. STEKOL, J. A., S. WEISS and C. SOMERVILLE: A Study of the Comparative Metabolism of Ethionine and Methionine in the Male and Female Rat. Arch. Biochem. Biophys. **100**, 86 (1963).

162. SUZUKI, U., S. ODAKE und T. MORI: Über einen neuen schwefelhaltigen Bestandteil der Hefe. Biochem. Z. **154**, 278 (1924).

163. SVIHLA, G. and F. SCHLENK: S-Adenosylmethionine in the Vacuole of *Candida utilis*. J. Bacteriol. **79**, 841 (1960).

164. TABOR, H., S. M. ROSENTHAL and C. W. TABOR: Role of Putrescine and Methionine in the Enzymic Synthesis of Spermidine in Fschorichiu coli Extracts. J. Amer. Chem. Soc. **79**, 2978 (1957).

165. — — — The Biosynthesis of Spermidine and Spermine from Putrescine and Methionine. J. Biol. Chem. **233**, 907 (1958).

166. Tabor, H. and C. W. Tabor: Spermidine, Spermine, and Related Amines. Pharmacol. Rev. **16**, 245 (1964).

167. Tait, G. H. and K. D. Gibson: The Enzymic Formation of Magnesium Protoporphyrin Monomethylester. Biochim. Biophys. Acta **52**, 614 (1961).

168. Tashjian, A. H., Jr., D. A. Ontjes and P. L. Munson: Alkylation and Oxidation of Methionine in Bovine Parathyroid Hormone: Effects on Hormonal Activity and Antigenicity. Biochemistry **3**, 1175 (1964).

169. Toennies, G.: Sulfonium Reactions of Methionine and their Possible Metabolic Significance. J. Biol. Chem. **132**, 455 (1940).

170. Toennies, G. and J. J. Kolb: Methionine Studies. VII. Sulfonium Derivatives. J. Amer. Chem. Soc. **67**, 849 (1945).

171. — — Methionine Studies. VIII. Regeneration of Sulfides from Sulfonium Derivatives. J. Amer. Chem. Soc. **67**, 1141 (1945).

172. — — Techniques and Reagents for Paper Chromatography. Analyt. Chemistry **23**, 823 (1951).

173. Verly, W. G.: Contribution à l'étude du métabolisme du groupe méthyle labile. Arch. internat. physiol. et biochim. **64**, 309 (1956).

174. Vigneaud, V. du: The Significance of Labile Methyl Groups in the Diet, and their Relation to Transmethylation. Harvey Lect. **38**, 39 (1942/43).

175. — A Trail of Research in Sulfur Chemistry and Metabolism, and Related Fields. Ithaca, N. Y.: Cornell Univ. Press. 1952.

176. Vigneaud, V. du and J. R. Rachele: The Concept of Transmethylation in Mammalian Metabolism and its Establishment by Isotopic Labeling through in vivo Experimentation. In: S. K. Shapiro and F. Schlenk (Edit.), Transmethylation and Methionine Biosynthesis, p. 1. Chicago and London: Univ. of Chicago Press. 1965.

177. Vigneaud, V. du, J. R. Rachele and A. M. White: A Crucial Test of Transmethylation *in vivo* by Intramolecular Isotopic Labeling. J. Amer. Chem. Soc. **78**, 5131 (1956).

178. Wagner, C. and E. R. Stadtman: Bacterial Fermentation of Dimethyl-β-propiothetin. Arch. Biochem. Biophys. **98**, 331 (1962).

179. Weiss, S., E. I. Anderson, P. T. Hsu and J. A. Stekol: An Adaptation of the Floyd-Lavine Procedure for the Isolation of Methionine to Tracer Work. J. Biol. Chem. **214**, 239 (1955).

180. Wendt, G.: Über den Thiozucker der Hefe. Z. physiol. Chem. **272**, 152 (1942).

181. Weygand, F., R. Junk und D. Leber: Adenylthiomethylpentose. Z. physiol. Chem. **291**, 191 (1952).

182. Weygand, F. und O. Trauth: Synthese der Adenylthiomethylpentose. Chem. Ber. **84**, 633 (1951).

183. Weygand, F., O. Trauth und R. Löwenfeld: Konstitutionsaufklärung des Thiozuckers der Adenylthiomethylpentose. Chem. Ber. **83**, 563 (1950).

184. Windmueller, H. G., C. J. Ackerman and R. W. Engel: Reaction of Ethylene Oxide with Histidine, Methionine, and Cysteine. J. Biol. Chem. **234**, 895 (1959).

185. Woods, D. D., M. A. Foster and J. R. Guest: Cobalamin-dependent and -independent Methyl Transfer in Methionine Biosynthesis. In: S. K. Shapiro and F. Schlenk (Edit.), Transmethylation and Methionine Biosynthesis, p. 138. Chicago and London: Univ. of Chicago Press. 1965.

(Received, December 18, 1964.)

Some Aspects of the Chemistry and Function of Human and Animal Hemoglobins.

By WALTER A. SCHROEDER, Pasadena, California
and RICHARD T. JONES, Portland, Oregon.

With 13 Figures.

Contents.

Acknowledgement. It is a pleasure to express our appreciation to Miss Lillian Casler for her patience and competence in drawing most of the figures. Her advice and suggestions played a major role in developing Fig. 10 into usable form.

Contribution No. 3207 from the Division of Chemistry and Chemical Engineering, California Institute of Technology.

References, pp. 181—194.

I. Introduction.

The concluding sentence of an earlier article on hemoglobin in 1959 (*213*) stated that "Many years of effort by many investigators will no doubt be required before the last amino acid residue [of hemoglobin] is placed in the sequence and before the nature of the heme-globin linkage becomes apparent, but it is not too much to expect that this goal will be achieved". Prophecy clearly is an uncertain art: within three years after the sentence was written, the complete amino acid sequence of human hemoglobins A and F was known and the heme-globin linkage appeared to be clearly defined. Yet a grain of truth was in the statement. The man-years of effort that were required to attain the goal within so short a time are not easily calculated.

Although the amino acid sequence and much of the three-dimensional arrangement of the hemoglobins are known, this review is not meant to be a summary of information about a problem that has been solved. Quite the opposite. The solution of the amino acid sequence and, to a degree at least, the determination of the three-dimensional arrangement of the molecule open new vistas to investigation. Now, one can more meaningfully plan and assess experiments on structure and function, and, if the older literature were carefully surveyed, perhaps draw worthwhile conclusions from results once obscure.

The hemoglobins offer items of interest to many disciplines. One cannot hope, therefore, to cover them all within the scope of this article. We have accordingly chosen to discuss certain aspects which interest us, which may have escaped recent review, or which offer opportunity for further investigation. Yet even within these limitations we do not propose to be encyclopedic. Rather, by key references we shall try to open the road to the reader who may wish to examine more fully the details of the original literature. Thus, for certain elementary facts about hemoglobin the reader may see the earlier article by one of us (*213*).

If the reader fails to find a discussion of a particular topic, the references below to various reviews may help him. Most of them have appeared relatively recently and contain a source of the literature. The arrangement is rather arbitrary because this wide-ranging subject does not lend itself to narrow compartmentalization.

General structure and function: (*18*), (*29*), (*43*), (*110*), (*112*), (*130*), (*163*), (*185*), (*213*), (*214*), (*230*).
Physical chemistry: (*204*), (*239*), (*255*), (*256*).
Fetal hemoglobin: (*248*).
Abnormal human hemoglobins: (*28*), (*88 a*), (*95*), (*112*).
Animal hemoglobins: (*78*), (*173*).
Comparative physiology: (*162*).
Hemoglobinopathies and thalassemia: (*25*), (*88 a*), (*159*), (*205*).
Genetics: (*18*), (*131*), (*132*), (*177 a*), (*180*), (*260*).

Biosynthesis: (*33*), (*77*), (*228*), (*250*).
Books: (*25*), (*130*), (*131*).
Symposia: (*52*), (*147*), (*238*).

II. Nomenclature.

The outpouring of literature on hemoglobin has necessarily led to some confusion of nomenclature. At times this confusion has resulted because several groups have simultaneously investigated the same substance or subject and have used different names. Even when identity has been established, usage is not necessarily uniform. In other instances, multiple nomenclatures for an apparently identical substance or phenomenon may persist because complete correlation of the data cannot be or has not been made. In perusing the original literature, therefore, the reader must be careful to determine what system of nomenclature is used. Some authors are shy in revealing this to the reader.

It will be more profitable to define various terms when a particular topic is discussed. Here let us define only the word "hemoglobin". "Hemoglobin" is the general term that will be used when neither the oxidation state of the heme iron nor the group attached to iron is vital to the discussion. In the main, "hemoglobin" then is oxyhemoglobin— that compound in which molecular oxygen is attached to ferrous iron.

Nomenclature with reference to particular topics is to be found on pp. 116—118, 120, 121, 123, 147, 155—157, 171, 172 for the most part in footnotes, or in italics, and, much is summarized in Section V, 2 (p. 132).

III. The Amino Acid Sequence of Human Hemoglobins A and F.

In all individuals whether young or old, "hemoglobin" is a more or less complex mixture. No individual possesses a single pure hemoglobin. Nevertheless, in normal new-born infants or in adults, a single hemoglobin component predominates. The new-born infant's hemoglobin (which for purposes of investigation is usually obtained from the umbilical cord) contains about 75% of so-called fetal hemoglobin or hemoglobin F*. The normal adult's hemoglobin consists to the extent of about 85% of adult hemoglobin or hemoglobin A**. *For the purposes of this article, hemoglobin F will be defined as the main component of the normal new-born infant and hemoglobin A as the main component of the normal adult.* The pure components may be isolated easily by chromatographic methods (see *114*, *142*, for example).

---

* This main component is referred to in various publications as F_{II}, F_0, and F_1. Sometimes cord blood hemoglobin without further definition is called "fetal hemoglobin" or "hemoglobin F".

** Here also the main component is variously referred to as A_{II}, A_1, A_0, etc., and the whole unfractionated hemoglobin as "adult hemoglobin" or "hemoglobin A".

1. Hemoglobin A.

Hemoglobin A contains 574 amino acid residues and 4 heme groups. The calculated molecular weight of 64,450 (*41*) agrees well with that of about 66,000 from most determinations of molecular weight.

The structure of the heme moiety (p. 152) has long been known and we shall, therefore, confine our attention to the globin for which the sequence is of such recent derivation.

The early work of SANGER had clearly shown that hemoglobin A contained several polypeptide chains. The investigations of RHINESMITH et al. (*197, 198*) and of BRAUNITZER (*40*) left no doubt that two pairs of chains were present. *The pair that is N-terminal* in the sequence val-leu** is called the α chains and the pair that is N-terminal in val-his-leu, the β chains.* Each chain is associated with a heme group. The gross structure of hemoglobin A is now commonly represented by $\alpha_2^A \beta_2^A$***.

The molecule thus is composed of identical halves (each half one $\alpha \beta$ unit), a conclusion to which X-ray data (*187*) and peptide patterns of tryptic peptides (*127*) had already led. However, the presence of two kinds of chains complicates the determination of sequence. This complication can be eliminated if the chains can be separated as SANGER, for example, did with the two chains of insulin.

The α and β chains may, indeed, be separated by chromatography (*249*), by counter-current distribution (*93*), or by fractional precipitation (*88, 249*) of globin. BUCCI and FRONTICELLI (*44a*) have recently described a separation of α and β chains with attached hemes by chromatography on carboxymethylcellulose after treatment with p-chloromercuribenzoate.

Because the two types of chains could be separated, the determination of the structure of hemoglobin A was greatly simplified, and the task for each chain was of the same magnitude as for ribonuclease or tobacco mosaic virus (TMV) protein. It is not germane at this point to discuss the methods of protein chemistry that went into the final determination of sequence. Suffice it to say that the advent of automatic amino acid analysis and the use of volatile buffers for the column chromatography of peptides added to the pace of the investigations.

---

* N-Terminal refers to the amino acid residue with the free α-amino group at the end of the polypeptide chain.

** The abbreviations are the first three letters of the name of the amino acid and are commonly used to designate a residue. The N-terminus of a sequence is always at the left.

*** The superscript "A" means that the chains have the sequence of hemoglobin A and the subscripts, in the usual chemical sense, indicate that two of each chain are in the molecule.

a. The Amino Acid Sequence* of the α^A Chain.

The data to be presented on the sequence of the α chain have resulted from the investigations of Braunitzer and collaborators, R. J. Hill, Konigsberg et al., and Schroeder and co-workers. Primary references which may be examined for detailed information and as a source of related references are (44, 92 and 217).

Thus, the α chain contains 141 residues and its empirical formula in residues of amino acid is as follows: lys_{11} his_{10} arg_3 asp_{12} thr_9 ser_{11} glu_5 pro_7 gly_7 ala_{21} $cys**_1$ val_{13} met_2 leu_{18} tyr_3 phe_7 try_1. Isoleucine is absent. These 141 residues have the sequence in *Scheme 1*.

NH$_2$***
| 10

val-leu-ser-pro-ala-asp-lys-thr-asp-val-lys-ala-ala-try-gly-lys-val-gly-ala-his-ala-gly-

      30   40

glu-tyr-gly-ala-glu-ala-leu-glu-arg-met-phe-leu-ser-phe-pro-thr-thr-lys-thr-tyr-phe-

NH$_2$
|

pro-his-phe-asp-leu-ser-his-gly-ser-ala-glu-val-lys-gly-his-gly-lys-lys-val-ala-asp-ala-

NH$_2$ NH$_2$
| 70 | 80

leu-thr-asp-ala-val-ala-his-val-asp-asp-met-pro-asp-ala-leu-ser-ala-leu-ser-asp-leu-his-

NH$_2$
|

ala-his-lys-leu-arg-val-asp-pro-val-asp-phe-lys-leu-leu-ser-his-cys-leu-leu-val-thr-leu-

110   120   130

ala-ala-his-leu-pro-ala-glu-phe-thr-pro-ala-val-his-ala-ser-leu-asp-lys-phe-leu-ala-ser-

140

val-ser-thr-val-leu-thr-ser-lys-tyr-arg

Scheme 1. Amino Acid Sequence of the α Chain in Human Hemoglobin A.

b. The Amino Acid Sequence of the β^A Chain.

Detailed information about the determination of the sequence of the β chain may be found in (41, 156 and 217).

The β chain with 146 residues is longer than the α chain and differs from it appreciably in amino acid composition as the following empirical formula shows: lys_{11} his_9 arg_3 asp_{13} thr_7 ser_5 glu_{11} pro_7 gly_{13} ala_{15} cys_2 val_{18} met_1 leu_{18} tyr_3 phe_8 try_2. Isoleucine is absent here also.

The amino acid sequence of the β chain† is shown in *Scheme 2*.

---

* The amino acid sequence is also termed the *primary structure*.

** This abbreviation is used throughout the present article to designate a cysteinyl residue.

*** The "NH$_2$" shows that the amide of the dicarboxylic acid is present.

† In this sequence the capitalized abbreviations of the amino acid residues show where a different type of residue is present in the γ chain as will be discussed below.

10
VAL-his-LEU-thr-PRO-glu-GLU-lys-SER-ALA-VAL-thr-ALA-leu-try-gly-lys-val-
NH$_2$
| 20 30
asp-val-ASP-GLU-VAL-gly-gly-glu-ALA-leu-gly-arg-leu-leu-val-val-tyr-pro-try-thr-
NH$_2$
| 40 50
glu-arg-phe-phe-GLU-ser-phe-gly-ASP-leu-ser-THR-PRO-ASP-ala-VAL-met-gly-
NH$_2$
| 60 70
asp-pro-lys-val-lys-ala-his-gly-lys-lys-val-leu-GLY-ALA-PHE-SER-asp-GLY-LEU-
NH$_2$
| 90
ALA-his-leu-asp-ASP-leu-lys-gly-thr-phe-ala-THR-leu-ser-glu-leu-his-cys-asp-lys-
NH$_2$ NH$_2$
100 | |
leu-his-val-asp-pro-glu-asp-phe-ARG-leu-leu-gly-asp-val-leu-val-CYS-val-leu-ala-
NH$_2$ NH$_2$
120 | 130 |
HIS-his-phe-gly-lys-glu-phe-thr-pro-PRO-val-glu-ala-ALA-TYR-glu-lys-VAL-val-
NH$_2$
| 140
ALA-gly-val-ala-ASP-ala-leu-ALA-HIS-LYS-tyr-his

Scheme 2. Amino Acid Sequence of the β Chain in Human Hemoglobin A.

2. Hemoglobin F.

Hemoglobin F like hemoglobin A contains 574 amino acid residues and 4 heme groups (which are identical in both proteins). The calculated molecular weight is about 64,700.

First investigations showed only two chains in hemoglobin F. If this had been correct, another item would have added to the long list of properties in which hemoglobins A and F differ. Actually, hemoglobin F like hemoglobin A has two pairs of chains [SCHROEDER and MATSUDA (218)]. These authors recognized that the N-terminal sequence val-leu of one pair is identical with that of the α chains of hemoglobin A and suggested that one pair of chains might be identical in hemoglobins A and F. The second pair is N-terminal in glycine (the actual sequence is gly-his-phe-) (223) and definitely distinct from the β chains. Consequently, the gross structure of hemoglobin F is represented by $\alpha_2^F \gamma_2^F$ (see footnote, p. 117). The α and γ chains may be separated by chromatography (249) or fractional precipitation (88, 221) and the aminoethylated* chains by countercurrent distribution (138).

a. The Amino Acid Sequence of the α^F Chain.

The suggestion of SCHROEDER and MATSUDA (218) that the α^A and α^F chains are identical received preliminary substantiation in the

---

* In aminoethylated globin, cysteinyl residues have been converted to β-amino-ethylcysteinyl residues by reaction with ethylenimine (192).

experiments of HUNT (*120*) and of JONES, SCHROEDER and VINOGRAD (*146*) and was confirmed by the determination of the complete sequence* [SCHROEDER, SHELTON, SHELTON and CORMICK (*220*)].

b. The Amino Acid Sequence of the γ^F Chain.

The detailed sequence of the γ^F chain has been presented by SCHROEDER, SHELTON, SHELTON, CORMICK and JONES (*221*).

Because the total number of amino acid residues in hemoglobins A and F is identical and because the α chains are identical, the β and γ chains must also be equal in length (146 residues). The empirical formula is lys_{12} his_7 arg_8 asp_{13} thr_{10} ser_{11} glu_{12} pro_4 gly_{13} ala_{11} cys_1 val_{13} met_2 $ileu_4$ leu_{17} tyr_2 phe_8 try_3. The most distinctive difference in amino acid composition between the β and γ chains is the presence of four isoleucyl residues in the γ chains.

The γ chain** has the sequence in *Scheme 3*.

```
                                    10
GLY-his-PHE-thr-GLU-glu-ASP-lys-ALA-THR-ILEU-thr-SER-leu-try-gly-lys-val-
NH₂
 |     20                                    30
asp-val-GLU-ASP-ALA-gly-gly-glu-THR-leu-gly-arg-leu-leu-val-val-tyr-pro-try-thr-
NH₂                                  NH₂
 |     40                             |         50
glu-arg-phe-phe-ASP-ser-phe-gly-ASP-leu-ser-SER-ALA-SER-ala-ILEU-met-gly-
NH₂
 |         60                                 70
asp-pro-lys-val-lys-ala-his-gly-lys-lys-val-leu-THR-SER-LEU-GLY-asp-ALA-ILEU-
                                             NH₂
          80                           |         90
LYS-his-leu-asp-ASP-leu-lys-gly-thr-phe-ala-GLU-leu-ser-glu-leu-his-cys-asp-lys-leu-
                   NH₂                 NH₂
         100       |                   |     110
his-val-asp-pro-glu-asp-phe-LYS-leu-leu-gly-asp-val-leu-val-THR-val-leu-ala-ILEU-
                                 NH₂                   NH₂
         120                     |       130  |
his-phe-gly-lys-glu-phe-thr-pro-GLU-val-glu-ala-SER-TRY-glu-lys-MET-val-THR-

          140
gly-val-ala-SER-ala-leu-SER-SER-ARG-tyr-his
```

Scheme 3. Amino Acid Sequence in the γ Chain of Human Hemoglobin F.

3. A Comparison of Hemoglobins A and F.

Hemoglobins A and F differ in UV-spectrum, rate of alkali denaturation, amino acid composition, and many other properties. The preceding discussion, however, has shown that not only are they similar in structure

---

* The gross structure of hemoglobin F accordingly becomes $\alpha_2^A \gamma_2^F$ or simply $\alpha_2 \gamma_2^F$. The superscript usually is omitted and, by definition, α, β, or γ without superscript denotes the sequence of the chains from hemoglobins A and F.

** The capitalized abbreviations are those residues that differ in the β chain.

but, more importantly, that half of each molecule (represented by the α chains) is identical in sequence. Without doubt, the differences in the β and γ chains are responsible for the dissimilarity in the properties of hemoglobins A and F.

It is easy to point out that there are 17 differences in amino acid composition and 39 differences in sequence between the β and γ chains. Although one can make a comparison, residue by residue, of the two chains and this, to a degree, has been done (219), little of importance derives therefrom because this linear comparison takes no account of the possible interaction of distantly separated residues. Nevertheless, a knowledge of amino acid sequence is valuable; it is necessary information that gives meaning to many other data. It, in conjunction with the three-dimensional structure, should eventually answer the question as to the manner in which hemoglobin functions. Some of these matters will be discussed in subsequent Sections (pp. 147—157).

4. A Comparison of the α and β Chains.

The α and β chains differ in length by only a few residues. Are there any identities in sequence? Actually, their similarity has been stressed repeatedly by BRAUNITZER (40, 41). If one sets the C-terminal* residues of the α and β chains side by side and then proceeds toward the N-terminus, 40 residues are identical between the C-termini and the 51st residue of the α chain (44% of 91 residues). Further identities require a shifting of one chain or the other with consequent formation of "gaps"** in one sequence or the other. Thus, in our comparison above, α_{141} was set beside β_{146} and 91 residues toward the N-terminus α_{51} equals β_{56}. The coincidences increase if α_{50} is set beside β_{50} by which shift a 5-residue gap** opens in the α chain. Likewise, if there is a one-residue gap between α_{46} and α_{47} and a two-residue gap between β_{19} and β_{20}, 22 more identities may be observed. There is much similarity, therefore, in the sequences of the α and β chains despite their difference in length.

IV. Other Hemoglobins in the Normal Human Individual.

We have already noted that no individual's hemoglobin is homogeneous. Hemoglobin as isolated from the erythrocyte contains a major component and thus by implication one or more other components must be present. Are these minor components normal constituents of the red blood cell

---

* That is, the residue at the opposite end from the N-terminus: it has a free α-carboxyl group.

** These "gaps" must not be thought of as actually present in the chains. They simply indicate where corresponding residues are "missing" in one or the other chain.

or do they arise from the procedure of isolation? Abundant data prove conclusively that at least some are authentic constituents of the erythrocyte. What then are these components? Their nature? Their quantity? Their function? Not all of these questions can be answered at present and others have answers that depend on the age of the individual.

1. The Hemoglobin Components of the New-born Infant.

It has long been known that hemoglobin from the umbilical cord of a new-born infant contains more than simply hemoglobin F. Approximately 15% is hemoglobin A (9, *167*) which cannot be considered a

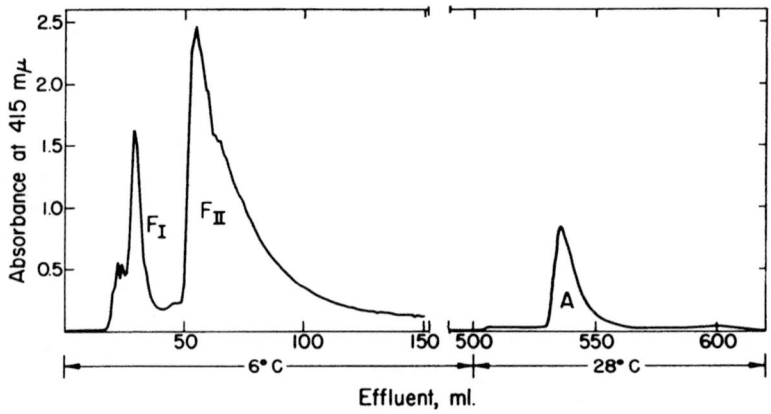

Fig. 1. The chromatography of umbilical cord blood hemoglobin on a 1 × 30 cm. column of the ion-exchange resin IRC-50 with Developer No. 5 (*51*). The temperature was 6° until 500 ml. of effluent had emerged and then was raised to 28°.

minor component in the normal sense. Actually, it results from the partial activation of the still unknown mechanism by which at birth the production of hemoglobin F in the normal individual is replaced by the production of hemoglobin A. Nevertheless, there are minor components in umbilical cord blood hemoglobin, and their presence is readily apparent upon column chromatography. Thus, *Fig. 1* depicts a typical chromatographic separation by the procedure of Allen, Schroeder and Balog (2). Zones F_I and F_{II} emerged relatively rapidly but Zone A was eluted only after the temperature was raised. Zone A contains hemoglobin A, and Zone F_{II} contains the major component, that is, the hemoglobin F that has already been discussed and of which the amino acid sequence is known.

Zone F_I, the most prominent minor component, normally amounts to about 10% of the hemoglobin in cord blood. It is easily separated by chromatography but is probably not detected in most electrophoretic systems, although NAIMAN and GERALD (177) have achieved a partial separation. Hemoglobin F_I cannot be distinguished from F by the rate of alkali denaturation, the spectrum, or the amino acid composition. Yet soon after its isolation, it was realized that F_I differed from F in the N-terminal residues. The specific chemical difference, as shown by SCHROEDER, CUA, MATSUDA and FENNINGER (215), is an acetyl group. It was concluded that the N-terminal residue of a single γ chain is acetylated. Hemoglobin F_I can, therefore, be represented as $\alpha_2 \gamma \gamma^{F_I}$ or more specifically* as $\alpha_2 \gamma \gamma^{N\text{-acetyl}}$. It was the first acetyl-containing hemoglobin to be detected. Its derivation and function are unknown.

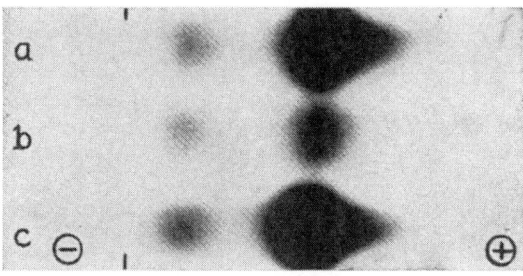

Fig. 2. Separation of components in normal adult hemoglobin (a and c) by starch block electrophoresis according to KUNKEL and WALLENIUS. The sample was applied at the position of the vertical lines. The smaller spot (left) is designated as hemoglobin A_2. The major spot is A_1 and its asymmetric part on the right, A_3. [From: Science 122, 288 (1955).]

Other minor hemoglobin components of unknown structure that precede F_I down the chromatographic column (Fig. 1) are present only to the extent of 1 to 2%. Because they are present in all cord bloods, it is unlikely that they are traces of hemoglobins H and Bart's that HORTON et al. (103) observed in about 30% of 300 samples of Negro cord blood.

2. The Hemoglobin Components of the Adult.

Minor components were observed in the hemoglobin of the adult almost simultaneously by electrophoretic and chromatographic methods [KUNKEL and WALLENIUS (158); MORRISON and COOK (171)]. Because it is well separated by most electrophoretic procedures, for example as in *Fig. 2*, the component designated as A_2 by KUNKEL and WALLENIUS

---

* As an extension of the nomenclature for gross structure, an abnormality in any chain is designated by a superscript equivalent to the name of the hemoglobin from which it derives. Thus, $\alpha_2 \gamma\gamma^{F_I}$ defines the two α chains and one γ chain as normal and one γ chain as different in an undefined way. The specific chemical alteration between the normal and abnormal chain can be defined also by a superscript. Thus, $\alpha_2 \gamma\gamma^{N\text{-acetyl}}$ means that F_I differs from F by the presence of acetyl in the N-terminal position of one chain. See GERALD and INGRAM (73) and Section V, 2 (p. 132), for further details.

(*158*) has received much attention. Chemical investigations have shown that hemoglobin A_2 also contains two α^A chains. The other pair is so different from the β or γ chains that they have been termed the δ chains (sometimes δ^{A_2}). The molecule is $\alpha_2 \delta_2$. Stretton and Ingram (*236*) have detected a minimum of eight differences in sequence between the β and the δ chain [substantiated by Hill and Kraus (*94*)], and Jones (*138*) has evidence for two more. *Table 1* compares the various residues in the β, γ and δ chains at those positions where the β and δ chains differ.

Table 1. Differences in Amino Acid Sequence in the β, γ and δ Chains.

Residue No.	β Chain	γ Chain	δ Chain
9	ser	ala	thr
12	thr	thr	aspNH$_2$
22	glu	asp	ala
50	thr	ser	ser
86	ala	ala	ser (?)
87	thr	gluNH$_2$	gluNH$_2$ (?)
116	his	ileu	arg
117	his	his	aspNH$_2$
124	pro	pro	
or			gluNH$_2$
125	pro	glu	
126	val	val	met

Hemoglobin A_2 thus is far more different from hemoglobin A than F_I is from F. There are many rather than a single difference. Again, the function of A_2 is unknown. Its quantity, which is about 2.5% of the hemoglobin of the normal individual, is increased or decreased in certain hemoglobinopathies. As a result, hemoglobin A_2 has received much attention from the standpoint of genetics and hemoglobin biosynthesis. Cord blood contains 0.2—0.3% of hemoglobin A_2 (*103*).

Other minor components are also present in adult hemoglobin. The asymmetry of the major spot in Fig. 2 (p. 123) suggests that a more rapidly moving component is also present. Although difficult to separate by electrophoresis, the more rapidly moving material is readily separated by chromatography as shown in *Fig. 3* from Schnek and Schroeder (*212*) who have correlated the various components as detected by starch block electrophoresis and chromatography.

Hemoglobin A_{Ic}*, the most prominent of these minor components, normally accounts for 5 to 7% of the total hemoglobin. According to

---

* Hemoglobin A_{Ic} apparently is incorrectly equated with A_3 by some authors and is probably identical with A_1^c of Huisman and Meyering (*117*).

References, pp. 181—194.

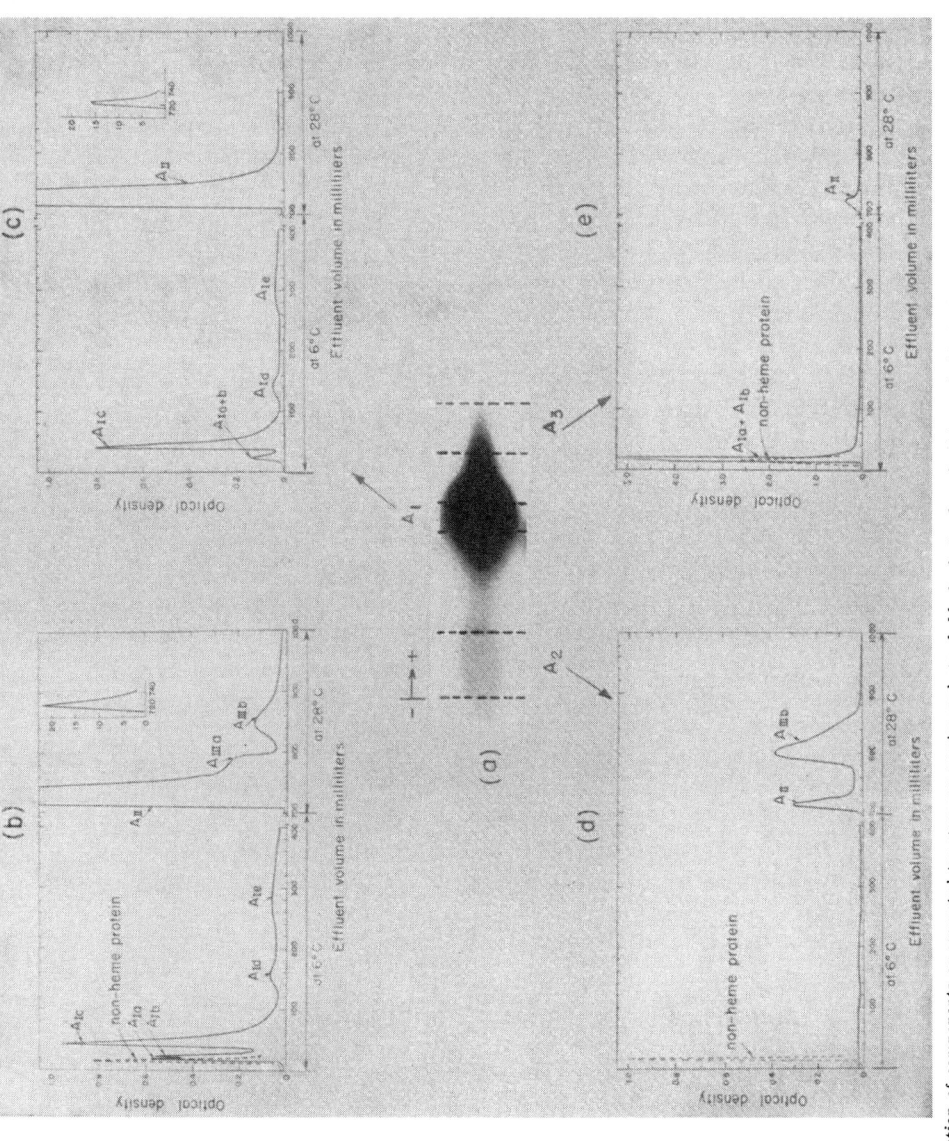

Fig. 3. Correlation of components as separated by chromatography and starch block electrophoresis: (a) starch block electrophoresis of whole adult hemoglobin at pH 8.6; (b) chromatogram of whole adult hemoglobin; (c), (d) and (e) chromatograms of A_1, A_2 and A_3 as isolated from a starch block. [From: SCHNEK and SCHROEDER. J. Amer. Chem. Soc. 83, 1472 (1961).]

Holmquist and Schroeder (*100*), hemoglobin A_{Ic} resembles hemoglobin A much as hemoglobin F_I resembles hemoglobin F. Thus, its gross structure is $\alpha_2 \beta \beta^{AIc}$: one of the normal β chains has a blocked N-terminal group. Although the blocking group has not yet been identified, an aldehyde or ketone (but not pyridoxal) probably has reacted with the amino group to give a Schiff's base type of compound. The group has a molecular weight of the order of 250 (*99*).

Horton and Huisman (*101a, 115*) have investigated the quantities of adult minor components in hemoglobins from a variety of disease states such as polycythemia vera, leukemia, etc., as well as in experiments in vitro that were designed to influence these components. Hemoglobin A_{Ic} was little influenced with certainty by various pathological states except by hemolytic anemias in which case there was a decrease. It was present in equal amount in young and old red cells and was not altered in quantity by storage of the cells at $25°$ for 7 days or at $4°$ for 49 days.

Hemoglobins A_{Ia} and A_{Ib} are small in amount and probably contain more than one component as well as non-heme proteins. Almost no investigation has been done on them. Zones A_{Id} and A_{Ie} were consistently observed in many chromatograms from the hemoglobin of numerous individuals (*51*). Despite the fact that there has been no obvious change in conditions of preparation or chromatography of hemoglobin, both zones have become virtually undetectable in the last several years (*144*). It is also known that aging or mistreatment of hemoglobin can produce zones that behave superficially like A_{Id} and A_{Ie}. Thus, convincing evidence for the existence of A_{Id} and A_{Ie} can no longer be adduced.

In conclusion, the examination of minor components has led to the discovery of the δ chain which, because of its existence in normal individuals, must be considered a normal chain. Hemoglobins F_I and A_{Ic}, though superficially similar, are very different in the nature of the group that blocks the N-terminal residue. Presumably, they derive in one way or another from F and A or vice versa or from common precursors, whereas the sequences of the β chains of A and δ chains of A_2 are determined independently.

V. The Abnormal Human Hemoglobins.

On the preceding pages, we have described the normal human hemoglobins A, F and A_2. They are made up of the α, β, γ and δ chains and the α chain is common to all. These three hemoglobins exemplify two of the three ways in which hemoglobins differ from each other in function or structure. Thus, hemoglobins F and A are related to the maturation of the individual while hemoglobins A and A_2 are related as major and minor components. These relationships, although not necessarily predicted or understood, must be considered to be normal. There is yet a third relationship when a hemoglobin is related to, but differs from, one of the normal hemoglobins; the different hemoglobin is termed an "abnormal" hemoglobin.

The story of the abnormal hemoglobins is so well known that it is almost superfluous to recall that the first one, now termed hemoglobin S, was detected in patients with sickle-cell anemia and reported by PAULING, ITANO, SINGER and WELLS in 1949 (*183*). ITANO and co-workers soon detected other abnormal hemoglobins and the world-wide search was on. Perhaps no other protein could have lent itself to such extensive study; none is so readily isolable and so apparent to the unaided eye. HUISMAN (*112*) has listed more than 80 reports of abnormal hemoglobins among which are many poorly characterized examples. Nevertheless, in other instances, the specific difference between abnormal and normal hemoglobin has been defined. It is pertinent, therefore, to consider briefly the methods by which this characterization of an abnormal hemoglobin may be done.

1. Methods.

For some years, hemoglobin S resisted efforts to determine in what way it differed from hemoglobin A. In part this inability to characterize the difference lay in the fact that methods for determining amino acid sequence were only then in the process of development. The experiment of INGRAM which limited the difference to a small fragment of the molecule was a notable advance and in conjunction with other data that appeared almost simultaneously made it possible soon to specify the abnormality in hemoglobin S.

Any investigation of the structure of an abnormal hemoglobin requires its isolation in pure form. Hemoglobin S was detected because of the electrophoretic difference between it and hemoglobin A. Indeed, electrophoretic techniques have been and continue to be the means by which most of the abnormal hemoglobins are detected. The experienced worker places much confidence in the interpretation of minute differences. Electrophoresis on supports such as paper or cellulose acetate is useful mainly for detection of abnormalities and for the examination of many samples. The scale is too limited to be useful for isolation, and, in addition, it is frequently difficult to elute the hemoglobin from the support. Increase in scale is accomplished by electrophoresis in blocks of such supports as starch granules, starch gel, agar gel, polyacrylamide gel, etc. That these supports are not inert media is shown by the difference in the quality of the separations: frequently marginal separations on one may be excellent on another. HUISMAN (*112*) has given a rather detailed discussion of electrophoretic methods.

Chromatographic techniques to separate the hemoglobins use ion-exchange materials such as Amberlite IRC-50, CM-cellulose, and DEAE-cellulose†. These methods are ideal for isolation because of their generally high capacity; for example, a 3.5×35-cm. column of Amberlite IRC-50 will commonly accommodate 2 to 3 g. of hemoglobin. By proper adjustment of conditions, a chromatographic procedure can probably be devised for the isolation of almost any hemoglobin. There is, therefore, no longer an excuse for investigating mixtures. Pertinent references to chromatographic procedures are (114) and (142).

If a reasonable quantity of a hemoglobin is available, the determination of the precise nature of the abnormality has become almost routine. Because hemoglobins usually have two types of chains, the first step is to locate the abnormality in one or the other or both types of chain. This may be done effectively through the so-called "hybridization" technique which will be discussed in detail in Chapter IX (p. 160). The principles, however, are these. When a mixture of an unknown and a reference hemoglobin is taken to an acidic or basic pH, each hemoglobin dissociates into subunits. After some time at the extreme pH, the hemoglobins may be reassociated by return to neutrality. During return to neutrality, random recombination and transfer of chains will occur as shown by the following equation†† in which radioactive hemoglobin A is the reference substance and the unknown hemoglobin is abnormal in the β chain

$$\alpha_2^* \beta_2^* + \alpha_2 \beta_2^X \rightarrow \alpha_2^* \beta_2^* + \alpha_2 \beta_2^X + \alpha_2^* \beta_2^X + \alpha_2 \beta_2^*$$
$$\text{(I)} \qquad \text{(II)} \qquad \text{(III)} \qquad \text{(IV)}$$

Now, (I) and (II) equal the starting products but (III) also equals (II), and (IV) equals (I) except for distribution of radioactivity. In essence, the α chains which are equal except for radioactivity have exchanged. Consequently, the specific activity of $\alpha_2^* \beta_2^*$ is less after the experiment and $\alpha_2 \beta_2^X$ has become radioactive. Proof that $\alpha_2 \beta_2^X$ had become radioactive in the α chains would require further experiments, but our example shows in principle how the abnormality can be limited to one type of chain.

CHERNOFF and PETTIT ($50a$) have described an electrophoretic procedure for separating the chains of various globins. This procedure is said to be sufficiently sensitive that an aberrant chain can be identified without resort to hybridization.

The position of the abnormality can be localized further by a study of the peptides in a tryptic digest. In this way INGRAM (127) showed that there was a chemical difference between hemoglobins A and S.

---

† CM = carboxymethyl; DEAE = diethylaminoethyl.

†† This equation does not imply a mechanism for the process (pp. 162—166).

References, pp. 181—194.

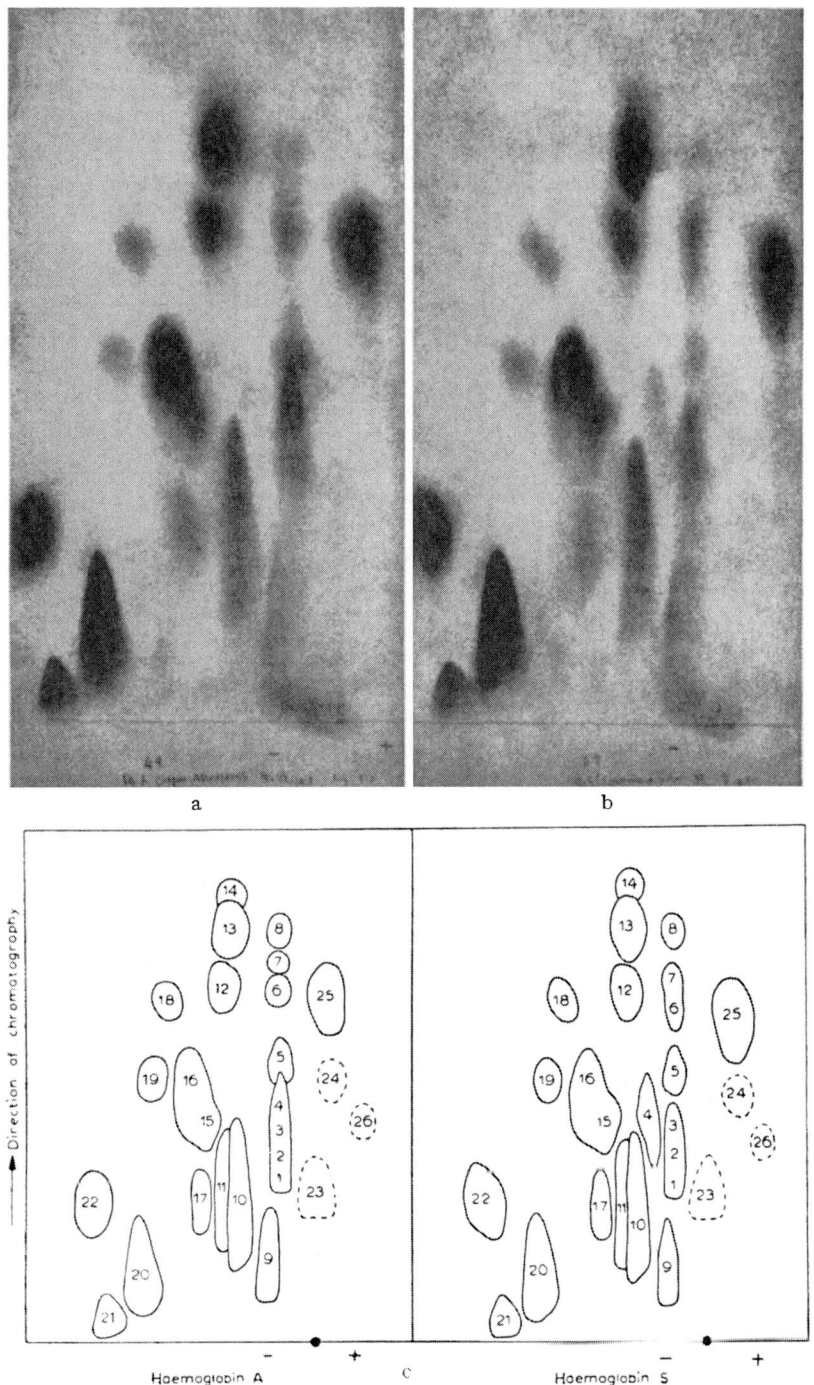

Fig. 4. Peptide patterns on paper of tryptic hydrolysates of hemoglobins A and S: a is from hemoglobin A, b is from hemoglobin S, and c is a tracing of the two. [From: INGRAM, Biochim. Biophys. Acta 28, 539 (1958).]

He first subjected individual tryptic hydrolysates of these two hemoglobins to paper electrophoresis in one dimension and then to paper
chromatography at 90° to the direction of electrophoresis. The array
of spots that was evident after staining with ninhydrin is called a "finger-
print" or, more formally, a "peptide map" or "pattern". *Fig. 4* shows
INGRAM's peptide patterns of hemoglobins A and S. Only one of the
spots (No. 4) on the pattern of hemoglobin A was out of position on the
pattern of hemoglobin S, and INGRAM concluded that the two hemoglobins differed at one position in one type of chain. At the present time,
the identity of the peptide in each spot on the pattern has been determined.
Because the entire amino acid sequence of the chains is known, the
position of each peptide in the chain is known also. If, then, the peptide
map of an abnormal hemoglobin shows the disappearance of a spot in
one position and its reappearance in another, the position of the
abnormality is limited to a specific peptide.

Peptide patterns may also be determined by column chromatography.
JONES (*138*) has modified the automatic amino acid analyzer of SPACKMAN,
STEIN and MOORE (*233*) to permit the automatic recording of the pattern
of enzymatic digests. Each column peptide pattern may be made with
hydrolysate from about 1 mg. of protein and requires about 24 hours.
The system is very reproducible and permits a quantitative evaluation
of the various peaks. A comparison of the pattern from tryptic digests
of hemoglobins A and S is depicted in *Fig. 5*. The peak marked with
an arrow on the chromatogram of hemoglobin S has moved with respect
to an equivalent peak from hemoglobin A. This designated peptide is
equivalent to spot No. 4 in Fig. 4.

The evaluation of the data is much simplified if the abnormality
has been limited to one chain or if the chains have been separated before
tryptic hydrolysis and the preparation of the peptide pattern.

The final identification of the abnormality requires usually the
isolation of the aberrant tryptic peptide and the determination of its
sequence. The isolation may be made by one of several column chromatographic methods (*44, 92, 156* and *216*) or by paper chromatography
(for example, see *15*). The amino acid composition of the abnormal
peptide by comparison with that of the corresponding normal peptide
may give a clue to the aberration which then must be confirmed by a
determination of sequence.

Accordingly, there may be different degrees of information about
the aberration in an abnormal hemoglobin: thus, the aberrant chain
only may be known; the aberrant tryptic peptide may have been
identified; the difference in the amino acid composition of the
aberrant peptide may be known; or the specific difference may have
been determined.

References, pp. 181—194.

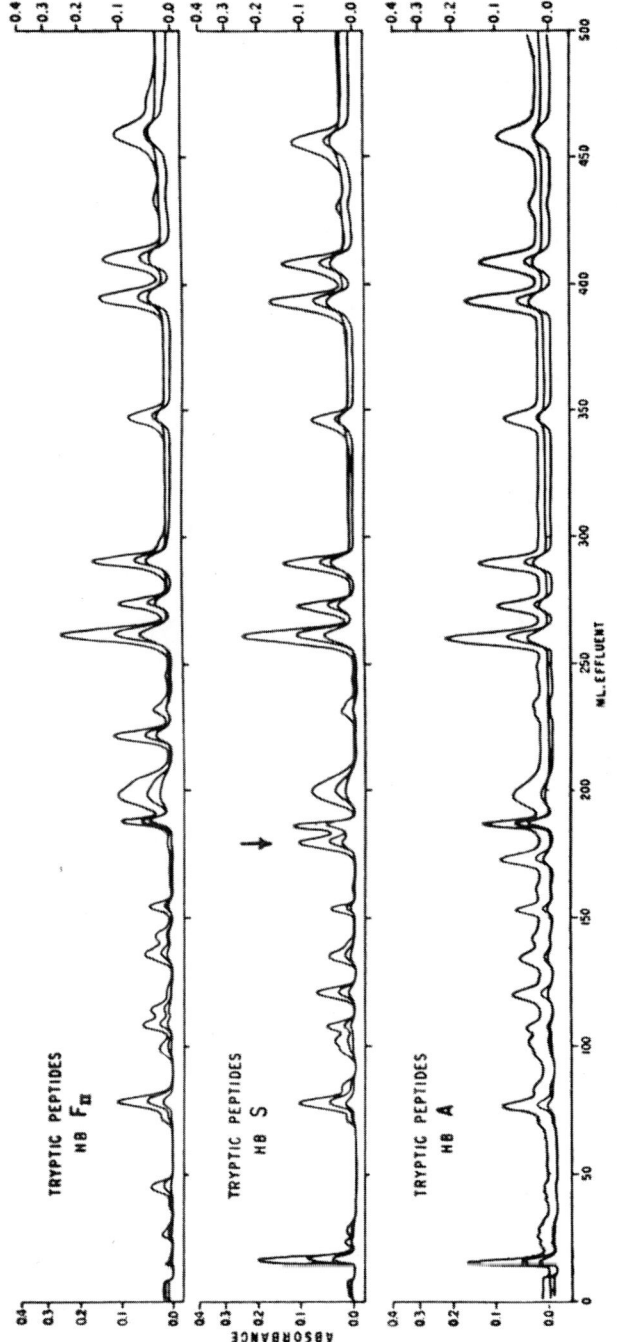

Fig. 5. Peptide patterns by automatic column chromatography of tryptic hydrolysates of hemoglobins A, S and F. [From: JONES, Cold Spring Harbor Symp. Quant. Biol. 29, 297 (1964).]

2. Nomenclature.

At this point, let us summarize and extend the nomenclature of hemoglobins especially as it relates to specifying the aberration in an abnormal hemoglobin.

After the discovery of sickle-cell hemoglobin, it was eventually called hemoglobin S, normal adult hemoglobin was designated as A, and normal fetal hemoglobin as F, with the proviso that other abnormal hemoglobins be denoted by capital letters. The pace of discovery soon almost exhausted the alphabet, and, indeed, a "new" hemoglobin could be differentiated only with difficulty from earlier discoveries. Finally, investigators refrained from assigning new letters to "new" hemoglobins, but assigned the letter of a previously discovered hemoglobin that corresponded best in properties and added a distinctive subscript (5). Thus, a particular hemoglobin that resembled hemoglobin G and was discovered in San Jose, California, was termed hemoglobin $G_{\text{San Jose}}$.

As we have seen, hemoglobin A is represented by $\alpha_2^A \beta_2^A$ where the superscripts denote the type of sequence in each chain and the subscripts show the number of each chain per molecule. When a normal chain is meant, the superscripts may be and usually are omitted, thus, $\alpha_2 \beta_2 =$ hemoglobin A. If then an abnormality is present in a β chain as in hemoglobin S, the formula $\alpha_2 \beta_2^S$ states this fact. $\alpha_2^I \beta_2$ shows that the α chains of hemoglobin I are abnormal.

As we have described above, the tryptic peptide in which an abnormality is present can be determined. Such information is incorporated in the nomenclature in the following way. The tryptic peptides are numbered from the N-terminus in the order in which they occur in the chain. Because of the specificity of trypsin, only those peptide bonds that are associated with the carboxyl groups of lysine and arginine will hydrolyze. Thus, the first tryptic peptide will extend from the N-terminus to and including the first lysyl or arginyl residue. For example, in the β chain (p. 119), the first tryptic peptide contains the residues 1 to 8 inclusive and is termed βT-1 where "β" is the chain, "T" refers to trypsin as the hydrolytic agent, and "1" denotes the position along the chain*. As a further example, residues 121 to 132 inclusive are βT-13 because this is the thirteenth tryptic peptide from the N-terminus. The abnormality of hemoglobin S is in βT-1 so that hemoglobin S $= \alpha_2 \beta_2^{\text{T-1}}$ (in which case, of course, the β is omitted in the superscript).

Let us now carry the nomenclature further. In hemoglobin S, a valyl residue replaces a glutamyl residue. This knowledge may be incorporated in this way: $\alpha_2 \beta_2^{\text{T-1 (glu} \rightarrow \text{val)}}$.

---

* We prefer this simple designation of tryptic peptides as given by Baglioni (14). Other systems are in use where βT-1 equals βTpI (73) equals βTp1 (98). Some designations have no relation to the order of the tryptic peptides (156).

Finally, because this substitution occurs in the sixth residue, hemo-globin S becomes $\alpha_2\beta_2^{6\ val}$.

3. Some General Genetic Considerations.

The study of the hemoglobins has added much to our knowledge of protein biosynthesis and genetics. Although these topics will be discussed in some detail in Chapter X (p. 169), brief mention of certain facts is necessary here for a meaningful consideration of the abnormal hemo-globins.

Although it may not be entirely correct, as a first approximation we may say that the normal α, β, γ and δ chains are independently synthesized under the control of genes that the individual has inherited*. The normal genes permit the individual to synthesize only the normal chains: he is a homozygote** for hemoglobins A, F, and A_2. The individual (a heterozygote) who as an adult synthesizes both hemoglobins A and S has inherited from both parents the ability to make normal α, γ, and δ chains but from one parent the ability to make β chains and from the other β^S chains. As another example, the individual who produces both hemoglobins A and I makes α, α^I, β, γ and δ chains. The individual who is homozygous for hemoglobin S makes α, β^S, γ, and δ chains***. A genetically very heterogeneous individual could produce a complex array of chains: α^W, $\alpha^X, \beta^Y, \beta^Z, \gamma^U, \gamma^V, \delta^S, \delta^T$.

Let us consider as typical of many hemoglobinopathies an adult with sickle-cell trait, that is, the heterozygote who produces mainly hemo-globins A and S in roughly equal amount. In adult life, α, β and β^S chains mainly will be formed. The main compounds are $\alpha_2\beta_2$ and $\alpha_2\beta_2^S$. Is $\alpha_2\beta\beta^S$ also present? Experience has shown that the abnormal hemo-

---

* The rate and quantity of production as well as the time sequence (thus, γ chains pre-natally but not post-natally) need not concern us here.

** The following definitions of some genetic terms may be helpful to the reader. *Chromosomes* are particulate structures present in the nucleus of cells. They serve as carriers of the genetic material. *Genes*, the units of inheritance, are those portions of a chromosome that determine given inherited characteristics. *Allelic genes* are different forms of the same gene that can occupy an identical site or locus on a chromosome. *Homologous* chromosomes are pairs of chromosomes that carry the same loci. One chromosome of each pair of homologous chromosomes is inherited from each parent. Most animals that possess hemoglobins have *diploid* cells, that is, cells which possess a duplicate set of homologous chromosomal pairs. An individual is *homozygous* for a given genetic characteristic if the two allelic genes on homologous chromosomes are identical. An individual is *heterozygous* if the two allelic genes are not identical.

*** The fact that every individual produces minor hemoglobin components of one type or another is not germane to this discussion where we deal only with the structure of the main components.

globin as isolated contains two *pairs* of chains with a few exceptions*. Accordingly, the abnormal hemoglobins to be described on the next few pages have the general formulas $\alpha_2 \beta_2^X$ or $\alpha_2^Y \beta_2$. Usually, an abnormal hemoglobin is isolated from an individual who is heterozygous for it and hemoglobin A. Homozygotes for an abnormal hemoglobin are rare except for hemoglobins S and C. Nevertheless, the literature provides abundant reference to many unusual heterozygotes (see *147* and *205* for many examples).

4. Hemoglobins with Abnormalities in the α Chain.

Fig. 6 represents the α chains as on p. 118, but, in addition, the tryptic peptides are designated and the kind and position of the alterations in hemoglobins with aberrant α chains are shown. Not only have many types of alteration been detected but alteration obviously is not limited to any one section of the chain. By the precise determination of the aberration, hemoglobins G$_{Honolulu}$, G$_{Singapore}$ and G$_{Hong Kong}$ have been shown to be identical, a conclusion that could not have been reached with certainty in any other way.

5. Hemoglobins with Abnormalities in the β Chain.

The position and kind of abnormalities that have been detected in the β chain are presented in *Fig. 7*. Here again the alterations are varied and by no means localized. At least four residues (Nos. 6, 7, 63 and 121) are altered differently in different hemoglobins.

6. Hemoglobins with Abnormalities in the γ Chain.

An abnormality in the γ chain is difficult to detect because of the transience of hemoglobin F in the infant. Such abnormalities are probably rare because only a few instances of possibly abnormal fetal hemoglobins have been suggested from among thousands of samples of umbilical cord hemoglobin that have been examined in one way or another. However, in at least three cases, there is convincing evidence that abnormal γ chains do exist.

Silvestroni and Bianco (*226*) have isolated a variant that they term F$_{Roma}$. Investigation by spectrum, alkali denaturation, and hybridization provides good evidence for an abnormal γ chain. Unfortunately, insufficient material was available for further characterization.

On much the same basis, Huisman (*113*) concludes that a fetal hemoglobin, termed hemoglobin Warren, which was present to the extent of about 13% at birth is abnormal in the γ chains. Hemoglobin Warren

---

* Further aspects of the question of unsymmetrical hemoglobins of the type $\alpha_2 \beta \beta^X$ will be discussed in Chapter IX (p. 160).

References, pp. 181—194.

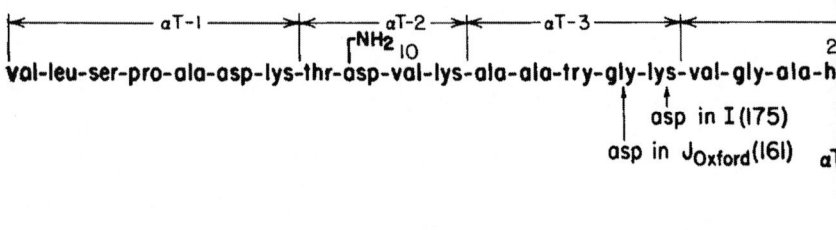

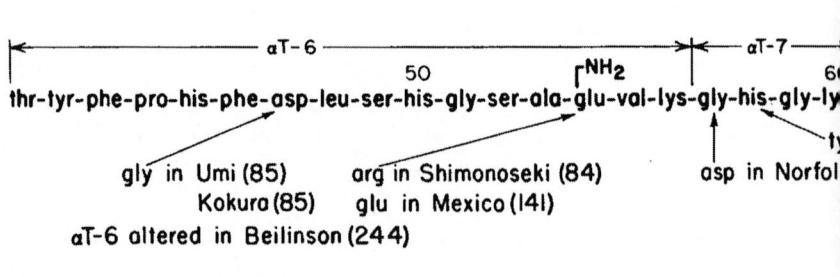

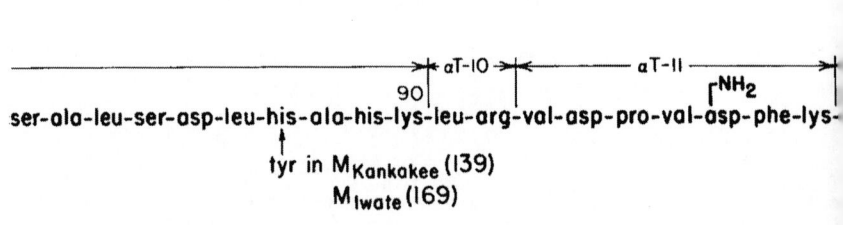

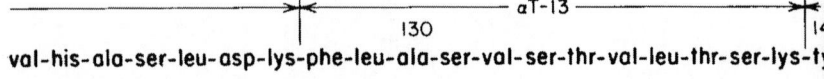

Fig. 6. The α chain of hemoglobin A. The numbering of the tryptic peptides and the aberratio

|← αT-4 ——————————————————→|←——— αT-5 ———————————→|

30   40

glu-tyr-gly-ala-glu-ala-leu-glu-arg-met-phe-leu-ser-phe-pro-thr-thr-lys-

in J$_{Medellin}$ (76)   gluNH$_2$ in G$_{Honolulu}$ (237)
in Da (34)   Singapore (237)
                                          Hong Kong (237)

——————————————————— αT-9 ———————————————————

$\lceil$NH$_2$ 70 $\lceil$NH$_2$ 80

ala-asp-ala-leu-thr-asp-ala-val-ala-his-val-asp-asp-met-pro-asp-ala-leu-

on (72)   lys in G$_{Philadelphia}$ (19)   αT-9 altered in Q (140)
                 G$_{Bristol}$ (58)
                 D$_{αSt. Louis}$ (81)
                 Stanleyville I (209)

——————————————————— αT-12 ———————————————————

110   120

-his-cys-leu-leu-val-thr-leu-ala-ala-his-leu-pro-ala-glu-phe-thr-pro-ala-

                                          lys in O$_{Indonesia}$ (21)

α chain abnormalities – G$_{Ibadan}$ (70)
                        G$_{Norfolk}$ (125)   L$_{Ferrara}$ (35)
                        Hopkins-2 (136)   M$_{Cologne}$ (152)
                        K (69)   Russ (118)

hemoglobins are shown. (The numbers in parentheses refer to the literature citations.)

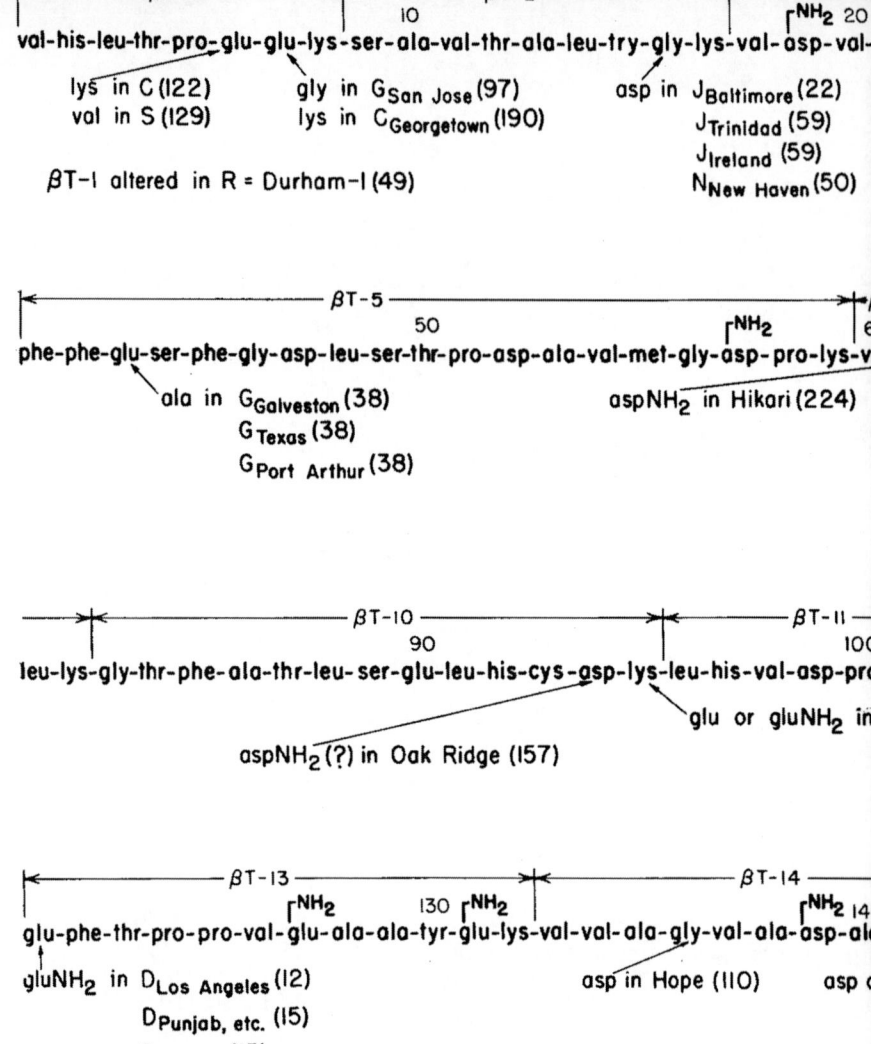

val-his-leu-thr-pro-glu-glu-lys-ser-ala-val-thr-ala-leu-try-gly-lys-val-asp-val-

lys in C (122)   gly in G$_{San\ Jose}$ (97)   asp in J$_{Baltimore}$ (22)
val in S (129)   lys in C$_{Georgetown}$ (190)   J$_{Trinidad}$ (59)

βT-1 altered in R = Durham-1 (49)   J$_{Ireland}$ (59)

N$_{New\ Haven}$ (50)

βT-5
50   NH₂   6

phe-phe-glu-ser-phe-gly-asp-leu-ser-thr-pro-asp-ala-val-met-gly-asp-pro-lys-v

ala in G$_{Galveston}$ (38)   aspNH$_2$ in Hikari (224)
G$_{Texas}$ (38)
G$_{Port\ Arthur}$ (38)

βT-10   βT-11
90   10C

leu-lys-gly-thr-phe-ala-thr-leu-ser-glu-leu-his-cys-asp-lys-leu-his-val-asp-pr

glu or gluNH$_2$ in

aspNH$_2$ (?) in Oak Ridge (157)

βT-13   βT-14
NH₂   130   NH₂   NH₂ 14

glu-phe-thr-pro-pro-val-glu-ala-ala-tyr-glu-lys-val-val-ala-gly-val-ala-asp-al

gluNH$_2$ in D$_{Los\ Angeles}$ (12)   asp in Hope (110)   asp c

D$_{Punjab,\ etc.}$ (15)

D$_{Cyprus}$ (15)

lys in O$_{Arabia}$ (21)

Fig. 7. The β chain of hemoglobin A. The numbering of the tryptic peptides and the aberrations i

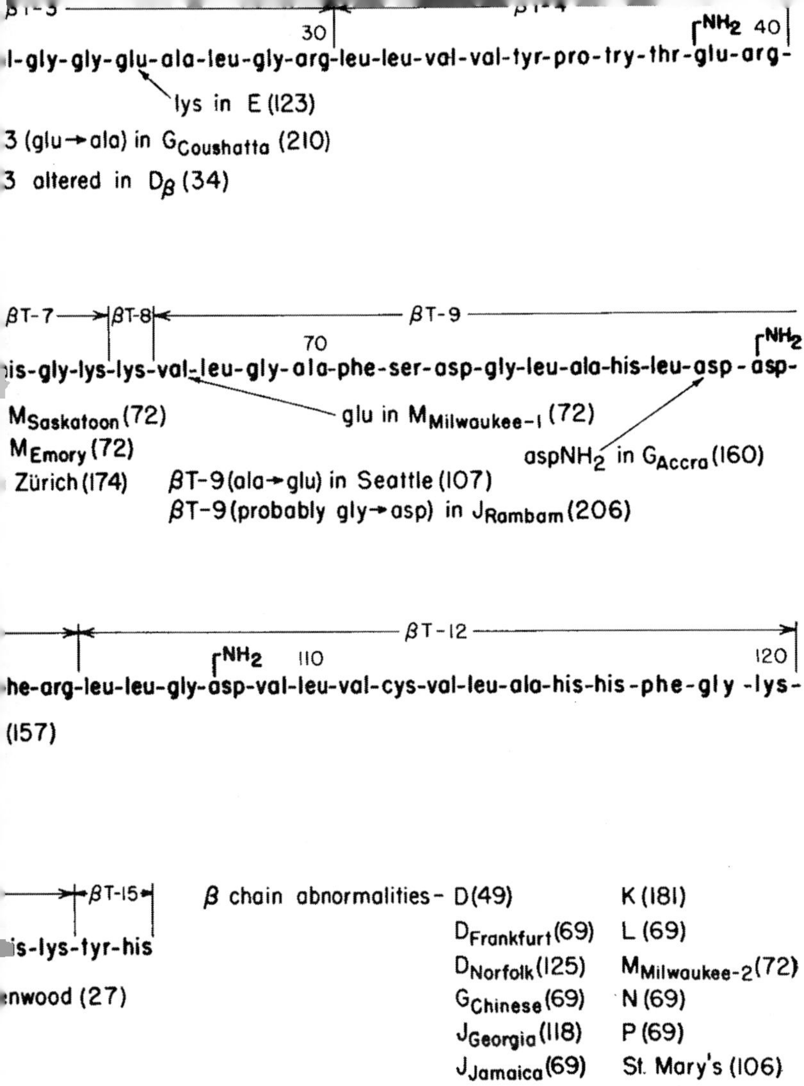

βT-3 ———— β1-4

```
                30│               ┌NH2 40│
-gly-gly-glu-ala-leu-gly-arg-leu-leu-val-val-tyr-pro-try-thr-glu-arg-
```
　　　　　　 ╲ lys in E (123)

3 (glu→ala) in G$_{Coushatta}$ (210)

3 altered in D$_β$ (34)

βT-7 ——→│βT-8│←——————— βT-9 ——————————————
```
                70
is-gly-lys-lys-val-leu-gly-ala-phe-ser-asp-gly-leu-ala-his-leu-asp-asp-
```
　　　　　　　　　　　　　　　　　　　　　　　　　　　　　　　┌NH2
M$_{Saskatoon}$ (72)

M$_{Emory}$ (72)　　　　　　 ╲ glu in M$_{Milwaukee-I}$ (72)

Zürich (174)　　βT-9 (ala→glu) in Seattle (107)　　　aspNH$_2$ in G$_{Accra}$ (160)

　　　　　　　　βT-9 (probably gly→asp) in J$_{Rambam}$ (206)

———→│←——————————— βT-12 ———————————————→│
```
       ┌NH2    110                                        120│
he-arg-leu-leu-gly-asp-val-leu-val-cys-val-leu-ala-his-his-phe-gly-lys-
```
(157)

——→│←βT-15→│　β chain abnormalities – D (49)　　　 K (181)

is-lys-tyr-his　　　　　　　　　　　　　　 D$_{Frankfurt}$ (69)　L (69)

nwood (27)　　　　　　　　　　　　　　　 D$_{Norfolk}$ (125)　M$_{Milwaukee-2}$ (72)

　　　　　　　　　　　　　　　　　　 G$_{Chinese}$ (69)　　N (69)

　　　　　　　　　　　　　　　　　　 J$_{Georgia}$ (118)　　P (69)

　　　　　　　　　　　　　　　　　　 J$_{Jamaica}$ (69)　　St. Mary's (106)

oglobins are shown. (The numbers in parentheses refer to the literature citations.)

is easily separated from hemoglobin F by chromatography. Its amino acid composition clearly relates it to hemoglobin F; tentatively, the content of glutamic acid appears to be lower than in hemoglobin F.

SCHNEIDER and JONES (211) have examined by column chromatography (138) the tryptic hydrolysates of hemoglobin F_{Texas} (208). The amino acid analysis of the aberrant peptide suggests that hemoglobin F_{Texas} is $\alpha_2 \gamma_2^{T-1 \ (glu \rightarrow lys)}$.

References to other fetal hemoglobins which may have abnormal γ chains may be found in (112).

Fetal hemoglobins with abnormal α chains are to be anticipated in offspring where at least one parent produces a hemoglobin with abnormal α chains. Therefore, the child may have not only $\alpha_2 \gamma_2$ but also $\alpha_2^X \gamma_2$. GERALD and EFRON (72), MINNICH et al. (168), and RANNEY et al. (195) have detected such abnormal fetal hemoglobins.

7. Hemoglobins with Abnormalities in the δ Chain.

Hemoglobin A_2 or $\alpha_2 \delta_2$, although its quantity in the adult is only 2 to 3%, contains normal chains. An abnormality of the δ chain is to be found in hemoglobin A_2' (termed B_2 in older literature) which was detected by CEPPELLINI (47). Hemoglobin A_2' has the structure $\alpha_2 \delta_2^{16 \ arg}$ [HORTON et al. (102); STRETTON quoted in (18); HUISMAN and JONES, unpublished]. RANNEY et al. (194) suggested that hemoglobin Flatbush is another variant of A_2, a conclusion which has received much support from the data of HUISMAN and LEE (116).

8. Hemoglobins with Other Abnormalities.

Because there are four normal hemoglobin chains, all abnormalities should be placed in one of the four preceding sections. Hemoglobins H and Lepore, however, do not fit into these categories.

Hemoglobin H is composed of four normal β chains and thus is β_4^A [JONES et al. (143, 145)]. It is abnormal not because of an amino acid substitution but because it is composed of a single type of chain. Hemoglobin Bart's has a related structure γ_4^F (124, 143), and β_4^S, β_4^C and δ_4 have also been reported (57, 111). The isolation of β_4^A and γ_4^F set off a search for α_4^A. FESSAS and LOUKOPOULOS (66) believe that they have isolated a human hemoglobin that contains only α chains [see also CHERNOFF (48)].

Hemoglobin Lepore$_{Boston}$ was detected by GERALD (71). BAGLIONI (17) concluded that hemoglobin Lepore contains normal α chains and another type whose N-terminal portion (about $1/2$ of the chain) is identical with the δ chain and whose C-terminal portion is identical with the β chain. This sequence presumably results from the break-up, crossing over, and recombination of genes. Accordingly, if the breaking point differs, different hemo-globins Lepore will result. Thus, the non-α chains of hemoglobin Lepore_Hollandia differ from those of Lepore_Boston (*26*). Extensive study by Curtain (*56*) with a New Guinean hemoglobin Lepore resulted in the conclusion that it was probably identical with Lepore_Hollandia but different from Lepore_Boston. Hemoglobin Pylos (*67*) may be identical with Lepore_Boston.

9. General Remarks.

One might remark about each abnormal or potentially abnormal hemoglobin and its significance. However, from the discussion in Section 3 (p. 133), the complexity of the hemoglobin mixture in any individual clearly depends upon the degree of heterozygosity and the particular chain or chains for which he is heterzygous. Because all hemo-globins (with few exceptions) have the α chain, the person with α-chain heterozygosity will produce abnormal hemoglobins related to A, F, and A$_2$, whereas β-chain heterozygosity, basically, will influence only A.

We must not lose sight of the fact that most individuals are normal in their hemoglobin. Consequently, those instances that delight the geneticist are very rare. As an example of an unusual hemoglobinopathy, consider the individual who was found by Atwater, Schwartz and Tocantins (*11*) to have four distinct hemoglobin components. Chemical investigation by Baglioni and Ingram (*19*) characterized them as $\alpha_2^A \beta_2^A$, $\alpha_2^G \beta_2^A$, $\alpha_2^A \beta_2^C$ and $\alpha_2^G \beta_2^C$. The individual was, therefore, heterozygous for hemoglobins A, G, C. The reader will find many more examples of rare combinations in (*147*) and (*205*).

VI. Animal Hemoglobins.

Many early investigations of hemoglobin used animal hemoglobins which sometimes crystallize very easily. During the era when the discovery of new abnormal human hemoglobins was especially common-place, the study of the animal hemoglobins lagged. Because the determination of the sequence can be made with relative ease, interest in the animal hemoglobins has revived.

The information about amino acid sequences that we have discussed in the preceding Sections tells us nothing about the three-dimensional arrangement of the molecule. On the other hand, much of what will be described in Chapter VII (p. 145) about the three-dimensional structure of hemoglobin comes from the X-ray diffraction work on horse hemo-globin. It is, therefore, of great interest to know the amino acid sequence of this animal hemoglobin at least.

A most important aspect of the study of hemoglobin asks the question: What features of structure give the molecule its particular and necessary properties? The abnormal human hemoglobins have yielded considerable information on this score. Yet they are, in most instances, very similar in properties and their most obvious difference—usually electrophoretic—is apparent once the aberration has been detected. It is not at all obvious why hemoglobin F is so much more resistant to alkali denaturation than hemoglobin A or why sickling occurs when hemoglobin S is in the erythrocyte [although MURAYAMA (176) has suggested an explanation]. The animal hemoglobins differ in properties from human hemoglobin A to a greater extent than do the abnormal human hemoglobins. Consequently, a study of animal hemoglobins should add much to our knowledge of those features of the molecule that appear to be essential for function and should give an insight into the effect of variation in structure on a given property. Although one also asks of other proteins how structure and function are related, hemoglobin, because it is easily isolated and because so many "derivatives" are present in abnormal human and animal hemoglobins, offers better opportunities than most proteins to achieve a solution to the problem.

GRATZER and ALLISON (78) have summarized in considerable detail the knowledge about animal hemoglobins as of 1960. At that time, information about amino acid sequences in any hemoglobin was scanty but many properties had been studied. Exact sequences of animal hemoglobins are still far from numerous but we propose to point out in the next few paragraphs some relationships that have become apparent.

1. Gross Structure of Animal Hemoglobins.

Do the animal hemoglobins have two pairs of polypeptide chains in the same way as the human hemoglobins? The first information came from PORTER and SANGER (191). Their results suggested six chains in horse and donkey hemoglobins, five in human hemoglobin, and four in cattle, sheep and goat hemoglobins. The latter three had both valine and methionine as N-terminal residues from which it must be inferred that at least two types of chains are present. From the work of RHINESMITH et al. (197, 198) and BRAUNITZER (40) (Chapter III, p. 116), human hemoglobin molecules obviously had two pairs of chains. However, BRAUNITZER (40) also concluded the same to be so for horse, sheep and cattle hemoglobins*.

The first separation of the individual chains of any hemoglobin was achieved by WILSON and SMITH (249) who separated the chains of horse globin. By their method, MULLER (173) has separated the chains of the hemoglobins of cattle, sheep, goats, rabbits, hare and chickens, and

---

* The qualitative data of OZAWA and SATAKE (182) disagree with the quantitative data of BRAUNITZER. They report that horse hemoglobin has *three* kinds of chains.

Riggs (*201*) has isolated the chains of mouse hemoglobins. The chains of several primate hemoglobins have also been separated [Zuckerkandl and Schroeder (*261*), and R. L. Hill et al. (*96*)]. The conclusion is inevitable that the mammalian hemoglobins contain two types of chains.

Avian, fish and other hemoglobins have received less attention. However, Konagaya (*154*) reports that hemoglobin of the blue-white dolphin has two pairs of chains with N-terminal sequences like the human chains. Buhler (*45*) has made an extended study of hemoglobin from the Chinook salmon and the rainbow trout.

2. Adult and Fetal Forms.

In mammals, where the placental barrier separates the fetal and maternal circulation, the fetus must obtain its oxygen through this barrier. Consequently, the presence of two types of hemoglobin is not surprising, and one might also anticipate that the fetal type would have a greater affinity in order to aid in the transfer of oxygen. In humans, the higher oxygen affinity of fetal blood evidently is associated largely with the intact erythrocyte rather than with a major difference in oxygen affinity between hemoglobins A and F in dialyzed hemolysates [Nechtman and Huisman (*179*)].

Information about fetal animal hemoglobins is fragmentary, but situations comparable to the human case are known largely due to Muller (*173*) who separated the chains of fetal and adult globins from man, cattle, sheep, goats and rabbits and compared the peptide patterns. As would be expected of the human hemoglobins, the α^A and α^F chains were chromatographically identical but the β^A and γ^F chains were not. In other species, likewise, one chain from each type of hemoglobin showed identical chromatographic characteristics, and these same chains had identical peptide patterns. The other types of chains showed definite differences.

Few as these examples are, they suggest that, in the animal hemoglobins as in human hemoglobin, one pair of chains is identical in the fetal and adult forms. On the other hand, Stockell et al. (*235*) were unable to detect a difference between fetal and adult horse hemoglobin. In some species, therefore, distinct fetal and adult types may not be present.

3. Nomenclature of the Animal Hemoglobins.

Much of the review of Gratzer and Allison (*78*) described the multiple hemoglobins that are found in certain animals. Although many species have more than one hemoglobin, they have been examined with far less thoroughness than the human hemoglobins. The nomenclature of these multiple animal hemoglobins follows no set pattern. Thus, two kinds of bovine hemoglobin are called A and B (23), of sheep hemoglobin A and B (65) or I and II (119), and of mouse hemoglobin "sharp" and "diffuse" (126). Unfortunately, the reader must determine the nomenclature of each paper and how it relates to that of other authors.

In their investigations of the N-termini of several animal hemoglobins, OZAWA and SATAKE (182) and BRAUNITZER (40) identified the sequence of two or three residues at the N-termini. In human, horse, pig, cattle, dog, goat, sheep, guinea pig, rabbit, snake and hen hemoglobins, the sequence val-leu was always present but a second sequence varied from species to species*. The human chains were originally designated α, β, and γ (footnote 6 of 145) because the sequences of two or three residues at the N-termini are different. Because so many animal hemoglobins have sequences N-terminal in val-leu, these are commonly referred to as the α chains of the animal hemoglobins. Some authors have called the other pair the "non-α" chains but it is more customary to refer to these chains in the adult form as β chains and in the fetal form as γ chains. It is desirable to define the source of the chain by superscripts thus, $\alpha_2^{\text{sheep I}} \beta_2^{\text{sheep I}}$.

4. Sequences in the Animal Hemoglobins.

The animal hemoglobins, thus, resemble human hemoglobin in gross structure and, frequently, in the presence of adult and fetal forms. Does the similarity end here or is it even more pronounced as one compares the sequences?

The N-terminal sequence val-leu is common to the species that were examined (40, 182). Val-gluNH$_2$-leu, the N-terminal sequence of the horse β chains (40), resembles the val-his-leu of the human β chains. In a more detailed study, ZUCKERKANDL, JONES and PAULING (259) compared the peptide patterns of several primate hemoglobins, porcine and bovine hemoglobins, and some fish hemoglobins. The peptide patterns of the primate hemoglobins were very similar to that of human hemoglobin, the porcine and bovine hemoglobins showed many similarities, and the fish hemoglobins almost none. MULLER (173) provided a more definitive comparison by studying the individual chains. R. L. HILL et al. (44b, 96) were able to see both similarities and dissimilarities in the peptide patterns of the chains from 13 primate hemoglobins. In more complete studies of animal hemoglobins then, one may expect to find peptides, and therefore portions of each chain, that are identical in hemoglobins from several species.

---

* The third sequence that OZAWA and SATAKE (182) detected in horse, pig, dog and guinea pig hemoglobins as well as some other sequences (such as met-gly in bovine hemoglobin) probably are incorrect.

Because many animal hemoglobins are now being investigated, complete sequences should be known in time. Much published information takes the form of the amino acid composition of the tryptic peptides from various animal hemoglobins. By comparison, then a minimum number of differences between the animal peptide and the corresponding human peptide can be deduced. As an example of this type of comparison, let us consider peptides from human hemoglobins A and F. These are βT-3 and γT-3 which differ in composition by only one residue: γT-3 has two valyl residues whereas βT-3 has three, and γT-3 contains one threonyl residue whereas βT-3 contains none. All other residues are equivalent in number in the two peptides. When γT-3 was first isolated, the sequence of βT-3 was known. One can so arrange the residues of γT-3 that they correspond with those of βT-3 except for the substitution of a threonyl residue for one of the three valyl residues. There is also potential danger in the above reasoning: although the compositions differ by one residue, the sequences of βT-3 and γT-3 are unlike at four places.

Sequences in animal hemoglobins are discussed below for various α, β and γ chains.

a. α Chains.

Matsuda, Gehring-Müller and Braunitzer (166) give the amino acid sequence of the α chains of the slow component of horse hemoglobin [most horses have two hemoglobins (24, 46)] as shown in Scheme 4.

$$NH_2$$
$$|\quad 10 \qquad\qquad\qquad\qquad\qquad\qquad\qquad 20$$
val-leu-ser-ALA*-ala-asp-lys-thr-asp-val-lys-ala-ala-try-SER-lys-val-gly-GLY-his-

$$30 \qquad\qquad\qquad\qquad\qquad\qquad 40$$
ala-gly-glu-tyr-gly-ala-glu-ala-leu-glu-arg-met-phe-leu-GLY-phe-pro-thr-thr-lys-thr-

$$NH_2$$
$$50 \qquad\qquad\qquad | \qquad\qquad\qquad 60$$
tyr-phe-pro-his-phe-asp-leu-ser-his-gly-ser-ala-glu-val-lys-ALA-his-gly-lys-lys-val-

$$70 \qquad\qquad\qquad\qquad\qquad\qquad 80$$
ala-asp-GLY-leu-thr-LEU-ala-val-GLY-his-LEU-asp-asp-LEU-pro-GLY-ala-leu-ser-

$$NH_2 \qquad\qquad\qquad\qquad\qquad\qquad NH_2$$
$$| \qquad\qquad\qquad 90 \qquad\qquad\qquad | \qquad\qquad 100$$
ASP-leu-ser-ASP-leu-his-ala-his-lys-leu-arg-val-asp-pro-val-asp-phe-lys-leu-leu-ser-

$$NH_2$$
$$110 \qquad\qquad\qquad | \qquad\qquad 120$$
his-cys-leu-leu-SER-thr-leu-ala-VAL-his-leu-pro-ASP-ASP-phe-thr-pro-ala-val-his-

$$130 \qquad\qquad\qquad\qquad 140$$
ala-ser-leu-asp-lys-phe-leu-SER-ser-val-ser-thr-val-leu-thr-ser-lys-tyr-arg

Scheme 4. Amino Acid Sequence of the α Chains of Horse Hemoglobin (slow component).

---

* The capitalized abbreviations designate the points of difference with the human α chain.

A minimum of 15 differences from human α chains was expected from the amino acid composition of the tryptic peptides, and 18 were found. The horse α chain has 141 residues just as the human α chain. If the nature of the differences is examined, it is found that, in 15 of the 18 changes, neutral amino acids have exchanged; in one, glutamic and aspartic acids are involved; and in the orther two, aspartic acid and asparagine that are only four residues apart (about one turn of an α helix) are in a sense interchanged. Thus, acidic, neutral and basic residues occupy essentially identical positions in both. Nevertheless, the character of the side-chains (hydrogen bonding vs. non-hydrogen bonding, for example) is such that the properties could be greatly influenced by the alterations in the neutral amino acids.

SATAKE and SASAKAWA (207) have presented a partial sequence of the bovine α chain. Their data do not yet permit an estimate of the number of differences between bovine and human α chains. However, some of their data are in conflict with fragmentary data of others. Thus, they list the sequence of bovine αT-7, 8 (residues 57–61) as -gly-his-ala-lys-lys-. MULLER (173), on the other hand, suggests that this sequence should be -gly-his-gly-ala-lys-, because free lysine was absent from peptide patterns of the bovine α chains and because the amino acid composition of the peptide was as given. The bovine α chain, according to SATAKE and SASAKAWA contains one cysteinyl residue, but BABIN et al. (13) have been unable to detect cysteic acid in oxidized α chains from adult bovine hemoglobins A and B and from fetal bovine hemoglobin. These disagreements may be due to the fact that hemoglobin from different breeds of cattle has been examined. Thus, MULLER used hemoglobin from Friesian stock, BABIN et al. from Hereford cows that have only bovine hemoglobin A and Jersey cows that have both types, and SATAKE and SASAKAWA (207) from an undesignated breed.

DIAMOND and BRAUNITZER (60) have listed the amino acid composition of the tryptic peptides of the rabbit α chain; the data of NAUGHTON and DINTZIS (178) agree with these for the most part. BRAUNITZER et al. (43) have reported the amino acid composition of the tryptic peptides of the α chain of pig and llama hemoglobin.

In the α chain of the several species discussed above, a minimum of 15 to 30 differences between a given chain and human α chains may be counted. Differences from species to species are also considerable. The gorilla α chains, however, appear to differ from human α chains only in an aspartyl instead of a glutamyl residue in position 23 [ZUCKERKANDL and SCHROEDER (261)].

Because the function of hemoglobin is the same in all species, it is natural to search out whether or not certain features of the sequence remain invariant and, thus, appear to be related to proper functioning.

Much caution, naturally, must be observed in trying to draw conclusions between sequence and function from amino acid composition of tryptic peptides alone. Nevertheless, if one examines the limited data on the α chains, several regions of the sequence seem to show little variation from species to species. We have already noted the presence of val-leu at the N-terminus of all of the α chains. The C-terminus with -lys-tyr-arg agrees also. The sequence from residues 36 to 46 may well be identical and, in fact, have only minor alteration for about five residues before and after. From residues 54 to 63, variation is of the type that is found between α and β chains, that is, an alanyl for a glycyl residue and vice versa. Finally, the portion from residues 89 to 106 may very well be identical in human, horse, pig, llama and rabbit α chains; it is somewhat surprising that SATAKE and SASAKAWA (207) report rather pronounced differences here in bovine α chains. Further comment on some of these apparent identities and their possible significance will be reserved until Chapter VII (p. 145) on the three-dimensional arrangement of the molecule.

b. β Chains.

SMITH (230) has placed 80% of the residues of the horse β chains in the sequence that *Scheme 5* shows.

10
val-GLU*-leu-SER-GLY-glu-glu-lys-ALA-ala-LEU-VAL-ala-leu-try-ASP-lys-val-
20   30
asp-GLU-GLU-glu-val-gly-gly-glu-ala-leu-gly-arg-leu-leu-val-val-tyr-pro-try-thr-glu-
40   50
arg-phe-(phe, glu, ser, phe, gly, asp, leu, ser, GLY, pro, asp, ala, val)**-met-gly-asp-pro-
60   70
lys-val-lys-ala-his-gly-lys-lys-val-leu-HIS-SER-phe-GLY-GLU-gly-VAL-HIS-his-
80   90
leu-asp-asp-leu-lys-gly-thr-phe-ala-ALA-leu-ser-glu-(leu, his, cys, asp, lys, leu, his, val,
100   110
asp, pro, glu, asp, phe)-arg-leu-leu-gly-asp-val-leu-ALA-LEU-val-VAL-ala-ARG-his-
120   130
phe-gly-lys-ASP-phe-thr-pro-GLU-LEU-glu-ala-SER-tyr-glu-lys-val-val-ala-gly-val-
140
ala-asp-ala-leu-ala-his-lys-tyr-his

Scheme 5. Amino Acid Sequence of the β Chain of Horse Hemoglobin.

A minimum of 25 differences between human and horse β chains may be observed.

Unpublished data from this laboratory (13) suggest that the sequence of the β chain of bovine hemoglobin B is as *Scheme 6* depicts.

---

* The capitalized abbreviations designate the points of difference from the human β chain.

** Abbreviations within parentheses and separated by commas indicate that the residue has not been placed definitely in sequence.

1   10
—MET*-leu-thr-ALA-glu-glu-lys, ALA-ala-val-thr-ala-PHE-(try, SER)**-lys, val-HIS-
                                                       NH_2

20   30 | 40
val-asp-glu-val-gly-gly-(glu, ala, leu, gly, arg), leu-leu-val-val-**tyr**-pro-try-thr-glu-arg,

                              50
phe-phe-glu-ser-phe-gly-(asp, leu, ser, **thr**, ALA, asp, ala, val, met, ASP, asp, pro, lys),

60   70
val- ys, ala-his-gly-lys, lys, val-leu-ASP-SER-phe-ser-(asp, gly, MET, LYS), his-leu-

  80   90
asp-ASP-leu-lys, gly-(ALA, phe, ala, thr, leu, ser, **glu**, leu, his, cys, asp, lys), leu-his-val-
      NH_2 NH_2
100 | 110 |
asp-pro-glu-asp-phe-lys, (leu, leu, gly, asp, val, leu, val, val, val, leu, ala, arg), ASP-

                                  130
phe-gly-(ASP, glu, phe, thr, pro, ASP, val, glu, ala, LEU, PHE, glu, lys), val-val-ala-
      NH_2
      | 140
gly-val-ala-asp-ala-leu-ala-his-ARG, tyr-his

Scheme 6. Amino Acid Sequence of the β Chain of Bovine Hemoglobin B.

Although the conclusion is still tentative, it is probable that these bovine β^B chains contain one residue less than human β chains as shown in Scheme 6 by the absence of a residue in position 1. All other positions have been numbered to correspond with obviously similar sequences of the human β chain. The data on the bovine β^A chain are less complete but indicate that this chain differs from the bovine β^B chain in at least three residues: glycine at residue 16, and lysine at residues 19 and 120.

The gorilla β chain may differ from the human β chain only in the exchange of an arginyl for an unidentified lysyl residue (261).

BRAUNITZER et al. (43) have summarized the amino acid composition of the tryptic peptides of pig, llama and rabbit β chains while R. L. HILL et al. (44b, 96) have determined the amino acid composition of many tryptic peptides from the β chains of eight primate hemoglobins.

A discussion of the significance of these data will be deferred until the end of the following Section on animal γ chains.

c. γ Chains.

The sequence of bovine γ chains as derived by BABIN et al. (13) is represented by *Scheme 7*.

---

* The capitalized abbreviations designate the points of difference from the human β chain.

** Abbreviations within parentheses and separated by commas indicate that the residue has not been placed definitely in sequence.

1   10

—MET*-LEU-SER-ALA-glu-GLU-lys-ala-ALA-VAL-thr-ser-leu-PHE-ALA-lys-val-

20   30

LYS-val-ASP-GLU-VAL-gly-gly-(glu, ALA)**-leu-gly-arg-leu-leu-val-val-tyr-pro-try-

$$\begin{array}{ll} \text{NH}_2 & \text{NH}_2 \\ | \quad 40 & | \end{array}$$

thr-glu-arg-phe-phe-GLU-ser-phe-gly-ASP-leu-ser-ser-ala-ASP-ala-ileu-LEU-gly-asp-

60   70

pro-lys-val-lys-ala-his-gly-lys-lys-val-leu-ASP-ser-PHE-CYS-GLU-GLY-LEU-lys-

$$\begin{array}{ll} \text{NH}_2 & \\ | \quad 80 & 90 \end{array}$$

GLU-leu-asp-asp-leu-lys-gly-ALA-phe-ala-SER-leu-ser-glu-leu-his-cys-asp-lys-leu-

$$\begin{array}{ll} \text{NH}_2 & \text{NH}_2 \\ 100 \quad | & | \quad 110 \end{array}$$

his-val-asp-pro-glu-asp-phe-ARG-(leu, leu)-gly-asp-val-leu-(val, GLU, val, leu)-ala-

$$\begin{array}{ll} \text{NH}_2 & \text{NH}_2 \\ 120 & | \quad\quad 130 \quad | \end{array}$$

ARG-ARG-phe-gly-SER-glu-phe-SER-pro-glu-LEU-glu-ala-ser-PHE-glu-lys-VAL-

$$\begin{array}{l} \text{NH}_2 \\ | \quad 140 \end{array}$$

val-thr-gly-val-ala-ASP-ala-leu-ALA-HIS-arg-tyr-his

Scheme 7. Amino Acid Sequence of the γ Chain of Bovine Fetal Hemoglobin.

In this instance also, it is probable that the chain is one residue shorter than the corresponding human chain. There are at least 39 differences in sequence. The bovine γ chains contain more basic amino acids than usual in β and γ chains, and a considerable number of the basic amino acids are in positions different from those in the human γ chain. Indeed, the differences between bovine γ chains and human β chains are less than those between bovine and human γ chains.

The more extensive data on the various β and γ chains suggest, even more strongly than in the case of the α chains, the virtual invariability of certain portions of the chain. Three sections correspond in many ways to the relatively invariant sections of the α chain. Most striking is the constancy between residues 28 and 42, and, indeed, few variations exist between residues 24 and 50. This region corresponds roughly to residues 36 to 46 of the α chain. Similarly, residues 56 to 68 approximate residues 54 to 63 of the α chain. Finally, residues 88 to 103 in the vicinity of a cysteinyl residue may well be as invariant as between 28 and 42.

---

  * The capitalized abbreviations designate the points of difference from the human γ chain.

  ** Abbreviations within parentheses and separated by commas indicate that the residue has not been placed definitely in sequence.

VII. Hemoglobin in Three Dimensions.

1. Introduction.

Because of the several independent determinations, the amino acid sequence of human hemoglobin probably is more certainly known than that of most proteins. In the above discussion, we have compared the known sequences of related chains of different hemoglobins and of the different chains of the same hemoglobin. Although the similarities are great, no sequence is so striking in contrast to that of other proteins that one can assuredly relate a certain part of the molecule to a specific property. Yet, in the case of hemoglobin, the multitude of examples and the apparent invariance of certain sequences suggest that these sequences must be vital for proper functioning. Nevertheless, the sequence alone presents the molecule essentially unidimensionally. For example, it has long been postulated that two histidyl residues are associated with the heme group [see WYMAN (255) for a detailed discussion]. But if this is correct, which two are they? Are they from two chains or one? If one, which of the histidyl residues are involved? Neither does the sequence show which other groups interact or how they are arranged in space. Although various physical measurements give some idea of the general size and shape of the molecule, we must turn to X-ray diffraction to learn the spatial arrangement of the atoms.

The X-ray investigation of hemoglobin and the sequence determinations reached climaxes at almost the same moment. For many years, even before the determination of sequence could be considered seriously, PERUTZ and co-workers had been attempting to elucidate the three-dimensional structure of horse hemoglobin through X-ray diffraction. This work as well as related work by KENDREW and others with myoglobin reached a very meaningful stage around 1960, and by the middle of 1961 the amino acid sequence of the human α, β and γ chains had been essentially solved. The correlation between these two types of data will now be described.

2. The Relationship between Hemoglobin and Myoglobin.

Myoglobin is a heme protein of muscle which stores oxygen that has been transported to the site by hemoglobin. Myoglobin contains only one heme group (identical with that of hemoglobin) and has a molecular weight of about 17,000. It is, therefore, roughly the equivalent of a single hemoglobin chain. Although the X-ray diffraction method can determine atomic dimensions, the size and complexity of a protein make it difficult to achieve this goal. Because thousands of spots on the X-ray photographs must be used and very many calculations made, the work has proceeded in stages which used increasingly more data to achieve greater resolution until atomic positions became apparent. When myoglobin was resolved to 6 Å, the outline of the molecule became evident. The observed convolutions were confirmed, and the positions of many atoms could be discerned when the resolution was increased to 2 Å. Indeed, α-helices in myoglobin as well as many side-chains of the amino acid residues could be identified.

At the same time, horse hemoglobin was resolved to 5.5 Å and was seen to have four subunits, each of which was almost identical in shape with myoglobin at an equivalent resolution. Because the individual chains of hemoglobin are so similar in convolution to those of myoglobin and because myoglobin is better resolved, the more precise location of individual residues in myoglobin has been used, in conjunction with the known sequence of hemoglobin A, to suggest how the residues in hemoglobin A are arranged in three dimensions.

3. Pertinent Aspects of the Structure of Myoglobin.

When sperm whale myoglobin was observed at 2 Å resolution (151), most of the molecule was seen to be arranged in right-handed α-helices of the type proposed by Pauling and Corey in 1951. Eight helical regions of different length are connected by non-helical regions of different length. The helical sections form a complicated and irregular array although the molecule is compact and interstices are virtually non-existent. The plane of the heme group is in a crevice between two helices. The bottom of the crevice is blocked by two other helices whose axes are roughly perpendicular to those that enclose the plane of the heme. The heme is rather tightly surrounded except on one side toward an

```
                  NH2
                   |      10                                  20
val-leu-ser-glu-gly-glu-try-glu-leu-val-leu-his-val-try-ala-lys-val-glu-ala-asp-val-ala-
                  NH2
                   |            30                            40
gly-his-gly-glu-asp-ileu-leu-ileu-arg-leu-phe-lys-ser-his-pro-glu-thr-leu-glu-lys-phe-
                       50                              60
asp-arg-phe-lys-his-leu-lys-thr-glu-ala-glu-met-lys-ala-ser-glu-asp-leu-lys-lys-his-gly-
                  70                                80
val-thr-val-leu-thr-ala-leu-gly-ala-ileu-leu-lys-lys-lys-gly-his-his-glu-ala-glu-leu-lys-
                       NH2
             90  |                      100
pro-leu-ala-glu-ser-his-ala-thr-lys-his-lys-ileu-pro-ileu-lys-tyr-leu-glu-phe-ileu-ser-glu-
                                        NH2                      NH2
  110                          120       |                        |      130
ala-ileu-ileu-his-val-leu-his-ser-arg-his-pro-gly-asp-phe-gly-ala-asp-ala-glu-gly-ala-
  NH2                                                            NH2
   |                           140                       150     |
met-asp-lys-ala-leu-glu-leu-phe-arg-lys-asp-ileu-ala-ala-lys-tyr-lys-glu-leu-gly-tyr-glu-
gly
```

Scheme 8. Amino Acid Sequence of Sperm Whale Myoglobin.

edge of its plane. Oxygen must arrive and depart from this side. The α and β chains at 5.5 Å resolution show very similar arrangement.

The amino acid sequence of sperm whale myoglobin has been determined chemically by EDMUNDSON (62) (*Scheme 8*).

The chain of 153 residues is 12 residues longer than the human α chain and 7 residues longer than the β and γ chains of hemoglobin.

If we compare the sequences of the α and β chains and of myoglobin as BRAUNITZER et al. (*41*) and CULLIS et al. (*55*) have done and as outlined for the α and β chains in Section III, 4 on p. 121, the three chains may be brought into coincidence in 26 positions. Because the general convolutions of the α and β chains and myoglobin are the same and because residues in the myoglobin chain can be positioned in space, this information has been used to arrange the residues of the α and β chains (especially in the helices) in space.

The probable three-dimensional shape of human hemoglobin will now be described with the realization that the extrapolation, though reasonable, goes from sperm whale myoglobin to horse hemoglobin to human hemoglobin.

4. The α and β Chains — Secondary and Tertiary Structure.

The amino acid sequence is commonly referred to as the "primary" structure, the arrangement in helical or non-helical regions as the "secondary" structure, and the arrangement of the secondary structures as the "tertiary" structure of a protein.

If, then, the sequences of the α and β chains and of myoglobin are placed side by side as already described, reasonable conclusions may be drawn as to which residues of the α and β chains are in helical and non-helical array. *Table 2* which summarizes this information is based upon the sequence of myoglobin and its correlation with helical and non-helical residues as presented by EDMUNDSON (*62*).

The eight helical regions can take their tortuous convolutions because of the interspersed non-helical regions. In some instances, the change of direction is almost complete, and consecutive helices are almost parallel. Each of the corners is different in length and type of residue. Helix D is absent in the α chain according to the arrangement of residues in Table 2; the X-ray data confirm this conclusion.

Fig. 8 pictures a model of the horse hemoglobin molecule. Many views of a similar model are presented by CULLIS et al. (*55*). It is difficult to gain an idea of the arrangement of such a complicated shape by the examination of two-dimensional pictures. Unfortunately, such a model cannot be provided for each reader! However, we encourage the reader to construct a paper model such as shown in *Fig. 9* from the drawings that are given in *Figs. 10a, b*. In Fig. 9, a single β chain has been removed from the model in Fig. 8 and placed beside a paper model. These models

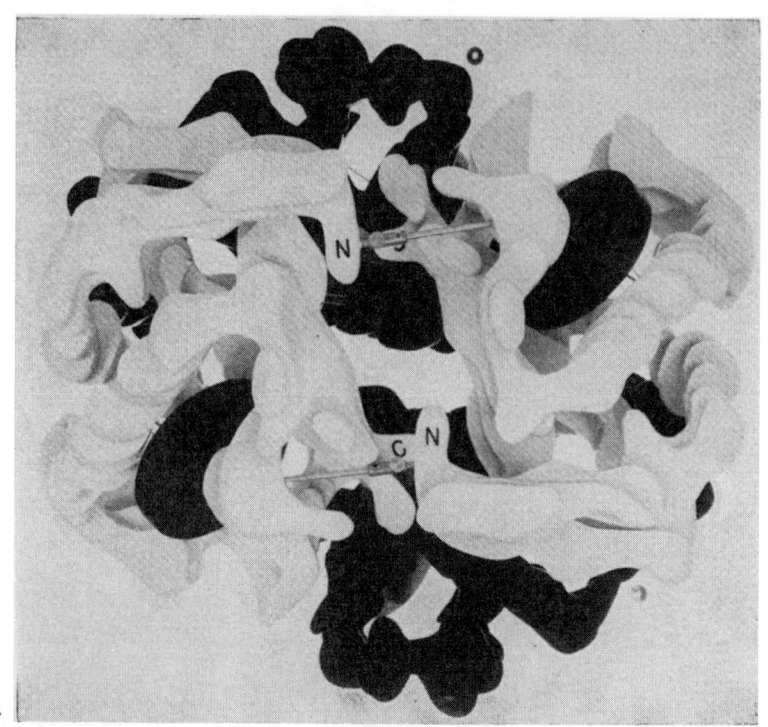

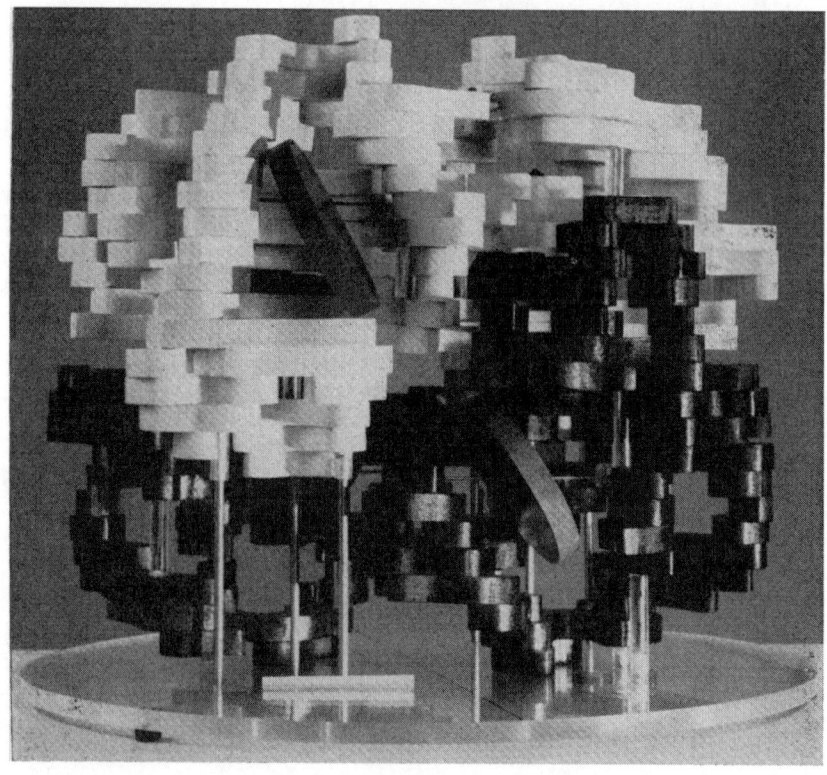

Fig. 8.

References, pp. 181—194.

were constructed on a scale of 0.25 inch (about 6 mm.) to the Ångstrom. In the paper model, the spiral on the surface of the cylinder has the pitch of an α-helix and a radius which approximates that of the main atoms of the polypeptide chain. At each 100° along the spiral, a dot shows the position at which a side-chain projects. This paper model probably shows the position of the helical residues rather correctly.

Table 2. Probable Residues in Helical and Non-Helical Arrangement in the α and β Chains of Hemoglobin and in Myoglobin.

Helical*	Non-Helical*	Residue Numbers			Length of helix (No. of residues)
		α chain	β chain	myoglobin	
	Pre-A	1–2	1–3	1–2	
A		3–18	4–19	3–18	16
	AB	19	—	19	
B		20–35	20–34	20–35	16
	BC	—	—	—	
C		36–42	35–41	36–42	7
	CD	43–51	42–49	43–50	
D		—	50–56	51–57	7
	DE	—	—	—	
E		52–71	57–76	58–77	20
	EF	72–79	77–84	78–85	
F		80–88	85–93	86–94	9
	FG	89–93	94–98	95–99	
G		94–112	99–117	100–118	19
	GH	113–118	118–123	119–124	
H		119–141	124–146	125–148	24
	Post-H	—	—	149–153	

* Beginning at the N-terminus, the helices are designated by capital letters and the interhelical regions by the letters of the two helices that they connect.

The non-helical residues are presented as a straight line rather than the zigzag path they would actually take. Therefore, this representation of the non-helical residues is meant to show only which residues are non-helical and approximately the positions they assume. The corners AB, BC, and DE which contain no residue appear to be odd contortions in this inflexible model: actually, rotation about several bond axes would readily permit the approximate course of the chain as shown.

The basic model of Fig. 10a may be elaborated on as *Fig. 10b* shows for helix A. The positions of the NH and CO groups and of the α-carbon atom of each residue are shown, and the identifying abbreviation for a

Fig. 8. Two views of a model of the horse hemoglobin molecule. Each piece is a contour of electron density at a given level in the molecule. (a) is a view along the dyad axis around which rotation of 180° produces superposition: the white units are α chains, the black β chains, and the discs hemes. (b) is a more general view of the model.

a

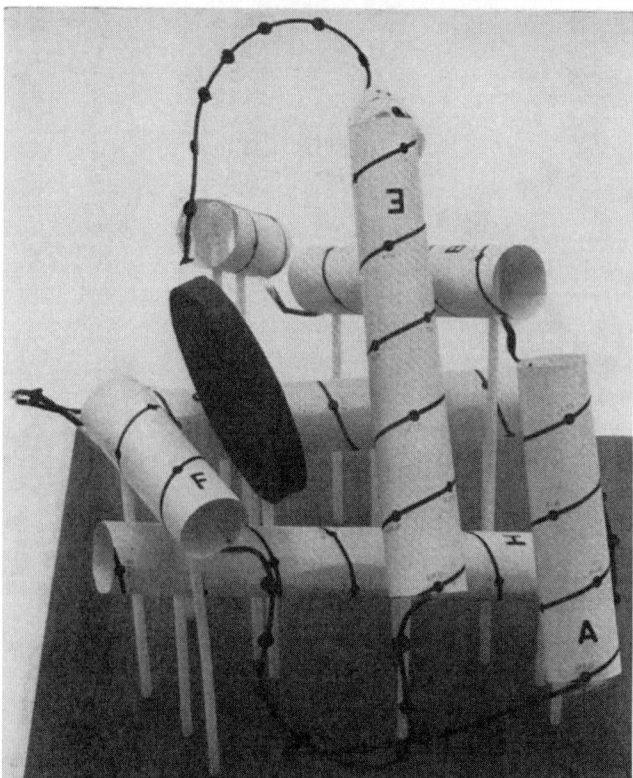

b

Fig. 9. Two models of the β chain of hemoglobin A. (Fig. 10a, b, pp. 152—153 gives directions for the construction of the paper model.)

References, pp. 181—194.

particular residue is placed in the position of the β-carbon atom. By the use of the data in Figs. 10a and b, models of the α and γ chains and of myoglobin may be made by altering the abbreviations for certain residues.

If a model is not constructed, *Fig. 11* may be helpful in visualizing some of the relationships. Fig. 11 is viewed from about the same angle as the models in Fig. 9.

The non-helical N-terminal sequence of several residues lies in the approximate position shown. At the N-terminal end of helix A and on the surface of the intact molecule are the abnormalities in human hemoglobins S, C, $G_{San\ Jose}$ and $C_{Georgetown}$ at residues 6 and 7.

Between residues 19 and 20, the direction of the chain changes abruptly.

Near corner BC, a prolyl residue (No. 36) is second from the N-terminus of helix C, and forces a change in direction of the helix which in this instance is almost 90°. As great a change of direction occurs at corner AB where no proline is near. However, the sequence from residues 28 to 42 is the most invariant that has been observed (or is presumed to be present) in the β and γ chains of various species. It includes all of helix C, the corner BC, and half of helix B. This region, then, probably is vital for the proper functioning of the hemoglobin molecule. Although the several α chains are more variable, but the prolyl residue that forces the corner is present both in the α chains and in myoglobin.

The longest non-helical region CD connects the shortest helices C and D. Indeed, the 5-residue "gap" at this point in the α chain eliminates helix D in the α chains, and, as mentioned above, its absence is confirmed by the X-ray data.

The prolyl residue at position 58 produces the corner DE just as that in position 36 produces BC.

The X-ray data place helices B and E very close together—so close, in fact, that a glycyl residue must be present at an appropriate position in both. This is residue 24 in helix B and residue 64 in helix E. Myoglobin and the α chains, likewise, have glycyl residues at the necessary positions. Surprisingly enough, the data of Satake and Sasakawa (*207*) place an alanyl instead of a glycyl residue at this position of the bovine α chains. If their data are correct, the spatial arrangement of helices B and E of the bovine α chains presumably is influenced.

Between helices E and F lies the heme group whose structure is as shown on p. 152.

Residue 92 in helix F is the so-called "proximal" histidine: a nitrogen atom of its imidazole group occupies the fifth coordination position of the heme iron. The "distal" histidine at position 63 of helix E is the other histidyl residue that is involved in the functioning of hemoglobin.

The non-polar side-chains of the heme are buried and the polar side-chains (that is, the propionic acid groups) are on the surface. The heme iron is accessible to oxygen and other ligands on the side of the distal histidine.

Hemin chloride.

The discussion of the β and γ chains of animal hemoglobins on p. 144 has pointed out that residues 88 to 103 vary little in sequence. This seems to be especially true of 88 to 94 which comprise most of helix F and include the proximal histidine. In the α chain, helix F shows greater variation in sequence although, the proximal histidine is present as it also is in myoglobin.

The sequence between residues 55 to 68 which at 63 has the distal histidine is again rather invariant. BRAUNITZER (42) has termed the section from residues 61 to 66 with its four basic residues the "basic

Fig. 10 a. Construction of a model of the β chain of hemoglobin A. The base may conveniently be made of $^3/_4$ inch (1.9 cm.) wood or plastic. The full size base is 10 inches square (25.4 × 25.4 cm.), and the grid lines are spaced one inch apart. Drill holes of equal depth at the positions that are indicated by + and are designated a and A, b and B, etc. to h and H, and he and HE. The diameter of the holes should equal that of the material for the supports which may be made of paper drinking straws, wood dowel, metal rod, etc. Beside each position a and A etc. on the base, the number in parenthesis is the length in inches of the support to be inserted into the adjacent hole. The cylinders that represent helices A to H have been unrolled. Enlarge the drawings of the helices to the proper size (the circumference is $3^{17}/_{32}$ inches = 9 cm. as shown by the dimension adjacent to helix C) on stiff paper or light cardboard, punch holes at (a) and (A), (b) and (B), etc. equal to the diameter of the supports (elongate if the helix is steeply tipped), form the cylinders, and paste together by overlapping the crosshatched portion. Slide hole (a) over the top of the support in hole a of the base, and slide hole (A) over the top of the support in hole A of the base until the opposite side of the cylinder rests on the top of each support. Place other cylinders in like manner. The helical sections are thereby arranged in proper relation. Cut wire for the non-helical sections pre-A, A-B, etc. to the indicated lengths and attach a circle of gummed tape to indicate the position of a residue. Attach pre-A to the N-terminal end of helix A by an adhesive tape. A-B, B-C, and D-E connect the appropriate helices. C-D forms an S-shaped loop and the others can be more or less obviously placed. The heme group is positioned by inserting the supports into the disc at the indicated angle. The length of the supports for the heme group is such that a one-half-inch (1.25 cm.) length should be inserted into the disc that represents the heme. In constructing a model from this figure, the reader may find it convenient to photograph the figure and then use a photographic enlarger to project the properly enlarged image onto paper where it may be traced easily. The letters (a), (A), (b) and (B) above represent the fully ancircled letters in this Figure.

References, pp. 181—194.

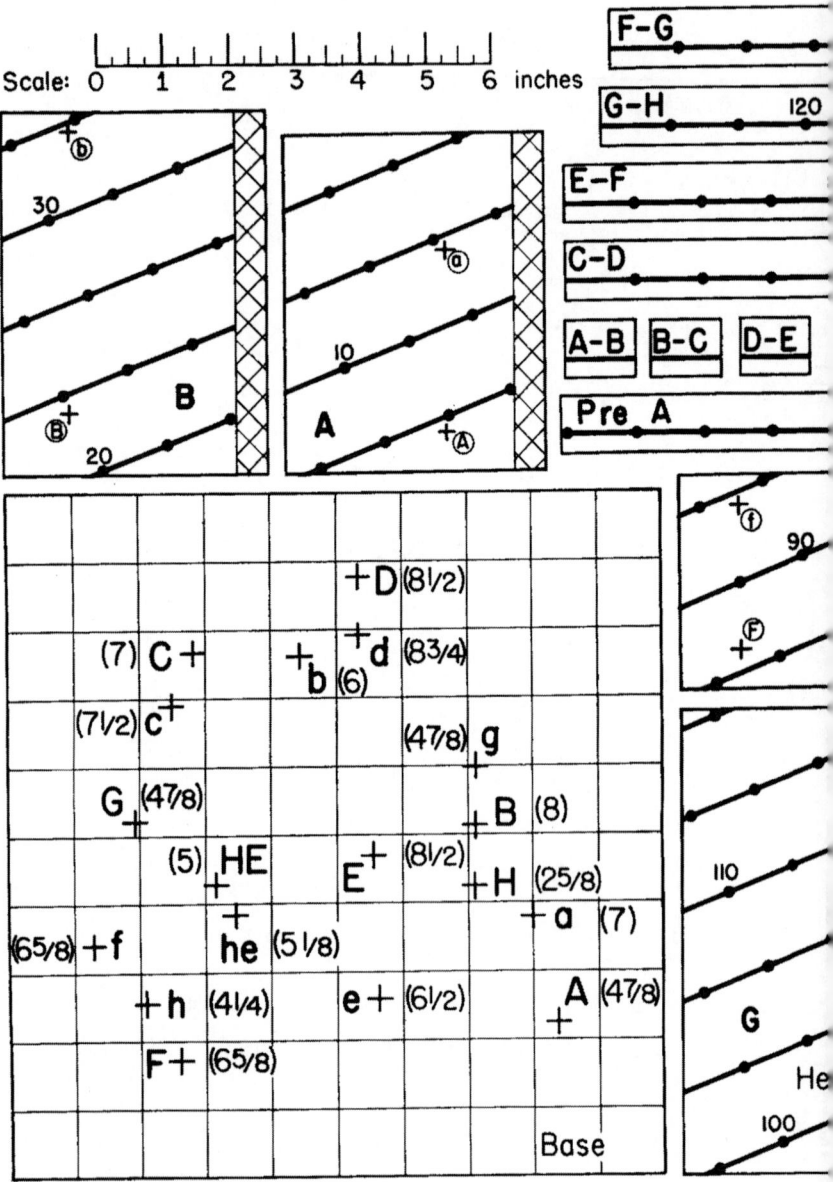

Fig. 10a (see Caption o

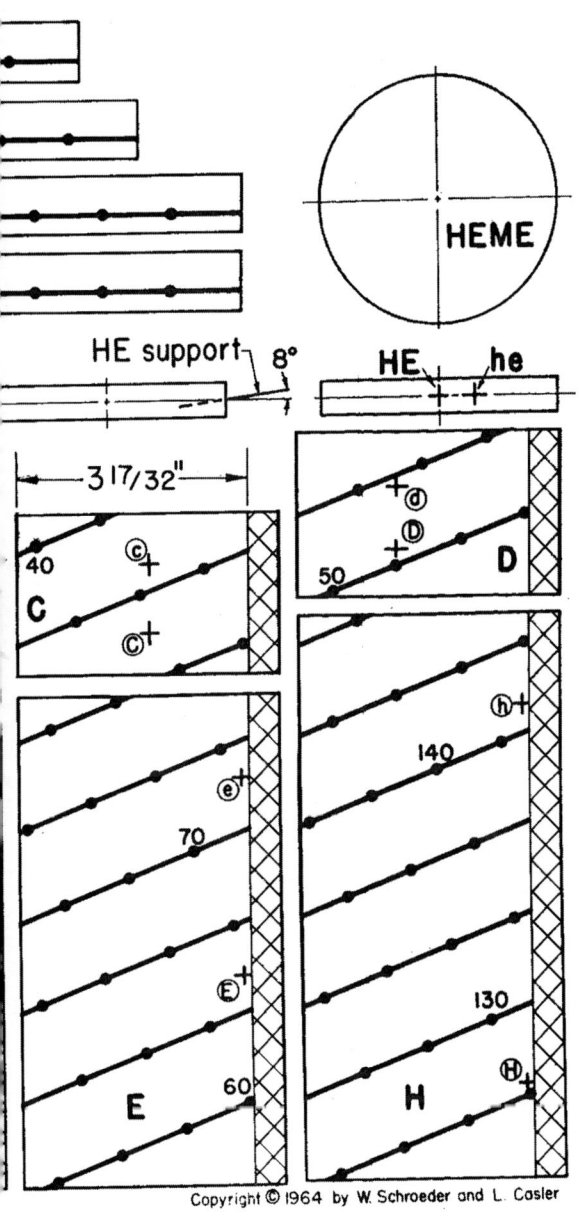

HEME

HE support 8°

HE he

3 17/32"

40 C

50 D

70 E

60

140 H

130

center". The importance of the distal histidine is apparent from the fact that in the improperly functioning hemoglobins Zürich and $M_{Saskastoon}$ the distal histidine is replaced by arginine and tyrosine, respectively, and in hemoglobin $M_{Milwaukee-1}$ glutamic acid replaces valine at residue 67. Hemoglobin M_{Iwate} ($= M_{Kankakee}$) is the first example of an alteration of the proximal histidine (in this instance of the α chain) (139). Needless

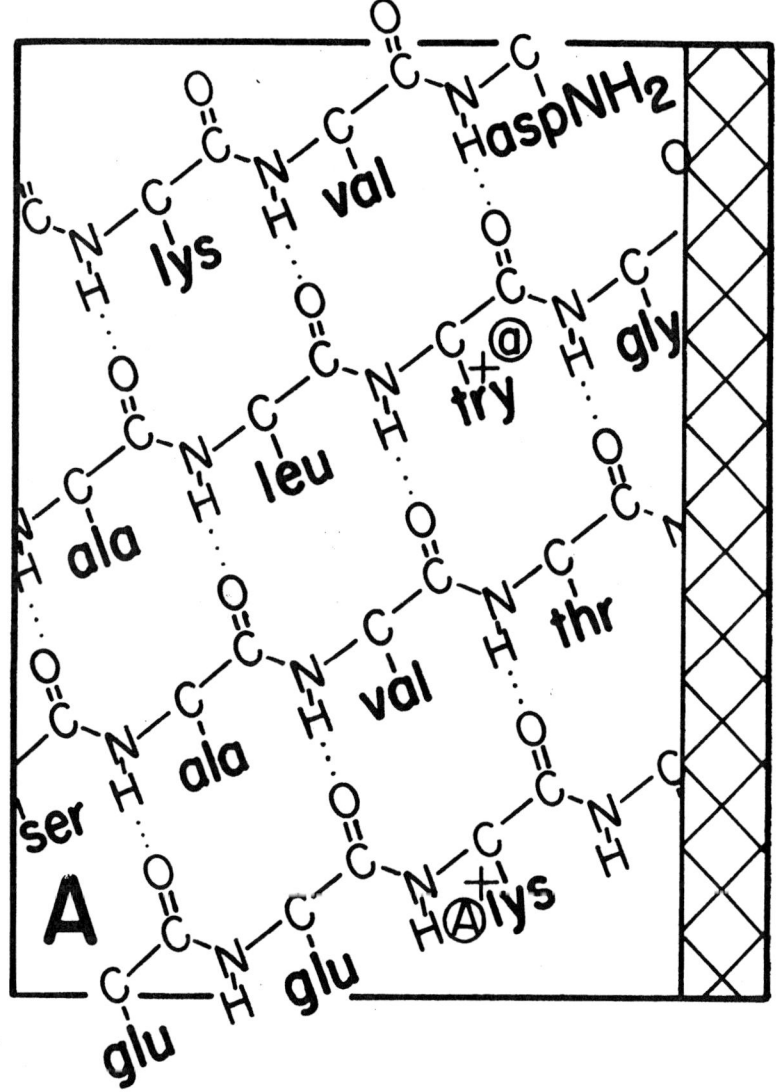

Fig. 10 b. An alteration of helix A as shown in Fig. 10 a. The main atoms of the polypeptide chain have been placed in position, and each residue has been identified. The α-carbon atom coincides with the dot that represents a residue in the helices of Fig. 10 a. This figure is full scale.

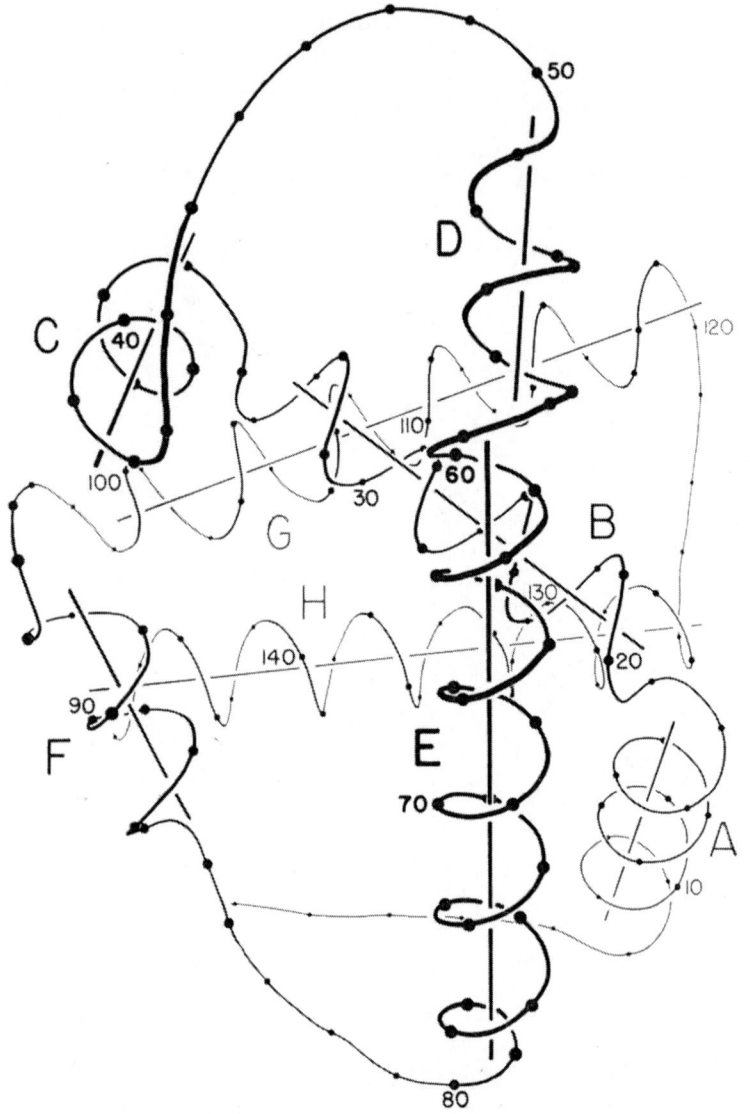

Fig. 11. A drawing of the β chain of hemoglobin A. The course of the helical and non-helical portions is shown in perspective. Each filled circle is placed at the approximate position of a residue. The helices are lettered to correspond with Table 2 (p. 149).

to say, myoglobin and the α chain have many similarities to the β chain near the distal histidine.

Helices G and H are furthest from the observer in Fig. 11. The C-terminus is at the left end of helix H. In both α and β chains, the

C-terminal residue is part of the helix, but in myoglobin five non-helical residues are present at the C-terminus. Both G and H are rather long helices which apparently are held together by a hydrogen bond that involves the tyrosyl residue at No. 145 (*150*). In the assembly of the four chains in hemoglobin, most of these helices are buried in the interior of the molecule.

By the use of a model or Fig. 11 and in conjunction with the data in Figs. 6 and 7 and Table 2 (pp. 134a, b; 149) the reader should be able to locate the position of the aberrations in the abnormal hemoglobins.

KENDREW (*150*) has described many features of the structure of myoglobin such as the environment of the heme, the interaction of side-chains, the positions of polar and non-polar side-chains, etc.

5. The Quaternary Structure.

The three-dimensional arrangement of the chains or subunits is termed the quaternary structure.

The two α and β units of hemoglobin are arranged in a tetrahedral form. Because of the dyad axis, the tetrahedron has a plane of symmetry although it is not a regular tetrahedron.

Examination of the entire model shows that none of the helices is entirely buried in the interior, although helices G and H are less exposed than the others. In general, with the exception of G and H, each helix exposes about half of its residues, more or less, at the surface of the normal four-chain molecule.

For the most part, there is relatively little contact between like chains, that is, between the two α chains or the two β chains. On the other hand, considerable areas of α chains touch β chains. Indeed, α and β chains meet in two different ways as examination of *Fig. 8 a* (p. 148) will reveal. Consider the α (white) chain to the right of the picture in its relation to the two β (black) chains at the top and bottom of the picture. The contact with the top β chain is different than the contact with the bottom β chain. This difference is also apparent in the distances between the iron atoms of the four heme groups as shown in *Table 3* [MUIRHEAD and PERUTZ (*172*)]. Thus, the distance from one α-chain iron atom to its nearer β-chain iron (α_1—β_1) is about 5 Å shorter than the distance to the farther β chain (α_1—β_2).

Table 3. Distances between Iron Atoms in Horse Oxyhemoglobin.

Pair of Atoms*	Distance (A)
α_1—α_2	36.0
β_1—β_2	33.4
α_1—β_1	25.2
α_1—β_2	30.4

* Each iron atom is designated here in association with a particular chain.

Although the four heme groups are far apart on an atomic scale, they interact according to the interpretation that is placed on the sigmoid oxygen dissociation curve. How this interaction occurs over a long distance is not evident.

6. The Three-dimensional Structure of Hemoglobins from Different Species.

It has been assumed above that there is such similarity and, in many cases, such identity that one can interpolate and extrapolate data among myoglobin, horse hemoglobin, and human hemoglobin. The relationships in amino acid sequence give confidence in this assumption, but it should be pointed out that X-ray evidence of a close relationship in the structure of several hemoglobins existed before sequence data became available. Thus, Bragg and Perutz (39) in 1952 concluded from an examination of horse, human and sheep hemoglobins that, with the exception of sheep fetal hemoglobin, the several molecules have the same external form. Blow's investigation (36) of pig and rabbit hemoglobins provided additional information in agreement. A few years later, Perutz et al. (189) decided that the similarities in the Patterson projections of horse and human hemoglobins implied a close relationship in the internal architecture of the two proteins. This statement has recently been strengthened by more detailed X-ray work on human hemoglobin which had as its object the solution of another problem, namely, the effect of oxygenation and deoxygenation on structure. These data will be discussed in the following Section.

7. Oxygenation and Structural Alterations.

The spatial configurations described in preceding Sections derive from horse oxyhemoglobin. Hemoglobin, of course, functions by combining with oxygen and later releasing it. More than 25 years ago, Haurowitz (87) proposed that this reversible reaction was accompanied by significant configurational alterations. Evidence to support this view has been summarized by Benesch and Benesch (29). Convincing proof of configurational change was first obtained in the X-ray investigation of Muirhead and Perutz (172) of human ferrohemoglobin*. At a resolution of 5.5 Å (with an accuracy somewhat less than that for horse oxy-hemoglobin because fewer data were used to derive the structure), the helices and configuration of the chains were seen to be essentially identical with those of horse oxyhemoglobin. Yet, the effect of deoxygenation manifests itself in the distances between iron atoms as shown in *Table 4*.

Within the limits of accuracy, neither the distance between α-chain iron atoms (α_1—α_2) nor the distance between an α-chain iron atom and its nearest β-chain iron atom (α_1—β_1) has been altered by deoxygenation. By contrast, the distance between β-chain iron atoms (β_1—β_2) and the α and β other iron atom (α_1—β_2) is 7 Å greater. The α chains thus are

---

* This form is frequently termed "reduced" hemoglobin or "deoxyhemoglobin".

References, pp. 181—194.

equidistant from each other in the two forms but β chains are not. Oxygenation brings the two β chains closer together, or conversely, deoxygenation causes them to separate. What less obvious alterations also occur must await investigation at greater resolution.

It was pointed out by MUIR-HEAD and PERUTZ (*172*) that it would have been better to compare horse rather than human ferrohemoglobin with horse oxyhemoglobin. This was not done in the first investigation because a crystalline form of horse ferrohemoglobin suitable for X-ray analysis was not available. In a later study (*186*), a satisfactory form of horse ferrohemoglobin was investigated, and results identical with those from human ferrohemoglobin were obtained.

Table 4. Distance between Iron Atoms in Horse Oxyhemoglobin and Human Ferrohemoglobin.

Pair of Atoms*	Distance (A)	
	Horse oxyhemoglobin	Human ferrohemoglobin
$\alpha_1-\alpha_2$	36.0	35.0
$\beta_1-\beta_2$	33.4	40.3
$\alpha_1-\beta_1$	25.2	25.0
$\alpha_1-\beta_2$	30.4	37.4

* As in Table 3 (p. 155) each iron atom is designated here in association with a particular chain.

Thus, a species difference was not responsible for the observed effect but, indeed, oxygenation and deoxygenation produce a structural change. A second conclusion is that the three-dimensional structures of horse and human hemoglobin must be very similar.

Although the evidence is less detailed [PERUTZ and MAZZARELLA (*188*)], oxyhemoglobin H (β_4^A) and ferrohemoglobin H have identical structure. Hemoglobin H behaves very differently than hemoglobin A in its reaction with oxygen: hemoglobin H shows no interaction between hemes and no Bohr effect* and its oxygen affinity is greater (*32*). Both α and β chains (perhaps merely two kinds of chains) apparently are necessary if hemoglobin is to have the properties so characteristic of the sigmoid oxygen dissociation curve. Further comparison of hemoglobins A and H may reveal much about the functioning of the molecules.

VIII. Chemical Modifications of Hemoglobin.

1. Introduction.

Hemoglobin and myoglobin are unique among the proteins in that their amino acid sequence is known, and their three-dimensional structure is known in much detail or can reasonably be guessed at. Thus, the relative positions of the polypeptide chains and of the side-chains of the various residues can be defined, and interactions between side-chains can be elucidated. The absence of disulfide bonds requires that the

---

* The Bohr effect is the dependence of oxygen affinity on pH.

entire molecule be held together by other than covalent bonds. The question naturally arises as to whether the molecule has the same configuration in solution and in the crystal. [See Perutz et al. (*186*) for comments on this point.] Although X-ray examination cannot answer this question, yet the X-ray work agrees that the molecular weight of hemoglobin is that of the ultracentrifugal studies. Despite the evidence that hemoglobin is in dynamic equilibrium with its parts (to be discussed in Chapter IX, p. 160), it may be considered to be $\alpha_2\beta_2$ (or $\alpha_2\gamma_2$) for practical purposes. If the configuration is different in solution than in the crystal, methods must be devised to discover these differences. That chemical methods for studying a protein in solution (especially when its sequence is known) can give meaningful results is apparent from the investigations of ribonuclease. In these experiments (*53, 54*), reaction of ribonuclease with iodoacetic acid modified one of two histidyl residues to form the carboxymethyl derivative. The reaction of either histidyl residue prevents the reaction of the other, and one product is enzymatically inactive and the other almost so. Both histidyl residues are presumed to be part of the active site. One is 12 and the other 119 residues from the N-terminus. Such experiments with hemoglobin are still few but their success with ribonuclease indicates that meaningful results should follow this approach.

2. Reactions of the Heme Group.

Let it suffice here to summarize some well-known data about the heme group and its reactions.

The heme may be removed from the globin by acidified ketones in the cold (*6, 241*); the globin precipitates while the heme remains in solution. Cleavage under such mild conditions led to the conclusion that the linkage was non-covalent—a conclusion that the X-ray data confirm.

The normally ferrous iron of the heme may be oxidized and reduced again. The ferrous form combines with many small molecules, such as CO and NO, in addition to oxygen. In the ferric form, hemoglobin no longer can carry oxygen, but the iron can combine with cyanide and other small ions. Because *t*-butyl isocyanide with a bulky group combined less strongly with hemoglobin than did ethyl isocyanide, St. George and Pauling (*234*) deduced that the heme group is buried—a conclusion again confirmed by the X-ray data.

3. Titration Curves and their Interpretation.

The differential and thermodynamic data that Wyman and associates (reviewed in detail in *255*) amassed by means of titration curves suggested that two imidazole groups from histidyl residues are linked to the heme iron. Again, the X-ray data substantiate this point.

References, pp. 181—194.

HERMANS (*89*) has concluded from titration studies that hemoglobin A has 8 tyrosyl side-chains with normal and 4 with abnormal ionization. In myoglobin (*150*) the hydroxyl group of the tyrosyl residue at residue 146 interacts with the carbonyl group of residue 99. The corresponding tyrosyl residues in hemoglobin are α_{140} and β_{145}, and the interacting residues are α_{93} and β_{98}. Accordingly, it may reasonably be inferred that these are the abnormal tyrosyl residues in hemoglobin A. However, HERMANS (*90*) has also examined the products of the reaction of hemoglobin A with carboxypeptidases A and B under conditions which remove residues α_{140} and α_{141}, or β_{145} and β_{146}, or all four of them [ANTONINI et al. (*7*)]. His data suggest that α_{140} and β_{145} are normal tyrosyl residues in carbonmonoxyhemoglobin, but the data are equivocal for ferrihemoglobin cyanide.

4. Reactions of the Sulfhydryl Group.

Before the sequence determinations showed one sulfhydryl group in α and γ chains and two in β chains, attempts to determine these numbers by titration with silver or mercury produced many conflicting reports (reviewed in *213*). Obviously, the sulfhydryl groups were not equally reactive. SMITH and PERUTZ (*231*) used this and other facts to identify the black chains in the crystallographic model of horse hemoglobin (Fig. 8, p. 148) with the β chains; they reacted horse hemoglobin with iodoacetamide and, after separation of the chains and hydrolysis, determined which contained S-carboxymethylcysteine. About the same time, RIGGS and WELLS (*200, 202*) showed that only one sulfhydryl group of the β chain of hemoglobin A would react with N-ethylmaleimide. It remained for GOLDSTEIN et al. (*75*) to prove that the reactive sulfhydryl group of the β^A chain is part of the cysteinyl residue at β_{93} (adjacent, therefore, to the proximal histidyl residue).

The profound effect of oxygenation and deoxygenation is also manifest in the reaction of this sulfhydryl group. BENESCH and BENESCH (*30*) found that the sulfhydryl group of β_{93} in ferrohemoglobin does not react with iodoacetamide whereas in oxyhemoglobin it does (*75*).

GUIDOTTI and KONIGSBERG (*80*) have investigated in detail the reactions of the sulfhydryl groups with iodoacetamide and N-ethylmaleimide.

KONIGSBERG (*155, 227*) by the use of a bifunctional maleimide derivative [bis-(N-methylmaleimido)-ether] has been able to react the sulfhydryl groups of two β chains at residue 93. This reaction, however, must have involved two molecules of hemoglobin; the two β_{93} residues in $\alpha_2\beta_2$ are on opposite sides of the surface of the molecule approximately 30 Å apart, and the reagent cannot span this distance.

5. Other Reactions.

Under certain conditions, N-ethylmaleimide will react with other than sulfhydryl groups. Thus, it reacts with the amino group of the N-terminal residue of the α chain but not with that of the β chain [GUIDOTTI and KONIGSBERG (80)]. On the other hand, the amino group of the N-terminal residue of the β chains of hemoglobin H(β_4^A) does react with N-ethylmaleimide. GUIDOTTI and KONIGSBERG conclude that the N-terminal group of the β chain is involved in some way in the interaction of α and β chains. We have pointed out earlier (214) and also in Section VII, 5 (p. 155) that a β chain may be associated with an α chain in two ways which can best be characterized by the distances between the heme iron atoms as in Table 3 (p. 155). Thus, we may speak about the "nearer" or the "farther" α chain in relation to a given β chain. If GUIDOTTI and KONIGSBERG's conclusion is correct, is the N-terminal group of the β chain associated with the nearer or the farther α chain? An inspection of the hemoglobin model demonstrates that any interaction of the N-terminus of the β chain must be with the farther α chain. Although it is difficult to judge with certainty, the model also suggests that a greater area of contact exists between a β chain and the farther α chain than between a β chain and the nearer α chain. Perhaps, in dissociation, the $\alpha\beta$ subunit involves the farther chains.

R. J. HILL (91) has found that the N-terminal amino group of the α chains is preferentially dinitrophenylated when dinitrofluorobenzene reacts with carbonmonoxyhemoglobin at pH 7.4 and 37° in homogeneous aqueous solution.

6. Conclusions.

Although other reactions of hemoglobin have been studied, these either are general such as acetylation or, if specific, the products have not been characterized so that the exact reaction can be written. Such information gives no clue to the differential reactivity of identical groups in different situations.

With the small beginnings that have been presented above, we may expect that much about hemoglobin will be learned by this approach.

IX. Reversible Dissociation and Subunit Hybridization of Hemoglobin.

1. Introduction.

The principal molecular form of hemoglobin A under physiological conditions is a specific aggregate of four chains and may, as we have seen, be represented by the formula $\alpha_2\beta_2$. Evidence for this degree of aggregation is based primarily upon molecular weight determinations which were first made by ADAIR (*1*) and PEDERSEN and ANDERSSEN (*184*). This four-chain aggregate can undergo reversible dissociation into double- and single-chain subunits depending upon the conditions of its aqueous environment (see *82* and *134a* for references). Extensive dissociation of hemoglobin A will occur under conditions of low and high pH (*68, 86*) and in strong salt solutions (*31, 153*). Even under physiological conditions one may conceive that hemoglobin is an equilibrium mixture of four-, two- and single-chain subunits. The mechanism of the dissociation of hemoglobin has received careful study since about 1956. These studies have been based upon direct physical-chemical measurement of dissociation and upon studies of the hybridization or recombination of subunits of different hemoglobins.

2. Possible Modes of Dissociation.

The dissociation of hemoglobin into two-chain subunits may possibly occur in three different ways. Non-identical dimer subunits can be formed from an asymmetrical dissociation, thus: $\alpha_2\beta_2 \to \alpha_2 + \beta_2$. Two types of symmetrical dissociation can theoretically occur as represented by $\alpha_2\beta_2 \to 2\,\alpha\beta$ and $\alpha_2\beta_2 \to 2\,\beta\,\alpha$ (*214* and p. 155). The two types of asymmetrical subunits* should differ from each other by the contact site(s) between the α and β chains.

Formation of single α and β chain subunits may occur either directly by dissociation of the four-chain molecule or indirectly by dissociation of any of the two-chain subunits. Combination of two or more of these reactions may account for the actual dissociation of hemoglobin under the various conditions which have been studied. All of these possible dissociations are combined in *Scheme 9*.

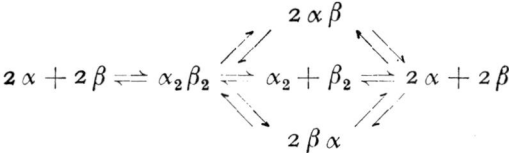

Scheme 9. Dissociation of Hemoglobin.

Because the dissociation of hemoglobin is reversible, each dissociation step shown in the Scheme has been represented as an equilibrium. However, not all of these steps need be reversible, nor need they all exist, provided there are enough reaction steps to account for the over-all reversible dissociation of hemoglobin. This point is discussed in more detail below (Section 6, p. 166).

---

* The reader should realize that *asymmetrical* dissociation of $\alpha_2\beta_2$ produces *symmetrical* double-chain subunits (α_2 and β_2) and that *symmetrical* dissociation forms *asymmetrical* subunits ($\alpha\beta$).

3. Evidence for Dissociation.

Direct evidence of dissociation of hemoglobin is based upon centrifugation, diffusion and light scattering measurements. Field and O'Brien (68) demonstrated by centrifugation studies that human hemoglobin dissociates into subunits of one-half normal molecular weight at pH 6 and below. Similarly, Hasserodt and Vinograd (86) detected a dissociation at an alkaline pH (10 and above). Dissociation in salt solutions at neutral pH has also been extensively studied by centrifugation (153, 31). The light scattering measurements of Rossi-Fanelli, Antonini and Caputo (203a) indicate that the salt-induced dissociation is very rapid. Apparently, in 3 M sodium chloride, most of the hemoglobin molecules exist in a dissociated form with a molecular weight approximately that of double-chain subunits. Other conditions such as concentrated mercaptoethanol (83) and urea (240) also result in dissociation of hemoglobin.

By means of thin-film dialysis, Guidotti and Craig (79) have shown that partial dissociation into single-chain subunits as well as double-chain subunits occurs at acid and alkaline pH. However, at high ionic strength the formation of single-chain subunits appears to be suppressed and the predominant dissociation is only into double-chain subunits.

4. Mechanism of Dissociation and Subunit Hybridization.

In addition to direct physical-chemical measurements, reversible dissociation of hemoglobin has been studied by subunit recombination or hybridization experiments. As we have pointed out in Chapter V, p. 128, this method is used to determine which chain of an abnormal hemoglobin is aberrant. It may also be applied to study the mechanism of the exchange of chains. Accordingly, these experiments depend upon the exchange of common subunits between human hemoglobins or the formation of new hybrid molecules from two hemoglobins which do not possess common subunits. After the hemoglobins have been dissociated and the common, or complementary, subunits have been allowed to equilibrate with one another, the four-chain hemoglobins are then reformed by conditions that favor reassociation. The various molecular species may then be separated by electrophoresis or chromatography.

Several mechanisms of reversible dissociation and subunit hybridization have been proposed from hybridization studies. These proposals which are discussed below in chronological order were made concurrently with many of the physical-chemical studies mentioned above.

Singer and Itano (229) were the first to propose a mechanism for hemoglobin dissociation and to suggest the type of subunits that result. These authors demonstrated that common subunits of hemoglobins A,

S and C can be exchanged from one hemoglobin to another by first dissociating a mixture of two hemoglobins at acid pH and then re-associating at neutral pH. Although common subunits exchanged, no "hybrid" molecules consisting of one-half hemoglobin A and one-half hemoglobin S or hemoglobin C such as $\alpha_2 \beta \beta^S$ or $\alpha_2 \beta \beta^C$ were detected. They concluded that the acid dissociation of hemoglobin must be asymmetric into α_2 and β_2 subunits (229), and that appreciable dissociation into single chains did not occur under their conditions (229). Therefore the predominant dissociation reaction as proposed by SINGER and ITANO can be written as

$$\alpha_2 \beta_2 \rightleftharpoons \alpha_2 + \beta_2,$$

where the position of the equilibrium is dependent upon pH.

VINOGRAD and HUTCHINSON (243) challenged this proposal of asymmetric dissociation when they demonstrated that the exchange of common subunits is not rapid at alkaline pH. The recombination of identical subunits requires a substantially longer time than that required for dissociation. They proposed that acid and alkaline dissociation is symmetrical with the formation of $\alpha \beta$ subunits. To account for the slow exchange of common subunits, they proposed that some α and β chains themselves must be in equilibrium with the $\alpha \beta$ subunits. The dissociation reactions which seemed most probable to VINOGRAD and HUTCHINSON are

$$\alpha_2 \beta_2 \rightleftharpoons 2 \alpha \beta \rightleftharpoons 2 \alpha + 2 \beta.$$

Because no hybrid molecules of the type $\alpha_2 \beta \beta^S$ could be detected among the products of hybridization of hemoglobins A and S, they further assumed that only two-chain subunits of like kind can recombine into four-chain molecules. Support for this assumption comes from the hybridization of hemoglobins H and S as follows:

$$3 \beta_4 + 3 \alpha_2 \beta_2^S \rightarrow 2 \beta_4 + \alpha_2 \beta_2^S + \beta_4^S + 2 \alpha_2 \beta_2.$$

Thus, the new products of this hybridization were $\alpha_2 \beta_2$ and β_4^S rather than $\alpha_2 \beta \beta^S$ and $\beta_2 \beta_2^S$ (145).

GUIDOTTI, KONIGSBERG, and CRAIG (82) re-examined the idea of symmetrical dissociation as formulated by VINOGRAD and HUTCHINSON. They suggest that one need not assume that only double-chain subunits of like kind can combine into four-chain molecules. Instead, they propose that the four-chain structures $\alpha_2 \beta_2$, $\alpha_2 \beta_2^X$, and $\alpha_2 \beta \beta^X$ are in equilibrium with the subunits $\alpha \beta$ and $\alpha \beta^X$ in the pH range 6 to 9; this is the pH range for separating the hybridization products. Thus, the four-chain structures are in an equilibrium which can be represented by

$$2 \alpha_2 \beta \beta^X \rightleftharpoons \alpha_2 \beta_2 + \alpha_2 \beta_2^X.$$

If such an equilibrium is attained more rapidly than the separation process can separate the forms, then any asymmetric hybrid such as $\alpha_2 \beta \beta^X$ will be converted into $\alpha_2 \beta_2$ and $\alpha_2 \beta_2^X$.

Strong support of these ideas came from thin film dialysis of dilute hemoglobin solutions (79) as well as from attempts to separate partially modified hemoglobins that resulted from incomplete reaction with iodoacetamide and N-ethylmaleimide (82). Accordingly, these authors have assigned equilibrium constants, K_1 and K_2, to the dissociation expression of VINOGRAD and HUTCHINSON thus:

$$\alpha_2 \beta_2 \overset{K_1}{\rightleftharpoons} 2\,\alpha\beta \overset{K_2}{\rightleftharpoons} 2\,\alpha + 2\,\beta.$$

Although the equilibrium constants have not been measured, an order of magnitude for each has been inferred from physical and chemical studies. The value of K_1 is small at neutral pH and is increased at high and low pH's. It also increases with salt concentration (153). The reaction rate of the dissociation into two-chain subunits is probably rapid (203a). Estimates of K_2 lack precision in large part because of the uncertainty in the value of K_1. The value of K_2 probably is very small at neutral pH but increases at extreme pH's. However, unlike K_1, K_2 becomes smaller with increasing salt concentration at all pH's. The rate of the dissociation into single chains is relatively rapid at least compared to the rate at which the chains dialyze through a semipermeable membrane (79).

The dissociation mechanism proposed by GUIDOTTI et al. (82) appears to fit well the experimental data on dissociation and subunit hybridization. In addition to presenting more data in support of symmetric dissociation, they stress the importance of considering hemoglobin as a system in mobile or "dynamic equilibrium" (82), as described above. The essential provision of their proposed mechanism is the conclusion that the rate at which the equilibrium between the different four-chain structures is reached is appreciably faster than the rate at which the products of hybridization and of asymmetric modification can be separated (82, 91a). Although the experimental evidence in support of this conclusion is indirect, the explanation is reasonable and probably valid. Nevertheless, it should be pointed out that the only method of separation which these authors used to study partially modified hemoglobins was column chromatography on IRC-50. The interaction between the hemoglobins and the resin which may occur during the separation procedure may influence the "dynamic equilibrium" by altering the dissociation constants, particularly K_1. Indeed, evidence of interaction between IRC-50 resin and hemoglobin has been obtained in studies of the chromatographic behavior of human hemoglobins (142). In these studies, hemoglobins were observed to chromatograph as double zones that may be explained by symmetrical dissociation during chromatography.

References, pp. 181—194.

At least two phenomena are unexplained and are in apparent conflict with the idea of a "dynamic equilibrium". The first is the observation that certain hemoglobins such as F_I (*215*) and A_{Ic} (*100*) are asymmetric: one of the two non-alpha chains of each is blocked at the N-terminus. The isolation of these hemoglobins by chromatography may be consistent with the dynamic equilibrium model either if the rate of dissociation into asymmetric half molecules $(\alpha_2 \gamma \gamma^{F_I} \to \alpha \gamma + \alpha \gamma^{F_I})$ is slow or if the molecular form represented by $\alpha_2 \gamma_2^{F_I}$ is unstable relative to the original asymmetric hemoglobin. Preliminary attempts at self-hybridization of hemoglobin A_{Ic} have been reported (*100*).

The second phenomenon, even more difficult to explain in terms of the proposed model, is the electrophoretic separation of components of a partially oxidized ($Fe^{II} \to Fe^{III}$) solution of hemoglobin (*133*). ITANO and ROBINSON (*133*) detected not only components which carried zero, two and four extra charges (relative to normal hemoglobin) but also those which carried either one or three extra charges. According to the "dynamic equilibrium" model, one would predict that these latter hemoglobins should rearrange during the separation procedure into symmetrical molecules which differ from one another by an even number of charges (zero, two, or four). The same possible explanation may be adduced as above: the rate of dissociation of partially oxidized hemoglobins is less than that for unoxidized hemoglobins, and the separation procedure is effective. However, if this explanation is correct, and if the model is applicable to the subunit recombination of the carbonmonoxyhemoglobin and ferrihemoglobin forms of hemoglobins A, S, and C as carried out by ITANO and SINGER (*135*), then hybrid molecules should have been detected. The absence of these hybrids, some of which can be formed by partial oxidation and detected by moving boundary electrophoresis (*133*), is unexplained by the idea of "dynamic equilibrium" alone. Again, one may ask if the conditions of chromatography may favor the rate at which equilibrium is attained. It is difficult to see how the procedure of moving boundary electrophoresis could effect the equilibrium rates. It would be interesting to know whether asymmetrically modified hemoglobins can be separated by the electrophoretic conditions which were used by ITANO and ROBINSON (*133*).

5. Other Factors which Influence Dissociation and Subunit Hybridization.

Other factors besides the pH, ionic strength and specific solutes have been found to effect dissociation and subunit hybridization. The state of oxygenation of hemoglobin which has already been shown to influence tertiary structure (Section VII, 7) appears to influence the extent of subunit dissociation in strong salt solutions (*31*). Ferrohemoglobin is less dissociated than oxyhemoglobin, at least at $5°$ in $2\,M$ sodium chloride. A similar difference in dissociation may also exist in acid pH. Ranney et al. (*193a*) have observed that hybridization of hemoglobins C and I at pH 4.7 is decreased upon deoxygenation. This decrease they attribute to be due to less dissociation.

Although the rate of dissociation into half-molecules may be rapid, the velocity of subunit hybridization is slow, as noted above. The rate of hybridization not only varies with pH, ionic strength and temperature but also with specific hemoglobins. Hybridization of hemoglobin F with different variants of adult hemoglobin and with hemoglobin H is slower than hybridization of hemoglobin A (*137, 143*). Hybridization of hemoglobin A_2 may be even slower than hemoglobin F (*108*). Huehns et al. (*105*) have investigated the reaction rates of hemoglobin α (prepared in vitro) with hemoglobins H (β_4), Bart's (γ_4), and δ_4 at neutral pH to form hemoglobins A, F and A_2. Although hemoglobins A and A_2 were formed rapidly at 37°, hemoglobin F formed at one-tenth the rate. These authors also found that the reaction to form hemoglobin A depended upon the duration of storage of the hemoglobin H which was used. The influence of aged samples on these and other hybridization studies is unknown but may be significant.

The relative affinities of complementary chains to form four-chain molecules is also a factor in hybridization reactions. Quantitative estimates of products from recombination of subunits of hemoglobins G and C (*11*) indicate differences in the stability of the various combinations of subunits relative to one another. Similar conclusions have been drawn from hybridization of hemoglobins A and γ_4 and hemoglobins F and β_4 under conditions of acid pH (*143*). The results were considered in terms of the following reaction:

$$2\,\alpha_2\gamma_2 + \beta_4 \overset{K}{\rightleftharpoons} 2\,\alpha_2\beta_2 + \gamma_4,$$

where K was concluded to be greater than one, and in the order of ten to twenty. The fact that K is not equal to one was interpreted to indicate unequal affinities of β and γ chains for α chains. The affinity of γ chains for one another relative to the affinity of β chains for one another should influence this equilibrium as well.

6. Comments on Dissociation and Hybridization.

On page 161 have been presented the various modes by which hemoglobin may possibly dissociate. These modes of dissociation have been given as reversible, and each dissociation has been shown as an equilibrium. Clearly, not all of these possible modes need exist. As we have seen, hybridization studies have been used to infer that one or more of these reversible steps do, in fact, occur in the dissociation of hemoglobin.

The first hybridizations of SINGER and ITANO (229) were interpreted to support an asymmetric dissociation

$$\alpha_2 \beta_2 \rightleftharpoons \alpha_2 + \beta_2.$$

If we consider the later studies of VINOGRAD and HUTCHINSON (243) and the "dynamic equilibrium" model of GUIDOTTI, KONIGSBERG and CRAIG (82), it is most reasonable to conclude that the principal dissociation of hemoglobin is symmetrical, thus

$$\alpha_2 \beta_2 \rightleftharpoons 2 \alpha \beta.$$

Although it is plausible to propose that the hybridization mechanism goes by way of further reversible dissociation into single chains,

$$\alpha \beta \rightleftharpoons \alpha + \beta$$

(let us term this the "single-chain mechanism"), other possible mechanisms have not been excluded. For example, assume that dissociation at low and high pH and in high salt concentrations, indeed, is predominantly into $\alpha \beta$ subunits but permit a small degree of dissociation into α_2 and β_2 subunits as follows:

$$\alpha_2 + \beta_2 \rightleftharpoons \alpha_2 \beta_2 \rightleftharpoons 2 \alpha \beta.$$

Such a scheme (the "partial asymmetric dissociation mechanism") in which small concentrations of α_2 and β_2 subunits are the active molecular species in hybridizations can explain the dissociation and hybridization data equally as well as the single-chain mechanism. This suggestion is a return to the ideas of SINGER and ITANO (229) but it explains the data of VINOGRAD and HUTCHINSON (243) in another way. The rapid rate of hybridizations with hemoglobin H, which is predominantly in the form of β_2 during hybridizations, certainly is consistent with this scheme, but, of course, hybridization by way of single chains is not ruled out. Likewise, the presence of single chains under conditions of acid dissociation as detected by thin film dialysis (79) does not exclude partial asymmetric dissociation as the principal mechanism of hybridization.

It is not unreasonable to postulate that the mode of dissociation of a complex, four-unit protein of unequal subunits may be different from the mode of reassociation of these subunits. Accordingly, another scheme that can be suggested for hybridization involves irreversible dissociation of $\alpha \beta$ subunits into single-chain subunits which re-associate first to α_2 and β_2 subunits and then to the four-chain hemoglobin molecule. This scheme may be called the "circular mechanism" and is illustrated by the expression

$$\alpha_2 \beta_2 \rightleftharpoons 2 \alpha \beta$$
$$\searrow \qquad\qquad \searrow$$
$$\alpha_2 + \beta_2 \leftarrow 2 \alpha + 2 \beta$$

Much evidence certainly points to a reversible dissociation into $\alpha\beta$ subunits. However, because a change in environment results upon dissociation, the secondary and tertiary structures of any double- or single-chain subunit may be different from the structural arrangements characteristic of these subunits in the four-chain molecule. If even small changes occur in the secondary and tertiary structure, then the sites which are responsible for binding the four-chain molecule together may be unfavorably disposed to take part in the reassociation of the separate subunits. Thus, it is possible that certain steps in the dissociation, especially into single chains, may not be reversible. Instead, re-association into the four-chain molecule from single chains may involve a different set of reaction steps which we suggest to be the formation of x_2 and β_2 subunits as intermediates in the formation of the whole molecule. Hence, we present the possibility that re-association of α and β chains occurs by going "around the circle".

In discussions of dissociation and re-association, emphasis is usually placed upon the interaction of the polypeptide chains, and the presence of the heme group in the molecule tends to be ignored. It may be fruitful to consider what role the heme groups do play in view of the fact that they do interact in the intact molecule and that globin itself has a molecular weight that is about half that of hemoglobin (*203*, *251a*).

As has been pointed out, hybridizations involving fetal hemoglobin appear to reach equilibrium more slowly than those involving hemoglobins that contain only normal and abnormal α and β chains. Thus, the hybridization of hemoglobins A and S appears to reach equilibrium with relative ease, although no extensive study of the rate of attaining equilibrium has been made (*137*, *146*). The slow rate of hybridization of hemoglobin F may, of course, be due to its less complete dissociation into the subunits which will exchange—presumably, α and γ chains (*104*) or, as indicated above, possibly α_2 and γ_2 subunits. However, another explanation of the slow rate would be that the configuration of the α chains which dissociate from hemoglobin F is different from that of the α chains which dissociate from hemoglobin A. Some data suggest that a greater affinity exists between α and β chains than between α and γ chains. This difference in affinity is not unreasonable in view of the considerable differences in amino acid sequences which exist between the β and γ chains. Therefore, is it not possible that in hemoglobin A or in its $\alpha\beta$ subunit the β chains may have imposed upon the associated α chains another configuration, however slightly different it may be, than the γ chains impose upon their associated α chains in hemoglobin F or its $\alpha\gamma$ subunit? Interconversion of these two hypothetical configurations of the α chain might take appreciably longer than the time required for dissociation and re-association. The possible reactions which would describe this scheme are shown below:

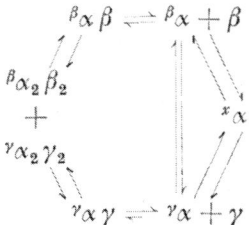

In these conversions the different configurations of the α chain are designated by β, γ and x superscripts to the left of the α. Configuration $^x\alpha$ may, conceivably, be equivalent either to $^\beta\alpha$ or $^\gamma\alpha$ or it may be some intermediate configuration. The reaction scheme has been illustrated above by the "single-chain mechanism" but equally valid schemes can be written for the "partial asymmetric dissociation" or "circular" mechanisms. The mechanism of interconversion of the different configurations of the α chain may involve interaction of the $^\gamma\alpha$ form with β chains and the $^\beta\alpha$ form with γ chains. Thus, the slow hybridization with hemoglobin F may be due either to a difference in the relative time required for transition of the $^\beta\alpha$ and $^\gamma\alpha$ to the $^x\alpha$ configuration, or to differences in the relative rates of reaction of $^\beta\alpha$ with β and γ chains versus the rates of reaction of $^\gamma\alpha$ with β and γ chains.

Although the "single-chain mechanism" of hybridization has gained the most acceptance, SINGER and ITANO's original suggestion of hybridization through asymmetric dissociation may be modified to explain the data equally well. In addition, we have proposed a third mechanism that requires re-association by a path that differs from that of dissociation. Unfortunately, none of these mechanisms explains convincingly why the main products of a hybridization take the form $\alpha_2\beta_2$ and $\alpha_2\beta_2^Z$, rather than $\alpha_2\beta\beta^X$, and we must resort to the idea of a "dynamic equilibrium" that places the onus upon the method of separation.

X. The Biosynthesis of Hemoglobin.

1. Introduction.

The rapid advances in our knowledge of protein biosynthesis that have occurred during the past ten years have resulted in part from investigations of the biological systems and chemical reactions that result in the synthesis of hemoglobin.

Hemoglobin biosynthesis may be divided into two parts: synthesis of the heme prosthetic group and synthesis of the polypeptide chains of globin. Although consideration will be given to both of these parts, emphasis will be placed upon the genetic and biochemical control of the synthesis of the polypeptide chains.

In humans and other mammals, heme and globin are synthesized intracellularly at certain stages in the development of erythroid cells before these cells become mature, circulating erythrocytes. Although most animals have nucleated erythrocytes in the circulation, the erythrocytes of mammals no longer contain nuclei. Hemoglobin synthesis does not take place in the unnucleated erythrocyte. The reticulocyte is an intermediate cell between the nucleated erythroid cell and the circulating erythrocyte. Although the reticulocyte lacks a nucleus, it contains sufficient remnants of the complete cell to carry out hemoglobin synthesis. Recently BORSOOK (37) has obtained evidence that nucleated erythroid cells may be the major site of hemoglobin synthesis. Nevertheless, the reticulocyte, particularly from rabbits and to a lesser extent from humans, has served as the principal biological system in which studies of hemoglobin synthesis have been made.

Although some in vivo studies of the synthesis of hemoglobin (particularly of heme synthesis) have been made, most of the studies of globin synthesis have been carried out in vitro either with intact reticulocytes or with cell-free systems that are derived partly or completely from reticulocytes. These reticulocytes are often obtained from rabbits after the animals have been made severely anemic either by repeated bleeding or by treatment with phenylhydrazine or similar agents. Thus, some caution must be exercised in interpreting much of the present data in terms of in vivo synthesis of hemoglobin in humans.

In the next Section we will direct attention to what is currently understood about the genetic and biochemical control of hemoglobin synthesis rather than to its course in various stages of the developing erythroid cell. For discussion of the latter topic the reader is referred to GRANICK and LEVERE (77).

2. Genetic Control of Hemoglobin Structure.

The current concept of protein biosynthesis states that the structure of proteins, at least their primary structure, is genetically determined (33). One of the first direct experimental illustrations that genes do indeed control the structure of proteins came when INGRAM (128) demonstrated a difference in the amino acid sequence of hemoglobins A and S. Chemical characterization of other abnormal human hemoglobins (see Chapter V; p. 126, and 130) supported this conclusion and led to the proposal that a linear correspondence exists between the DNA of the gene and the sequence of amino acids in the polypeptide which the gene determines. Although verification of this proposal has not been possible from genetic and chemical studies of hemoglobin alone, experimental support has come from studies of mutants of microorganisms [YANOFSKY (257)].

The genes which control the structure of the abnormal hemoglobins in most cases appear to follow simple Mendelian inheritance (180 and 18). Allelism* of the genes determining two abnormal hemoglobins was first postulated by RANNEY (193) from the inheritance of hemoglobins C and S in a given family. The fact that the sixth residue of the β chain is differently substituted in hemoglobins S and C [HUNT and INGRAM (121)] is confirmatory evidence that these two abnormal hemoglobins are controlled by allelic genes (see 130 for further discussion). This was the first example in which a comparison of protein sequence was used to conclude allelism. Similar chemical evidence has been employed to conclude allelism of all of the β-chain variants. Although genetic studies of allelism of α-chain variants have not yet been reported, chemical studies have been used in strong support of allelism.

Soon after the chemical abnormality in hemoglobin S (128) had been identified and two types of polypeptide chains had been detected in hemoglobin A (198), ITANO (132) postulated that separate genetic control of the synthesis of the two chains should result in the presence of four hemoglobins in an individual who was heterozygous* for both genetic factors (i. e., for both α and β chain genes). The first evidence of non-allelism of two abnormal hemoglobins was reported in 1958 by SMITH and TORBERT (232), who found independent segregation** of the genes controlling hemoglobins S and Hopkins-2. ITANO and ROBINSON (134) later demonstrated that the α chain of Hopkins-2 was abnormal and that four hemoglobins (A $= \alpha_2^A \beta_2^A$, Hopkins-2 $= \alpha_2^{Ho-2} \beta_2^A$, S $= \alpha_2^A \beta_2^S$ and Ho-2/S $= \alpha_2^{Ho-2} \beta_2^S$) were present in individuals who were doubly heterozygous for these genes. Chemical and genetic studies of other individuals with four hemoglobins (11, 20, 196) support the conclusion that two separate genetic loci control the structure of hemoglobin A. Thus, it has been concluded that the sequence of the α chain is under the control of the so-called "α-chain gene" which segregates independently from the "β-chain gene".

The chemical identity of the α chains of hemoglobins A, F and A_2 suggests a common genetic control. Genetic proof of this conclusion comes from instances where one of the major hemoglobins in adult life is an α chain variant. Thus, the presence of this abnormal α chain may be demonstrated as a variant of hemoglobin F of the infant or in the hemoglobin analogous to A_2 [reviewed by BAGLIONI (18)]. For example, MINNICH et al. (168) found a variant of hemoglobin F in cord blood

---

* For definitions of genetic terms, see p. 33.
** Independent segregation of two genes indicates that they occupy separate loci on different chromosomes.

of two infants; they postulated that the variant contained the abnormal
α^DSt. Louis chain because one parent of each child was heterozygous for
hemoglobin $D_{\alpha \text{ St. Louis}}$, an α-chain variant. Now, hemoglobins $D_{\alpha \text{ St. Louis}}$
and $G_{\text{Philadelphia}}$ are identical (Fig. 6, p. 134a) and it remained for WEATHERALL
and BAGLIONI (247) to demonstrate by chemical means the presence
of hemoglobin A, G, F and the variant of F in the affected individual.
Likewise, HUEHNS and SHOOTER (108) were able to demonstrate that
a previously detected variant (225), designated hemoglobin G_2, was
comprised of normal δ chains and abnormal $\alpha^{G\text{Ibadan}}$ chains.

Variation in the δ chain of hemoglobin A_2 such as occurs in A_2' (see
Section V, 7, p. 135) does not appear to alter the sequence of α or β chains
and vice versa. This chemical evidence suggests that the genetic factors
which control the sequence of the γ and δ chains are independent of the
genes which determine the structure of the α and β chains. Thus it is
postulated that at least four genes exercise the genetic control of the
sequences of human hemoglobins (see 18 and 131 for further discussion).

Limited genetic data suggest a linkage* of the loci for the β-chain
gene and δ-chain gene (18, 101). Some indirect evidence also links the
γ-chain gene with the genes for the β and δ chains (177a, 258).

3. Biosynthesis of the Polypeptide Chains of Globin.

The chemical components required for the synthesis of the polypeptide chains of globin are the same as those required for the synthesis
of any protein. They include several types of ribonucleic acids (RNA),
proteins, free amino acids, and an energy source (see 228 for review).
Synthesis takes place at ribosomes, which are particles composed of
ribosomal RNA and protein. Two other types of RNA, messenger RNA
(m-RNA) and transfer or soluble RNA (s-RNA), are also known to be
required for protein synthesis.

The genetic information which controls the sequence of amino acid
residues in proteins is believed to be contained in the sequence of
pyrimidine and purine bases of the deoxyribonucleic acid (DNA) that
is present in chromosomes of the nucleus. The DNA genetic information
is transcribed into RNA structure in the form of m-RNA which passes
from the nucleus to sites in the cytoplasm where protein synthesis takes
place. Under the influence of ribosomes with which the m-RNA associates,
the genetic information in the m-RNA directs the sequential synthesis
of the polypeptide chains. Translation of this genetic information from
DNA structure to protein structure results from the interaction of the
m-RNA-ribosome complex with specific aminoacyl-s-RNA complexes.

---

* The presence of two non-allelic genes on the same chromosome constitutes
"linkage".

Some of the important biochemical steps in protein biosynthesis are schematically represented in *Fig. 12* to outline the formation of the β^A chains. This figure will be the basis for our discussion of the biosynthesis of proteins in general and of hemoglobin specifically.

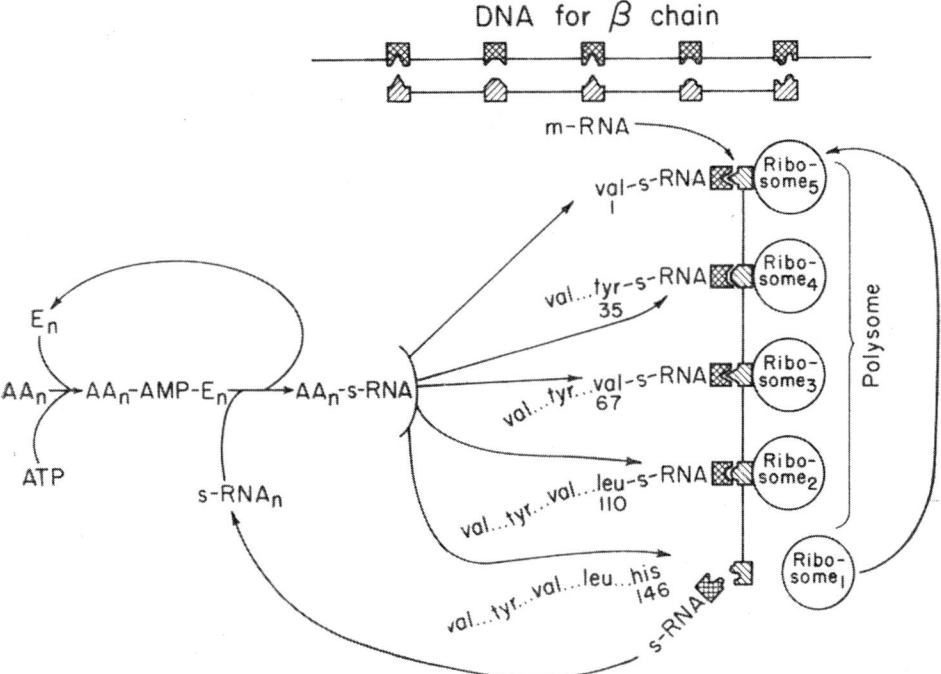

Fig. 12. Protein biosynthesis as exemplified by the formation of the β chain.

a. The Activation of Amino Acids for Protein Biosynthesis.

Amino acids enter into the biosynthesis of proteins through an activation and coupling with s-RNA. Thus, as shown at the left of Fig. 12 a given amino acid (here represented in general as AA_n) reacts with adenosine triphosphate (ATP) under the influence of its own specific activating enzyme E_n (an aminoacyl-RNA synthetase) to form an aminoacyl-AMP* intermediate. The chemical energy which is ultimately required for the formation of the peptide bonds of the polypeptide chain is derived from the ATP at this activation step. The aminoacyl-AMP intermediate is bound to the enzyme until the next reaction step. At this point, the aminoacyl-AMP-enzyme intermediates react with s-RNA to form aminoacyl-s-RNA's. A specific activating enzyme, of course, is used over again to activate another molecule of its specific amino acid (*228*).

---

* AMP represents adenosine monophosphate.

The s-RNA's are a group of low molecular weight RNA molecules which serve as specific carriers of amino acids in the synthesis of proteins. Each specific s-RNA has the potential to react with its own specific aminoacyl-AMP-enzyme intermediate to form a specific aminoacyl-s-RNA. This specificity between amino acid, activating enzyme, and s-RNA is indicated by the subscript "n" in Fig. 12, Although there is at least one specific s-RNA for each amino acid found in proteins, there are two or more specific and chemically different s-RNA's for at least some amino acids (33). The studies of EHRENSTEIN et al. (63) lead to the conclusion that the structure of the nucleic acid portion of the amino-acyl-s-RNA rather than the structure of the amino acid which it carries determines where the given amino acid is finally incorporated into the peptide chain. Thus, they converted the cysteine of cysteine-s-RNA to alanine by reduction with Raney nickel. The resultant alanine-s-RNA, however, was directed into the position of cysteine in the α chain of rabbit hemoglobin and not into positions that are occupied by alanine.

b. DNA and the Role of m-RNA.

The sequence of bases in the ribosomal RNA was once believed to be the genetic information which determined the sequence of amino acids in the biosynthesis of the polypeptide chain (165, 228), and each type of protein was thought to be determined by a specific type of ribosome. This idea no longer appears valid because of the detection and partial characterization of m-RNA (228). The m-RNA from bacterial systems has been studied most extensively. Chemical studies indicate that its base composition is similar to that of DNA. It is synthesized in the nucleus and at a faster rate than either ribosomal RNA or s-RNA (10, 228).

Although the presence and function of m-RNA in animal cells has, in large part, been postulated from the studies of microorganisms, ARNSTEIN, COX and HUNT (10) have recently obtained evidence for a high molecular weight RNA in rabbit reticulocytes which appears to function as m-RNA. Indirect evidence for m-RNA in animal cells also is found in studies of the aggregations of microsomes into polyribosome structures (165).

Presumably the genetic information of the β chain gene of the cell nucleus is transcribed from the DNA structure into the structure of an m-RNA under the influence of an RNA polymerase enzyme (254). This is shown schematically at the top of Fig. 12 (p. 173) where portions of the genetic information are shown as geometric figures in the representation of DNA. Complementary figures illustrate the transcription of the genetic information into the m-RNA structure. These symbols represent the bits of genetic information or "code" units which correspond to specific

References, pp. 181—194.

amino acid residues in the protein. An m-RNA for the α chain is produced in a similar way. In the case of microorganisms, several adjacent or closely linked genes may be transcribed into one molecule of m-RNA (3); in like manner, the m-RNA which carries the β chain information may also, in another part of the messenger molecule, carry information for the structure of the δ and possibly of the γ chain.

c. The Function of the Ribosomes.

Several groups (74, 164, 165, 245) have found that the ribosomes of reticulocytes are aggregated into clusters of 2 to 6 ribosomes. Most frequently, these clusters which are termed polyribosomes or polysomes contain 4 or 5 ribosomes (165). Active hemoglobin synthesis takes place at these polysomes rather than at single ribosomes (74, 165, 245). There is good chemical (165) and morphological (199) evidence that the ribosomes are held together by m-RNA to form the polysome. RICH et al. (199) have shown a relative movement between the m-RNA and the ribosomes in the polysomes from reticulocytes. Each ribosome appears to be making a polypeptide chain as it moves along the m-RNA. Thus, several polypeptide chains, one for each ribosome, presumably are synthesized simultaneously in the polysome aggregate from different portions of the m-RNA (246a). A functional representation of a polysome is illustrated to the right of Fig. 12.

DINTZIS (61) and NAUGHTON and DINTZIS (178) have demonstrated by means of labeling with radioactive amino acids that the amino acid residues of rabbit hemoglobin are incorporated into the polypeptide consecutively. Fig. 13 and its legend from DINTZIS (61) present schematically the reasoning behind and the course of the experiments. Thus, as samples are taken at times t_2, t_3, and t_4, it would be anticipated that the chains would become more and more radioactive throughout their length. Accordingly, the tryptic peptides (represented by the symbols a to g) near one end of the chain would become radioactive first, and this radioactivity would move toward the other end of the chain as the time of reaction is prolonged. That the incorporation occurred in this way was studied by isolation of the hemoglobin, separation of the chains, preparation of a tryptic hydrolysate, and the isolation of the individual tryptic peptides. By a correlation of the specific radioactivity of these peptides with their position in the polypeptide chain, it could be deduced that the C-terminal sequences are the *last* to be synthesized before release of the complete chain from the ribosome. Thus, synthesis is initiated at the N-terminal end of the chains and progresses sequentially towards the C-terminal end. The studies of DINTZIS (61, 178) also indicate that the α and β chains are synthesized independently of one another.

Thus, Rich et al. (*199*) have concluded from the study of polysomes and from Dintzis' experiments that the biosynthetic process for proteins is associated with the movement of ribosomes along the m-RNA molecule in such a way that the genetic information in the m-RNA is translated sequentially into the polypeptide chain.

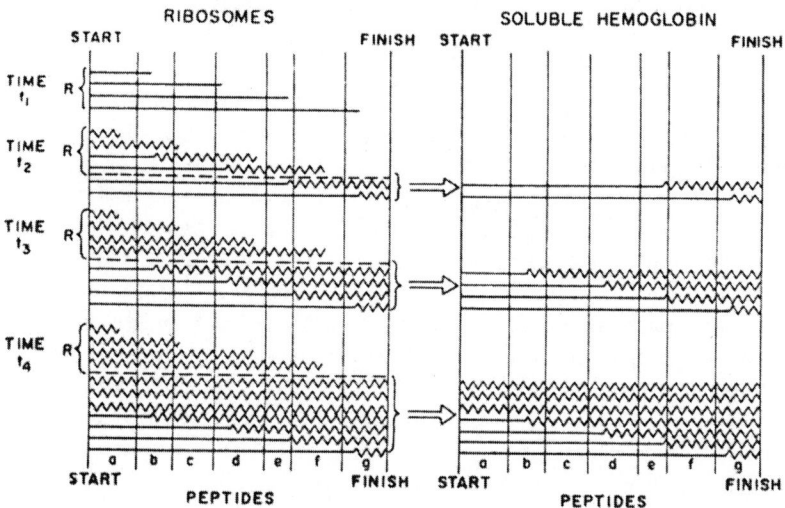

Fig. 13. Model of sequential chain growth. The straight lines represent unlabeled peptide chain. The zigzag lines represent radioactively labeled peptide chain formed after the addition of radioactive amino acid at time t_1. The groups of peptides labeled R are those unfinished bits attached to the ribosome at each time; the rest, having reached the finish line, are assumed to be present in the soluble hemoglobin. In the ribosome at time t_2, the top two completely zigzag lines represent peptide chains formed completely from amino acids during the time interval between t_1 and t_2. The middle two lines represent chains which have grown during that time interval, but have not reached the finish line and are therefore still attached to the ribosomes. The bottom two chains represent those which have crossed the finish line, left the ribosomes and are to be found mixed with other molecules of soluble hemoglobin. The symbols a-g indicate the tryptic peptides to be expected. [From: Dintzis, Proc. Nat. Acad. Sci. (USA) **47**, 247 (1961).]

d. The Assembling of the Polypeptide Chain.

The final step in translating genetic information from nucleic acid structure (m-RNA) into the primary structure (p. 147) of hemoglobin chains apparently takes place as a result of specific interaction between the bases of a "code" unit in the m-RNA and the homologous bases of specific aminoacyl-s-RNA's. The exact mechanism of this very precise interaction and role of the ribosome are not yet completely understood. The process has been schematically represented in Fig. 12 (p. 173). There, the placing of the N-terminal valyl residue is shown at the moment that the synthesis of a β chain begins at the site of ribosome No. 5. The interaction between an adapter portion of the valyl-s-RNA with the region of the m-RNA, which codes for the N-terminal valyl residue, is visualized as a pairing of the geometric figures. Ribosome No. 4 had

References, pp. 181—194.

begun the synthesis earlier and is shown as a tyrosyl residue is placed in the 35th position of the β chain. Similarly, valyl and leucyl residues are being introduced at positions 67 and 110, respectively. Although the code unit for the valyl residue at position 67 is represented by the same geometric figure as that for the first valyl residue, the codes for these two valyl residues may be different. There is evidence that some amino acids may be represented by two or more codes, each requiring a different specific s-RNA for translation (33).

A completed β chain and a ribosome (No. 1) are shown as they detach from the m-RNA just after the hystidyl, the last amino acid residue, is placed in the chain*. The ribosomes, at least in reticulocytes, appear to "recirculate" by attaching to another molecule of m-RNA (199). PHILIPPS (189a) suggests another mechanism.

The exact chemistry which is involved in the transfer of the amino-acyl group from the s-RNA to the growing polypeptide chain with the subsequent formation of the peptide bond is not as precisely known as that of other reactions in protein biosynthesis.

The chemical constituents which are required in addition to s-RNA's, m-RNA, and ribosomes include guanosine triphosphate (GTP), reduced glutathione (GSH), and one or more soluble enzymes. ARLINGHAUS et al. (8) propose that the reaction has two stages of which the first, that is, the binding of the aminoacyl-s-RNA to the ribosome, is stimulated by GTP and is catalyzed by a "binding" enzyme. The second step is the formation of the peptide bond for which a "peptide synthetase" is proposed. The peptide synthetase apparently does not require an energy source other than that already incorporated in the structure of amino-acyl-s-RNA. In the in vitro studies of the system, GSH stimulates the step at which the peptides are formed. Independent observations by WARNER and RICH (246) and ARLINGHAUS et al. (8) suggest that there are two binding sites for aminoacyl-s-RNA on each ribosome of polysomes from reticulocytes.

e. The Final Release of the Protein.

MORRIS (170) has concluded that the final step in the biosynthesis of hemoglobin involves a specific release reaction that is independent of the process of forming the polypeptide chains. GTP again appears to be required for this release. Chemical characterization of the soluble products of the protein-release reaction indicates that they are α and β chains without heme groups and thus not yet in the form of finished hemoglobin. Because of the requirement of GTP and the enhancement

---

* The representation of Fig. 12 is schematic and is not meant to imply that only five ribosomes may be associated with m-RNA or that they are equally spaced along the m-RNA.

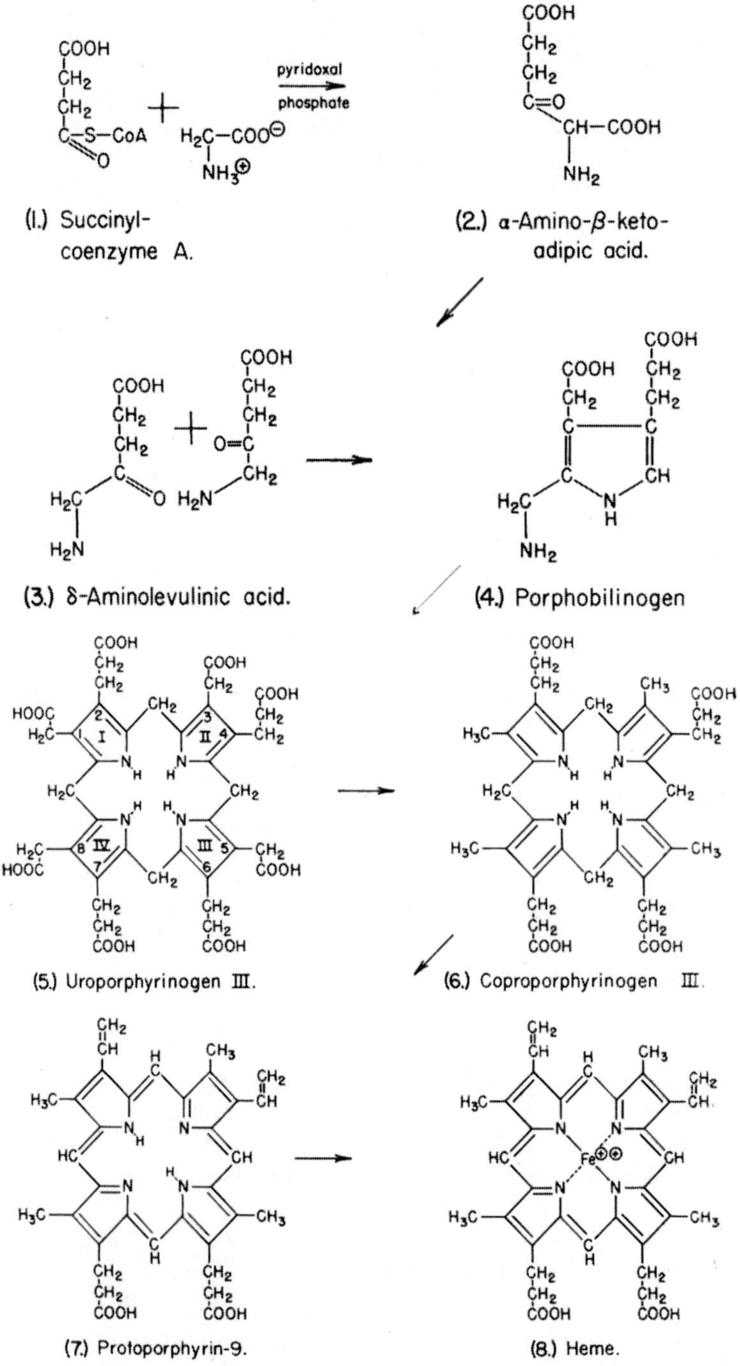
pyridoxal
phosphate
(1.) Succinyl-coenzyme A.
(2.) α-Amino-β-keto-adipic acid.
(3.) δ-Aminolevulinic acid.
(4.) Porphobilinogen
(5.) Uroporphyrinogen III.
(6.) Coproporphyrinogen III.
(7.) Protoporphyrin-9.
(8.) Heme.

Scheme 10. Steps in the biosynthesis of heme from succinyl-coenzyme A and glycine.

References, pp. 181—194.

by GSH, the protein-release reaction described by Morris (170) may be related to the final step of synthesis that is proposed by Arlinghaus et al. (8) and has been described above.

4. Biosynthesis of Heme.

Heme, the prosthetic group of hemoglobin, is synthesized independently of but in parallel with globin in erythroid cells. Most of the reaction steps in the biosynthesis of heme have been delineated. The reader is referred to other reviews (77, 148, 242, 253) for more complete details of these reactions and for references to the original literature. A summary of the steps in the biosynthesis of heme is illustrated in *Scheme 10*. The first step is the condensation of glycine with succinyl-coenzyme A to form α-amino-β-ketoadipic acid which is immediately decarboxylated to δ-aminolevulinic acid. Pyridoxal phosphate is required for the activation of the condensation reaction. Two molecules of δ-aminolevulinic acid then undergo an enzymatically controlled condensation to form porphobilinogen. Under the influence of two enzymes, a deaminase and an isomerase, by a mechanism which is not yet completely understood, four molecules of porphobilinogen combine to form a molecule of uroporphyrinogen*. The four acetic acid side-groups of uroporphyrinogen are next enzymatically decarboxylated to form the methyl groups of coproporphyrinogen III which on dehydrogenation and decarboxylation yields protoporphyrin-9. The last step in the biosynthesis is the enzymatic introduction of one atom of iron into protoporphyrin to form heme.

5. The Attachment of Heme to Globin Chains.

Although it is not known yet at what stage the heme is attached to the globin (110, 250), the attachment may be enzymatically mediated (222). A heme synthetase enzyme [Schwartz et al. (222)] which has been partially purified from chicken erythrocytes and may be detected in human erythrocytes will catalyze the formation of hemoglobin from iron, protoporphyrin and globin by an unknown mechanism. Because free $\alpha\beta$ subunits have been reported to be present in erythrocytes (251) and because they will rapidly combine with heme to form a hemoglobin chemically identical with normal hemoglobin A (250), it has been proposed that the α and β chains are synthesized and released from the ribosomes before the heme group is added. On the other hand, Granick and Levere (77) suggest that the addition of heme to the globin chain on

---

* The uncertain feature of this step is the mechanism by which the arrangement of side-chains is achieved in uroporphyrinogen III. At positions 1 to 6 of uroporphyrinogen (Scheme 10), the acetic acid and propionic acid side-chains alternate, but at positions 7 and 8, the relative positions are reversed. The isomerase is invoked to explain the final product.

the ribosome may bring about the appropriate folding of the chain which stabilizes it and allows it to leave the ribosome. It may be of interest to note that about two-thirds of the chain would have to be synthesized before the attachment would be complete.

These contradictory views bring up another problem in protein and hemoglobin biosynthesis: when, where, and how does the protein achieve its secondary, tertiary and quaternary structures?

We have been stressing the way in which the amino acid residues are put together biosynthetically but we have also pointed out in other comments that the sequence as yet tells us little about the spatial arrangement of the molecule. There is relatively little information from any source for any protein about the manner in which secondary and tertiary structure is achieved. It is uncertain whether the final spatial arrangement requires genetic control as such or is under genetic control only through the arrangement of amino acid residues in the chain. Some of the most definite information has come from a study of ribonuclease. Thus, Epstein et al. (4, 64) have reduced the four disulfide bridges of ribonuclease and then reoxidized them to recover virtually complete enzymatic activity. Presumably, the entire structure has been re-established through the information in the sequence. However, ribonuclease has disulfide bonds and a small percentage of helical structure whereas hemoglobin has no disulfide bonds and a high percentage of helical structure. It remains to be seen to what extent such differences bear upon the manner in which the three-dimensional structure is achieved.

6. Assembly of Chains into the Whole Hemoglobin Molecule.

Because the point at which heme and a given chain are assembled is unknown, the point and mechanism of assembly into the four-chain molecule is also obscure at present. However, several features about the in vivo association of the chains into molecules with two and four chains may bear on these points.

Because the α chains of hemoglobins A, F and A_2 are chemically and genetically identical, it is reasonable to conclude that these α chains are derived from a common source or pool. Furthermore, because the abnormal tetramer hemoglobins, β_4, γ_4 and δ_4 can exist, the β, γ and δ chains presumably can be synthesized and released independently of the α chain. The absence of a hemoglobin of α chains alone, except perhaps as formed as a result of unstable hemoglobins abnormal in β chains, has led to the postulate that the release of these chains from ribosomes is dependent upon the availability of β, γ, or δ chains. Huehns and Schooter (109) have proposed that the α chain is removed from its synthetic site on the ribosome by direct interaction with one of its complementary chains to form $\alpha\beta$, $\alpha\gamma$, or $\alpha\delta$. The subunit $\alpha\beta$ will combine spontaneously with heme to form a hemoglobin apparently indistinguishable from hemoglobin A (252). Two points are consistent with this idea: (1) a hemoglobin of α chains alone may never be present except through the deterioration of unstable hemoglobin abnormal in the β chain (106), and (2) small amounts of $\alpha\beta$ globin may be detected in erythrocytes (252). Perhaps this $\alpha\beta$ globin is identical with the non-heme protein which KARPATKIN (149) observed to become labeled rapidly in studies of the incorporation of amino acids into erythrocyte proteins.

At the time of the first dissociation and subunit recombination studies, it was proposed that the aggregation of α_2 and β_2 subunits was the final step in hemoglobin biosynthesis. This idea was developed to explain the absence of hemoglobins of the type $\alpha_2\beta^A\beta^S$ (136). The specificity of identical chains for each other to form α_2 or β_2 subunits has been explained in two ways: (1) either both chains are somehow synthesized together under the influence of a single gene so that the dissimilar subunit $\beta^A\beta^S$ is not formed, or (2) the attraction of identical chains to form subunits is much greater than that between chains that differ by as little as one amino acid (250). A more plausible explanation may be that a mixed molecule such as $\alpha_2\beta^A\beta^S$ does exist but is converted by the separatory procedures into the other molecules $\alpha_2\beta_2^A$ and $\alpha_2\beta_2^S$ with which it presumably is in equilibrium (82).

In conclusion, hemoglobin has played an important role in the study of protein biosynthesis and has therefore been the source of much of the information that has been presented in the preceding paragraphs. It is not too much to suggest that it will continue in this role and that it will aid in illuminating those paths that still are in the dark.

References.

1. ADAIR, G. S.: A Theory of Partial Osmotic Pressures and Membrane Equilibria with Special Reference to the Application of Dalton's Law to Haemoglobin Solutions in the Presence of Salts. Proc. Roy. Soc. (London) A 120, 573 (1928).
2. ALLEN, D. W., W. A. SCHROEDER and J. BALOG: Observations on the Chromatographic Heterogeneity of Normal Adult and Fetal Human Hemoglobin: A Study of the Effects of Crystallization and Chromatography on the Heterogeneity and Isoleucine Content. J. Amer. Chem. Soc. 80, 1628 (1958).
3. AMES, B. N. and R. G. MARTIN: Biochemical Aspects of Genetics: The Operon. Annu. Rev. Biochem. 33, 235 (1964).
4. ANFINSEN, C. B.: Macromolecular Considerations of Cellular Organization. Canadian Cancer Conference, Vol. 5, p. 3. New York: Academic Press. 1963.
5. Anonymous: Nomenclature of Abnormal Hemoglobins. Blood 17, 125 (1961).
6. ANSON, M. L. and A. E. MIRSKY: Protein Coagulation and its Reversal. The Preparation of Insoluble Globin, Soluble Globin, and Heme. J. Gen. Physiol. 13, 469 (1929/30).

7. Antonini, E., J. Wyman, R. Zito, A. Rossi-Fanelli and A. Caputo: Studies of Carboxypeptidase Digests of Human Hemoglobin. J. Biol. Chem. **236**, PC 60 (1961).

8. Arlinghaus, R., J. Shaeffer and R. Schweet: Mechanism of Peptide Bond Formation in Polypeptide Synthesis. Proc. Nat. Acad. Sci. (USA) **51**, 1291 (1964).

9. Armstrong, D. H., W. A. Schroeder and W. D. Fenninger: A Comparison of the Percentage of Fetal Hemoglobin in Human Umbilical Cord Blood as Determined by Chromatography and by Alkali Denaturation. Blood **22**, 554 (1963).

10. Arnstein, H. R. V., R. A. Cox and J. A. Hunt: The Function of High Molecular-Weight Ribonucleic Acid from Rabbit Reticulocytes in Haemoglobin Biosynthesis. Biochem. J. **92**, 648 (1964).

11. Atwater, J., I. R. Schwartz and L. M. Tocantins: A Variety of Human Hemoglobin with Four Distinct Electrophoretic Components. Blood **15**, 901 (1960).

12. Babin, D. R., R. T. Jones and W. A. Schroeder: Hemoglobin $D_{Los Angeles}$: $\alpha_2^A \beta_2^{121 gluNH_2}$. Biochim. Biophys. Acta **86**, 136 (1964).

13. Babin, D. R., W. A. Schroeder, J. R. Shelton, J. B. Shelton and B. Robberson: Unpublished results.

14. Baglioni, C.: An Improved Method for the Fingerprinting of Human Hemoglobin. Biochim. Biophys. Acta **48**, 392 (1961).

15. — Abnormal Human Haemoglobins. VIII. Chemical Studies on Haemoglobin D. Biochim. Biophys. Acta **59**, 437 (1962).

16. — A Chemical Study of Hemoglobin Norfolk. J. Biol. Chem. **237**, 69 (1962).

17. — The Fusion of Two Peptide Chains in Hemoglobin Lepore and its Interpretation as a Genetic Deletion. Proc. Nat. Acad. Sci. (USA) **48**, 1880 (1962). See also Biochim. Biophys. Acta **97**, 37 (1965).

18. — Correlations between Genetics and Chemistry of Human Hemoglobins. In: J. H. Taylor (Edit.), Molecular Genetics, Part 1, p. 405. New York: Academic Press. 1963.

19. Baglioni, C. and V. M. Ingram: Abnormal Human Haemoglobins. V. Chemical Investigation of Haemoglobins A, G, C, X from one Individual. Biochim. Biophys. Acta **48**, 253 (1961).

20. — — Four Adult Haemoglobin Types in One Person. Nature **189**, 465 (1961).

21. Baglioni, C. and H. Lehmann: Chemical Heterogeneity of Haemoglobin O. Nature **196**, 229 (1962).

22. Baglioni, C. and D. J. Weatherall: Abnormal Human Haemoglobins. IX. Chemistry of Hemoglobin $J_{Baltimore}$. Biochim. Biophys. Acta **78**, 637 (1963).

23. Bangham, A. D.: Distribution of Electrophoretically Different Haemoglobins among Cattle Breeds of Great Britain. Nature **179**, 467 (1957).

24. Bangham, A. D. and H. Lehmann: "Multiple" Haemoglobins in the Horse. Nature **181**, 267 (1958).

25. Bannerman, R. M.: Thalassemia — A Survey of Some Aspects. New York: Grune and Stratton. 1961.

26. Barnabas, J. and C. J. Muller: Haemoglobin $Lepore_{Hollandia}$. Nature **194**, 931 (1962).

27. Bayrakci, C., A. Josephson, L. Singer, P. Heller and R. D. Coleman: A New Fast Hemoglobin. Xth Congr. Int. Soc. Haematology, Stockholm, 1964, Abstract L: 6.

28. Beaven, G. H. and W. B. Gratzer: A Critical Review of Human Haemoglobin Variants. J. Clin. Pathol. **12**, 1, 101 (1959).

29. BENESCH, R. and R. E. BENESCH: Some Relations between Structure and Function in Hemoglobin. J. Mol. Biol. 6, 498 (1963).
30. BENESCH, R. E. and R. BENESCH: The Influence of Oxygenation on the Reactivity of the —SH Groups of Hemoglobin. Biochemistry 1, 735 (1962).
31. BENESCH, R. E., R. BENESCH and G. MACDUFF: The Dissociation of Hemoglobins A and H in Concentrated Sodium Chloride. Biochemistry 3, 1132 (1964).
32. BENESCH, R. E., H. M. RANNEY, R. BENESCH and G. M. SMITH: The Chemistry of the Bohr Effect. II. Some Properties of Hemoglobin H. J. Biol. Chem. 236, 2926 (1961).
33. BENNETT, J. C. and W. J. DREYER: Genetic Coding for Protein Structure. Annu. Rev. Biochem. 33, 205 (1964).
34. BENZER, S., V. M. INGRAM and H. LEHMANN: Three Varieties of Human Haemoglobin D. Nature 182, 852 (1958).
35. BIANCO, I., G. MODIANO, E. BOTTINI and R. LUCCI: Alteration in the α-Chain of Haemoglobin L$_{Ferrara}$. Nature 198, 395 (1963).
36. BLOW, D. M.: An X-ray Examination of Some Crystal forms of Pig and Rabbit Haemoglobin. Acta Crystallogr. 11, 125 (1958).
37. BORSOOK, H.: DNA, RNA and Protein Synthesis after Acute Severe Blood Loss: a Picture of Erythropoiesis at the Combined Morphological and Molecular Level. Ann. New York Acad. Sci. 199, 523 (1964).
38. BOWMAN, B. H., C. P. OLIVER, D. R. BARNETT, J. E. CUNNINGHAM and R. G. SCHNEIDER: Chemical Characterization of Three Hemoglobins G. Blood 23, 193 (1964).
39. BRAGG, W. L. and M. F. PERUTZ: The External Form of the Haemoglobin Molecule. II. Acta Crystallogr. 5, 323 (1952).
40. BRAUNITZER, G.: Vergleichende Untersuchungen zur Primärstruktur der Proteinkomponente einiger Hämoglobine. Z. physiol. Chem. 312, 72 (1958).
41. BRAUNITZER, G., R. GEHRING-MÜLLER, N. HILSCHMANN, K. HILSE, G. HOBOM, V. RUDLOFF und B. WITTMANN-LIEBOLD: Die Konstitution des normalen adulten Humanhämoglobins. Z. physiol. Chem. 325, 283 (1961).
42. BRAUNITZER, G., N. HILSCHMANN, V. RUDLOFF, K. HILSE, B. LIEBOLD and R. MÜLLER: The Haemoglobin Particles. Chemical and Genetic Aspects of their Structure. Nature 190, 480 (1961).
43. BRAUNITZER, G., K. HILSE, V. RUDLOFF and N. HILSCHMANN: The Hemoglobins. Adv. Protein Chem. 19, 1 (1964).
44. BRAUNITZER, G., V. RUDLOFF und N. HILSCHMANN: Hämoglobine. X. Die Analyse der α- und β-Ketten des adulten normalen Humanhämoglobins aus seinen tryptischen Spaltprodukten. Z. physiol. Chem. 331, 1 (1963).
44a. BUCCI, E. and C. FRONTICELLI: A New Method for the Preparation of α and β Subunits of Human Hemoglobin. J. Biol. Chem. 240, PC 551 (1965).
44b. BUETTNER-JANUSCH, J. and R. L. HILL: Molecules and Monkeys. Science 147, 836 (1965).
45. BUHLER, D. R.: Studies on Fish Hemoglobins. Chinook Salmon and Rainbow Trout. J. Biol. Chem. 238, 1665 (1963).
46. CABANNES, R. et CH. SERAIN: Étude électrophorétique des hémoglobines des Mammifères domestiques d'Algérie. C. R. séances soc. biol. 149, 1193 (1955).
47. CEPPELLINI, R.: Ciba Foundation Symp., Biochemistry of Human Genetics, p. 133. Boston: Little, Brown & Co. 1959.
48. CHERNOFF, A. I.: Hemoglobin A₄, a Naturally Occurring Hemoglobin Possessing only α Chains. Xth Congr. Int. Soc. Haematology, Stockholm, 1964, Abstract L: 3.

49. Chernoff, A. I. and J. C. Liu: The Amino Acid Composition of Hemoglobin. II. Analytical Technics. Blood 17, 54 (1961).
50. Chernoff, A. I. and P. E. Perillie: The Amino Acid Composition of HGB New Haven 2 (HGB N$_{New\ Haven}$). Biochem. Biophys. Res. Comm. 16, 368 (1964).
50 a. Chernoff, A. I. and N. M. Pettit, Jr.: The Amino Acid Composition of Hemoglobin. III. A Qualitative Method for Identifying Abnormalities of the Polypeptide Chains of Hemoglobin. Blood 24, 750 (1964).
51. Clegg, M. D. and W. A. Schroeder: A Chromatographic Study of the Minor Components of Normal Adult Human Hemoglobin Including a Comparison of Hemoglobin from Normal and Phenylketonuric Individuals. J. Amer. Chem. Soc. 81, 6065 (1959).
52. Conference on Hemoglobin. Nat. Acad. Sci. (USA), Nat. Research Council, Publ. No. 557, 1958.
53. Crestfield, A. M., W. H. Stein and S. Moore: Alkylation and Identification of the Histidine Residues at the Active Site of Ribonuclease. J. Biol. Chem. 238, 2413 (1963).
54. — — — Properties and Conformation of the Histidine Residues at the Active Site of Ribonuclease. J. Biol. Chem. 238, 2421 (1963).
55. Cullis, A. F., H. Muirhead, M. F. Perutz, M. G. Rossmann and A. C. T. North: The Structure of Haemoglobin. IX. A Three-dimensional Fourier Synthesis at 5.5 Å Resolution: Description of the Structure. Proc. Roy. Soc. (London) A 265, 161 (1962).
56. Curtain, C. C.: A Structural Study of Abnormal Haemoglobins Occurring in New Guinea. Austral. J. Exp. Biol. Med. Sci. 42, 89 (1964).
57. Dance, N., E. R. Huehns and G. H. Beaven: The Abnormal Haemoglobins in Haemoglobin-H Disease. Biochemic. J. 87, 240 (1963).
58. Dance, N., E. R. Huehns and E. M. Shooter: The Chemical Investigation of Haemoglobins G$_{Bristol}$ and G$_{Bristol/C}$. Biochim. Biophys. Acta 86, 144 (1964).
59. — — — Personal communication.
60. Diamond, J. M. and G. Braunitzer: α-Chain of Rabbit Haemoglobin. Nature 194, 1287 (1962).
61. Dintzis, H. M.: Assembly of the Peptide Chains of Hemoglobin. Proc. Nat. Acad. Sci. (USA) 47, 247 (1961).
62. Edmundson, A. B.: Amino-acid Sequence of Sperm Whale Myoglobin. Nature 205, 883 (1965).
63. Ehrenstein, G. v., B. Weisblum and S. Benzer: The Function of sRNA as Amino Acid Adaptor in the Synthesis of Hemoglobin. Proc. Nat. Acad. Sci. (USA) 49, 669 (1963).
64. Epstein, C. J., R. F. Goldberger, D. M. Young and C. B. Anfinsen: A Study of the Factors Influencing the Rate and Extent of Enzymic Reactivation during Reoxidation of Reduced Ribonuclease. Arch. Biochem. Biophys. Suppl. 1, 223 (1962).
65. Evans, J. V., J. W. B. King, B. L. Cohen, H. Harris and F. L. Warren: Genetics of Haemoglobin and Blood Potassium Differences in Sheep. Nature 178, 849 (1956).
66. Fessas, Ph. and D. Loukopoulos: Alpha-Chain of Human Hemoglobin: Occurrence in vivo. Science 143, 590 (1964).
67. Fessas, Ph., G. Stamatoyannopoulos and A. Karaklis: Hemoglobin "Pylos": Study of a Hemoglobinopathy Resembling Thalassemia in the Heterozygous, Homozygous and Double Heterozygous State. Blood 19, 1 (1962).
68. Field, E. O. and J. R. P. O'Brien: Dissociation of Human Haemoglobin at Low pH. Biochem. J. 60, 656 (1955).

69. GAMMACK, D. B., E. R. HUEHNS, H. LEHMANN and E. M. SHOOTER: The Abnormal Polypeptide Chains in a Number of Haemoglobin Variants. Acta Genet. Statist. Med. 11, 1 (1961).

70. GAMMACK, D. B., E. R. HUEHNS, E. M. SHOOTER and P. S. GERALD: Identification of the Abnormal Polypeptide Chain of Haemoglobin G_{Ib}. J. Mol. Biol. 2, 372 (1960).

71. GERALD, P. S.: Starch Electrophoresis of Hemoglobin: Findings in Thalassemia Syndromes. In: Conference on Hemoglobin, p. 212. Washington, D. C.: Nat. Acad. Sci., Nat. Research Council, Publication No. 557, 1958.

72. GERALD, P. S. and M. L. EFRON: Chemical Studies of Several Varieties of Hb M. Proc. Nat. Acad. Sci. (USA) 47, 1758 (1961).

73. GERALD, P. S. and V. M. INGRAM: Recommendations for the Nomenclature of Hemoglobins. J. Biol. Chem. 236, 2155 (1961).

74. GIERER, A.: Function of Aggregated Reticulocyte Ribosomes in Protein Synthesis. J. Mol. Biol. 6, 148 (1963).

75. GOLDSTEIN, J., G. GUIDOTTI, W. KONIGSBERG and R. J. HILL: The Amino Acid Sequence around the "Reactive Sulfhydryl" Group of the β Chain from Human Hemoglobin. J. Biol. Chem. 236, PC 77 (1961).

76. GOTTLIEB, A. J., A. RESTREPO and H. A. ITANO: HbJ$_{Medellin}$: Chemical and Genetic Study. Federat. Proc. (Amer. Soc. Exp. Biol.) 23, 172 (1964).

77. GRANICK, S. and R. D. LEVERE: Heme Synthesis in Erythroid Cells. Progr. Hematology 4, 1 (1964).

78. GRATZER, W. B. and A. C. ALLISON: Multiple Haemoglobins. Biol. Rev. 35, 459 (1960).

79. GUIDOTTI, G. and L. C. CRAIG: Dialysis Studies. VIII. The Behavior of Solutes which Associate. Proc. Nat. Acad. Sci. (USA) 50, 46 (1963).

80. GUIDOTTI, G. and W. KONIGSBERG: The Characterization of Modified Human Hemoglobin. I. Reaction with Iodoacetamide and N-Ethylmaleimide. J. Biol. Chem. 239, 1474 (1964).

81. — — Personal communication.

82. GUIDOTTI, G., W. KONIGSBERG and L. C. CRAIG: On the Dissociation of Normal Adult Human Hemoglobin. Proc. Nat. Acad. Sci. (USA) 50, 774 (1963).

83. GUTTER, F. J., H. A. SOBER and E. A. PETERSON: The Effect of Mercaptoethanol and Urea on the Molecular Weight of Hemoglobin. Arch. Biochem. Biophys. 62, 427 (1956).

84. HANADA, M. and D. L. RUCKNAGEL: The Characterization of Hemoglobin Shimonoseki. Blood 24, 624 (1964).

85. HANADA, M., D. L. RUCKNAGEL and M. M. COHEN: Hemoglobin Abnormalities Characterized by Amino Acid Substitutions in the Sixth Tryptic Peptide of the α-Chain. Abstracts, Amer. Soc. Human Genetics. New York, July 1963.

86. HASSERODT, U. and J. (R.) VINOGRAD: Dissociation of Human Carbonmonoxyhemoglobin at High pH. Proc. Nat. Acad. Sci. (USA) 45, 12 (1959).

87. HAUROWITZ, F.: Das Gleichgewicht zwischen Hämoglobin und Sauerstoff. Z. physiol. Chem. 254, 266 (1938).

88. HAYASHI, H.: A Simple Method for the Fractionation of Globins into their α- and β Chains. J. Biochemistry (Tokyo) 50, 70 (1961).

88a. HELLER, P.: The Molecular Basis of the Pathogenicity of Abnormal Hemoglobins. Some Recent Developments. Blood 25, 110 (1965).

89. HERMANS, J., Jr.: Normal and Abnormal Tyrosine Side-chains in Various Heme Proteins. Biochemistry 1, 193 (1962).

90. — Spectrophotometric Titration Curves of Human Hemoglobin and Its Carboxypeptidase Digests. Biochemistry 2, 453 (1963).

91. Hill, R. J.: Personal communication.

91a. — Discussion following Reference *134a.*

92. Hill, R. J. and W. Konigsberg: The Structure of Human Hemoglobin. IV. The Chymotryptic Digestion of the α Chain of Human Hemoglobin. J. Biol. Chem. **237**, 3151 (1962).

93. Hill, R. J., W. Konigsberg, G. Guidotti and L. C. Craig: The Structure of Human Hemoglobin. I. The Separation of the α and β Chains and their Amino Acid Composition. J. Biol. Chem. **237**, 1549 (1962).

94. Hill, R. J. and A. P. Kraus: Studies on the Amino Acid Sequence of HbA₂. Federat. Proc. (Amer. Soc. Exp. Biol.) **22**, 597 (1963).

95. Hill, R. L.: The Abnormal Human Hemoglobins. Lab. Invest. **10**, 1012 (1961).

96. Hill, R. L., J. Buettner-Janusch and V. Buettner-Janusch: Evolution of Hemoglobin in Primates. Proc. Nat. Acad. Sci. (USA) **50**, 885 (1963).

97. Hill, R. L., R. T. Swenson and H. C. Schwartz: Characterization of a Chemical Abnormality in Hemoglobin G. J. Biol. Chem. **235**, 3182 (1960).

98. Hilschmann, N. und G. Braunitzer: Über Hämoglobine. XII. Die Sequenzanalyse des Humanhämoglobins. Die Analyse des β-Core-Peptids und des Peptids α Tp 6. Z. physiol. Chem. **335**, 21 (1964).

99. Holmquist, W. R.: Personal communication.

100. Holmquist, W. R. and W. A. Schroeder: Properties and Partial Characterization of Adult Human Hemoglobin A₁c. Biochem. Biophys. Acta **82**, 639 (1964).

101. Horton, B. F. and T. H. J. Huisman: Linkages of the β-Chain and δ-Chain Structural Genes of Human Hemoglobins. Amer. J. Human Genet. **15**, 394 (1963).

101a. — — Studies on the Heterogeneity of Haemoglobin. VII. Minor Haemoglobin Components in Haematological Diseases. Brit. J. Haematol. (in press.)

102. Horton, B. (F.), R. A. Payne, M. T. Bridges and T. H. J. Huisman: Studies on an Abnormal Minor Hemoglobin Component (Hb-B₂). Clin. Chim. Acta **6**, 246 (1961).

103. Horton, B. F., R. B. Thompson, A. M. Dozy, C. M. Nechtman, E. Nichols and T. H. J. Huisman: Inhomogeneity of Hemoglobin. VI. The Minor Hemoglobin Components of Cord Blood. Blood **20**, 302 (1962).

104. Huehns, E. R., G. H. Beaven and B. L. Stevens: Recombination Studies on Haemoglobin at Neutral pH. Biochemic. J. **92**, 440 (1964).

105. — — — Reaction of Haemoglobin α^A with Haemoglobin β₄^A, γ₄^F, and δ^A₂. Biochemic. J. **92**, 444 (1964).

106. Huehns, E. R., A. Buchanan, P. Barkhan, P. E. Crome and P. L. Hollison: Personal communication.

107. Huehns, E. R., H. Hartman, F. Hecht and A. G. Motulsky: Personal communication.

108. Huehns, E. R. and E. M. Shooter: Polypeptide Chains of Haemoglobin A₂. Nature **189**, 918 (1961).

109. — — Reaction of Haemoglobin α^A with Haemoglobin H. Nature **193**, 1083 (1962).

110. — — Haemoglobin. Science Progr. **52**, 353 (1964).

111. Huisman, T. H. J.: Genetic Aspects of Two Different Minor Haemoglobin Components Found in Cord Blood Samples of Negro Babies. Nature **188**, 589 (1960).

112. — Normal and Abnormal Human Hemoglobins. Adv. Clin. Chem. **6**, 231 (1963).

113. — Personal communication.

114. Huisman, T. H. J. and A. M. Dozy: Studies on the Heterogeneity of Hemoglobin. IV. Chromatographic Behavior of Different Human Hemoglobins on Anion-exchange Cellulose (DEAE-cellulose). J. Chromatogr. **7**, 180 (1962).

115. Huisman, T. H. J. and B. F. Horton: Studies on the Heterogeneity of Hemoglobin. VIII. Chromatographic and Electrophoretic Investigations of Various Minor Hemoglobin Fractions Present in Normal and *in vitro* Modified Red Blood Cell Hemolysates. J. Chromatogr. (in press).

116. Huisman, T. H. J. and R. C. Lee: Two δ-Chain Abnormal Hemoglobins in One Individual. (To be published.)

117. Huisman, T. H. J. and C. A. Meyering: Studies on the Heterogeneity of Hemoglobin. I. The Heterogeneity of Different Human Hemoglobin Types in Carboxymethylcellulose and in Amberlite IRC-50 Chromatography: Qualitative Aspects. Clin. Chim. Acta **5**, 103 (1960).

118. Huisman, T. H. J. and V. P. Sydenstricker: Difference in Gross Structure of two Electrophoretically Identical "Minor" Haemoglobin Components. Nature **193**, 489 (1962).

119. Huisman, T. H. J., G. van Vliet and T. Sebens: Sheep Haemoglobins. Nature **182**, 171 (1958).

120. Hunt, J. A.: Identity of the α-Chains of Adult and Foetal Human Haemoglobins. Nature **183**, 1373 (1959).

121. Hunt, J. A. and V. M. Ingram: Allelomorphism and the Chemical Differences of the Human Haemoglobins A, S and C. Nature **181**, 1062 (1958).

122. — — Abnormal Human Haemoglobins. IV. The Chemical Difference between Normal Human Haemoglobin and Haemoglobin C. Biochim. Biophys. Acta **42**, 409 (1960).

123. — — Abnormal Human Haemoglobins. VI. The Chemical Difference between Haemoglobins A and E. Biochim. Biophys. Acta **49**, 520 (1961).

124. Hunt, J. A. and H. Lehmann: Haemoglobin "Bart's": a Foetal Haemoglobin without α-Chains. Nature **184**, 872 (1959).

125. Huntsman, R. G., M. Hall, H. Lehmann and P. K. Sukumaran: A Second and a Third Abnormal Haemoglobin in Norfolk. Haemoglobin $G_{Norfolk}$ and Haemoglobin $D_{Norfolk}$. Brit. Med. J. **1963**, I, 720.

126. Hutton, J. J., J. Bishop, R. Schweet and E. S. Russell: Hemoglobin Inheritance in Inbred Mouse Strains. II. Genetic Studies. Proc. Nat. Acad. Sci. (USA) **48**, 1718 (1962).

127. Ingram, V. M.: A Specific Chemical Difference between the Globins of Normal Human and Sickle-cell Anaemia Haemoglobin. Nature **178**, 792 (1956).

128. — Gene Mutations in Human Haemoglobin: The Chemical Difference Between Normal and Sickle Cell Haemoglobin. Nature **180**, 326 (1957).

129. — Abnormal Human Haemoglobins. III. The Chemical Difference between Normal and Sickle Cell Haemoglobins. Biochim. Biophys. Acta **36**, 402 (1959).

130. — Hemoglobin and Its Abnormalities. Springfield, Ill.: Charles C. Thomas. 1961.

131. — The Hemoglobins in Genetics and Evolution. New York: Columbia Univ. Press. 1963.

132. Itano, H. A.: The Human Hemoglobins: Their Properties and Genetic Control. Adv. Protein Chem. **12**, 215 (1957).

133. Itano, H. A. and E. (A.) Robinson: Electrophoretic Separation of Intermediate Compounds in Two Reactions of Ferrihemoglobin. Biochim. Biophys. Acta **29**, 545 (1958).

134. — — Genetic Control of the α- and β-Chains of Hemoglobin. Proc. Nat. Acad. Sci. (USA) **46**, 1492 (1960).

134a. Itano, H. A., E. A. Robinson and A. J. Gottlieb: Dissociation and Association of Hemoglobin Subunits. In: Subunit Structure of Proteins. Biochemical and Genetic Aspects. Brookhaven Sympos. Biol., No. 17, p. 194. Washington: Office of Technical Serv., Dept. Commerce. 1964.

135. Itano, H. A. and S. J. Singer: On Dissociation and Recombination of Human Adult Hemoglobins A, S, and C. Proc. Nat. Acad. Sci. (USA) **44**, 522 (1958).

136. Itano, H. A., S. J. Singer and E. (A.) Robinson: Chemical and Genetical Units of the Haemoglobin Molecule. Ciba Foundation Symp., Biochemistry of Human Genetics, p. 96. Boston: Little, Brown & Co. 1959.

137. Jones, R. T.: Chromatographic and Chemical Studies of Some Abnormal Human Hemoglobins and Some Minor Hemoglobin Components. Ph. D. Thesis, Calif. Instit. of Technology, Pasadena, 1961.

138. — Structural Studies of Aminoethylated Hemoglobins by Automatic Peptide Chromatography. Cold Spring Harbor Sympos. Quant. Biol. **29**, 297 (1964).

139. Jones, R. T., R. D. Coleman and P. Heller: The Chemical Structure of M_{Iwate} ($M_{Kankakee}$). Federat. Proc. (Amer. Soc. Exp. Biol.) **23**, 173 (1964).

140. Jones, R. T. and R. L. Hill: Personal communication.

141. Jones, R. T., R. D. Koler and R. Lisker: The Chemical Structure of Hemoglobin Mexico by Automatic Peptide Chromatography and Subunit Hybridization. Clin. Res. **11**, 105 (1963).

142. Jones, R. T. and W. A. Schroeder: Chromatography of Human Hemoglobin. Factors influencing Chromatography and Differentiation of Similar Hemoglobins. J. Chromatogr. **10**, 421 (1963).

143. — — Chemical Characterization and Subunit Hybridization of Human Hemoglobin H and Associated Compounds. Biochemistry **2**, 1357 (1963).

144. — — Unpublished.

145. Jones, R. T., W. A. Schroeder, J. E. Balog and J. R. Vinograd: Gross Structure of Hemoglobin H. J. Amer. Chem. Soc. **81**, 3161 (1959).

146. Jones, R. T., W. A. Schroeder and J. R. Vinograd: Identity of the α Chains of Hemoglobins A and F. J. Amer. Chem. Soc. **81**, 4749 (1959).

147. Jonxis, J. H. P. and J. F. Delafresnaye (Edit.): Abnormal Haemoglobins - A Symposium. Springfield, Ill.: Charles C. Thomas. 1959.

148. Karlson, P.: Introduction to Modern Biochemistry. (Translated by C. H. Doering.) New York: Academic Press. 1963.

149. Karpatkin, S.: Globin Synthesis in Human Reticulocytes. J. Lab. Clin. Med. **62**, 121 (1963).

150. Kendrew, J. C.: Side-chain Interactions in Myoglobin. In: Enzyme Models and Enzyme Structure. Brookhaven Symp. in Biol., No. 15, p. 216 (1962). Washington: Office of Technical Serv., Dept. Commerce. 1962.

151. Kendrew, J. C., R. E. Dickerson, B. E. Strandberg, R. G. Hart, D. R. Davies, D. C. Phillips and V. C. Shore: Structure of Myoglobin. A Three-dimensional Fourier Synthesis at 2 Å Resolution. Nature **185**, 422 (1960).

152. Kersten, H. G. and E. Kleihauer: Hemoglobin M Cologne, a New Hemoglobin Variant. The Differential Diagnosis of Cyanosis. Med. Welt **1964**, 1607 [Chem. Abstr. **61**, 10901 (1964)].

153. Kirshner, A. G. and C. Tanford: The Dissociation of Hemoglobin by Inorganic Salts. Biochemistry **3**, 291 (1964).

154. Konagaya, S.: N-Terminal Amino Acid Sequences of Hemoglobin of Blue-White Dolphin *(Prodelphinus coeruleo-albus)*. J. Biochemistry (Tokyo) **54**, 189 (1963).

155. Konigsberg, W.: The Reaction of Hemoglobin with a Bifunctional Maleimide Derivative. Federat. Proc. (Amer. Soc. Exp. Biol.) **23**, 173 (1964).

156. KONIGSBERG, W., J. GOLDSTEIN and R. J. HILL: The Structure of Human Hemoglobin. VII. The Digestion of the β Chain of Human Hemoglobin with Pepsin. J. Biol. Chem. **238**, 2028 (1963).

157. KRAUS, L. M.: Personal communication.

158. KUNKEL, H. G. and G. WALLENIUS: New Hemoglobin in Normal Adult Blood. Science **122**, 288 (1955).

159. LEHMANN, H. and J. A. M. AGER: The Hemoglobinopathies and Thalassemia. In: J. B. Stanbury, J. B. Wyngaarden and D. S. Frederickson (Edit.), The Metabolic Basis of Inherited Disease, p. 1086. New York: McGraw-Hill. 1960.

160. LEHMANN, H., D. BEALE and F. S. BOI-DOKU: Haemoglobin G$_{Accra}$. Nature **203**, 363 (1964).

161. LIDDELL, J., D. BROWN, D. BEALE, H. LEHMANN and R. G. HUNTSMAN: A New Haemoglobin J$_{\alpha\,Oxford}$ Found during a Survey of an English Population. Nature **204**, 269 (1964).

162. MANWELL, C.: Comparative Physiology: Blood Pigments. Annu. Rev. Physiol. **22**, 191 (1960).

163. MARGOLIASH, E.: Porphyrins and Hemoproteins. Annu. Rev. Biochem. **30**, 549 (1961).

164. MARKS, P. A., E. R. BURKA and D. SCHLESSINGER: Protein Synthesis in Erythroid Cells. I. Reticulocyte Ribosomes Active in Stimulating Amino Acid Incorporation. Proc. Nat. Acad. Sci. (USA) **48**, 2163 (1962).

165. MATHIAS, A. P., R. WILLIAMSON, H. E. HUXLEY and S. PAGE: Occurrence and Function of Polysomes in Rabbit Reticulocytes. J. Mol. Biol. **9**, 154 (1964).

166. MATSUDA, G., R. GEHRING-MÜLLER und G. BRAUNITZER: Die vollständige Sequenz der α-Kette der langsamen Komponente des Pferdehämoglobins. Biochem. Z. **338**, 669 (1963).

167. MATSUDA, G., R. T. JONES and W. A. SCHROEDER: Characterization of Hemoglobin A from Human Umbilical Cord Blood. Biochim. Biophys. Acta **82**, 180 (1964).

168. MINNICH, V., J. K. CORDONNIER, W. J. WILLIAMS and C. V. MOORE: Alpha, Beta and Gamma Hemoglobin Polypeptide Chains During the Neonatal Period with Description of a Fetal Form of Hemoglobin D$_{\alpha\,St.Louis}$. Blood **19**, 137 (1962).

169. MIYAGI, T., I. IUCHI, S. SHIBATA, I. TAKEDA and A. TAMURA: Possible Amino Acid Substitution in the α Chain ($\alpha^{87\,tyr}$) of HbM$_{Iwate}$. Acta Haem. Jap. **26**, 538 (1963).

170. MORRIS, A. J.: Terminal Stages in the Biosynthesis of Haemoglobin. The Release of Protein from Reticulocyte Ribosomes. Biochem. J. **91**, 611 (1964).

171. MORRISON, M. and J. L. COOK: Chromatographic Fractionation of Normal Adult Oxyhemoglobin. Science **122**, 920 (1955).

172. MUIRHEAD, H. and M. F. PERUTZ: Structure of Haemoglobin. A Three-Dimensional Fourier Synthesis of Reduced Human Haemoglobin at 5.5 Å Resolution. Nature **199**, 633 (1963).

173. MULLER, C. J.: Molecular Evolution. Assen: Van Gorcum. 1961.

174. MULLER, C. J. and S. KINGMA: Hemoglobin Zürich: $\alpha_2^A\beta_2^{63\,Arg}$. Biochim. Biophys. Acta **50**, 595 (1961).

175. MURAYAMA, M.: Chemical Difference between Normal Human Haemoglobin and Haemoglobin-I. Nature **196**, 276 (1962).

176. — A Molecular Mechanism of Sickled Erythrocyte Formation. Nature **202**, 258 (1964).

177. Naiman, J. L. and P. S. Gerald: Fetal Hemoglobin: Improved Separation by a Modified Agar Gel Electrophoresis. J. Lab. Clin. Med. **61**, 508 (1963).

177a. Nance, W. E.: Genetic Control of Hemoglobin Synthesis. Science **141**, 123 (1963).

178. Naughton, M. A. and H. M. Dintzis: Sequential Biosynthesis of the Protein Chains of Hemoglobin. Proc. Nat. Acad. Sci. (USA) **48**, 1822 (1962).

179. Nechtman, C. M. and T. H. J. Huisman: Comparative Studies of Oxygen Equilibria of Human Adult and Cord Blood Red Cell Hemolyzates and Suspensions. Clin. Chim. Acta **10**, 165 (1964).

180. Neel, J. V.: The Genetics of Human Haemoglobin Differences: Problems and Perspectives. Ann. Human Genet. **21**, 1 (1956/57).

181. O'Gorman, P., H. Lehmann, K. M. Allsopp and P. K. Sukumaran: Sickle-cell Haemoglobin K Disease. Brit. Med. J. **1963**, II, 1381.

182. Ozawa, H. and K. Satake: On the Species Difference of N-Terminal Amino Acid Sequence in Hemoglobin. I. J. Biochemistry (Tokyo) **42**, 641 (1955).

183. Pauling, L., H. A. Itano, S. J. Singer and I. C. Wells: Sickle Cell Anemia, a Molecular Disease. Science **110**, 543 (1949).

184. Pedersen, K. O. and K. J. I. Anderssen: Unpublished data, cited by T. Svedberg and K. O. Pedersen, in: The Ultracentrifuge, p. 407. Oxford: Clarendon Press and New York: Johnson Reprint Corp. 1940.

185. Perutz, M. F.: Proteins and Nucleic Acids: Structure and Function. Amsterdam: Elsevier. 1962.

186. Perutz, M. F., W. Bolton, R. Diamond, H. Muirhead and H. C. Watson: Structure of Haemoglobin. An X-Ray Examination of Reduced Horse Haemoglobin. Nature **203**, 687 (1964).

187. Perutz, M. F., A. M. Liquori and F. Finch: X-Ray and Solubility Studies of the Haemoglobin of Sickle-cell Anemia Patients. Nature **167**, 929 (1951).

188. Perutz, M. F. and L. Mazzarella: A Preliminary X-Ray Analysis of Haemoglobin H. Nature **199**, 639 (1963).

189. Perutz, M. F., I. F. Trotter, E. R. Howells and D. W. Green: An X-Ray Study of Reduced Human Haemoglobin. Acta Cristallogr. **8**, 241 (1955).

189 a. Philipps, G. R.: Haemoglobin Synthesis and Polysomes in Intact Reticulocytes. Nature **205**, 567 (1965).

190. Pierce, L. E., C. E. Rath and K. McCoy: A New Hemoglobin Variant with Sickling Properties. New England J. Med. **268**, 862 (1963).

191. Porter, R. R. and F. Sanger: The Free Amino Groups of Haemoglobins. Biochemic. J. **42**, 287 (1948).

192. Raftery, M. A. and R. D. Cole: Tryptic Cleavage at Cysteinyl Peptide Bonds. Biochem. Biophys. Res. Comm. **10**, 467 (1963).

193. Ranney, H. M.: Observations on the Inheritance of Sickle Cell Hemoglobin and Hemoglobin C. J. Clin. Invest. **33**, 1634 (1954).

193a. Ranney, H. M., R. E. Benesch, R. Benesch and A. S. Jacobs: Hybridization of Deoxygenated Human Hemoglobin. Biochim. Biophys. Acta **74**, 544 (1963).

194. Ranney, H. M., A. S. Jacobs, T. B. Bradley and F. A. Cordova: A "New" Variant of Haemoglobin A_2 and its Segregation in a Family with Haemoglobin S. Nature **197**, 164 (1963).

195. Ranney, H. M., C. O'Brien and A. S. Jacobs: An Abnormal Human Fetal Haemoglobin with an Abnormal Alpha-Polypeptide Chain. Nature **194**, 743 (1962).

196. Raper, A. B., D. B. Gammack, E. R. Huehns and E. M. Shooter: Four Haemoglobins in One Individual. A Study of the Genetic Interaction of Hb-G and Hb-C. Brit. Med. J. **1960**, II, 1257.

197. RHINESMITH, H. S., W. A. SCHROEDER and N. MARTIN: The N-Terminal Sequence of the β Chains of Normal Adult Human Hemoglobin. J. Amer. Chem. Soc. **80**, 3358 (1958).

198. RHINESMITH, H. S., W. A. SCHROEDER and L. PAULING: A Quantitative Study of the Hydrolysis of Human Dinitrophenyl(DNP)globin: The Number and Kind of Polypeptide Chains in Normal Adult Human Hemoglobin. J. Amer. Chem. Soc. **79**, 4682 (1957).

199. RICH, A., J. R. WARNER and H. M. GOODMAN: The Structure and Function of Polyribosomes. Cold Spring Harbor Sympos. Quant. Biol. **28**, 269 (1963).

200. RIGGS, A.: The Binding of N-Ethylmaleimide by Human Hemoglobin and its Effect upon the Oxygen Equilibrium. J. Biol. Chem. **236**, 1948 (1961).

201. — The Amino Acid Composition of Some Mammalian Hemoglobins: Mouse, Guinea Pig and Elephant. J. Biol. Chem. **238**, 2983 (1963).

202. RIGGS, A. and M. WELLS: The Identification of the Oxygenation-linked Acid Groups with the β-Chain of Human Hemoglobin. Federat. Proc. (Amer. Soc. Exp. Biol.) **19**, 78 (1960).

203. ROSSI-FANELLI, A., E. ANTONINI and A. CAPUTO: Studies on the Structure of Hemoglobin. I. Physico-chemical Properties of Human Globin. Biochem. Biophys. Acta **30**, 608 (1958).

203a. — — — Studies on the Relations between Molecular and Functional Properties of Hemoglobin. I. The Effect of Salts on the Molecular Weight of Human Hemoglobin. J. Biol. Chem. **236**, 391 (1961).

204. — — — Hemoglobin and Myoglobin. Adv. Protein Chem. **19**, 73 (1964).

205. RUCKNAGEL, D. L. and J. V. NEEL: The Hemoglobinopathies. Progr. Med. Genetics **1**, 158 (1961).

206. SALOMON, H., I. TATARSKI, N. DANCE, E. R. HUEHNS and E. M. SHOOTER: A New Hemoglobin Variant Found in a Beduin Tribe: Hemoglobin "Rambam". Asian Congr. of Pathology, Israel, August 1964, and personal communication.

207. SATAKE, K. and S. SASAKAWA: Studies on Hemoglobin. VIII. The Tryptic Peptides of Bovine Globin α. J. Biochemistry (Tokyo) **53**, 201 (1963).

208. SCHNEIDER, R. G., F. ARAT and M. E. HAGGARD: An Inhomogeneous Foetal Haemoglobin Variant (the Texas Type). Nature **202**, 1346 (1964).

209. SCHNEIDER, R. G. and B. H. BOWMAN: Personal communication.

210. SCHNEIDER, R. G., M. E. HAGGARD, C. W. McNUTT, J. E. JOHNSON, Jr., B. H. BOWMAN and D. R. BARNETT: Hemoglobin $G_{Coushatta}$: A New Variant in an American Indian Family. Science **143**, 697 (1964).

211. SCHNEIDER, R. G. and R. T. JONES: Hemoglobin F_{Texas}: Gamma-Chain Variant. Science **148**, 240 (1965).

212. SCHNEK, A. G. and W. A. SCHROEDER: The Relation between the Minor Components of Whole Normal Human Adult Hemoglobin as Isolated by Chromatography and Starch Block Electrophoresis. J. Amer. Chem. Soc. **83**, 1472 (1961).

213. SCHROEDER, W. A.: The Chemical Structure of the Normal Human Hemoglobins. Fortschr. Chem. organ. Naturstoffe **17**, 322 (1959).

214. — The Hemoglobins. Annu. Rev. Biochem. **32**, 301 (1963).

215. SCHROEDER, W. A., J. T. CUA, G. MATSUDA and W. D. FENNINGER: Hemoglobin F_I, an Acetyl-containing Hemoglobin. Biochim. Biophys. Acta **63**, 532 (1962).

216. SCHROEDER, W. A., R. T. JONES, J. CORMICK and K. McCALLA: Chromatographic Separation of Peptides on Ion Exchange Resins. Separation of Peptides from Enzymatic Hydrolyzates of the α, β, and γ Chains of Human Hemoglobins. Anal. Chem. **34**, 1570 (1962).

217. Schroeder, W. A., R. T. Jones, J. R. Shelton, J. B. Shelton, J. Cormick and K. McCalla: A Partial Sequence of the Amino Acid Residues in the γ Chain of Human Hemoglobin F. Proc. Nat. Acad. Sci. (USA) 47, 811 (1961).
218. Schroeder, W. A. and G. Matsuda: N-Terminal Residues of Human Fetal Hemoglobin. J. Amer. Chem. Soc. 80, 1521 (1958).
219. Schroeder, W. A., J. R. Shelton, J. B. Shelton and J. Cormick: Further Sequences in the γ Chain of Human Fetal Hemoglobin. Proc. Nat. Acad. Sci (USA) 48, 284 (1962).
220. — — — — The Amino Acid Sequence of the α Chain of Human Fetal Hemoglobin. Biochemistry 2, 1353 (1963).
221. Schroeder, W. A., J. R. Shelton, J. B. Shelton, J. Cormick and R. T. Jones: The Amino Acid Sequence of the γ Chain of Human Fetal Hemoglobin. Biochemistry 2, 992 (1963).
222. Schwartz, H. C., R. Goudsmit, R. L. Hill, G. E. Cartwright and M. M. Wintrobe: The Biosynthesis of Hemoglobin from Iron, Protoporphyrin and Globin. J. Clin. Invest. 40, 188 (1961).
223. Shelton, J. R. and W. A. Schroeder: Further N-Terminal Sequences in Human Hemoglobins A, S and F by Edman's Phenylthiohydantoin Method. J. Amer. Chem. Soc. 82, 3342 (1960).
224. Shibata, S., T. Miyaji, I. Iuchi, S. Ueda and I. Takeda: Hemoglobin Hikari ($\alpha_2 \beta_2^{61\ asp\ NH_2}$): A Fast Moving Hemoglobin Found in two Unrelated Japanese Families. Clin. Chim. Acta 10, 101 (1964).
225. Shooter, E. M., E. R. Skinner, J. P. Garlick and N. A. Barnicot: The Electrophoretic Characterization of Haemoglobin G and a New Minor Haemoglobin, G_2. Brit. J. Haematol. 6, 140 (1960).
226. Silvestroni, E. and I. Bianco: A New Variant of Human Fetal Hemoglobin: HbF$_{Roma}$. Blood 22, 545 (1963).
227. Simon, S. R. and W. H. Konigsberg: Investigation of Relations Among Molecular Structure, Oxygenation, and Dissociation with Cross-Linked Hemoglobin. Xth Congr. Int. Soc. Haematology, Stockholm, 1964, Abstr. L: 21.
228. Simpson, M. V.: Protein Biosynthesis. Annu. Rev. Biochem. 31, 333 (1962).
229. Singer, S. J. and H. A. Itano: On the Asymmetrical Dissociation of Human Hemoglobin. Proc. Nat. Acad. Sci. (USA) 45, 174 (1959).
230. Smith, D. B.: Some Aspects of the Structure of Hemoglobin. Canad. J. Biochem. Physiol. 42, 755 (1964), and personal communication.
231. Smith, D. B. and M. F. Perutz: Identification of the Black Sub-Unit of the Crystallographic Model of Horse Haemoglobin with the Valyl-Glutaminyl Polypeptide Chain. Nature 188, 406 (1960).
232. Smith, E. W. and J. V. Torbert: Study of Two Abnormal Hemoglobins with Evidence for a New Genetic Locus for Hemoglobin Formation. Bull. John Hopkins Hosp. 101, 38 (1958).
233. Spackman, D. H., W. H. Stein and S. Moore: Automatic Recording Apparatus for Use in the Chromatography of Amino Acids. Analyt. Chemistry 30, 1190 (1958).
234. St. George, R. C. C. and L. Pauling: The Combining Power of Hemoglobin for Alkyl Isocyanides, and the Nature of the Heme-Heme Interactions in Hemoglobin. Science 114, 629 (1951).
235. Stockell, A., M. F. Perutz, H. Muirhead and S. C. Glauser: A Comparison of Adult and Foetal Horse Haemoglobins. J. Mol. Biol. 3, 112 (1961).
236. Stretton, A. O. W.: Personal communication.
237. Swenson, R. T., R. L. Hill, H. Lehmann and R. T. S. Jim: A Chemical Abnormality in Hemoglobin G from Chinese Individuals. J. Biol. Chem. 237, 1517 (1962).

238. Symposium on Molecular Heterogeneity of Hemoglobin. Federat. Proc. (Amer. Soc. Exp. Biol.) **16**, 740 (1957).

239. TANFORD, C.: The Interpretation of Hydrogen Ion Titration Curves of Proteins. Adv. Protein Chem. **17**, 69 (1962).

240. — Cohesive Forces and Disruptive Reagents. In: Subunit Structure of Proteins. Biochemical and Genetic Aspects. Brookhaven Sympos. Biol. No 17, p. 154. Washington: Office of Technical Serv., Dept. Commerce. 1964.

241. TEALE, F. W. J.: Cleavage of the Haem-Protein Link by Acid Methylethylketone. Biochim. Biophys. Acta **35**, 543 (1959).

242. TSCHUDY, D. P.: Porphyrin Biosynthesis. In: F. W. Sunderman and F. W. Sunderman, Jr. (Edit.), Hemoglobin, its Precursors and Metabolites, p. 159. Philadelphia: Lippincott. 1964.

243. VINOGRAD, J. R. and W. D. HUTCHINSON: Carbon-14 Labelled Hybrids of Haemoglobin. Nature **187**, 216 (1960).

244. VRIES, A. DE, H. JOSHUA, H. LEHMANN, R. L. HILL and R. E. FELLOWS: The First Observation of an Abnormal Haemoglobin in a Jewish Family: Haemoglobin Beilinson. Brit. J. Haematol. **9**, 484 (1963).

245. WARNER, J. R., P. M. KNOPF and A. RICH: A Multiple Ribosomal Structure in Protein Synthesis. Proc. Nat. Acad. Sci. (USA) **49**, 122 (1963).

246. WARNER, J. R. and A. RICH: The Number of Soluble RNA Molecules on Reticulocyte Polyribosomes. Proc. Nat. Acad. Sci. (USA) **51**, 1134 (1964).

246a. — — The Number of Growing Polypeptide Chains on Reticulocyte Polyribosomes. J. Mol. Biol. **10**, 202 (1964).

247. WEATHERALL, D. J. and C. BAGLIONI: A Fetal Hemoglobin Variant of Unusual Genetic Interest. Blood **20**, 675 (1962).

248. WHITE, J. C. and G. H. BEAVEN: Foetal Haemoglobin. Brit. Med. Bull. **15**, 33 (1959).

249. WILSON, S. and D. B. SMITH: Separation of the Valyl-leucyl- and Valyl-glutamyl-polypeptide Chains of Horse Globin by Fractional Precipitation and Column Chromatography. Canad. J. Biochem. Physiol. **37**, 405 (1959).

250. WINTERHALTER, K. H.: Hemoglobin Synthesis. Pathol. Microbiol. **27**, 508 (1964).

251. WINTERHALTER, K. H. and E. R. HUEHNS: Free Globin in Red Cells. J. Clin. Invest. **42**, 995 (1963).

251a. — — Preparation, Properties, and Specific Recombination of $\alpha\beta$-Globin Subunits. J. Biol. Chem. **239**, 3699 (1964).

252. WINTERHALTER, K. H., E. R. HUEHNS and C. A. FINCH: Free Globin in Normal Red Cells (quoted in *250*).

253. WINTROBE, M. M.: Clinical Hematology. 5th ed. Philadelphia: Lea and Febiger. 1961.

254. WOOD, W. B. and P. BERG: Studies on the "Messenger" Activity of RNA Synthesized with RNA Polymerase. Cold Spring Harbor Sympos. Quant. Biol. **28**, 237 (1963).

255. WYMAN, J., Jr.: Heme Proteins. Adv. Protein Chem. **4**, 410 (1948).

256. — Linked Functions and Reciprocal Effects in Hemoglobin: a Second Look. Adv. Protein Chem. **19**, 223 (1964).

257. YANOFSKY, C.: Discussion to W. Gilbert, Protein Synthesis in *Escherichia coli*. Cold Spring Harbor Sympos. Quant. Biol. **28**, 287, 296 (1963).

258. ZUCKERKANDL, E.: Controller-Gene Diseases: The Operon Model as Applied to β-Thalassemia, Familial Fetal Hemoglobinemia and the Normal Switch from the Production of Fetal Hemoglobin to that of Adult Hemoglobin. J. Mol. Biol. **8**, 128 (1964).

259. Zuckerkandl, E., R. T. Jones and L. Pauling: A Comparison of Anima
    Hemoglobins by Tryptic Peptide Pattern Analysis. Proc. Nat. Acad. Sci. **46**,
    1349 (1960).
260. Zuckerkandl, E. and L. Pauling: Molecular Disease, Evolution, and Genic
    Heterogeneity. In: M. Kasha and B. Pullman (Edit.), Horizons in Bio-
    chemistry, p. 189. New York: Academic Press. 1962.
261. Zuckerkandl, E. and W. A. Schroeder: Amino-Acid Composition of the
    Polypeptide Chains of Gorilla Haemoglobin. Nature **192**, 984 (1961).

(Received, January 13, 1965.)

Kollagen.
(Collagen.)

Von W. GRASSMANN,

unter Mitarbeit von

J. ENGEL, K. HANNIG, H. HÖRMANN, K. KÜHN und A. NORDWIG,
München.

Mit 24 Abbildungen.

Inhaltsübersicht.

I. Einleitung.

Der Anlaß für unseren Arbeitskreis, sich mit dem Problem der Kollagenstruktur zu befassen, war ein äußerer. Die Kollagenfaser ist das hauptsächliche Bauelement des Bindegewebes und damit auch der Haut und des Leders. Zunächst haben wir uns — und dies gilt schon von meinem Vorgänger Max Bergmann — das Problem der chemischen Struktur des Kollagens viel zu einfach vorgestellt: Die Molekülkette, von der es zunächst offen schien, ob sie in der Länge begrenzt sei oder „fakultativ endlos" wie die der Cellulose, sollte aufgebaut sein aus Sequenzen von je 3 bzw. 6 Aminosäuren, unter denen Glycin, Prolin und Hydroxyprolin regelmäßig wiederkehren sollten (27, 12). Im Gegensatz zu den rein spekulativen Vorstellungen über die Aminosäureanordnungen anderer Proteine, die damals auf ähnlicher Grundlage entwickelt worden sind, hat sich im Falle des Kollagens die Bergmannsche Vorstellung einer periodischen Wiederkehr kurzer, gleichartiger Sequenzen als fruchtbar erwiesen. Sie war wegleitend für die späteren 3-Ketten-Helix-Modelle nach Pauling (316), Ramachandran (334, 335, 337) und Rich und Crick (345), und sie hat in den chemischen Untersuchungen, über die unten berichtet wird, zum mindesten für die „apolaren" Bereiche eine Bestätigung gefunden.

Als man dann erkannt hatte, daß das Molekül weit größer und seine Struktur weit komplizierter sei als ursprünglich erwartet, haben wir uns oft gefragt, ob der für seine Aufklärung notwendige Arbeitsaufwand lohnend sei, und auch wir haben gefürchtet, „die richtige Struktur könne ausdruckslos sein, d. h. keine weiteren Folgerungen zulassen und uns ebenso wenig anregen wie irgendeine inerte Substanz" [J. D. WATSON (402)].

Heute glauben wir immerhin, daß der Arbeitsaufwand sich zu lohnen beginnt. Das Kollagen, das durch die strukturellen Besonderheiten seines Moleküls eine Sonderstellung unter den Proteinen innehat, hat zwar keine so zentrale Bedeutung für die Lebensvorgänge wie die Nukleinsäuren, und seine Funktionen scheinen zunächst weit weniger interessant als die der chemisch viel weniger aufgeklärten Faserproteine der Muskeln. Die einzige Aufgabe, die dem Kollagen zufällt, scheint die zu sein, rasch im extracellulären Raum Fibrillen und Fasern zu bilden, die hoch reißfest, quellbar und für die normalen Körperfermente unangreifbar sein sollen — eine Kombination, die keinem anderen Faserprotein zukommt — und mit ihrer Hilfe in einem bisher noch kaum verstandenen, aber für die Formbildung des tierischen Organismus offenbar entscheidenden Vorgang der Zusammenordnung, übergeordnete makroskopische Strukturen, wie Haut, Sehnen, Knorpel, Knochen und Zähne, aufzubauen. Für die *Mineralisierung* der zuletzt genannten Strukturen, deren Mechanismus sich gegenwärtig zu entschleiern beginnt (vgl. Abschnitt VIII/4, S. 289), ist ohne Zweifel das Kollagen das auslösende und dirigierende Element. Die Bildung der Kollagenfibrillen ist der wesentliche Vorgang im Prozeß der *Wundheilung*. Änderungen des kollagen-bildenden Apparates und der Kollagenfibrillen selbst sind das wichtigste und bisher wohl am meisten verständliche Symptom des *Alterns* eines Individuums und eines Gewebes. Störungen und Fehlleitungen des Kollagenaufbaus sind offenbar mitentscheidend für eine Reihe von *Krankheitsbildern* (Scorbut, Lathyrismus, rheumatische Erkrankungen, Sclerodermie (vgl. Abschnitt VIII/5, S. 290). Wohl die interessanteste Eigenschaft der Kollagenfibrille ist ihre im Jahre 1942 unabhängig voneinander und ziemlich gleichzeitig von HALL, JAKUS und SCHMITT (*146*) und von WOLPERS (*412, 413*) entdeckte und inzwischen von einer Generation von Forschern intensiv untersuchte, periodisch wiederkehrende *Querstreifung*. In den Fibrillen des nativen Typs kann man bis zu 15 innerhalb einer Periode von etwa 650 Å wiederkehrende (*294*), in den bisher nur im Laboratorium aufgefundenen Long-spacing-Strukturen mit einer Periodenlänge von etwa 2800 Å 40 und mehr (*178*) dunkle und ebenso viele helle Querstreifen erkennen. Die Abmessungen der kleinsten erkennbaren Einzelheiten, die bei etwa 10—15 Å liegen, sind dabei so, daß ein Zusammenhang mit der Aminosäureanordnung, also der Primärstruktur, zwingend angenommen werden muß (*294*). Die Primärstruktur enthält, wie im VI. und VII. Kapitel gezeigt wird, alle Informationen zum Aufbau des Tripelhelix-Moleküls und der hochstrukturierten Fibrillen und Long-spacing-Strukturen. Aufgeklärt ist sie gegenwärtig nur in den wesentlichen Zügen; die vollkommene Aufklärung erscheint als eine mit den heutigen Mitteln allenfalls gerade noch lösbare Aufgabe, von der noch nicht feststeht, ob sie den Arbeitsaufwand lohnen wird. Aber schon der heutige Stand reicht aus, um die Bildung der optisch sichtbaren Strukturen und ihres elektronenmikroskopischen Bildes auf molekularer Basis zu verstehen. Der Reiz für den Forscher auf diesem Gebiet liegt darin, daß von den Untereinheiten des Moleküls über das Stäbchenmolekül selbst bis zu den im Elektronenmikroskop und schließlich mit freiem Auge sichtbaren Strukturen im Zusammenwirken von Chemie, Röntgenstrukturforschung, Elektronenmikroskopie und anderen physikalischen Methoden alle Schritte in vitro verfolgt werden können und tatsächlich zum größten Teil auch im Grundsätzlichen ihres Mechanismus aufgeklärt sind. Gerade im Falle des Kollagens ist die Grenze zwischen *chemischer* und *morphologischer Struktur* fließend, und so mag es gerechtfertigt sein, wenn im folgenden das Wort „Struktur" immer wieder in beiden Bedeutungen gebraucht wird.

II. Übersicht.

Die Kollagenfaser ist der hauptsächliche Faserbestandteil des *Bindegewebes*. Als Bindegewebe bezeichnet man das zwischen den Organen und Muskeln sich ausbreitende und sie umhüllende Gewebe. Dazu gehören beispielsweise die Haut und die inneren Membranen des Körpers, die Knorpel, die organische Grundsubstanz des Knochens und der Zähne, das Synovialgewebe, große Teile der Lymphdrüsen und der Gefäßwände. Das Bindegewebe besteht neben wenigen *Zellen*, die zu seiner Erhaltung und zu seinem Aufbau notwendig sind (Fibroblasten, Mastzellen), aus *Fasern*, unter denen *Kollagenfasern*, die wesentlich dünneren, aber im chemischen Aufbau weitgehend mit Kollagenfasern übereinstimmenden *Retikulinfasern* und die meist mengenmäßig stark zurücktretenden, chemisch und physikalisch vollkommen verschiedenen *Elastinfasern* unterschieden werden. Fasern und Zellen sind eingebettet in die gallertige *Zwischen- oder Intercellularsubstanz*, die durch einen hohen Gehalt an Glykoproteiden und Mucopolysacchariden gekennzeichnet ist und außerdem neben einer größeren Zahl in der Hauptsache Serumproteinen entsprechenden Eiweißstoffen (*152, 76, 3*) eine lösliche Vorstufe des Kollagens (Tropokollagen*) enthält. Die Bildung von Kollagenvorstufen erfolgt innerhalb der Fibroblasten, die Bildung der Fibrillen extracellulär inner-

---

* Die Bezeichnung *Tropokollagen* wurde ursprünglich von F. O. SCHMITT und Mitarb. (*133a*) für das Kollagenmonomere vorgeschlagen. Sie wurde später häufig ausschließlich für neutralsalzlösliches Kollagen verwendet (*100, 134*). In der vorliegenden Arbeit soll nur mehr von löslichem Kollagen oder dem Kollagenmonomeren die Rede sein.

Literaturverzeichnis: SS. 293—314.

halb der Zwischensubstanz, wobei die Einzelheiten dieses Übergangs noch nicht völlig abgeklärt sind (*90, 371*) (vgl. Abschnitt VIII/1, S. 283 f.). Den saueren Mucopolysacchariden der Zwischensubstanz kommt offenbar eine wesentliche, aber bisher unbekannte Funktion für die Entstehung der Fibrillen zu. Das ergibt sich daraus, daß einer erhöhten Bildung der Fibrillen immer eine gesteigerte Synthese von Mucopolysacchariden — im Isotopenversuch leicht am Einbau von Schwefel erkennbar — parallel oder wahrscheinlich vorausgeht (*331, 89, 55a*).

Als charakteristische und ausführlich untersuchte (*280*) Mucopolysaccharide seien Chondroitin, die drei Chondroitin-Schwefelsäuren A, B und C, Keratosulfat, Hyaluronsäure und Heparin genannt, die ihrerseits aus N-Acetyl-glucosamin, N-Acetyl-galactosamin, Glucuronsäure und Iduronsäure aufgebaut sind und zum großen Teil Schwefelsäure in esterartiger Bindung enthalten.

Tabelle 1. Aminosäurezusammensetzung von Kollagenen unterschiedlicher Herkunft sowie von Elastin.

(% Aminosäure von Gesamtaminosäuren; Mol%.)

	Humanhaut (*33*)	Kalbshaut (*108*)	Rattenhaut (*321*)	Karpfen-Schwimmblase (*321*)	Dogfischhaut (*321*)	Regenwurmhaut (*403, 275*)	Nemathelminthen (*404*)	Anneliden (*404*)	Mollusken (*407*)	Spongin B (*297*)	Elastin (*391*)
Gly	33,0	32,7	33,1	33,3	33,8	25,5	28,6	32,4	32,1	32,3	21,9
Ala	11,0	10,8	10,6	12,5	10,6	8,9	6,9	10,3	7,2	9,4	5,7
Val	2,4	2,3	2,4	1,8	2,5	3,6	1,3	1,7	2,2	2,4	16,7
Leu	2,4	2,6	2,4	2,0	2,5	3,8	1,8	2,9	2,4	2,4	9,9
Ileu	1,0	1,2	1,1	1,1	1,5	2,4	1,4	1,5	1,2	1,7	4,6
Ser	3,6	2,9	4,3	3,7	6,1	8,9	2,3	10,5	6,9	2,4	1,1
Thr	1,8	1,9	1,9	2,7	2,3	6,8	1,8	5,2	2,8	2,7	1,4
Met	0,6	0,5	0,8	1,6	1,8	0,4	1,2	0,0	0,1	0,3	0,1
Phe	1,2	1,5	1,1	1,3	1,3	2,1	0,8	0,5	1,0	1,0	8,2
Tyr	0,3	0,4	0,3	0,3	0,3	1,9	0,1	0,0	0,9	0,4	2,6
Lys	2,7	2,9	2,8	2,6	2,6	2,8	3,7	1,5	0,8	2,4	0,7
OH—Lys	0,6	0,5	0,5	0,7	0,5	—	0,0	—	0,8	2,4	—
Arg	5,1	5,4	5,1	5,2	5,1	2,7	4,2	2,1	5,1	4,3	2,0
His	0,5	0,6	0,5	0,4	1,2	0,4	0,8	0,0	0,3	0,3	0,1
Asp	4,5	4,8	4,6	4,7	4,3	7,6	7,6	5,6	6,7	9,7	0,8
Glu	7,3	7,6	7,1	7,1	6,8	8,4	10,1	8,1	9,9	8,6	5,0
Pro	12,8	13,2	12,1	11,6	10,6	2,1	28,0	1,3	10,4	7,3	18,5
OH—Pro	9,3	8,2	9,2	7,6	5,9	9,9	2,4	16,5	10,0	9,4	0,0
Amid—N	(4,0)	(4,3)	(4,1)	(4,0)	(3,6)	(13,3)	(4,4)	(9,7)	(4,6)	(9,0)	—

In den einzelnen Bindegewebsformen sind diese Komponenten nach Art und Mengenverhältnis verschieden vertreten. Dies gilt zunächst für das Verhältnis zwischen Fasern und Zwischensubstanz: So sind besonders Knorpel reich an Zwischensubstanz, während Sehnen und Haut überwiegend aus Fasern bestehen, die in den Sehnen parallel, in der Haut der Säugetiere als unregelmäßiges, dreidimensionales Flechtwerk vorliegen. Aber auch die Art der Mucopolysaccharide wechselt: So finden sich in Haut, Sehnen und Nackenband Chondroitinsulfat B; in Knorpel, Knochen und Aorta Chondroitinsulfat A.

$$
\begin{array}{c}
H \\
| \\
H_2N-C-COOH \\
| \\
(CH_2)_l \\
| \\
CH_2
\end{array}
$$

Desmosin $(k + l + m = 4)$.

(I)

Die Fasern des *Elastins* unterscheiden sich von den Kollagenfasern durch ihre vollkommen andere Aminosäurezusammensetzung *(Tabelle 1)*, durch die seit langem bekannte, sehr unterschiedliche mikroskopische Anfärbbarkeit *(153, 405)*, durch die hohe Dehnbarkeit, aber auch durch das Verhalten beim Erhitzen in Wasser oder gegenüber Fermenten. Im nativen Zustand werden Kollagen und Elastin von den üblichen Gewebsproteinasen nicht, wohl aber von zwei verschiedenen spezifischen Enzymen, *Kollagenase* aus *Clostridium histolyticum* und *Elastase*, einem in geringer Menge in der Pankreasdrüse und in Mikroorganismen *(270, 306)* gebildeten Enzym, abgebaut. Nach Denaturierung durch Hitze und chemische Reagenzien ist Kollagen spielend für alle Proteinasen verdaulich (s. unten), während die Widerstandsfähigkeit des Elastins sich nicht ändert. Charakteristisch für Elastin und verantwortlich für seine hohe Resistenz ist die vor kurzem durch Partridge *(390)* entdeckte Quervernetzung durch die Desmosine (1), (2), die durch einen Ring-schluß der Seitenketten von Lysinresten benachbarter Ketten entstanden sein dürften. Die Annahme einer Umwandlung von Elastin in Kollagen (*213*) ist sicher unrichtig.

Die Kollagenfasern können im Bindegewebe lichtmikroskopisch durch ihre spezifische *Anfärbbarkeit* erkannt werden. Besondere Bedeutung hat dabei die Anfärbung mit Pikrofuchsin nach Hansen und die Van-Gieson-Färbung (*351*) erlangt. Für denselben Zweck haben sich auch verschiedene Fluoreszenzfarbstoffe bewährt (*384*). Im Elektronenmikroskop können die Kollagenfasern eindeutig durch ihre charakteristische Querstreifung erkannt werden.

(2)

Isodesmosin ($w + x + y = 4$).

Charakteristisch für die kollagenen Fasern sind weiterhin:

a) Die sich von anderen Proteinen erheblich unterscheidende *Aminosäurenzusammensetzung* (*34*). Sie ist gekennzeichnet durch einen hohen Gehalt an Glycin, welcher etwa einem Drittel der Menge aller Aminosäuren entspricht, weiters durch einen hohen Gehalt an Prolin und Hydroxyprolin. Hydroxyprolin kommt, von geringen Ausnahmen im Tier- und Pflanzenreich abgesehen, nur in Kollagen vor. Die Bestimmung von Hydroxyprolin ist somit eine viel verwendete Methode zur Ermittlung des Kollagengehaltes eines Gewebes (*296, 385, 332, 410*) und zur Verfolgung des Kollagenstoffwechsels. Neben Hydroxyprolin enthält Kollagen in geringer Menge auch Hydroxylysin. Auch diese Aminosäure wird in anderen Proteinen nur sehr selten angetroffen.

b) Die hohe *Quellbarkeit* und das Verhalten bei der *Denaturierung*. Bei letzterem Vorgang schrumpft die ursprünglich nur wenig dehnbare Faser auf etwa $1/_3$ ihrer Länge zusammen. Die geschrumpfte Faser ist gummelastisch (vgl. Abschn. VI/2, S. 254).

c) Die hohe *Widerstandsfähigkeit* des nativen Kollagens *gegenüber Proteasen*. Lediglich Kollagenasen, die eine ausgesprochene Sonderstellung einnehmen, vermögen natives Kollagen anzugreifen. Im Gegensatz zu nativem Kollagen wird die geschrumpfte Faser von Proteasen rasch verdaut (*252, 99, 114*). Der Angriff von Pepsin auf Kollagen bei

40° ist damit zu erklären, daß im sauren Wirkungsbereich des Pepsins die Denaturierungstemperatur infolge der Quellung herabgesetzt ist (*99, 346*). Über besondere Veränderungen des Kollagens durch enzymatischen Angriff bei tieferer Temperatur s. Abschnitt IV/3, S. 237.

d) Das *Röntgenbeugungsdiagramm* des Kollagens (s. Abschnitt V/1, S. 238).

Die Kollagenfasern setzen sich aus Bündeln von Fibrillen zusammen. Diese Fibrillen sind nur aus einer einzigen Art von monomeren Molekülen, den Kollagenmolekülen, aufgebaut. Das *Kollagenmolekül* ist ein steifes Stäbchen von etwa 3000 Å Länge und 14,5 Å Durchmesser. An seinem Bau sind drei Peptidketten von je etwa 1000 Aminosäuren beteiligt. Jede Kette bildet eine linksgewundene Schraube. Die drei Ketten sind zusätzlich im Rechtssinn nochmals zu einer übergeordneten Schraube verdrillt. Wasserstoffbindungen quer zur Faserachse sorgen für den Zusammenhalt der drei Peptidketten (*343, 337*).

Die Existenz des aus drei Peptidketten aufgebauten Stäbchenmoleküls ist auf drei Wegen bewiesen:

a) Durch Auswertung der Röntgendiagramme (Abschnitt V/1, S. 238).

b) Auf Grund des Molekulargewichts und der Dimensionen des Moleküls, die an löslichem Kollagen ermittelt wurden (Abschnitt V/2, S. 245).

c) Auf Grund des Zerfalls bei der Denaturierung, wobei die aus einer Kette bestehenden α-, die aus zwei Ketten bestehenden β- und die aus drei Ketten aufgebauten γ-Komponenten erhalten werden.

Von den drei Ketten sind nach Untersuchungen von Piez (*323*) zwei, die mit $\alpha 1$ bezeichnet werden, in ihrer Aminosäurezusammensetzung identisch. Die dritte, $\alpha 2$ genannt, unterscheidet sich dagegen von den anderen beiden. Neue Befunde zeigen, daß zumindest in einzelnen Fällen alle *drei* Peptidketten unterschiedlich aufgebaut sind (*160a, 320*). In vielen Fällen sind zwei, in seltenen alle drei Ketten durch covalente „*intramolekulare*" Quervernetzungen miteinander verbunden. Besteht eine Vernetzung zwischen zwei $\alpha 1$-Ketten, dann bezeichnet man das entstandene Dimere mit β_{11}-Komponente; liegt die Vernetzung zwischen einer $\alpha 1$- und einer $\alpha 2$-Kette, so liegt eine β_{12}-Komponente vor. Im γ-Kollagenmolekül sind alle drei Ketten miteinander verbunden. Kollagen ohne intramolekulare Vernetzung, bestehend aus drei α-Ketten, wird mit Typ I, solches bestehend aus einer α- und einer β-Komponente als Typ II und vollständig vernetztes γ-Kollagen mit Typ III bezeichnet (*104*). Kollagen vom Typ III wird in Kollagenen verschiedener Herkunft immer nur in geringer Menge gefunden (5—10%).

Vorhandensein und Zahl kovalenter intermolekularer *Quervernetzungen* sind entscheidend für den Unterschied zwischen löslichem und unlöslichem Kollagen, aber auch zwischen dem reifen jugendlichen und älteren

Literaturverzeichnis: SS. 293—314.

Kollagen (*394, 395*) (s. Abschnitt VIII/2, S. 285 f.). Für die *Gerbung* ist eine Einführung zusätzlicher intermolekularer Quervernetzungen, für die *Gelatine-Herstellung* umgekehrt eine Auflösung der vorhandenen Quervernetzungen charakteristisch.

Nach neueren Untersuchungen (s. Abschn. IV/2, S. 232) bestehen die Einzelketten des Kollagenmonomeren aus Untereinheiten, die durch besondere Bindungen — diskutiert werden Ester und esterähnliche Bindungen — miteinander verknüpft sind.

Das bisher gezeichnete Bild erlaubt noch keine Aussage über die Ursache der im Elektronenmikroskop sichtbaren *hochstrukturierten Querstreifung*. Zu ihrer Erklärung sind in der Hauptsache zwei Theorien zu diskutieren, die möglicherweise bei entsprechender Interpretierung beide richtig sein können:

a) Abwechseln von Bereichen mit leichten apolaren und von solchen mit schweren polaren Aminosäuren;

b) Abwechseln von kristallin geordneten und weniger geordneten Bereichen.

Sicher ist, daß dieses Abwechseln eine Eigenschaft der Molekülkette selbst sein muß; sicher ist weiters (s. Abschnitt IV/1, S. 217 f.), daß die apolaren Bereiche, die den Hellteilen des elektronenmikroskopischen Bildes zugeordnet sind, im wesentlichen aus Tripeptid-Sequenzen der allgemeinen Formel $(—G—P—R)_n$ aufgebaut sind (G = Glycin, sehr selten Alanin; P = Prolin; R = jede beliebige Aminosäure einschließlich Hydroxyprolin, aber nicht Prolin). Fest steht weiter, daß in den polaren Bereichen Anhäufungen basischer und Anhäufungen saurer Aminosäuren vorhanden sind und daß sie Prolin nur in geringerem Umfange enthalten. Unentschieden ist aber, welcher Grad von Ordnung in chemischem und kristallographischem Sinn diesen Bereichen zukommt. Völlig ungeordnet können sie nicht sein, weil sonst die Steifheit des Molekülstäbchens unverständlich wäre. Die Wiederkehr von Glycin oder Alanin an jeder dritten Stelle, die für die Ausbildung einer den „geordneten" Bereichen analogen Struktur postuliert worden ist (*335, 336, 343*), ist mit dem größten Teil der bisherigen Befunde vereinbar. Es ist immerhin wahrscheinlich, daß auch diesen Bereichen eine ähnliche Struktur zukommt wie den apolaren Bereichen, wenn sie auch möglicherweise weniger stabil ist.

Die Zusammenlagerung der Kollagenmoleküle zu den verschiedenen *im Elektronenmikroskop sichtbaren Strukturformen* erfolgt im wesentlichen durch elektrostatische Kräfte. Folgende Strukturformen sind gegenwärtig bekannt. Ihr molekularer Bau und der Mechanismus ihrer Entstehung werden im VII. Kapitel behandelt (S. 264 ff.).

a) Die native Fibrille. Sie ist die einzige im adulten Wirbeltierorganismus bisher aufgefundene Strukturform des Kollagens. [HALL, JAKUS und SCHMITT (*146, 365*) 1942 und WOLPERS (*412, 413*) 1943.]

Hochdifferenzierte Aufnahmen zeigen heute bis zu 10—14 Querstreifen innerhalb einer Periode von etwa 650 Å. Das Querstreifungsmuster ist asymmetrisch *(Abb. 1)*. Fibrillen des nativen Typs, die mit der nativen

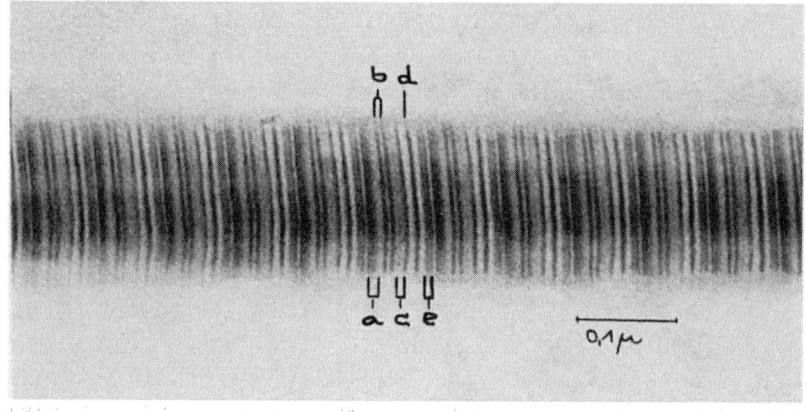

Abb. 1. Kollagenfibrillen des nativen Typs mit einem asymmetrischen Querstreifungsmuster, Querstreifungsperiode 670 Å; 12 Querstreifen pro Periode, angefärbt mit Phosphorwolframsäure und Uranylacetat.

Fibrille in der Querstreifung völlig übereinstimmen, werden auch aus sauren Kollagenlösungen durch Dialyse gegen Wasser, verdünnte NaCl-Lösung, neutralen Phosphatpuffer oder durch Erwärmen neutraler Lösungen der Ionenstärke 0,15—0,4 auf etwa 35° erhalten.

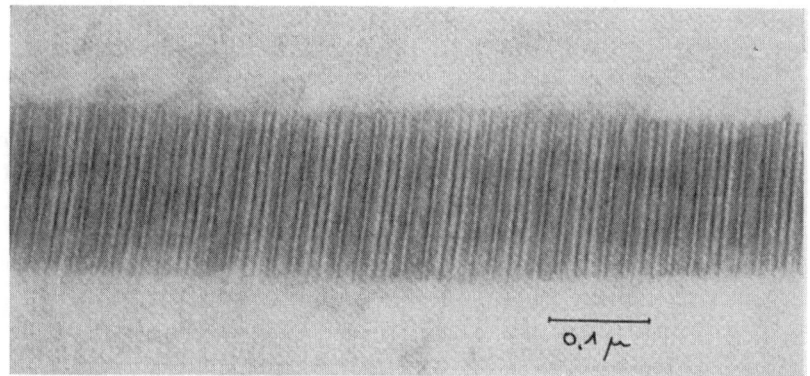

Abb. 2. Kollagenfibrillen mit symmetrischem Querstreifungsmuster aus durch Pepsinbehandlung hochgereinigtem Kollagen, Periode etwa 650 Å, angefärbt mit Phosphorwolframsäure und Uranylacetat.

b) Fibrillen mit einem symmetrischen Querstreifungsmuster und einer Periode von 650 Å. Von Kühn bei der Abscheidung aus hochgereinigten Kollagenlösungen beobachtet *(Abb. 2) (244)*.

Literaturverzeichnis: SS. 293—314.

c) Fibrillen mit einer Querstreifung von 100—120 Å. Von Kühn bei der Abscheidung aus hochgereinigten Kollagenlösungen beobachtet *(Abb. 3) (244)*.

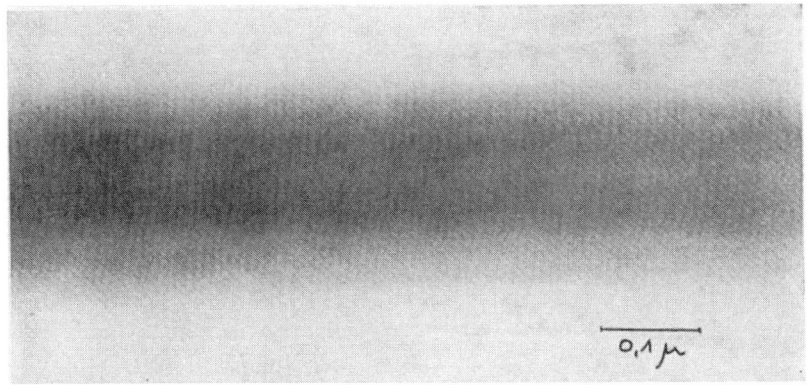

Abb. 3. Kollagenfibrille mit einer Querstreifungsperiode von etwa 100 Å aus durch Pepsinbehandlung hochgereinigtem Kollagen, angefärbt mit Phosphorwolframsäure und Uranylacetat.

d) Long-spacing-Segmente entdeckt 1953 durch Schmitt, Gross und Highberger *(363)*. Länge 2800 Å, 40—45 Querstreifen, Querstreifungsmuster asymmetrisch. Abscheidung aus sauren Kollagenlösungen durch ATP *(Abb. 4) (363)*.

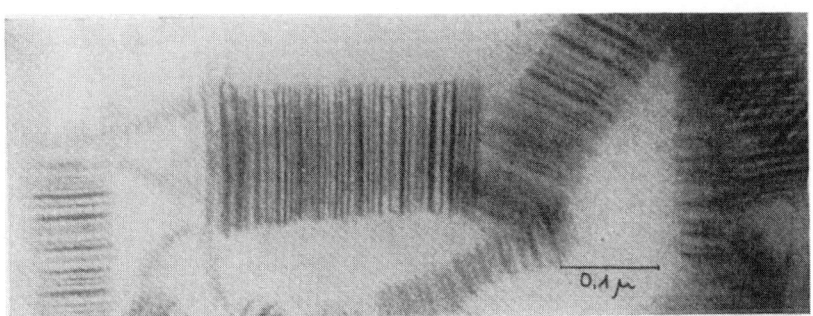

Abb. 4. Long-spacing-Segmente mit einem asymmetrischen Querstreifungsmuster und über 40 dunklen Querstreifen, angefärbt mit Uranylacetat, Länge 2800 Å. In dem mittleren Segment befindet sich das „A-Ende" links, das „B-Ende" rechts.

e) Long-spacing-Segmente mit symmetrischer Querstreifung. Aufgefunden durch Hörmann *(190)*, mitgeteilt durch Grassmann (Scheveningen 1963) und Kühn *(Abb. 5) (244)*.

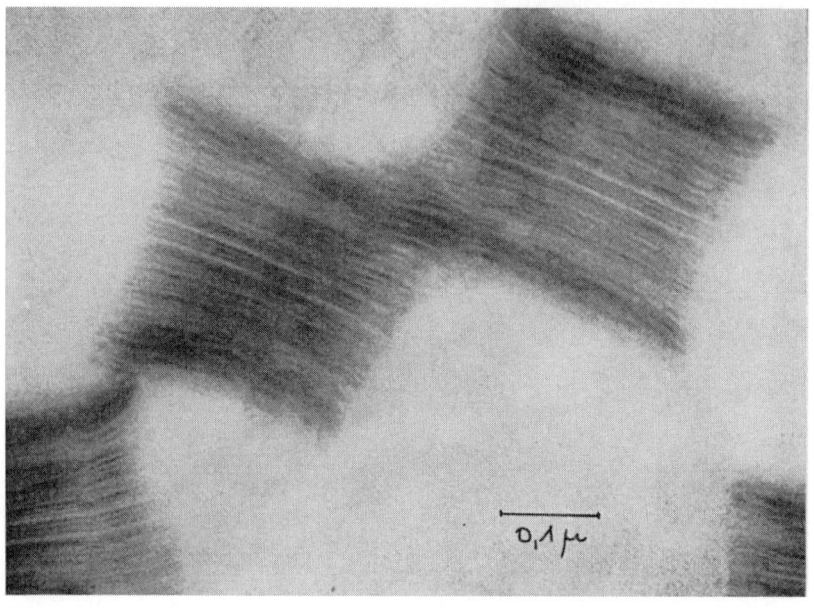

Abb. 5. Long-spacing-Segmente mit einem symmetrischen Querstreifungsmuster, hergestellt aus hochgereinigtem Kollagen, angefärbt mit Phosphorwolframsäure und Uranylacetat.

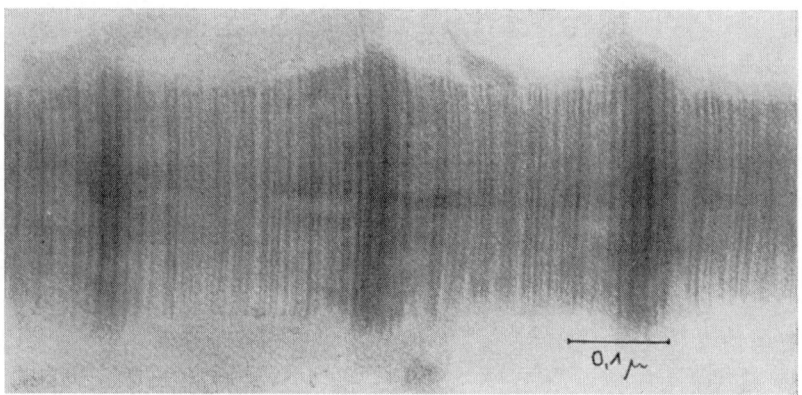

Abb. 6. Long-Spacing-Fibrille Typ I mit einem symmetrischen Querstreifungsmuster und einer Periode von 2600 Å, angefärbt mit Phosphorwolframsäure und Uranylacetat.

f) Long-spacing-Fibrillen Typ I. Entdeckt durch Highberger, Gross und Schmitt 1950, Periodenlänge ca. 2500 Å, Muster symmetrisch *(Abb. 6)* *(167)*.

Literaturverzeichnis: SS. 293—314.

g) Long-spacing-Fibrillen Typ II. Entdeckt durch KÜHN und ZIMMER 1961, Periodenlänge ca. 2500 Å, Muster symmetrisch *(Abb. 7) (249)*.

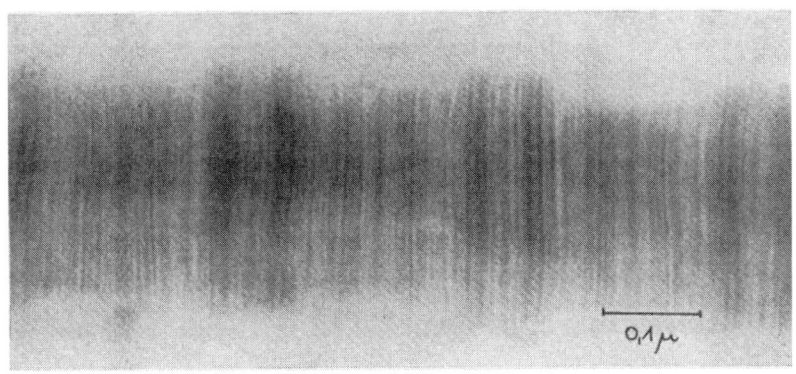

Abb. 7. Long-spacing-Fibrille Typ II mit einem symmetrischen Querstreifungsmuster und einer Periode von 2600 Å, angefärbt mit Phosphorwolframsäure und Uranylacetat.

III. Aminosäurezusammensetzung.

1. Kollagene der Wirbeltiere.

Chemisch ist Kollagen gekennzeichnet durch einen hohen Gehalt an Glycin und den Aminosäuren Prolin und Hydroxyprolin, durch die Anwesenheit von Hydroxylysin und einen relativ geringen Gehalt an aromatischen und schwefelhaltigen Aminosäuren. Neben dem mengenmäßig bei weitem überwiegenden L-4-Hydroxyprolin sind neuerdings auch kleine Mengen L-3-Hydroxyprolin aufgefunden worden *(309)*. Die Aminosäurezusammensetzung der *Säugetier-Kollagene* und — soweit bekannt — auch deren elektronenmikroskopisches Bild, Röntgendiagramm und physikalisch-chemisches Verhalten, stimmen weitgehend überein. Ungefähr ein Drittel der gesamten Aminosäurereste entfällt auf Glycin, 8—9% auf Hydroxyprolin und 14% auf Prolin. Tyrosin, Histidin und die schwefelhaltigen Aminosäuren machen jeweils weniger als 1% aus. Ein Teil des in unreinen Präparaten gefundenen Tyrosins gehört sicher Verunreinigungen an, doch ist eine kleine Menge Tyrosin ein wesentlicher Bestandteil des Kollagens (vgl. Tabelle 1, S. 199 und Tabelle 3, S. 216—217). Die Aminosäurezusammensetzung der *Fisch-Kollagene* variiert in einem weiteren Bereich als die der Säugetier-Kollagene. Der Gesamtgehalt an Iminosäuren ist geringer als bei Säugetier-Kollagen. Andererseits sind die Hydroxyaminosäuren Serin, Threonin und manchmal auch Hydroxylysin erhöht, so daß die gesamte Zahl der Hydroxylgruppen in

Säugetier- und Fischkollagen etwa übereinstimmt. Abgesehen vom Kollagen der Lungen-Fische, deren Aminosäurezusammensetzung derjenigen der Säugetiere am nächsten steht, ist der Methionin-Gehalt höher als bei Säugetieren. Der Gehalt an Iminosäuren (*40, 207, 388, 142, 276, 143*) und die Denaturierungstemperatur gehen bei Kollagen parallel. Dies stützt die Ansicht, daß den unter vorwiegender Beteiligung von Prolin und Hydroxyprolin aufgebauten „apolaren" Bereichen eine besondere Bedeutung für die Stabilität des Kollagenmoleküls zukommt. Das Kollagen von Fischen, die in warmen Gewässern leben, zeigt höhere Temperaturbeständigkeit und höheren Gehalt an Iminosäuren als dasjenige von Bewohnern kalter Gewässer (*140*).

Am Aufbau des Kollagens ist neben Aminosäuren eine kleine Menge von Hexosen beteiligt, die kovalent an das Kollagen gebunden ist (*120, 113, 132, 233, 244, 23*). Es handelt sich dabei um Galaktose und Glucose (*120, 30*); gelegentlich wurden auch Spuren von Mannose (*133*) und Fucose (*92*) gefunden. Ichthyocoll (Kollagen der Karpfenschwimmblase) enthält insgesamt 6,3 Galaktose- und 4,2 Glucosereste auf 3000 Aminosäuren (*30*). Ähnliche Mengen wurden auch in löslichem und unlöslichem Rinderkollagen festgestellt (*120, 113, 233, 132*). Die Hexosen sind einer Osazonbildung nicht zugänglich, was für eine O-glykosidische Bindung spricht (*111*). Für Ichthyocoll konnten Blumenfeld und Mitarb. (*30*) zeigen, daß die Hexosen monofunktionell gebunden vorliegen, also als einzelne Reste an das Protein gebunden sind. Bezüglich einer Beteiligung von Hexosen an den Quervernetzungen des unlöslichen Kollagens s. Abschnitt IV/3, S. 236.

2. Kollagene im niederen Tierreich.

Außer in Wirbeltieren sind Kollagene im Regenwurm, in den Rundwürmern (Anneliden) und Nemathelminthen, den Echinodermen, Coelenteraten und Mollusken gefunden (*403, 275, 276, 407*). Auch das Spongin B der Schwämme ist zu den Kollagenen zu rechnen (*404*). Auffallend ist dabei der meist weitgehend konstante Glycingehalt bei mitunter erheblicher Variabilität der übrigen Aminosäuren. Der Gesamtgehalt an Iminosäuren schwankt zwischen 11 und 30%, wobei die Differenzen der beiden einzelnen Iminosäuren noch wesentlich höher sind (Prolin zwischen 1,3 und 28%; Hydroxyprolin zwischen 2,4 und 16,5%).

IV. Primärstruktur.

1. Aminosäuresequenzen des Kollagens.

Anfangs der dreißiger Jahre hatte Astbury (*12*) in einem Vortrag geäußert, „es könnten die Aminosäuren des Kollagens irgendwie in Dreiergruppierungen angeordnet sein"; jede dritte Aminosäure sollte dabei Glycin, jede neunte Hydroxyprolin sein. Bergmann (*27, 28*) griff diesen Gedanken auf und entwickelte ihn zunächst für das Kollagen weiter. Darüber hinaus sollte seine „Frequenzregel" genannte Theorie auch auf andere Proteinstrukturen anwendbar sein.

Literaturverzeichnis: SS. 293—314.

Die ersten sequenz-analytischen Studien am Kollagen sind später in der Absicht unternommen worden, die Gültigkeit dieser ein wenig an Zahlenmystik erinnernden Gedankengänge zu prüfen. Es wird im folgenden jedoch gezeigt werden, daß die These eines periodischen Aufbaus aus kleinen Einheiten, die sich in der übrigen Proteinchemie nicht bewährt hat [eine Ausnahme bildet möglicherweise das Seidenfibroin (262, 400a)], für Teile des Kollagenmoleküls in der Tat Gültigkeit besitzt.

Neben diese Fragestellung sind in der Folgezeit zwei weitere getreten, die die volle Sequenzermittlung des Kollagens zu einem brennenden Problem machen: In der Primärstruktur der Polypeptidketten des Kollagens sind, wie im VI. und VII. Kapitel eingehend belegt werden wird, alle „Informationen" zum Aufbau der Tertiärstruktur des Tripelhelix-Moleküls und der übergeordneten elektronenmikroskopisch sichtbaren Kollagenstrukturen enthalten (vgl. 101a). Von einer vollständigen Aufklärung der Primärstruktur wird man also vertiefte Einblicke in den Mechanismus des geordneten Zusammentritts der drei Ketten zum Molekül und der Moleküle zu den sichtbaren Strukturen erwarten dürfen.

Der vollständigen Ermittlung der Aminosäuresequenz kommt aber im Falle des Kollagens auch deswegen eine grundsätzliche Bedeutung zu, weil sie die Voraussetzung für eine exakte chemische Interpretierung des elektronenmikroskopischen Bildes der Fibrillen und Segmente bildet. Zwar kennen wir schon heute, wie im folgenden näher gezeigt werden soll, die grundsätzlichen chemischen Ursachen für das Auftreten von Hell- und Dunkelstreifen, aber von einer vollständigen Lösung dieses Problems wird man erst dann sprechen können, wenn man wirklich jedem einzelnen der hellen und dunklen Querstreifen bestimmte Aminosäuresequenzen innerhalb der Polypeptidketten zuordnen kann. Von diesem Ziel sind wir noch weit entfernt. Das hat vor allem drei Gründe:

a) Verglichen mit den bisher zur Konstitutionsermittlung herangezogenen Proteinen [Myoglobin (60), Hämoglobin (38), Cytochrom c (273), Proteineinheit des Tabakmosaik-Virus (8), Ribonuklease (10), Trypsin (401), Chymotrypsin (214, 157)] ist das Molekulargewicht des Kollagens mit etwa 320000 um ein Vielfaches größer. Die Zahl der Aminosäurereste einer einzelnen Kette beträgt also etwa 1100, die des Gesamtmoleküls das Dreifache.

b) Die Einzelketten sind im chemischen Aufbau außerordentlich ähnlich und zugleich recht empfindlich; ihre Trennung ist wesentlich schwieriger als bei anderen großen, aus Untereinheiten aufgebauten Proteinen und in größerem präparativem Maßstabe gegenwärtig noch nicht gelungen. Im Prinzip scheint dieses Problem jedoch nun gelöst zu sein (321).

c) Auch der periodische Aufbau weiter Teile des Moleküls aus Tripeptidsequenzen erweist sich, zunächst paradoxerweise, als Nachteil für eine vollständige Sequenzaufklärung, wie unten (S. 225 f.) gezeigt werden wird.

Das Problem der vollständigen Sequenzanalyse des Kollagens würde sich stark vereinfachen, wenn die These von Hodge (324, 177) sich bestätigen sollte, nach der die α1-Ketten aus fünf, die α2-Ketten aus sieben untereinander *identischen* Untereinheiten aufgebaut sein sollen. Es scheinen aber gegenwärtig eine Reihe von Tatsachen (vgl. vor allem S. 262) gegen diese aus einer Analyse des elektronenmikroskopischen Bildes der Kollagensegmente abgeleiteten und durch Analogien mit anderen Proteinen höheren Molekulargewichts (z. B. Tabakmosaik-Virus) nahegelegte These zu sprechen. Man muß heute zwar, wie weiter unten näher ausgeführt wird, Untereinheiten der α-Ketten annehmen (im englischen Sprachgebrauch als ,,sub-subunits" bezeichnet); sie dürften jedoch untereinander keineswegs identisch sein.

Eine systematische Sequenzanalyse des Kollagens sollte ausgehen von den einheitlichen α1-, α2- (und eventuell α3-) Ketten und nach Möglichkeit von größeren einheitlichen Bruchstücken derselben, die durch eine spezifische chemische Spaltung an bestimmten Stellen der Kette erhalten werden können. Gegenwärtig kommen drei, zumindest weitgehend spezifische Methoden für diesen Zweck in Frage:

1. Die Spaltung mit Hydroxylamin. Der unter relativ gelinden Bedingungen erfolgende Abbau des Kollagens durch H_2NOH (83) war zunächst auf eine Spaltung esterartiger Quervernetzungen bezogen worden (181, 182). Eine Spaltung von Quervernetzungen durch Hydroxylamin erfolgt auch tatsächlich; der quantitativ bedeutsamere Vorgang ist aber, wie heute feststeht, eine Spaltung der Hauptkette in kleinere Bruchstücke (vgl. Abschnitt IV/2, S. 232).

2. Behandlung von denaturiertem Kollagen mit Alkali führt, wie in unserem Laboratorium gefunden wurde (359), zu einem Zerfall in Bruchstücke von hohem Molekulargewicht. Auch diese Spaltungsmöglichkeit wird in Abschnitt IV/2 ausführlich behandelt.

3. Die kürzlich beschriebene, für Methionin-Bindungen spezifische Spaltung von Proteinen mit BrCN (128) läßt sich mit Erfolg auch auf Kollagen anwenden, wenn man es vorher denaturiert (304, 33a). (Dagegen wird das native Molekül auch bei sehr stark saurer Reaktion, pH = 1, nicht gespalten.) Das mittlere Molekulargewicht der Bruchstücke im nicht aggregierten Zustand ist zu etwa 15—16000 gefunden worden (303). Dies entspricht der Bildung von etwa 18 Bruchstücken auf das Gesamtmolekül oder 6 Teilstücken je Einzelkette, was mit dem Methioningehalt in Übereinstimmung steht.

Die erwähnten Spaltungsmöglichkeiten werden in den kommenden Jahren für die Sequenzermittlung des Kollagens zunehmende Bedeutung gewinnen. Die bisher vorliegenden Teilergebnisse zur Aminosäuresequenz bauen aber auf wesentlich weniger spezifischen chemischen oder enzym-chemischen Spaltungsreaktionen auf, wobei durchwegs Gesamtkollagen verwendet wurde. Folgende Wege sind dabei bisher beschritten worden:

a. Partielle Säure- und Alkali-Hydrolyse.

Im Gegensatz zu den im vorausgehenden besprochenen Methoden ist diese Spaltung weitgehend unspezifisch, wenn auch gewisse Bindungen, z. B. diejenigen des Serins und der Asparaginsäure, durch besonders hohe, andere, z. B. die des Valins und Isoleucins, durch besonders geringe Spaltbarkeit ausgezeichnet sind. Die Reaktion ist deshalb zu einem willkürlich festgelegten oder empirisch ermittelten Zeitpunkt zu beenden, die Zahl der Bruchstücke meist sehr groß, die Aufarbeitung schwierig.

SCHROEDER (368, 369) und KRONER (221, 222) wandten diese Methode auf das Kollagen an und beschrieben in den Jahren 1953—1955 über 60 Di-, Tri- und Tetrapeptide, die sie so erhalten hatten (Tabelle 2). Die wichtigste sequenzanalytische Erkenntnis aus ihren Ergebnissen war, daß nur die wenigsten der gefundenen Bruchstücke exakt zu dem von ASTBURY (13) bevorzugten Strukturvorschlag $(P—G—R)_n$ passen wollten (P = Pro oder Hypro; G = Gly; R = eine der anderen Aminosäuren). So ist Glycin auf der Aminoseite häufig mit Aminosäuren verknüpft, die von Prolin verschieden sind; Prolin ist seinerseits mit seiner Carboxylgruppe häufig an Aminosäuren gebunden, die nicht Glycin sind. Zu dem erwähnten Vorschlag paßte ferner nicht das Vorkommen von Gly—Gly als solches oder in Form von Arg—Gly—Gly, vor allem aber die Existenz von Gly—Pro—Hypro und das bemerkswert häufige Auftreten des Dipeptids Gly—Pro, was eher auf ein Element Gly—Pro—R hinzuweisen schien. Jedenfalls waren solche Tripeptide mit einer gewissen Variabilität von R ebenfalls isoliert worden. SCHROEDER schloß aus diesen Ergebnissen und der Tatsache, daß auch Hypro—Gly in großer Menge auftrat, auf eine regel-mäßige Anordnung aus vier Aminosäuren. Die Sequenz sollte R—P—P—R sein (P = Pro oder Hypro; R = beliebige Aminosäure), ihr wichtigstes Beispiel Gly—Pro—Hypro—Gly (369). In seiner Arbeit taucht zum ersten Mal der Gedanke auf, daß die Periodizität nur einen Teil der Kollagen-struktur beherrschen könnte, ein Gedanke, der im Zusammenhang mit späteren Ergebnissen in den Vordergrund rückte (s. unten).

b. Enzymatische Methoden.

α) Abbau des denaturierten Kollagens mit Trypsin.

Unter den klassischen Proteinasen bietet nur das Trypsin Aussichten, beim Abbau eines so großen Moleküls zu definierten Abbauprodukten zu

Tabelle 2. Peptide aus Partialhydrolysaten von Gelatine und Kollagen [größtenteils aus *(221, 222; 369, 370)*]. Die Zahlen geben die Ausbeute in Mikromolen Peptid je 250 mg Gelatine bzw. Kollagen an [entnommen aus *(100)*].

Neutrale Peptide	Basische Peptide	Saure Peptide	Peptide mit sauren und basischen Aminosäuren	Prolin- und hydroxyprolinhaltige Peptide
Ala—Gly 13,0	Ala—Arg 3,0	Gly—Asp 1,0	Asp—Arg 1,2	Gly—Pro 61,8
Ala—Ala 4,6	Ala—(Arg, Gly) ... 0,2	Gly—Glu 7,0	Asp—(Arg, Gly) 0,7	Gly—Pro—Gly 0,4
Gly—Ala 9,0	Ala—Lys 1,9	Ala—Asp 1,9	Glu—Arg 1,8	Gly—Pro—Ala 3,5
Val—Gly 4,1	Arg—Gly—Gly 0,4	Val—Glu 0,5	Glu—Arg—Gly 1,0	Gly—Pro—Glu 0,7
Ser—Gly 18,4	Lys—Gly 1,8	Leu—Glu 0,4	4,7	Ser—Pro—Gly 0,5
Ser—Ala 1,5	Ser—Arg 0,9	Glu—Gly 4,5		Ala—Pro—Gly 0,3
Thr—Gly 17,4	8,2	Glu—Ala 6,6		Lys—Pro—Gly .... —,—
Thr—Ala 1,1		Gly—Asp—Gly 0,5		Pro—(Gly, Lys) .. 1,0
Gly—Gly .... —,—		22,4		Pro—Ser 1,0
Leu—Ala .... —,—				Pro—Thr 0,7
Ala—Gly—Ala.... —,—				Hypro—Gly 35,6
Ala—Ala—Gly .... —,—				Ala—Hypro—Gly . 3,6
69,1				Ala—Hypro 1,0
				Leu—Hypro 1,7
				Ser—Hypro—Gly . 1,4
				Glu—Hypro 0,9
				Glu—Hypro—Gly . 2,4
				Gly—Pro—Hypro . 4,2
				Gly—(Hypro, Pro)—Gly 3,1
				123,8

gelangen. Denn im Gegensatz beispielsweise zu Chymotrypsin, Pepsin und Papain ist nur die Spezifität des Trypsins scharf begrenzt. Trypsin spaltet Bindungen, an denen basische Aminosäuren, nämlich Arginin und Lysin mit ihren COOH-Gruppen beteiligt sind. Nach Spaltung dieser Bindungen bleibt der Angriff scharf stehen.

Wir haben schon 1955 damit begonnen (*103*), die Produkte des Abbaues von hitzedenaturiertem Kollagen mit kristallisiertem Trypsin zu untersuchen. Es erwies sich nach den Ergebnissen der ersten Abbauversuche als notwendig, das noch Spuren von Chymotrypsin enthaltende Trypsin elektrophoretisch zu reinigen (*107*). Erst dann erfolgte der Abbau streng nach der Spezifitätsregel. Der tryptische Abbau des hitzedenaturierten Kollagens wurde titrimetrisch nach der Methode von WALDSCHMIDT-LEITZ verfolgt (*Abb. 8*). Aus der Kurve ist

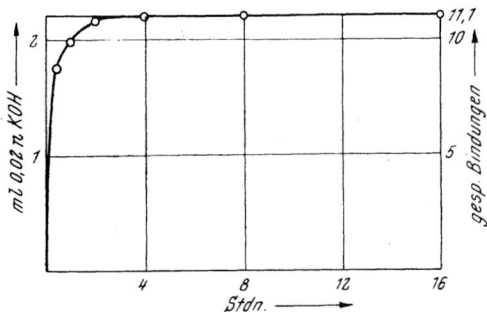

Abb. 8. Abbau des lösl. Kollagens mit gereinigtem Trypsin. Titration der freigesetzten Carboxylgruppen nach WALDSCHMIDT-LEITZ. pH 8,2 eingestellt mit NaOH, 84 mg Kollagen pro Titrationsprobe; linke Ordinate: ml 0,02-n alkohol. KOH gegen Thymolphthalein; rechte Ordinate: Zahl der gespaltenen Bindungen, berechnet auf 216 Aminosäuren.

zu ersehen, daß bereits nach vierstündiger Inkubation bei 37° der Abbau zu einem scharfen Endpunkt kommt; trotzdem wurde die Enzymeinwirkung über 16 Stunden weitergeführt.

Aus dem Aciditätszuwachs in Verbindung mit der Aminosäureanalyse ist zu errechnen, daß nur etwa 60% der auf Grund des Gehalts an basischen Aminosäuren errechneten Angriffsstellen geöffnet wurden. Es liegt nahe, dieses Ergebnis mit dem hohen Gehalt des Kollagens an den Iminosäuren Prolin und Hydroxyprolin in Verbindung zu bringen. GRIMM (*127*) konnte nämlich an den synthetischen Substraten Leu-–Lys—Pro-—Gly und Leu—Lys—Hypro—Gly, die sich als völlig trypsinresistent erwiesen, belegen, daß Bindungen zwischen der Carboxylgruppe einer basischen Aminosäure und dem Stickstoff der beiden Iminosäuren für Trypsin unangreifbar sind. Das gleiche ergibt sich aus der Tatsache, daß beim Abbau des Kollagens mit Trypsin und den meisten anderen Proteinasen strenge Äquivalenz zwischen freigesetzten Carboxyl- und Aminogruppen gefunden wird (s. S. 223). In diesem Zusammenhang ist erwähnenswert, daß GRASSMANN und RIEDERLE (*119*) schon 1936 das Vorkommen von Lys—Pro—Gly in Gelatine wahrscheinlich gemacht haben. Auch in jüngerer Zeit gelangen Anreicherungen von basischen Fraktionen aus Gelatine, die zu etwa $^2/_3$ aus den Aminosäuren Gly, Pro und Lys bestanden (*185*).

Aus der Zahl der gelösten Peptidbindungen errechnet sich für die Produkte des tryptischen Abbaues eine mittlere Kettenlänge von etwa 19—20 Aminosäuren. Bei einem Molekulargewicht von 320000 sind also etwa 170 Abbauprodukte zu erwarten. Die Auftrennung so komplizierter Gemische derartiger Peptide ist aber schwierig und bisher noch kaum versucht worden.

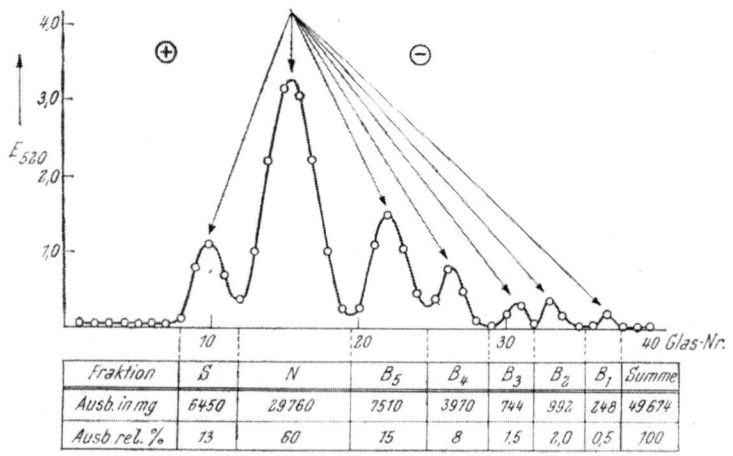

Fraktion	S	N	B_5	B_4	B_3	B_2	B_1	Summe
Ausb. in mg	6450	29760	7510	3970	744	992	248	49674
Ausb. rel. %	13	60	15	8	1,5	2,0	0,5	100

Abb. 9. Verteilung und Ausbeute der Fraktionen nach der elektrophoretischen Trennung des tryptischen Abbaugemisches von lösl. Kollagen in der Elphor-VaP-Apparatur; Anfärbung mit Ninhydrin (nach alkalischer Hydrolyse). Pyridin-Essigsäure-Puffer; pH 4,9; 1200 V, 175 mA, Filtrierkarton Binzer 230, Durchsatz: 3 g Peptidgemisch/Tag, insgesamt 62 g.

Als überraschend aussichtsreicher Weg dafür hat sich die von uns entwickelte kontinuierliche Elektrophorese erwiesen (*148*). *Abb.* 9 zeigt die Trennung der Spaltstücke bei neutraler Reaktion, wobei eine größere Zahl basischer Fraktionen, eine der Menge nach überwiegende Neutralfraktion und einige saure Fraktionen erhalten werden. Das Trennbild ist für Kollagen aus Haut oder Sehne und für lösl. Kollagen identisch. Die Neutralfraktion kann weiter aufgetrennt werden, wenn man sie bei saurer Reaktion als Kationen wandern läßt.

Durch diese Vortrennung erhielten wir schließlich zwölf elektrophoretisch einheitliche Fraktionen, die aber zum größten Teil noch aus einer Vielzahl von Komponenten bestehen (*108*). Zu ihrer weiteren Auftrennung bewährte sich die Chromatographie an Dowex 1 × 2 (*Abb. 10*). Als endgültig einheitlich sehen wir dabei eine Fraktion erst dann an, wenn sie weder elektrophoretisch noch chromatographisch weiter zerlegt werden kann und bei der quantitativen Aminosäureanalyse ein sauberes, ganzzahliges Verhältnis der Aminosäuren ergibt.

In der *Tabelle 3* sind 20 der über 100 erhaltenen und näher untersuchten Peptide mit ihren Ausbeuten und den gefundenen Aminosäure-

Literaturverzeichnis: SS. 293—314.

verhältnissen zusammengestellt. Die Kettenlänge der erhaltenen Peptide liegt zwischen 3 und 110 Aminosäuren. In fast allen Fällen macht Glycin ein Drittel der gesamten Menge aller gefundenen Aminosäuren aus, was zunächst für eine gleichmäßige Verteilung dieser Aminosäure innerhalb der Ketten spricht. Die zunächst naheliegende Vorstellung, daß Glycin an jeder dritten Stelle der Kette wiederkehre, läßt sich trotzdem für das Gesamtkollagen nicht aufrechterhalten. Dem widerspricht schon die Isolierung von Peptiden, in denen zwei Glycinreste unmittelbar mit-

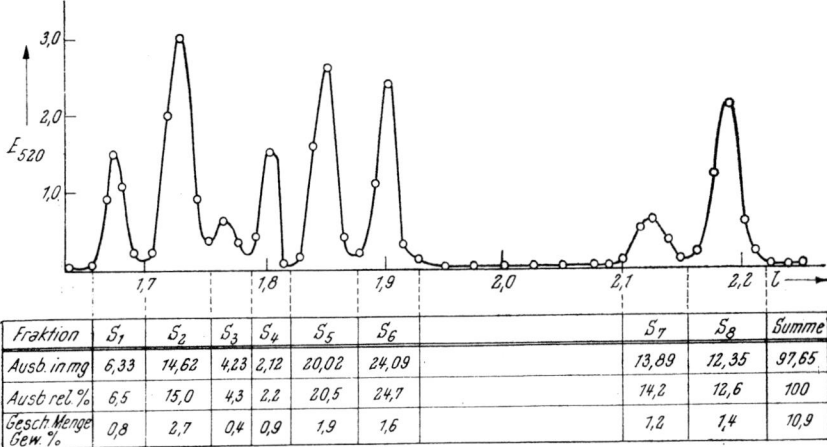

Fraktion	S_1	S_2	S_3	S_4	S_5	S_6		S_7	S_8	Summe
Ausb. in mg	6,33	14,62	4,23	2,12	20,02	24,09		13,89	12,35	97,65
Ausb. rel. %	6,5	15,0	4,3	2,2	20,5	24,7		14,2	12,6	100
Gesch. Menge Gew. %	0,8	2,7	0,4	0,9	1,9	1,6		1,2	1,4	10,9

Abb. 10. Verteilung und Ausbeuten der Fraktionen nach der säulenchromatographischen Trennung von Fraktion S_{II} an Dowex 1×2. Anfärbung mit Ninhydrin (nach alkalischer Hydrolyse). pH-Gradienten-Elution: pH 8,0 → pH 3,0.

einander verknüpft oder nur durch eine Aminosäure getrennt vorliegen (vgl. dazu S. 211 f.), wie durch SCHROEDER (369), KRONER (221, 222) und uns (108) gezeigt werden konnte. Anderseits wurden aber auch Peptide gefunden, in denen wenigstens 4 oder 5 Sequenzen ohne Glycin auftraten (s. unten).

Am Aminoende wurde bisher ohne Ausnahme in allen Peptiden Glycin gefunden. Am Carboxylende traten erwartungsgemäß ausschließlich Lysin und Arginin auf. In den trypsinspaltbaren Bindungen sind also die beiden basischen Aminosäuren ausschließlich an Gly gebunden. Ein großer Teil der Peptide enthält basische Aminosäuren auch in der Mitte der Kette. Dieser Befund ist im Hinblick auf die tryptische Unspaltbarkeit von Bindungen des Prolin- und Hydroxyprolin-Stickstoffs oben bereits diskutiert worden.

Ein Peptid mit der Bezeichnung S_f aus einem älteren Abbauversuch, in dem ein noch Spuren von Chymotrypsin enthaltendes Trypsinpräparat für den Abbau verwendet wurde, ist ganz rechts in der Tabelle 3 mit

Tabelle 3. Aminosäurezusammensetzung der in

Fraktion	B₁	S₈ G	B₃	S₁	N_IV 1	S₅	N_IV 3	N_IV 2	S₄
Gew.-% im Gemisch	0,50	1,70	0,5	0,80	2,20	1,9	2,40	3,90	0,90
Gly	1	5	5	19	11	12	12	20	20
Ala.........	—	—	—	2	2	6	2	6	4
Val.........	—	—	—	—	1	—	2	1	1
Leu	—	—	—	—	—	—	—	—	—
Ileu	—	—	—	—	—	—	—	—	2
Phe	—	—	—	—	—	—	—	—	—
Ser	—	—	—	1	3	—	2	1	1
Thr	—	1	—	—	—	—	1	—	2
Tyr	—	—	—	—	—	—	—	—	—
Met	—	—	—	—	—	—	—	—	—
Pro	—	—	—	7	1	7	5	10	9
Hypro	—	—	—	8	2	2	1	2	6
Asp	—	2	—	15	2	3	2	3	1
Glu	—	3	—	2	3	6	5	6	11
Lys	—	—	1	—	2	—	1	4	3
Hylys	—	1	—	—	—	—	—	—	—
His.........	1	—	—	—	—	—	—	—	—
Arg	1	1	4	4	1	2	2	2	—
Kettenlänge .	3	13	10	58	28	38	35	55	60
Mol-% Gly..	33	38	50	33	39	32	34	36	33
Mol-% polare Aminosäure	66	54	50	36	29	29	29	27	25
Mol-% Pro + Hypro	—	—	—	26	11	24	17	22	25
NH₂-Endgruppen	Gly	Gly	3 Gly	3 Gly	3 Gly	Gly	3 Gly	3 Gly	Gly
COOH-Endgruppen	Arg	Arg	Arg	3 Arg	2 Lys 1 Arg	Arg	1 Lys 2 Arg	2 Lys 1 Arg	Lys
MG aus AS Analyse .	368	1321	1055	6528	2718	3468	3584	4915	5528
MG aus Endgruppen .	350	1200	1000	2300	870	3350	1200	1700	5000

aufgeführt; am Carboxylende fanden wir Asparaginsäure. Das auch in anderer Hinsicht bemerkenswerte Peptid wird unten noch diskutiert werden.

Wenn man die bisher untersuchten einheitlichen Peptide, so wie es in der Tabelle 3 geschehen ist, nach ihrem Gehalt an stark polaren Aminosäuren ordnet, fällt auf, daß sich der Gehalt an Iminosäuren im allgemeinen

Literaturverzeichnis: SS. 293—314.

einheitlicher Form gewonnenen Abbaupeptide (*108*).

$N_{VI\,g\,1}$	$N_{VI\,v}$	$N_{VI\,w}$	$N_{VI\,i\,1}$	$N_{VI\,f}$	$B_{5\,r}$	$N_{IV\,4}$	$N_{VI\,1}$	$N_{VI\,g\,2}$	$N_{VI\,c}$	S_f
0,60	0,20	0,40	0,60	0,50	2,50	3,00	0,20	0,30	0,30	3,1
38	22	15	20	28	18	16	33	17	23	14
14	8	6	9	10	5	2	10	4	7	3
2	2	3	—	2	1	1	2	2	5	—
3	1	2	3	2	1	2	4	2	6	1
—	—	—	—	1	—	1	—	—	1	—
—	2	1	—	2	—	3	2	—	1	1
3	4	8	2	1	3	3	4	4	7	2
1	2	6	—	2	1	—	—	2	7	1
—	—	1	—	—	—	—	—	—	—	2
2	—	1	1	1	—	—	—	—	3	—
8	5	3	1	12	9	6	14	5	6	10
13	4	4	10	8	4	7	12	7	4	5
3	4	3	3	3	2	1	3	—	2	2
16	6	1	7	8	2	3	11	4	—	2
2	4	4	—	3	2	4	2	—	3	—
—	—	—	—	—	—	—	—	—	—	—
—	—	—	—	—	—	1	—	—	—	—
5	1	3	2	2	4	—	2	2	2	—
110	65	63	58	85	52	50	99	49	77	43
34	34	24	34	33	34	32	33	34	30	33
24	23	21	21	19	19	18	18	12	9	9
19	14	11	19	24	25	26	26	24	13	35
3 Gly	3 Gly	Gly	3 Gly	3 Gly	3 Gly	3 Gly	3 Gly	Gly	3 Gly	Gly
3 Arg	2 Lys 1 Arg	Lys	3 Arg	2 Lys 1 Arg	2 Lys 1 Arg	3 Lys	1 Lys 2 Arg	Arg	3 Arg	Asp
10035	5840	5913	5234	7722	4670	4630	9013	4354	6936	3916
3450	1700	5800	1600	2300	1420	1200	3000	4000	2400	3700

gegenläufig zu der Menge der polaren Aminosäuren verhält. Aus diesem Ergebnis kann man bereits auf das Vorliegen apolarer Bereiche schließen, die reich an Iminosäuren sind, während andere Bereiche arm an Iminosäuren, jedoch reich an polaren Säuren sein müssen. Die an zahlreichen isolierten Einzelpeptiden durch enzymatische und chemische Abbaumethoden vom Amino- und Carboxylende her ganz oder teilweise durchgeführte

Sequenzanalyse bestätigt dieses Ergebnis (vgl. *Tabelle 4*). Auch diese Befunde zeigen in ihrer Gesamtheit ein wechselweises Auftreten prolin- und hydroxyprolin-reicher (apolarer) und an Iminosäuren armer polarer Bereiche. In letzteren werden teils vorwiegend basische, teils vorwiegend saure Bezirke angetroffen.

So befinden sich z. B. im sequenzmäßig nicht erfaßten Mittelstück des Peptids N_{IV_3} sechs Prolin- und Hydroxyprolinreste, die zusammen mit zwölf weiteren neutralen kurzkettigen Aminosäuren (Glycin

Tabelle 4. Konstitution von Abbaupeptiden des lösl. Kollagens.

1. Prolin-freie stark polare Peptide.

B_1:   H—Gly—His—Arg—OH

S_{86}:   H—Gly—Asp—Glu—Gly—Hylys—(3 Gly,   Thr,   2 Glu,   Asp)—Arg—OH

2. Prolin- und hydroxyprolin-haltige Peptide.

S_f: H—Gly—Ser—Ala—(Thr, Leu, Gly)— $\begin{bmatrix} 10\,\text{Pro}\ 5\,\text{Hypro, } 12\,\text{Gly,} \\ 2\,\text{Ala, Ser, }2\,\text{Tyr, Glu} \end{bmatrix}$ —Phe—
Glu—Leu—Asp—Asp—COOH

S_1: $\left.\begin{array}{l} \text{H—Gly—} \\ \text{H—Gly—} \\ \text{H—Gly—} \end{array}\right(\text{Gly, Ser, 2 Asp, Glu} \left.\right)$ — $\begin{bmatrix} 7\,\text{Pro, }8\,\text{Hypro, }15\,\text{Gly,} \\ 2\,\text{Ala, Glu, }13\,\text{Asp, Arg} \end{bmatrix}$ $\begin{array}{l} \text{—Arg—OH} \\ \text{—Arg—OH} \\ \text{—Arg—OH} \end{array}$

S_5: $\text{H—Gly—}\left(\text{Asp, Glu, Ala}\right)$ — $\begin{bmatrix} 7\,\text{Pro, }2\,\text{Hypro, }11\,\text{Gly,} \\ 5\,\text{Ala, }2\,\text{Asp, }5\,\text{Glu, Arg} \end{bmatrix}$ —Arg—OH

N_{IV_1}: $\left.\begin{array}{l} \text{H—Gly—} \\ \text{H—Gly—} \\ \text{H—Gly—} \end{array}\right(\text{Gly, Ser, Glu, Ala} \left.\right)$ — $\begin{bmatrix} \text{Pro, }2\,\text{Hypro, }7\,\text{Gly, Ala} \\ \text{Val, }2\,\text{Ser, }2\,\text{Asp, }2\,\text{Glu} \end{bmatrix}$ $\begin{array}{l} \text{—Lys—OH} \\ \text{—Lys—OH} \\ \text{—Arg—OH} \end{array}$

N_{IV_2}: $\left.\begin{array}{l} \text{H—Gly—} \\ \text{H—Gly—} \\ \text{H—Gly—} \end{array}\right(\begin{array}{l}\text{4 Gly, 4 Ala, 2 Glu,} \\ \text{2 Asp}\end{array} \left.\right)$ — $\begin{bmatrix} 10\,\text{Pro, }2\,\text{Hypro, }13\,\text{Gly,} \\ 2\,\text{Ala, Val, Ser, Asp,} \\ 4\,\text{Glu, }2\,\text{Lys, Arg} \end{bmatrix}$ $\begin{array}{l} \text{—Lys—OH} \\ \text{—Lys—OH} \\ \text{—Arg—OH} \end{array}$

N_{IV_3}: $\left.\begin{array}{l} \text{H—Gly—} \\ \text{H—Gly—} \\ \text{H—Gly—} \end{array}\right(\begin{array}{l}\text{Gly, Ala, Ser, Thr,} \\ \text{3 Glu, Asp}\end{array} \left.\right)$ — $\begin{bmatrix} 5\,\text{Pro, Hypro, }8\,\text{Gly, Ala} \\ \text{Ser, }2\,\text{Val, Asp, }2\,\text{Glu} \end{bmatrix}$ $\begin{array}{l} \text{—Lys—OH} \\ \text{—Arg—OH} \\ \text{—Arg—OH} \end{array}$

N_{VI_1}: $\left.\begin{array}{l} \text{H—Gly—} \\ \text{H—Gly—} \\ \text{H—Gly—} \end{array}\right(\text{Gly, Ala, 2 Glu, Asp} \left.\right)$ — $\begin{bmatrix} 14\,\text{Pro, }12\,\text{Hypro, }29\,\text{Gly,} \\ 9\,\text{Ala, }2\,\text{Val, }4\,\text{Leu, }2\,\text{Phe,} \\ 4\,\text{Ser, }9\,\text{Glu, }2\,\text{Asp, Lys} \end{bmatrix}$ $\begin{array}{l} \text{—Lys—OH} \\ \text{—Arg—OH} \\ \text{—Arg—OH} \end{array}$

N_{VI_v}: $\left.\begin{array}{l} \text{H—Gly—} \\ \text{H—Gly—} \\ \text{H—Gly—} \end{array}\right(\text{Gly, Glu, Ala, Ser} \left.\right)$ — $\begin{bmatrix} 5\,\text{Pro, }4\,\text{Hypro, }18\,\text{Gly,} \\ 7\,\text{Ala, }2\,\text{Val, Leu, }2\,\text{Phe,} \\ 3\,\text{Ser, }2\,\text{Thr, }4\,\text{Asp,} \\ 5\,\text{Glu, }2\,\text{Lys} \end{bmatrix}$ $\begin{array}{l} \text{—Lys—OH} \\ \text{—Lys—OH} \\ \text{—Arg—OH} \end{array}$

N_{VI_w}: $\text{H—Gly—}\left(\text{Ala, Ser, Glu}\right)$ — $\begin{bmatrix} 3\,\text{Pro, }4\,\text{Hypro, }14\,\text{Gly,} \\ 5\,\text{Ala, }3\,\text{Val, }2\,\text{Leu, Phe,} \\ 7\,\text{Ser} \\ 6\,\text{Thr, Tyr, Met, }3\,\text{Asp,} \\ 3\,\text{Lys, }3\,\text{Arg, }2\,\text{Glucos-} \\ \text{amin} \end{bmatrix}$ —Lys—OH

Literaturverzeichnis: SS. 293—314.

Alanin, Serin, Valin) einem apolaren Bereich angehören, während sich unter den ersten acht Aminosäuren nach dem Aminoende vier Dicarbonsäuren befinden, so daß dieser Bereich des Peptids ausgesprochen sauren Charakter besitzt. Auch bei den übrigen in Tabelle 4 angeführten Peptiden besitzt das Aminoende stark sauren Charakter, während im nicht aufgeklärten Mittelstück neben einer starken Anreicherung von Prolin und Hydroxyprolin auch noch saure bzw. basische Aminosäuren anzutreffen sind. Isolierte prolin- und hydroxyprolinfreie Peptide, wie die Peptide B_1 und S_{86}, zeigen ausgesprochen polaren Charakter.

Besonders aufschlußreich war die Untersuchung des aus 44 Aminosäuren bestehenden Peptides S_f. In diesem Peptid entfallen rund $^1/_3$ der Aminosäuren auf Prolin und Hydroxyprolin. Aber diese beiden Iminosäuren sind keineswegs gleichmäßig über die Kette verteilt; man kann vom Aminoende insgesamt sechs, vom Carboxylende fünf Aminosäuren abspalten, unter denen sich keine der beiden Iminosäuren befindet. Diese gehören vielmehr einem in seiner Sequenz noch nicht völlig aufgeklärten Mittelstück aus 33 Aminosäuren an, in dem außerdem noch Glykokoll, Serin und Alanin, also Aminosäuren mit geringer Raumbeanspruchung, angehäuft sind.

In einigen Fällen waren die Ergebnisse der quantitativen Endgruppenbestimmungen überraschend. Bezogen auf das Molekulargewicht wiesen sehr viele Peptide drei Aminoendgruppen (und zwar immer Glycin) und drei Carboxylendgruppen auf (Arginin und Lysin im Verhältnis 2 : 1 oder umgekehrt). Zum Beispiel ergibt sich für das Peptid N_{VII} aus der Aminosäureanalyse ein Mindestmolekulargewicht von 9013, während die Aminoendgruppenbestimmung nach SANGER (354) unter Zugrundelegung dieses Molekulargewichtes drei aminoendständige Glycinreste ergibt. Die C-terminale Aminosäurebestimmung nach AKABORI (4) ergab für dieses Peptid pro angenommenes Mindestmolekulargewicht 2 Argininreste und 1 Lysin. Auf Grund dieses Befundes kann man annehmen, daß die Fraktion N_{VII} aus drei Ketten besteht, die durch intramolekulare Bindungen miteinander verknüpft sind. Ähnliche Ergebnisse wurden bei den Peptiden S_1, N_{IV1}, N_{IV2}, N_{IV3}, N_{VIc}, N_{VIf}, N_{VIg1}, N_{VII}, N_{VIv} und B_{5f} erhalten (vgl. Tabelle 3 und 4). Die Annahme des Vorliegens von Dreierketten in den Abbaupeptiden wird durch weitere Befunde gestützt.

Gegen die Annahme, daß in den isolierten Abbaupeptiden die ursprüngliche Helix des Kollagens noch erhalten sei, muß eingewendet werden, daß schon bei der Hitzedenaturierung des Kollagens vor dem tryptischen Abbau eine Entfaltung, also eine Trennung der Dreier-Peptidkette, eintritt. Die Möglichkeit, daß bei derartigen prolin- und hydroxyprolin-reichen Peptiden eine Tendenz zu einem sekundären Wiederzusammentritt zur Tripelhelix besteht, wird durch die Rena-

turierungsversuche am Kollagen nahegelegt. Dagegen spricht aber, daß
der Drehwert dieser Peptide, soweit ermittelt, niedrig und etwa in der
Gegend des vollständig denaturierten Kollagens gelegen ist und sich beim
Erwärmen und Abkühlen nur unwesentlich ändert. Der Zusammenhalt
der Dreikettenpeptide könnte daher zunächst nur durch kovalente
Quervernetzungen erklärt werden, wobei die Ketten auch statistisch
geknäuelt vorliegen können [vgl. dazu auch Hodge (*324*, *177*)].

Nachdem Kühn (*247*) gezeigt hat, daß in den dunklen Teilen Anhäufungen saurer und basischer Aminosäuren, also polare Bereiche,
vorliegen, und nachdem die Massendicke der vorwiegend aus leichten
Aminosäuren aufgebauten „apolaren" Bereiche gering ist, erscheint es ein-
leuchtend, die dunklen Querstreifen den ersteren, die Hellteile den letzteren
zuzuordnen. Weitere Belege für diese Auffassung werden unten (S. 229)
gebracht werden. Da die native Fibrille unter Versetzung der Kollagenmoleküle um jeweils $1/_4$ ihrer Länge gebildet wird (vgl. Kapitel VII,
S. 266 ff.) und nur das elektronenmikroskopische Bild der Long-spacing-
Segmente das wirkliche Verteilungsmuster der Aminosäuren entlang
des Moleküls richtig widerspiegelt, ist die Zuordnung im einzelnen nur
an elektronenmikroskopischen Aufnahmen von Long-spacing-Segmenten
möglich, die bei einer Länge von 2800 Å 40 und mehr dunkle Querstreifen
erkennen lassen. Bei der Kompliziertheit des Moleküls ist es jedoch noch
verfrüht, die isolierten Peptide bestimmten Teilen des elektronenmikroskopischen Bildes zuordnen zu wollen.

β) Kollagenase — der Schlüssel für die apolaren Bereiche.

Bei der Kollagenase aus *Clostridium histolyticum* handelt es sich um
ein Enzym, dessen Spezifität bisher in der Proteasen-Chemie einzig da-
steht. Kollagen und sein Denaturierungsprodukt Gelatine sind die
einzigen natürlich vorkommenden Substrate dieses Enzyms, und umgekehrt vermag Kollagenase als einzige Protease Kollagen im nativen
Zustand bei annähernd neutraler Reaktion abzubauen. Die umfangreiche
Literatur zu diesem Thema findet man in zwei Übersichtsartikeln aufgeführt (*268*, *302*).

Die komplizierte Spezifität des Enzyms ist von den Arbeitskreisen
von Nagai und Noda (*289*), Heyns (*166*) und von uns (*112*) geklärt
worden. Nach den aus Abbaustudien an Kollagen selbst und aus Untersuchungen an synthetischen Substraten gewonnenen Ergebnissen muß
die den Spezifitätsansprüchen der Kollagenase genügende Anordnung

$$\overset{\downarrow}{X—P—R—G—P—Y}$$

lauten. (Dabei bedeuten: P = Prolin, Hydroxyprolin oder auch Sarkosin;
G = Glycin oder Alanin; R = eine, soweit man weiß, beliebige Amino-

Literaturverzeichnis: SS. 293—314.

säure; X und Y können Aminosäuren, aber auch Schutzgruppen sein, z. B. X der Carbobenzoxyrest und Y eine Amid- oder Estergruppe; ↓ bedeutet die Spaltstelle.) Bei der Spezifität der Kollagenase liegt also der bisher in der Peptidasen-Chemie unbekannte Fall vor, daß eine größere Sequenz bestimmter Aminosäuren für den Angriff des Enzyms Voraussetzung ist.

Für die photometrische Bestimmung des Enzyms haben sich vor allem drei Substrate bewährt:

$$\overset{\downarrow}{\text{Cbo—Gly—Pro—Gly—Gly—Pro—Ala}} \; (116),$$

$$\overset{\downarrow}{\text{Cbo—Gly—Pro—Leu—Gly—Pro}} \; (290) \text{ und}$$

$$\overset{\downarrow}{\text{PZ—Pro—Leu—Gly—Pro—D—Arg}} \; (422)$$

(Cbo = Carbobenzoxy-; PZ = p-Phenylazobenzyloxycarbonyl-; ↓ = = Spaltstelle).

Zur Reinigung der Kollagenase siehe (373, 123, 387a).

Der Anwendung der Kollagenase sind entscheidende Einblicke in die Sequenz der apolaren Bereiche, aber auch wichtige Aufschlüsse über die polaren Bereiche des Kollagenmoleküls zu verdanken. Bei Einwirkung von Kollagenase auf Kollagen entstehen keine freien Aminosäuren (373), sondern Peptide. Aminoendständig findet sich in Übereinstimmung mit der oben angegebenen Spezifitätsregel nahezu ausschließlich Glycin, am Carboxylende dagegen eine Vielzahl verschiedener neutraler, saurer und basischer Aminosäuren, wobei Hydroxyprolin und Alanin überwiegen (166, 282, 288, 125). Das bestätigte die erste präparative Arbeit auf diesem Gebiet; SCHROHENLOHER et al. (370) fraktionierten das entstehende Peptidgemisch und wiesen nach, daß bei einem solchen Abbau u. a. Tripeptide entstehen. Sie konnten in hoher Ausbeute die Bruchstücke Gly—Pro—Ala und Gly—Pro—Hypro isolieren, die beide schon früher als Produkte partieller Säurehydrolyse bekannt waren (Tabelle 2, S. 212). Der Anteil der in dritter Position stehenden Aminosäuren belief sich auf jeweils etwa 23% des Gehalts von Alaninbzw. Hydroxyprolin im Kollagen.

In eigenen Arbeiten (117) wurde zunächst versucht, ohne Fraktionierung des Peptidgemisches, nämlich durch Edman-Abbau (59) in Verbindung mit quantitativen Aminosäureanalysen, zu quantitativen Angaben über die mit Kollagenase entstehenden Abbauprodukte zu gelangen. Am Aminoende der Spaltstellen wurde, neben sehr geringen Mengen an Alanin, ausschließlich Glycin, in zweiter Position zu 82% Prolin und zu 18% Alanin festgestellt. Es wurde gefunden, daß der Anteil der Dreiereinheiten Gly—Pro—R an der Primärstruktur des Kollagens mindestens 33 bis 35% betragen muß. Die Sequenz Gly—Pro macht also allein 22—24% aller Aminosäuren des Kollagens aus und dürfte folglich das wichtigste Strukturelement der Kollagenhelix sein. Vom Prolin sind nach unseren Ergebnissen 75% oder mehr in dieser Bindungsweise vorhanden.

Noda (81) hat den Anteil der als Tripeptide anfallenden prolinhaltigen Neutralbereiche mit 40,6% angegeben, und Franzblau (75) hat eine sogar noch höhere Menge geschätzt (50—60%). Wir haben aber unseren Wert inzwischen in den weiter unten wiedergegebenen Versuchen bestätigen können. Nagai (288) gibt an, daß die Summe von Gly—Pro—Hypro und Gly—Pro—Ala allein etwa 25% des ganzen Kollagens erfaßt.

Aus diesen Ergebnissen in Verbindung mit den Spezifitätsregeln der Kollagenase geht hervor, daß die apolaren Teile eine sich ständig wiederholende Folge von Dreiereinheiten nach dem Muster

$$—Gly—Pro—R_1—Gly—Pro—R_2—Gly—Pro—R_3— \text{ usw.}$$

sein müssen. Die aus Röntgendaten abgeleiteten Modellvorstellungen (vgl. S. 238 ff.) werden damit für diesen Teil des Moleküls chemisch bestätigt.

Welche Aminosäuren können außer Hydroxyprolin und Alanin noch die Position R im Tripeptid Gly—Pro—R einnehmen? Vor allem wohl Glycin (222), wenn auch etwas weniger, als wir (117) ursprünglich wegen eines Versagens der Edman-Methodik errechnet haben. Nach Angaben von Mandl (267) beträgt sein Anteil an den Aminosäuren der dritten Position etwa 15%; Greenberg und Mitarb. fanden 16% (125). Ferner sind direkt oder indirekt Arg (166, 81, 125, 308, 305), Glu (222, 105, 125), Asp (125), Phe (265), Ser (105) und Thr (105) als in der Position R stehend identifiziert worden.

Es erhebt sich die Frage, ob die Neutralbereiche ausschließlich Gly—Pro—R-Polymere sind. Dies scheint nicht ganz streng der Fall zu sein. Zum Beispiel kann nach Untersuchungen sowohl an synthetischen Substraten (291) als auch am Kollagen selbst (117, 125, 265) Alanin in Position 1 (statt Glycin) erscheinen. Das gilt bei synthetischen Peptiden auch für Sarcosin (291). Auch in Position 2 kann Alanin Prolin ersetzen (117, 125); Ogle et al. (308) haben Gly—Ala—Hypro sogar in Substanz isolieren können.

Im Moment herrscht noch Unklarheit darüber, ob mit dem Tripeptid Gly—Pro—Hypro der Hauptanteil des Hypro im Kollagen erfaßt wird. Nagai (288) fand in ausgezeichneter Übereinstimmung mit Logan et al. (370), daß das Hydroxyprolin dieser Sequenz nur etwa 24% des gesamten Kollagen-Hypro ausmacht. Unsere eigenen Resultate (105) sprechen bereits für einen größeren, bei 40% liegenden Anteil. Möglicherweise ist er aber noch höher anzusetzen, wie aus einer soeben erschienenen Arbeit von Greenberg et al. (125) hervorgeht.

Literaturverzeichnis: SS. 293—314.

Die Autoren haben unsere oben beschriebene Methodik (*117*) nachgearbeitet; im Zuge des Edman-Abbaus wurde dabei eine andere Cyclisierungs-Technik verwendet. Während die Menge des gefundenen Prolins in Position 2 gut mit dem von uns angegebenen Wert übereinstimmt, ist das Durchschnitts-Molekulargewicht des Peptidgemisches wesentlich kleiner (Kettenlänge: 4,5 Aminosäuren). Offensichtlich wurde für den Abbau kein reines Enzym benutzt; das findet auch darin seinen Ausdruck, daß die Anteile des Gly der 1. Position und des Pro der 2. Position sich wie 200 zu 107 verhalten, also auch nicht annähernd äquivalent sind. Für das Hydroxyprolin in der 3. Position des Abbaugemisches wird ein Wert angegeben, der 78% des gesamten Hypro im Kollagen deckt. Dieser Prozentsatz liegt ganz wesentlich höher als die oben angeführten Ausbeuten. Die Autoren rechnen selbst mit der Möglichkeit, daß nicht die ganze Menge dieses Hydroxyprolins Tripeptiden des Typs Gly—Pro—R entstammt.

In Anbetracht der Tatsache, daß Hydroxyprolin im Tierreich ausschließlich im Kollagen vorkommt und sicher eine wichtige Funktion bei der Ausbildung seiner Struktur hat, erscheinen diese Diskrepanzen besonders klärungsbedürftig.

γ) Isolierung der apolaren Bereiche.

Noch schärfere und weitergehende Einblicke in den Bau der apolaren Bereiche kann man gewinnen, wenn man diese zunächst isoliert bzw. anreichert und erst dann dem Abbau mit Kollagenase unterwirft. Eine solche Isolierung ist möglich auf Grund der Tatsache, daß die prolinreichen Bezirke nur von Kollagenase, nicht aber von den meisten bekannten Proteinasen und Peptidasen angegriffen werden.

Durch nacheinanderfolgende Einwirkung von Trypsin, Chymotrypsin, Carboxypeptidase A und B und Aminopeptidase kann man die iminosäurefreien Bereiche weitgehend abbauen und dann präparativ leicht von den intakt gebliebenen iminosäurereichen abtrennen [HANNIG (*149*), GRASSMANN (*101*)]. Diese prolinreichen Anteile können durch Kollagenase weiter gespalten werden bis zu einer Kettenlänge von 3—4 Aminosäuren. Bei allen diesen Spaltversuchen (auch mit Kollagenase) werden Amino- und Carboxylgruppen in gleicher Menge in Freiheit gesetzt, d. h. es werden keine Bindungen gespalten, an denen der Stickstoff von Prolin oder Hydroxyprolin beteiligt ist. Läßt man aber auf das zuletzt erhaltene Peptidgemisch Peptidasen eines Nierenextraktes einwirken, so wird es weiter abgebaut, diesmal praktisch ohne Auftreten freier Aminogruppen (*101*) *(Abb. 11)*. Bekanntlich enthält der Nierenextrakt Peptidasen, welche Peptidbindungen des Prolins und Hydroxyprolins zu spalten vermögen (vgl. z. B. *122, 379*). Vor der Einwirkung von Kollagenase ist das Nierenenzym unwirksam. Es werden von ihm nur die niedermolekularen Prolinpeptide, nicht aber die längerkettigen apolaren Zwischenbereiche angegriffen. Dieser Versuch zeigt in Übereinstimmung mit der oben gegebenen Spezifitätsregel, daß Kollagenase Bindungen des

Prolinstickstoffs zwar nicht angreift, aber einem nachfolgenden Angriff durch Nierenpeptidase zugänglich macht (Abb. 11).

Wir haben (105) das nach erschöpfender Spaltung mit Pepsin, Trypsin, Chymotrypsin, Leucin-Aminopeptidase, Carboxypeptidase A und B erhaltene Abbaugemisch zunächst elektrophoretisch in saure, basische und neutrale Fraktionen getrennt, wobei der Anteil der Neutralfraktion mit über 70% bei weitem überwog. Durch Chromatographie an Sephadex G-25 konnte daraus ein Peptidgemisch mit einem Iminosäuregehalt von 39,6 Mol-% erhalten werden. Die Ausbeute, bezogen auf eingesetztes

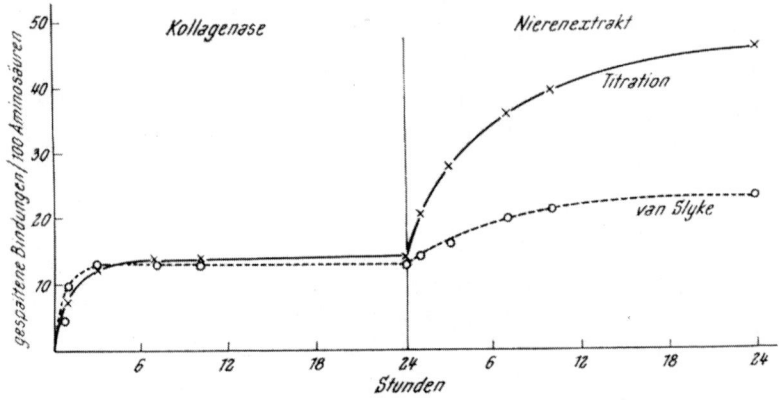

Abb. 11. Restpeptidgemisch nach erschöpfendem nacheinanderfolgendem Abbau von Kollagen mit Trypsin, Chymotrypsin, Aminopeptidase und Carboxypeptidase, angereichert durch Fällung mit Alkohol (Prolin + Hydroxyprolin 28,5 Mol./100 Mol. Aminosäuren, Glycin 31,8 Mol./100 Mol. Aminosäuren). Aufeinanderfolgender Abbau mit Kollagenase und Nierenextrakt.

Kollagenmaterial, betrug 44,3 Gew.-%, d. h. diese Fraktion enthielt etwa 80% der eingesetzten Iminosäuren. Auch die basischen und sauren Fraktionen der Gruppentrennung enthielten noch Peptide mit Prolin und Hydroxyprolin, allerdings ist ihr Anteil gegenüber dem Gesamt-kollagen wesentlich geringer.

Abb. 12 zeigt eine Fraktionierung dieses Peptidgemisches an einer 1,5 m langen Sephadex-G-25-Säule. Unter den einzelnen Gipfeln sind die jeweiligen mittleren Molekulargewichte angegeben, wie sie auf Grund der quantitativen Umsetzung mit Dinitrofluorbenzol (118), sowie die damit praktisch übereinstimmenden, die in der Ultrazentrifuge durch Gleichgewichtslauf bei 0,8 mm Schichtdicke in der Multichannel-Zelle (428) erhalten wurden. Im untersuchten Peptidgemisch sind demnach iminosäurereiche Peptide aus den apolaren Bereichen mit einer Ketten-länge von 6—30 Aminosäuren anzunehmen. Das entspricht unter Zu-grundelegung des Rich-Crick-Modells einer Helixlänge von 17—85 Å.

Die einzelnen Fraktionen dieses Trennversuches sind jedoch noch nicht einheitlich, wie sich unter anderem auch aus der Endgruppen-

Literaturverzeichnis: SS. 293—314.

bestimmung ergibt. Sie konnten papierchromatographisch sowie an einer Ionenaustauschersäule mit Dowex 1 × 2 noch weiter aufgetrennt werden. Man erhält auf diese Weise etwa 40 Peptide, von denen der größte Teil

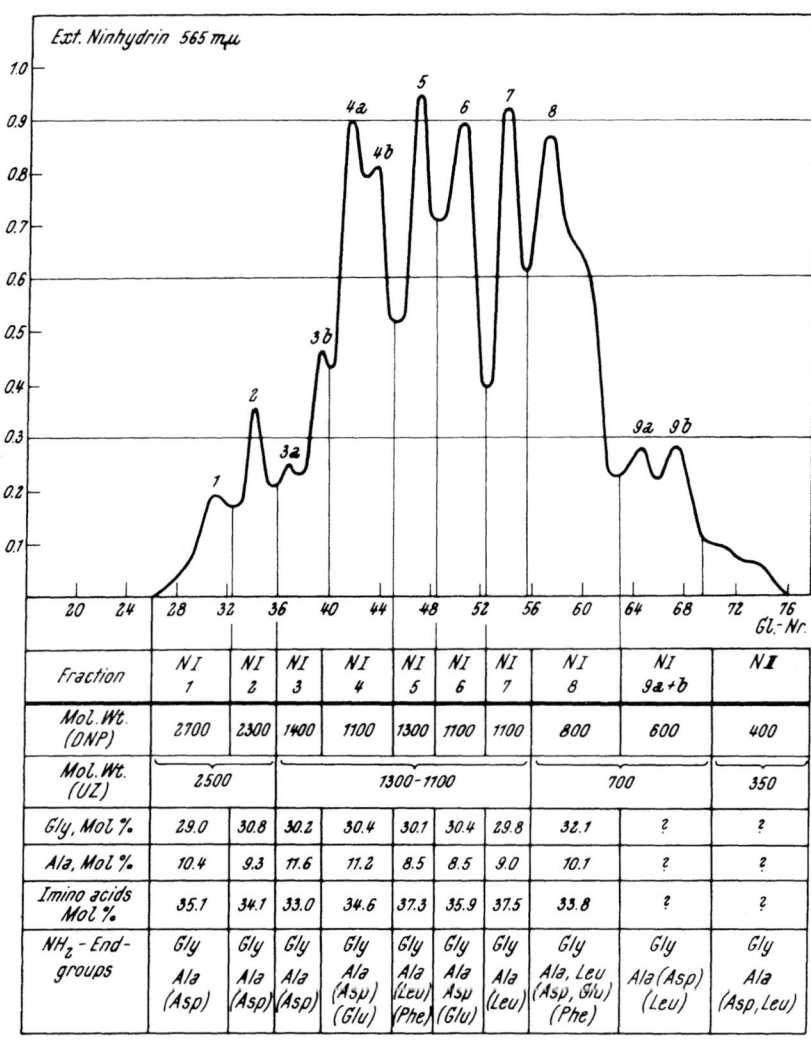

Fraction	NI 1	NI 2	NI 3	NI 4	NI 5	NI 6	NI 7	NI 8	NI 9a+b	NI
Mol.Wt. (DNP)	2700	2300	1400	1100	1300	1100	1100	800	600	400
Mol.Wt. (UZ)	2500		1300-1100					700		350
Gly, Mol %	29.0	30.8	30.2	30.4	30.1	30.4	29.8	32.1	?	?
Ala, Mol %	10.4	9.3	11.6	11.2	8.5	8.5	9.0	10.1	?	?
Imino acids Mol %	35.1	34.1	33.0	34.6	37.3	35.9	37.5	33.8	?	?
NH₂-End-groups	Gly Ala (Asp)	Gly Ala (Asp)	Gly Ala (Asp)	Gly Ala (Asp) (Glu)	Gly Ala (Leu) (Phe)	Gly Ala Asp (Glu)	Gly Ala (Leu)	Gly Ala, Leu (Asp, Glu) (Phe)	Gly Ala(Asp) (Leu)	Gly Ala (Asp, Leu)

Abb. 12. Auftrennung des Peptidgemisches der Neutralfraktion an Sephadex G-25-Säule: 150 cm × 2 cm; Elutionsmittel: Wasser 20°, 20 ml/Stunde; eingesetzt: 255 mg N-Fraktion (105).
[Aus: Z. physiol. Chem. 333, 154 (1963).]

als einheitlich anzusehen ist. Ihre vollständige Sequenzermittlung, die wir in Angriff genommen haben, dürfte bis zu Kettenlängen von drei Tripeptideinheiten ohne allzu große Schwierigkeiten mit Hilfe der Kollagenasespaltung möglich sein. Für die längeren Peptide, die nach den Ergebnissen der Abb. 12 schätzungsweise die Hälfte ausmachen, dürfte aber gerade der regelmäßige Bau aus sich wiederholenden, außerordentlich ähnlichen Tripeptideinheiten die Konstitutionsermittlung er-

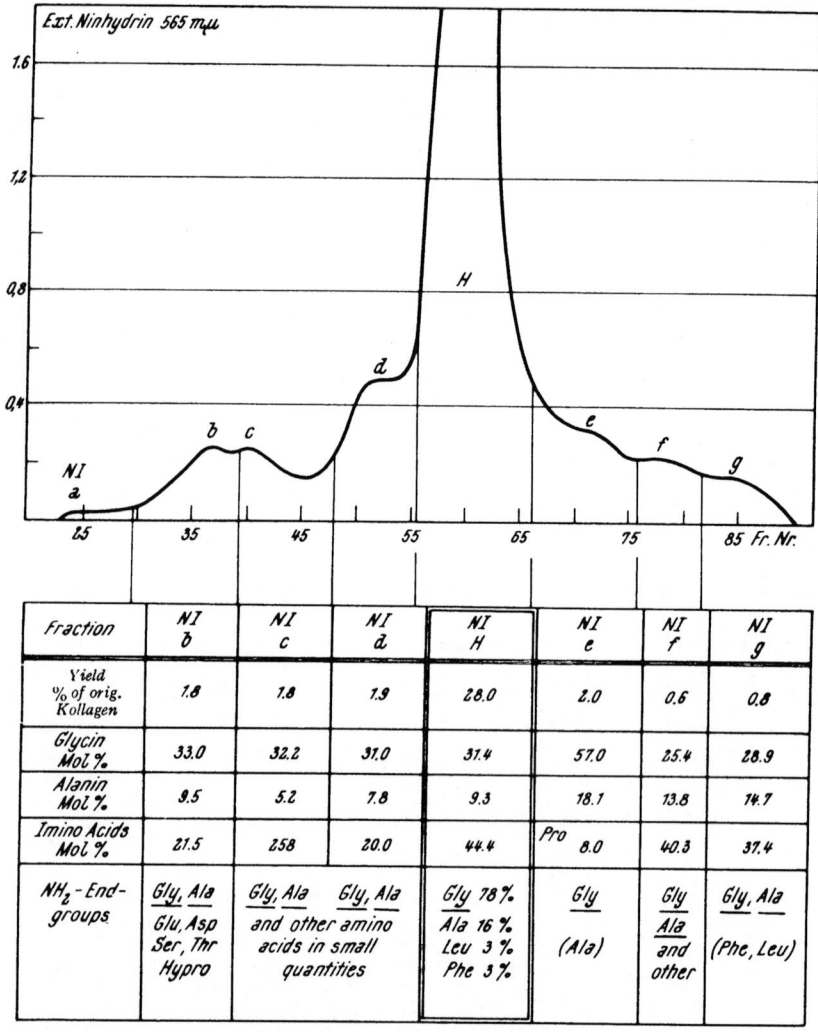

Fraction	NI b	NI c	NI d	NI H	NI e	NI f	NI g
Yield % of orig. Kollagen	1.8	1.8	1.9	28.0	2.0	0.6	0.8
Glycin Mol %	33.0	32.2	31.0	31.4	57.0	25.4	28.9
Alanin Mol %	9.5	5.2	7.8	9.3	18.1	13.8	14.7
Imino Acids Mol %	21.5	25.8	20.0	44.4	Pro 8.0	40.3	37.4
NH₂-End-groups	Gly, Ala Glu, Asp Ser, Thr Hypro	Gly, Ala	Gly, Ala and other amino acids in small quantities	Gly 78% Ala 16% Leu 3% Phe 3%	Gly (Ala)	Gly Ala and other	Gly, Ala (Phe, Leu)

Abb. 13. Auftrennung der Neutralfraktion nach Kollagenaseabbau an Sephadex G-25-Säule: 170 cm × 2 cm; Elutionsmittel: Wasser 20°, 40 ml/Stunde (105).
[Aus: Z. physiol. Chem. 333, 154 (1963).]

schweren. Da außer der Kollagenase kein anderes Enzym in der Lage zu sein scheint, Strukturen dieses Typus anzugreifen, kann das bei anderen Proteinen so erfolgreich angewandte Prinzip der „überlappenden Spaltung" mit mehreren Enzymen [BRAUNITZER (37)] hier bisher nicht

Literaturverzeichnis: SS. 293—314.

angewandt werden. Wir hoffen indessen, Wege zur Lösung dieses Problems zu finden, das als ein Schlüsselproblem in der Sequenzanalytik des Kollagens gelten kann.

Im Hinblick auf diese Schwierigkeiten wurde in einer weiteren Versuchsreihe zunächst das gesamte neutrale Peptidgemisch vor der Auftrennung an Sephadex G-25 mit Kollagenase abgebaut. Nach Kollagenasespaltung ergab sich an Sephadex G-25 das in *Abb. 13* dargestellte Trennbild. Ein Vergleich mit Abb. 12 zeigt deutlich, daß die Vielzahl der Peptide mit unterschiedlicher Kettenlänge zu Bruchstücken aufgespalten wurde, die im wesentlichen in einem Hauptgipfel H enthalten sind. Die Molekulargewichtsbestimmung der Peptide im Gipfel H ergab einen Wert von etwa 300, was einer Kettenlänge von durchschnittlich drei Aminosäuren entspricht. Der Anteil des Gipfels H beträgt etwa 28 Gew.-% vom eingesetzten Kollagen. Er besteht zu 44,4 Mol-% aus den Aminosäuren Prolin und Hydroxyprolin und umfaßt somit 58% der insgesamt eingesetzten Iminosäuren.

Es ist hervorzuheben, daß der Gipfel H nicht weniger als 97,6 Mol-% neutrale Aminosäuren enthält, wovon allein 85 Mol-% auf Glycin, Alanin, Prolin und Hydroxyprolin entfallen; der Rest setzt sich aus anderen neutralen Aminosäuren (12,6 Mol-%) sowie 1,6 Mol-% Glutaminsäure und nur 0,8 Mol-% weiteren polaren Aminosäuren zusammen. Eine weitere Auftrennung des Gipfels H an Dowex 1 × 2 liefert reines Gly—Pro—Hypro sowie Gly—Pro—Ala (10% bzw. 3% bezogen auf eingesetztes Kollagen). Weiter gesichert sind die Tripeptide Gly—Pro—Glu, Gly—Pro—Ser und Gly—Pro—Thr. Ersteres war schon von den Säurepartialhydrolysaten her bekannt, die beiden anderen hatten, wegen des häufigen Vorkommens von Pro—Ser und Pro—Thr einerseits (*369*) und Ser—Gly und Thr—Gly (*369*) anderseits, als Bruchstücke von Gly—Pro—Ser—Gly—Pro—R bzw. Gly—Pro—Thr—Gly—Pro—R ohnehin eine hohe Existenzwahrscheinlichkeit.

NODA und Mitarb. (*81*) haben inzwischen in ähnlicher Arbeitsweise gleichfalls die apolaren Bereiche des Kollagens isoliert. Denaturiertes Kollagen wurde erst mit Trypsin, dann mit Carboxypeptidase und Leucinaminopeptidase verdaut. Das Abbauprodukt wurde durch Chromatographie an Ionenaustauschersäulen, aber nicht an Sephadex fraktioniert. Für die Neutralsequenzen wurde eine durchschnittliche Kettenlänge von ungefähr 35 Å entsprechend 4,3—4,6 Tripeptideinheiten gefunden.

δ) Isolierung der polaren Bereiche.

Nach dem Gesagten kann es nicht überraschen, daß man, sozusagen in Umkehrung des oben geschilderten Weges, den Abbau mit Kollagenase auch benutzen kann, um die prolinreichen apolaren Bereiche zu zerstören und die polaren „amorphen" zu isolieren bzw. anzureichern. Voraussetzung für einwandfreie Ergebnisse ist allerdings dabei die Verwendung von Kollagenasepräparaten, die frei von beigemengten unspezifischen Peptidasen sind, was für die käuflichen Präparate nicht zutrifft (*269, 387a*).

Schon früher war aufgefallen, daß nach Verdauung von Kollagen mit Kollagenase das mittlere Molekulargewicht des Peptidgemisches über dem von Tripeptiden liegt. Michaels et al. (*282*) hatten den Wert 500, Nagai (*288*) und wir (*117*) 600—700 gefunden. In mehreren Arbeitskreisen wurde versucht, auch die längeren Bruchstücke, die offenbar der Anlaß für diese Erhöhung des Durchschnitts-Molekulargewichts sind, zu isolieren. Die Gruppe von Seifter und Mitarb. (*372, 75*) fand eine Fraktion, die aus dem Verdauungsansatz von Kollagen mit Kollagenase nur langsam abdialysiert (sogenannte „nichtdialysierbare Fraktion"). Sie enthält, wie die Aminosäureanalyse zeigt, noch etwa 7 Mol-% Prolin und 5—6 Mol-% Hydroxyprolin, während der Glycingehalt unverändert bei etwa 32% liegt. Der Gehalt an polaren Aminosäuren und der Kohlenhydratgehalt ist aber, wie unsere Analysen (*302*) bestätigten, zum Teil sehr stark gestiegen. Die Annahme erscheint naheliegend, daß es sich hierbei um die längeren von den polaren Querstreifen der Struktur handeln könnte. Der Anteil dieser schwer dialysierbaren Bruchstücke beträgt etwa 10—15 N-% von Kollagen. Noda et al. (*298*) bestätigten diese Resultate.

Auf Grund der Spezifität der Kollagenase ist zu erwarten, daß sich am Aminoende der verbliebenen größeren Bruchstücke, soweit sie aus dem Inneren der ursprünglichen Ketten stammen, die Gruppierung Gly—Pro— (bzw. Ala—Pro—) und am Carboxylende —Pro—X (beide eigentlich noch den apolaren Bereichen zugehörend) finden sollten. Die Ergebnisse von Manahan und Mandl (*265*) bestätigen diese Erwartung. Diese Autoren haben zwei Peptide aus Kollagen-Kollagenase-Verdauungsansätzen mit einer Kettenlänge von 21 bzw. 24 Aminosäuren isolieren können, die ihrer Zusammensetzung nach zweifellos hierher gehören:

$$\text{Ala—Pro—(Gly}_8\text{, Ala}_2\text{, Glu}_2\text{, Asp}_2\text{, Phe}_2\text{, Thr)—Pro—Ala und}$$
$$\text{Gly—Pro—(Gly}_6\text{, Ala}_3\text{, Val}_2\text{, Leu}_2\text{, Lys}_2\text{, Glu}_2\text{, Asp, Phe, Ser)—Pro—Phe.}$$

Interessant ist ferner, daß die optische Drehung mit $[\alpha]_{405\,m\mu} = -160°$ ganz beträchtlich unter dem für denaturiertes Kollagen gemessenen Wert, $[\alpha]_{405\,m\mu} = -360°$, liegt (*302, 81*). Das könnte bedeuten, daß die (in diesem Material fehlenden) Gly—Pro—R-Polymeren auch im Zustand der Denaturierung des Moleküls eine bestimmte Konfiguration einnehmen, was dann zu dem relativ hohen Drehwert von —360° führt.

Literaturverzeichnis: SS. 293—314.

Schließlich soll noch erwähnt werden, daß man die Einwirkung von Kollagenase auf Kollagen sichtbar machen kann. Zwei Arbeiten, die gleichzeitig und unabhängig voneinander durchgeführt wurden (298, 305), benutzten zu diesem Zweck die sogenannten SLS (s. S. 205, 266 f.). Elektronenoptisch konnte nachgewiesen werden, daß die Kollagenase wirklich nur die Hellteile angreift, anders ausgedrückt: daß die Hellteile wirklich mit den iminosäurereichen Bezirken der Faser identisch sind (vgl. *Abb. 14a—e*). Bei geeigneter Arbeitsweise kann man beobachten, wie mit fortschreitender Enzymeinwirkung die Hellteile verschwinden, während die „bands" auch nach längerer Zeit unangegriffen bleiben.

Zusammenfassend kann man nach dem derzeitigen Stand der Kenntnisse folgendes über die Primärstruktur des Kollagens sagen:

a. Die Primärstruktur ist diskontinuierlich aufgebaut; es existieren durch Kollagenase spaltbare Sequenzabschnitte, die der Sequenz Gly—Pro—R entsprechen und vorwiegend aus neutralen Aminosäuren zusammengesetzt sind („apolare" oder „geordnete" Teile), sowie Bereiche mit Anhäufungen polarer Aminosäuren. Diese Tatsache erklärt die elektronenmikroskopisch beobachtete Querstreifung. Eine wirklich scharfe Trennung liegt aber nicht vor: Die apolaren Bereiche enthalten vereinzelt auch polare Aminosäuren *(Tabelle 5)*, und umgekehrt sind z. B. Prolin und Hydroxyprolin auch Bestandteile polarer Sequenzen.

Tabelle 5. Die heute bekannten Tripeptide der Struktur Gly—Pro—R aus den Neutralbereichen des Kollagens.

Gly—Pro—Ala	Gly—Pro—Arg
Gly—Pro—Gly	Gly—Pro—Asp
Gly—Pro—Glu	Gly—Pro—Phe*
Gly—Pro—Hypro	Gly—Pro—Ser*
	Gly—Pro—Thr*

* Nicht isoliert, aber nach analytischen Daten wahrscheinlich.

b. Innerhalb der Sequenz Gly—Pro—R kann nach dem heutigen Stand R = Hypro, Ala, Gly, Glu, Arg, Asp, Phe, Thr und Ser sein (Tabelle 5). Es ist sehr wahrscheinlich, daß auch noch andere Aminosäuren die Position R einnehmen können. Der Anteil von Gly—Pro—R an der Kollagenstruktur ist durch den Prolingehalt limitiert und kann wegen anderer Prolylbindungen (Punkt *f*) nur wenig über dem gefundenen Wert von 33% liegen.

c. Gly—Pro und Hypro—Gly sind die wichtigsten Dipeptide; Gly—Pro—Hypro ist das wichtigste Tripeptid der Neutralbereiche und darüber hinaus wohl der gesamten Kollagenstruktur. Die Modellvorstellungen für Kollagen haben durch diese Befunde eine starke Unterstützung erfahren.

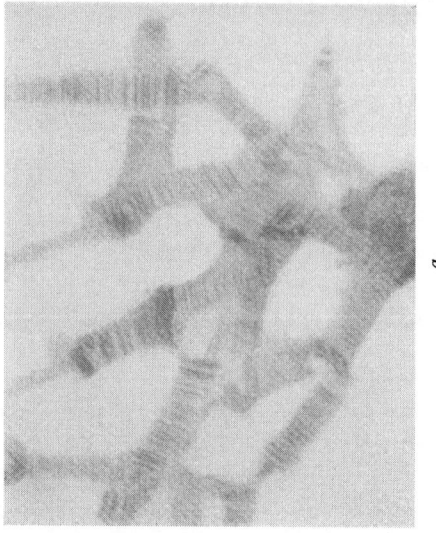

b

a

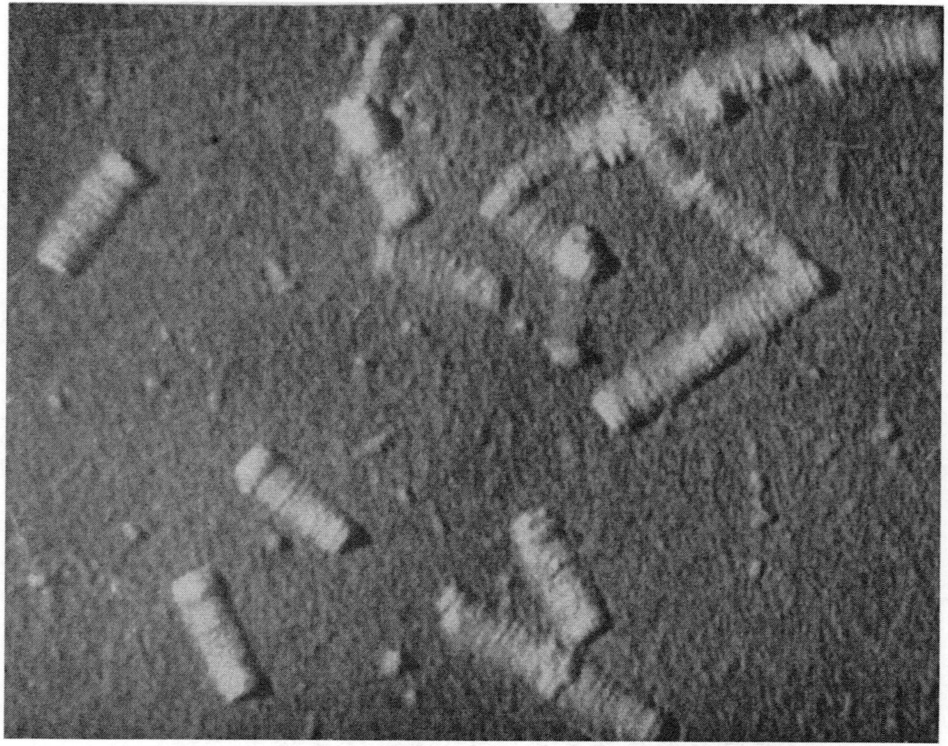

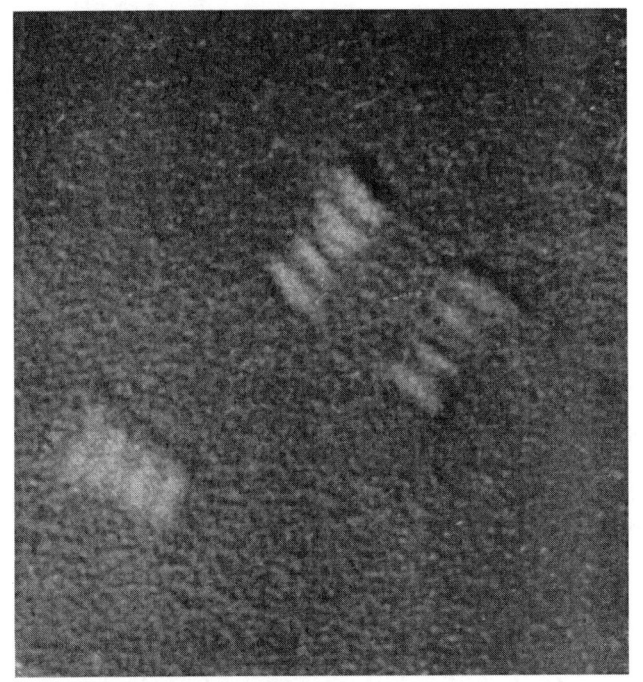

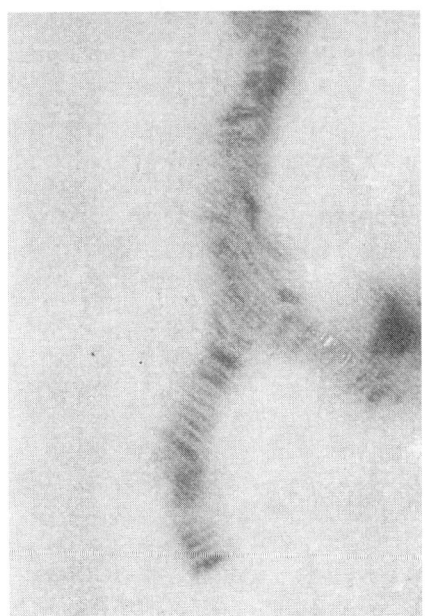

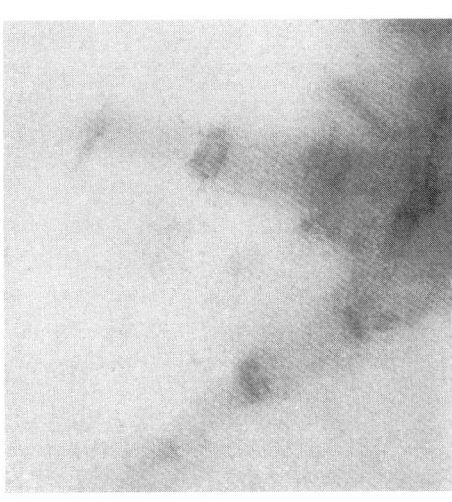

Abb. 14. *a* Long-spacing-Segmente, behandelt mit Phosphorwolframsäure. *b* wie *a*, 1 Minute lang abgebaut mit Kollagenase. *c* wie *a*, 3 Minuten lang abgebaut mit Kollagenase. *d* Long-spacing-Segmente aus Rattenschwanzsehne. *e* wie *d*, 20 Minuten mit Kollagenase behandelt, *a — c* : nach Nordwig, Hörmann, Kühn und Grassmann [aus: Z. physiol. Chem. **325**, 242 (1961)], *d* und *e* : nach Nishigai, Nagai und Noda [aus: J. Biochem. (Tokyo) **48**, 152 (1960)].

d. Glycin ist nicht ganz so gleichmäßig über die gesamte Struktur verteilt, wie man früher angenommen hat. Es sind Häufungen nachgewiesen worden (vgl. Tabelle 2, S. 212 und *108*). Dementsprechend gibt es auch Sequenzen mit einem Glycingehalt unter 33%, so etwa das 22 Aminosäuren lange Peptid B_3d_2 mit 14% Glycin u. a. (*108*). Über große Bereiche allerdings beträgt der Glycin-Anteil stets etwa $^1/_3$, z. B. auch in den „nichtdialysierbaren Anteilen".

e. Prolin kommt zur Hauptsache innerhalb der Sequenzen Gly—Pro—R (nachgewiesen: etwa 80%) sowie an basische Aminosäuren gebunden vor. Die zweite Iminosäure, Hydroxyprolin, spielt wider Erwarten eine andere Rolle; sie kann die Position des Prolins in den Tripeptiden Gly—Pro—R nicht einnehmen (*117*).

f. Ein großer Teil der basischen Aminosäuren, nämlich 40%, ist mit seinen Carboxylgruppen an Iminosäuren, der Rest an Glycin (*108*) gebunden. Arginin steht möglicherweise zum Teil am Übergang zwischen den neutralen und den polaren Bereichen, wie man aus Nodas Ergebnissen (*81*) schließen könnte.

2. Alkali- und Hydroxylamin-empfindliche Bindungen, Endgruppen und Acetylgruppen.

Behandelt man denaturiertes lösliches Kollagen mit Hydroxylamin unter alkalischen Bedingungen, dann bilden sich proteingebundene Hydroxamsäuren, deren Menge kolorimetrisch bestimmt werden kann, nachdem vorher Anteile niedrigmolekularer Hydroxamsäuren — z. T. wohl aus der Spaltung von Beimengungen herrührend — durch Dialyse entfernt worden sind (*83, 181, 189*). Die Reaktion ist stark von der Hydroxylaminkonzentration und vom Ionenmilieu abhängig. Sie erfolgt zunächst rasch, erreicht aber keinen scharfen Endpunkt und schreitet, wohl unter unspezifischer Spaltung von Peptid- und Amidbindungen, langsam fort *(Abb. 15)*. Während des ersten Reaktionsabschnittes werden α-Aminogruppen nur in unbedeutender Menge frei (*29, 191, 408*). Die Kinetik der Hydroxamsäurebildung entspricht in diesem Bereich derjenigen, wie man sie bei der Reaktion von Carbonsäureestern mit Hydroxylamin beobachtet (*184*). Man schließt daraus, daß es sich bei den hydroxylamin-spaltbaren Bindungen um Ester oder esterähnliche Bindungen handelt. Die Menge der proteingebundenen Hydroxamsäuren beträgt bei Anwendung einer optimalen Hydroxylaminkonzentration nach Abschluß der raschen Reaktion etwa 6 Mol pro 1000 Mol Aminosäuren (*83, 181*).

Die Hydroxamsäurebildung ist mit einem Abbau der Kollagenketten verbunden. In 4 M-LiCl bei 40° und pH 9,5 fand Hörmann (*191*) nach 7 Stunden ein mittleres Mol-Gewicht der Abbauprodukte von etwa

15500. Dieser Wert ist zu erwarten, wenn — wie oben gefunden — 6 hydroxylamin-empfindliche Bindungen in eine Kollagenkette von etwa 1000 Aminosäuren Länge eingebaut sind. Ähnliche Mol-Gewichte (21000 bzw. 14000) haben BAILEY und HODGE (15) nach Hydroxyl-aminabbau der isolierten α1- und α2-Ketten des Kollagens in Gegenwart von 1,2 M-KCNS erhalten. Unter anderen Spaltbedingungen, in Abwesenheit von Denaturierungsmitteln, sinkt das Mol-Gewicht weniger weit ab. GALLOP und Mitarb. (83) fanden in diesem Fall einen mittleren Wert von etwa 30000. Nach diesem Ergebnis würden

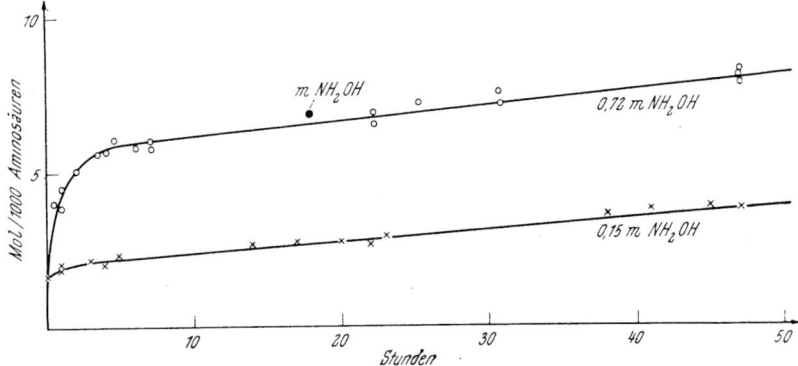

Abb. 15. Hydroxamsäurebildung von löslichem Kollagen in 4 M-LiCl bei verschiedenen Konzentrationen an Hydroxylamin, 37°, pH 9,5.

die Peptidketten des Kollagens nur in Viertel zerfallen. Das Vorliegen von mindestens sechs hydroxylaminspaltbaren Bindungen wird daher von den Autoren damit gedeutet, daß sie annehmen, die einzelnen Untereinheiten der Kollagenketten seien durch doppelte Esterbindungen aneinandergekettet (82). Stützen für diese Vorstellung leiten sie aus den im folgenden angeführten Versuchsergebnissen ab:

Ähnlich wie Hydroxylamin spaltet auch Hydrazin die Peptidketten des Kollagens (29). Dabei bilden sich Säurehydrazide, die nach Umsetzung mit p-Dimethylaminobenzaldehyd kolorimetrisch bestimmt werden können. Ihre Menge und Bildungsgeschwindigkeit entspricht derjenigen der Hydroxamsäuren. BLUMENFELD et al. (29) spalteten das hydrazinbehandelte Kollagen weiter mit Kollagenase und trennten aus dem erhaltenen Hydrolysat die Hauptmenge der hydrazidhaltigen Peptide ab. In ihnen bestand Äquivalenz zwischen Asparaginsäureresten und gebundenem Hydrazin, woraus geschlossen wurde, daß Hydrazin an Asparaginsäurereste gebunden vorliegt. Auf zwei Asparaginsäurereste traf weiterhin ein Mol aminoendständiges Glycin und ein Mol Prolin. Gleiche Mengen an aminoendständigem Glycin und an Prolin sind in

Peptiden aus Kollagenasehydrolysaten wegen der Spezifität dieses Enzyms (vgl. dazu S. 220 ff.) nur dann zu erwarten, wenn diese Peptide carboxylendständigen Sequenzen entstammen. (Mittelständige Peptide müßten auf ein Mol N-terminales Glycin zwei Mole Prolin aufweisen, eines in Nachbarschaft des Aminoendes, das andere in Nachbarschaft des Carboxylendes. Aminoendständige Peptide dürften wegen des Fehlens freier Aminoendgruppen in Kollagen keine freien α-Aminogruppen enthalten.) Aus diesen Analysenergebnissen schlossen die Autoren, daß in der Nähe der Carboxylenden von Untereinheiten der Peptidketten des Kollagens jeweils zwei Asparaginsäurereste in kurzem Abstand voneinander angeordnet sein sollen, welche mit je einer ihrer Carboxylgruppen esterartig mit der nachfolgenden Untereinheit verbunden sind (82). Die Untereinheiten sollen somit durch paarweise angeordnete Esterbindungen aneinandergekettet sein.

Die Alkoholkomponente der Esterbindungen stellt nach den Untersuchungen von Gallop (82) ein noch unbekannter Polyhydroxydialdehyd dar, der mit den bereits in Kollagen nachgewiesenen Hexosen nicht identisch ist. Spaltet man Kollagen bei pH 8 mit Hydrazin, dann bilden sich neben den Säurehydraziden der Asparaginsäurereste auch Aldehydhydrazone. Die Kinetik ihrer Bildung stimmt mit derjenigen der Hydrazide überein. Daraus wird geschlossen, daß die entsprechenden Aldehydgruppen erst bei der Hydrazinspaltung frei werden (82).

Wenn die Kollagenketten durch eine beachtliche Zahl von Esterbindungen unterbrochen sind, dann erhebt sich die Frage nach den Aminoendgruppen der durch diese Ester zusammengehaltenen Untereinheiten der Peptidketten. Im Kollagen sind, wie von zahlreichen Arbeitskreisen festgestellt wurde (35, 110, 206, 188), freie α-Aminogruppen nicht nachweisbar. Schon beim Aufbau aus durchlaufenden Peptidbindungen hätte man eine, bei einem Aufbau aus esterartig verknüpften Untereinheiten etwa 6—7 α-Aminogruppen auf 1000 Aminosäuren zu erwarten. Eine Bestimmung mit einer auf hohe Empfindlichkeit gebrachten DNP-Methode ergab jedoch nur 0,13 Mol auf 1000 Aminosäuren (188, 408).

Das Problem der α-Aminoenden der Untereinheiten des Kollagens dürfte eine Lösung gefunden haben durch die neue Feststellung von Hörmann und Joseph (187, 188), wonach in Kollagen gebundene Acetylreste gefunden wurden. Von diesen sind 6—7 Mol berechnet auf 1000 Mol Aminosäuren durch 0,72 M-Hydroxylamin nicht vom Protein abspaltbar, was ihre Bindung an Aminogruppen beweist. Nachdem sämtliche ε-Aminogruppen einer Substitution durch FDNB (1-Fluor-2,4-dinitrobenzol) zugänglich sind (182), kommen nur mehr α-Aminogruppen für die Bindung der Acetylreste in Frage. Die Anzahl der an α-Aminogruppen gebundenen Acetylreste stimmt recht gut mit der der hydroxylaminempfindlichen

Bindungen im Kollagen überein. Man kann daraus schließen, daß die Kollagenketten aus etwa 6—7 Untereinheiten bestehen, die mit Esterbindungen aneinander gekettet sind und deren Aminoenden durch Acetylreste substituiert sind.

Neben hydroxylamin- und hydrazin-spaltbaren Bindungen finden sich im Kollagen auch besonders *alkali-labile Verknüpfungen*, deren Natur bis heute unbekannt ist. Bei pH 11 und 40° erhält man nach 20—30 Min. zwei Hauptkomponenten, die als A und B bezeichnet wurden. (Längere Reaktionszeiten bewirken einen wesentlich langsameren unspezifischen Weiterzerfall in kleinere Abbauprodukte.) Als Aminoendgruppe tritt im wesentlichen Glycin neben Spuren von Asparaginsäure, Glutaminsäure, Alanin und Serin auf (*292*).

Die Bruchstücke A und B konnten durch Chromatographie an CM-Cellulose getrennt werden (*359*). Es konnte gezeigt werden, daß die A-Komponente aus der α1-Kette, B aus der α2-Kette hervorgeht. Beide Komponenten sind jedoch nach neueren Untersuchungen noch nicht einheitlich. Ihre Charakterisierung gelang bisher nur unvollständig (*151*). BEIER (*67*) konnte zeigen, daß die A- und B-Komponenten auch durch saure Behandlung von α- und β-Komponenten (pH 3,7; 40°) gewonnen werden können.

Die bei der beschriebenen milden alkalischen Spaltung erfaßten Bindungen scheinen mit den hydroxylaminspaltbaren nicht identisch zu sein, denn man findet die Zahl der hydroxylaminspaltbaren Bindungen nach dem alkalischen Abbau praktisch unverändert (*292*). Natives Kollagen, sowohl als Faser als auch in gelöstem Zustand, ist interessanterweise diesem Abbau unter gelinden Bedingungen nicht zugänglich. Nach mehrtägiger Einwirkung von 5%iger NaOH in der Kälte findet man zwar die Quervernetzungen gelöst und den größten Teil des Amid-N abgespalten, aber keine Verkürzung der Ketten. Aus den Lösungen können wieder normale Segmente gewonnen werden (*250*).

3. Quervernetzungen.

Die Peptidketten des Kollagens können, wie bereits eingangs ausgeführt, durch Quervernetzungen miteinander verbunden sein. Man unterscheidet intramolekulare und intermolekulare Vernetzungen je nachdem, ob die Brücken zwischen den Peptidketten des gleichen Kollagenmoleküls oder zwischen denen verschiedener Moleküle ausgebildet sind. Die Quervernetzungen können durch Alkali oder Hydroxylamin gespalten werden. Bei längerer Einwirkung von Natronlauge auf natives, reifes Kollagen wird dasselbe in sauren Puffern löslich (*53, 250*). Nachdem das gelöste Material in der Ultrazentrifuge ähnlich wie gelöstes Kollagen sedimentiert und mit ATP (Adenosin-triphosphat) Segmente gefällt werden können (53, 250), ist das Löslichwerden auf eine Spaltung von Quervernetzungen zurückzuführen. Das denaturierte Material zeigt in der Ultrazentrifuge nur einen einzigen Gipfel ähnlich dem der α-Komponente von Kollagen, woraus man schließen kann, daß auch die intramolekulare Quervernetzung durch Alkali angegriffen wurde. Die Spaltung der Quervernetzungen des Kollagens spielt für den Aufschluß von Haut zum Zwecke der Gelatineherstellung eine maßgebliche Rolle [(189), dort weitere Literatur].

Auch Hydroxylamin öffnet die Quervernetzungen des Kollagens, im nativen Protein allerdings erst bei stark alkalischer Reaktion. Unlösliches Kollagen geht dabei allmählich in Lösung (183). Es gelingt, einen Teil des gelösten Kollagens in Form von Segmenten oder auch quergestreiften Fibrillen abzuscheiden. Denaturiertes Kollagen wird durch Hydroxylamin bereits bei pH 9,5 aufgelöst (181).

Das Verhalten gegen Alkali und Hydroxylamin legt die Annahme nahe, daß es sich bei den Quervernetzungen um Esterbindungen, etwa zwischen Asparagin- und Glutaminsäure einerseits und Serin oder Threonin anderseits, handelt. Eine amidartige Bindung zwischen carboxyltragenden Seitenketten und den ε-Aminogruppen des Lysins haben Levy und Mitarb. (259) auf Grund ihres Befundes angenommen, daß etwa 30% der ε-Aminogruppen des Lysins der Umsetzung durch Nitrosylchlorid unzugänglich seien. Von Hörmann (182, 188) ist aber gezeigt worden, daß innerhalb einer Fehlergrenze von wenigen Prozenten alle ε-Aminogruppen reagieren, wenn man die Umsetzung unter denaturierenden Bedingungen (z. B. in Gegenwart von NaClO$_4$) durchführt. Ihre Unzugänglichkeit im nativen Zustand entspricht vielfach analogen Erfahrungen der Proteinchemie. Es kann also höchstens ein sehr kleiner Bruchteil des Lysins an derartigen Quervernetzungen beteiligt sein. Die Isolierung eines Peptids, in dem dieser Bindungstypus vorliegt, aus Gelatine (277) — die Ausbeute entspricht etwa 1% des Lysins — kann ebensogut darauf zurückzuführen sein, daß unter den Bedingungen längerer alkalischer Behandlung im Laufe der Gelatineherstellung Esterbindungen zu Amidbindungen umgelagert werden.

Arbeiten von Hörmann (181, 182) messen den Hexosen eine Bedeutung für die Ausbildung der Quervernetzungen in reifem Kollagen bei. Ihnen liegt der Befund zugrunde, daß unlösliches Kollagen durch Behandlung mit saurem Perjodat unter denaturierenden Verhältnissen aufgelöst werden kann (115), ohne daß eine nennenswerte Anzahl von Peptidbindungen gespalten wird (408). Dabei werden die Hexosen des Kollagens nahezu vollständig abgebaut (233, 181, 182, 17). Die übrigen durch Perjodat in ihren Seitenketten veränderten Aminosäuren Hydroxylysin, Tyrosin oder Methionin kommen für eine Beteiligung an Quervernetzungen nicht in Frage, da sie nur dann von Perjodat angegriffen werden können, wenn sie in ihren Seitenketten nicht substituiert sind. Beim partiellen enzymatischen Abbau, z. B. mit Trypsin und Kollagenase, reichern sich die Kohlenhydrate in den am schwersten angreifbaren Anteilen an (*181, 182*).

Eine Beteiligung der Hexosen an den Quervernetzungen setzt eine zumindest bifunktionelle Bindung derselben in reifem Kollagen voraus. Ein Hexosenrest sollte demnach mit seiner reduzierenden Gruppe O-glykosidisch an eine hydroxylgruppentragende Seitenkette der einen Peptidkette und mit einer oder mehreren OH-Gruppen esterartig an carboxylgruppentragende Seitenketten benachbarter Peptidketten gebunden sein. Diese Estergruppen stellen nach Vorstellungen von Hörmann (*181, 182*) die hydroxylamin-empfindlichen Bindungen in den Quervernetzungen des Kollagens dar. Die Tatsache, daß in Ichthyocoll lediglich eine monofunktionelle Bindung der Hexosen an Kollagen festgestellt wurde (*30*) (s. S. 208), steht diesen Vorstellungen nicht entgegen, denn in Ichthyocoll sind keine oder nur sehr wenige Quervernetzungen, auch diese nur intramolekular, vorhanden.

Die exakte Ermittlung von Zahl und Art der Quervernetzungen stößt auf die Schwierigkeit, daß beim Abbau sowohl mit NH_2OH wie mit Alkali unter schonenden Bedingungen auch eine Spaltung der Hauptkette an mehreren Stellen erfolgt (s. oben). Dies gilt auch schon für die reduzierende Spaltung mit $LiBH_4$, aus deren Ergebnissen Grassmann und Mitarb. erstmals auf ein Vorkommen von Esterbindungen geschlossen hatten (*102*).

Die Spaltung der intra- und intermolekularen Quervernetzungen ist auch auf enzymatischem Wege möglich. In neuerer Zeit ist eine Art des enzymatischen Angriffs auf Kollagen beschrieben und eingehend untersucht worden, bei der die native Form des Kollagen-helixmoleküls erhalten bleibt. Das Wesentliche für diesen Abbau, für den unter anderem die Enzyme Pepsin (*299, 300, 236*), Papain (*228, 228 a*), Ficin (*228, 228 a*) und Pronase (*78*) in Betracht kommen, ist, daß der Angriff des Enzyms unter Bedingungen erfolgt, bei denen eine Denaturierung vermieden wird, also insbesondere bei niederer Temperatur (25° oder darunter). Die Einwirkung auf unlösliches reifes Kollagen unter diesen Bedingungen führt zur Auflösung, wobei Lösungen von Kollagen erhalten werden, die sich von den auf anderem Wege erhaltenen Kollagenlösungen (s. unten) nur unwesentlich unterscheiden und wie diese die normalen Tripelhelix-Stäbchenmoleküle vom Molekulargewicht 320000 enthalten. Diese Befunde sowie Untersuchungen des gelösten Materials in der Ultrazentrifuge zeigen, daß durch die Enzymbehandlung inter- und intramolekulare Vernetzungen gelöst werden (*228*), und zwar, je nach der Art des angewandten Enzyms, beide in ungleichem Maße. Trypsin spaltet die Quervernetzungen des Kollagens nur in geringem Maße.

Für die Auflösung von älterem, hoch verfestigten Kollagen ist häufig eine mehrmalige Anwendung von Pepsin, Papain oder Ficin oder eine kombinierte Anwendung mehrerer Enzyme notwendig (300, 228). Bei kurzer Einwirkung von Pepsin wird Kollagen aus Rinderhaut nur in einen hoch gequollenen Zustand übergeführt, der erst bei einer zweiten Pepsinbehandlung in echte Lösung übergeht. Offenbar ist für die Auflösung von verfestigtem Kollagen die Spaltung von Bindungen verschiedener enzymatischer Angreifbarkeit notwendig. Für eine nähere Diskussion dazu, die sich auf elektronenmikroskopische Untersuchungen gründet, s. Abschnitt VII/4, S. 276).

V. Sekundär- und Tertiärstruktur.

1. Ergebnisse der Röntgenstruktur-Untersuchung.

Die Peptidketten des Kollagens liegen in der nativen Fibrille in einer Konformation vor, die sich von den Faltungen der meisten übrigen Eiweißkörper (α-Schraube, pleated sheet-Struktur) erheblich unterscheidet. Bestimmend für die Kollagenstruktur sind der hohe Gehalt an den Iminosäuren Prolin und Hydroxyprolin sowie die regelmäßige Verteilung von Glycin, welches wenigstens in den allermeisten Bereichen jede dritte Aminosäurenposition besetzt. In prolinhaltigen Ketten ist nicht nur wie in allen Peptidketten aus Mesomeriegründen die freie Drehbarkeit um die Peptidbindungen eingeschränkt, sondern darüber hinaus wegen der Ringstruktur auch die Drehbarkeit zwischen α-C- und N-Atom aufgehoben. Dies bedeutet, daß Prolin und Hydroxyprolin weder in der trans-Anordnung der gestreckten Peptidkette bzw. in der pleated-sheet-Struktur noch in der α-Schraube untergebracht werden können.

Die heutige Kenntnis über die räumliche Anordnung der Peptidketten des Kollagens beruht zur Hauptsache auf einer sorgfältigen Analyse der Röntgeninterferenzen. Bei diesen unterscheidet man zwischen Kleinwinkel- und Weitwinkelinterferenzen. Die nahe am Primärstrahl auftretenden Kleinwinkelinterferenzen entsprechen langen Perioden in der Faserachse. Zum Beispiel erscheint ein Reflex für die Identitätsperiode von 640 Å in diesem Bereich. Weitwinkelinterferenzen dagegen werden von Abständen in der Größenordnung von Aminosäuren erzeugt. Aus ihnen ist im wesentlichen die räumliche Struktur der Kollagenketten abgeleitet worden.

Die ersten Aufnahmen der Weitwinkel-Reflexe des Kollagens liegen bereits weit zurück [s. z. B. Herzog und Gonnell 1924 (162, 358, 161, 163, 164, 165)]. Sie wurden in der Folgezeit stark verbessert. Die besten

Literaturverzeichnis: SS. 293—314.

Aufnahmen wurden von Cowan, North und Randall (51) an gestreckten Fasern erhalten. Charakteristisch für die Röntgenaufnahmen des Kollagens ist ein starker Meridionalreflex von 2,89 Å (2,86—2,95 Å). Daneben fallen noch Reflexe bei 4,0 und 9,5 Å auf. Diese stellen jedoch keine wirklichen Meridionalreflexe dar, sondern erscheinen im Abbild der gestreckten Faser in zwei Teilreflexe rechts und links von der Faserachse aufgespalten. Die äquatorialen Reflexe für trockenes Kollagen entsprechen Abständen von 4,5 und 10,5 Å. Im feuchten oder gequollenen Zustand kann sich der Abstand von 10,5 Å bis auf 16 Å und mehr vergrößern [Küntzel (253)]. Gerngross und Mitarb. (88) fanden diesen Reflex erstmals in gedehnter und ungedehnter Gelatine.

Die ersten Vorschläge über die Anordnung der Polypeptidketten in Kollagen stammen von Astbury (13) (1940), Huggins (195) (1943) sowie von Ambrose und Elliott (7) (1951). Ihre Modelle stellen gestreckte Ketten dar. Eine prinzipielle Neuerung erfuhr die Entwicklung durch die Arbeiten von Pauling, Corey und Mitarb. (315, 317), die erstmals eine genaue Untersuchung sämtlicher Strukturelemente anstellten, die für die Ausbildung der Konformation von Peptidketten bestimmend sind. Nach einer genauen Bestimmung der Atomabstände und Winkel stellten sie fest, daß die Peptidgruppe auf Grund ihrer Mesomerie eine planare Anordnung besitzt und forderten, daß in einer stabilen Proteinstruktur eine maximale Anzahl von Wasserstoffbindungen ausgebildet sein müsse. In Ausarbeitung dieser Voraussetzungen schlugen sie für Kollagen eine Struktur aus drei Peptidketten vor (316), die sich um eine gemeinsame Achse winden. Der Abstand von 2,86 Å entsprach der Höhe einer Aminosäure. Dieser gegenüber anderen Proteinstrukturen verminderte Abstand wurde durch Annahme von cis-trans-cis-Anordnungen der aufeinanderfolgenden Peptidbindungen erreicht. Die Wasserstoffbindungen liegen in der Richtung der Faserachse.

Wenn dieses Strukturmodell auch den späteren Untersuchungen nicht standhielt, so bestätigten doch die Fourier-Transformationen von Cochran und Mitarb. (44) den schraubenförmigen Charakter der Tertiärstruktur des Kollagens. Als beweisend werden die oben genannten, vom Meridian symmetrisch abstehenden Reflexe bei 4,0 und 9,5 Å angesehen. Die Annahme von drei Einzelketten ist inzwischen auf verschiedenen Wegen voll bestätigt worden (104). Gegen das Modell von Pauling sprechen die Ergebnisse über den Dichroismus der IR-Absorption der Wasserstoffbrücken, aus denen auf eine Lage quer zur Faserachse geschlossen werden muß (340). Auch erscheint die Annahme von cis-Peptidbindungen wegen des höheren Energiegehaltes weniger wahrscheinlich (14, 47, 46). Strukturvorschläge von Randall (339), Bear (20, 19, 46) und Huggins (196) versuchten die damit zusammenhängenden Schwierigkeiten zu überwinden.

Eine Weiterentwicklung wurde jedoch erst erreicht, als Ramachandran und Kartha (*334*) 1954 forderten, daß die drei Peptidketten des Kollagens nicht miteinander verdrillt sein sollten, sondern daß jede für sich getrennt eine linksgewundene Schrauben-Anordnung mit einer Ganghöhe von drei Aminosäuren einnehmen soll. Die einzelnen Peptidschrauben sind durch Wasserstoffbindungen zwischen NH- und C=O-Gruppen quer zur Faserachse miteinander verbunden. Die Struktur setzt eine regelmäßige Anordnung von —Gly—Pro—X— voraus, wobei X jede beliebige Aminosäure mit Ausnahme von Prolin oder Hydroxyprolin sein soll. Eine solche Dreiereinheit besitzt zwei NH-Gruppen, von denen beide als Donatoren für Wasserstoffbrücken zu C=O-Gruppen benachbarter Peptidketten auftreten. Die Glycinreste nehmen die dem Innern der Dreikettenstruktur zugewandten Positionen der Peptidketten ein, während die Pyrrolidinringe und andere Seitenketten nach außen ragen. Eine Verbesserung erfuhr dieser Vorschlag ein Jahr später durch dieselben Autoren (*335*), indem sie annahmen, daß die drei Ketten zusätzlich noch im Rechtssinn miteinander verdrillt seien, was zur sogenannten coiled coil-Struktur führte. In ihr beträgt der Abstand einer Aminosäure 2,85 Å. Identität wird erst nach $3^1/_3$ Umdrehungen, also 10 Aminosäuren, erreicht, wenn die nächste Kette in die Position der vorhergehenden gerückt ist. Nach 10 Umdrehungen bzw. 30 Aminosäuren oder 85,8 Å ist eine volle Windung der übergeordneten Schraube vollzogen. Der röntgenographisch gemessene Äquatorialabstand von 4,5 Å entspricht dem Durchmesser einer Einzelschraube, der Seitenabstand von 10,5 Å dem Durchmesser der Tripelhelix.

In der Zwischenzeit hatten Schroeder und Mitarb. (*369*) auf Grund chemischer Ergebnisse (vgl. Kapitel II, S. 211) die Vorstellung entwickelt, daß in weiten Bereichen des Kollagens die Sequenz —Gly—Pro—Hypro— vorliegt. Diese Folge ist nur zur Ausbildung einer Wasserstoffbindung auf drei Aminosäuren befähigt und nicht von zweien, wie sie die Struktur von Ramachandran und Kartha (*335*) fordert. Dies führte zu einem neuen Vorschlag von Rich und Crick (*343, 344*). Diese Autoren hatten zuvor eine Modifikation von Polyglycin (Polyglycin II) untersucht (*52*) und festgestellt, daß in ihr die Peptidketten schraubenförmig angeordnet und durch Wasserstoffbindungen quer zu den Schraubenachsen zusammengehalten werden. Die Ganghöhe beträgt 3 Aminosäuren, von denen jede 3,1 Å einnimmt. Wenn man in einer linksgewundenen Polyglycinschraube in regelmäßiger Weise zwei von drei Glycinresten durch L-Iminosäuren ersetzt, so entfallen zwei von drei möglichen Wasserstoffbindungen. Die verbleibenden Bindungen sind so gerichtet, daß jeweils drei Peptidketten im Dreiecksverband miteinander vernetzt bleiben, wodurch das unendliche Gitter von Polyglycin II auf eine Dreikettenstruktur reduziert wird. Auch dieses Modell wurde dadurch verfeinert, daß man eine zusätzliche Verdrillung der drei Ketten, also eine coiled coil-Struktur, annahm. Dadurch verringert sich der Abstand für eine Aminosäure von 3,1 auf 2,9 Å. Die untergeordneten Peptidschrauben sind links, die übergeordnete Verdrillung der drei Ketten ist rechts gewunden. Auf unabhängigem Wege, nämlich durch Untersuchung von Poly-L-prolin, sind COWAN, McGAVIN und NORTH (50) zur gleichen Kollagenstruktur gelangt. Frisch synthetisiertes Poly-L-prolin weist in Wasser oder organischen Säuren eine Mutarotation auf und wandelt sich dabei in eine stark linksdrehende Modifikation um (Poly-L-prolin II, $[\alpha]_D^{20} = = -540°$) (209), in der die Peptidkette eine ähnliche Schraubenstruktur wie in Polyglycin II einnimmt. Ersatz jedes dritten Prolinrestes durch Glycin führt zur Möglichkeit der Ausbildung von Wasserstoffbrücken zu Nachbarketten, von denen sich drei zu einer Tripelhelix zusammenlagern.

Für die Zusammenlagerung der drei Peptidketten bestehen nach den Angaben von RICH und CRICK (343, 344) zwei Möglichkeiten. Entweder können die Wasserstoffbindungen N—H ... O=C im Uhrzeigersinn oder im Gegenuhrzeigersinn angeordnet sein. Letztere Struktur wird Kollagen I oder auch plus-Struktur [RAMACHANDRAN 1956 (333)], erstere Kollagen II oder minus-Struktur genannt. Die beiden unterscheiden sich in der gegenseitigen Anordnung der Aminosäuren verschiedener Ketten zueinander. Eine spannungsfreie Ausbildung beider Strukturen ist nur dann gegeben, wenn in den Peptidketten jede dritte Aminosäurenposition durch Glycin besetzt ist. Diese Reste kommen in der Dreikettenstruktur nach innen zu liegen. Bei geringen Deformierungen ist es allerdings möglich, insbesondere in Kollagen I, daß auch andere Aminosäuren mit kleinen Seitenketten Position 1 einnehmen. Die Pyrrolidinringe von Prolin und Hydroxyprolin in den Stellungen 2 und 3 ragen nach außen. Diese Positionen können aber auch durch andere seitenkettentragende Aminosäuren besetzt werden (Tabelle 6).

Eine detaillierte Beschreibung ihrer Kollagenstrukturen mit genauen Atompositionen geben RICH und CRICK 1961 (345). Abb. 16 bringt eine vereinfachte schematische Wiedergabe der Struktur von Kollagen II auf der Basis einer immer wiederkehrenden Sequenz Gly—Pro—Hypro. Man erkennt, daß die Pyrrolidinringe nach außen ragen, während die Glycinreste im Innern der Tripelhelix angeordnet sind. Die Ganghöhe der untergeordneten Linksschraube beträgt drei Aminosäurereste zu je 2,86 Å, also 8,58 Å. Innerhalb dieses Abstandes verschiebt sich die Kette um 36° nach rechts, so daß nach 10 Windungen = 30 Aminosäuren die volle Ganghöhe der übergeordneten Schraube erreicht ist. Das entspricht einem Abstand von 85,8 Å.

Nach neueren Untersuchungen von Rogulenkova et al. (*350*) stimmen die Röntgenreflexe von orientiertem synthetischem (Gly—Pro—Hypro)$_n$ mit denen von Kollagen recht gut überein. Dasselbe gilt für (Pro— Gly—Pro)$_n$ (*69, 70*), das sich von obigem Polymeren durch den Ersatz von Hypro durch Pro unterscheidet.

Die Anbringung der Seitenketten ist in den Positionen 2 und 3 von Kollagen II leichter möglich als in Kollagen I, weshalb Rich und Crick dem Modell II den Vorzug geben. Auch stimmen die mit einem optischen Diffraktometer an Modellen ermittelten Reflexe mit den Röntgendiagrammen von nativem Kollagen am besten überein, wenn man die Kollagenstruktur II zugrunde legt (*21, 345*). In den Strukturmodellen von Rich und Crick besteht auch die Möglichkeit, daß eine der beiden Peptidketten in antiparalleler Anordnung zu den beiden anderen vorliegt. Bezüglich weiterer Strukturmodelle s. (*283, 284*).

In Verteidigung seiner ursprünglichen Vorschläge entwickelte Ramachandran (*336, 337*) eine Verbesserung seines Modells, von dem Rich und Crick (*345*) sagen, daß es bei Einhaltung der vorgeschriebenen Valenzwinkel, Van-der-Waals-Kontakte und Bindungslängen für die Wasserstoffbindungen nicht aufgebaut werden könne. Ramachandran dagegen steht auf Grund verfeinerter Röntgenuntersuchungen an Kollagen (*254*) auf dem Standpunkt, daß die Toleranz für diese Werte wesentlich größer sei und daß tatsächlich in Kollagen erheblich von der Norm abweichende Werte gefunden werden. Danach beträgt in der gestreckten Faser die Länge für eine Aminosäure bis zu 3,05 Å. Infrarotmessungen lassen weiterhin auf Wasserstoffbindungen mit Längen bis zu 3,0—3,05 Å schließen. In dem verbesserten Modell gehen ähnlich wie im älteren zwei allerdings gewinkelte Wasserstoffbrücken von drei Aminosäureresten aus. Die Energien, die für die starke Deformation der Bindungen notwendig sind, werden durch die Ausbildung der größeren Anzahl von Wasserstoffbindungen aufgebracht.

Tabelle 6. Die möglichen Positionen der Seitenketten in den Kollagenmodellen I und II von Rich und Crick (*344*)

Position	Kollagen I		Kollagen II
	nicht deformiert	deformiert	
1	nur Glycin	andere Reste sind möglich, Prolin und Hydroxyprolin ausgeschlossen	nur Glycin
2	jeder Rest einschließlich Prolin und Hydroxyprolin		
3	nur Glycin	jeder Rest einschließlich Prolin und Hydroxyprolin, nicht jedoch Valin und Isoleucin	jeder Rest einschließlich Prolin und Hydroxyprolin

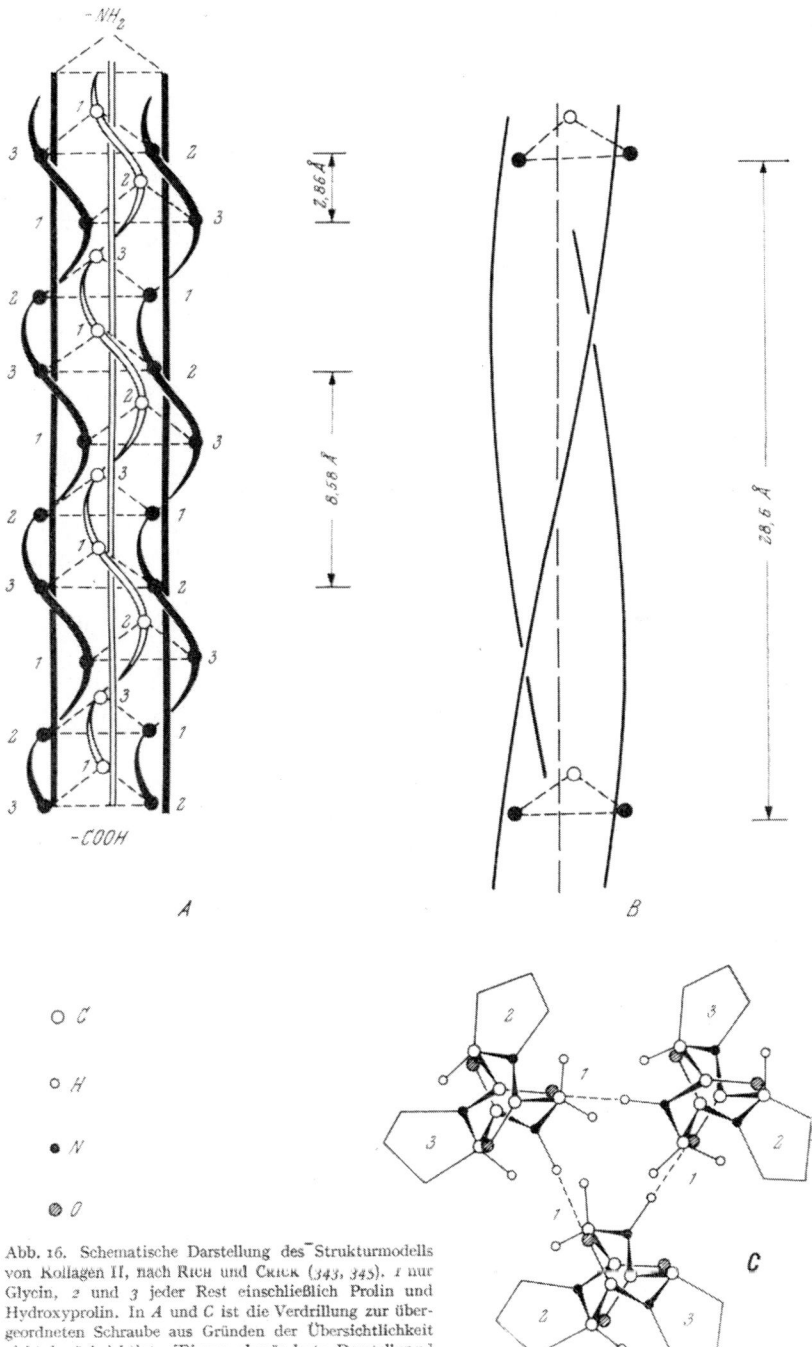

Abb. 16. Schematische Darstellung des Strukturmodells von Kollagen II, nach Rich und Crick (343, 345). 1 nur Glycin, 2 und 3 jeder Rest einschließlich Prolin und Hydroxyprolin. In A und C ist die Verdrillung zur übergeordneten Schraube aus Gründen der Übersichtlichkeit nicht berücksichtigt. [Eigene abgeänderte Darstellung.]

Das Modell von RAMACHANDRAN ist in Bereichen, in denen die Sequenz Gly—Pro—Hypro kumuliert ist, wegen der zu geringen Anzahl von NH-Gruppen nicht zu verwirklichen. Diese Bereiche sind aber nicht so häufig, wie angenommen wurde. Anderseits ist in den Bereichen der Form Gly—Pro—X (X = jede Aminosäure außer Prolin und Hydroxyprolin) und in den iminosäurefreien polaren Regionen das Modell von RAMACHANDRAN durchaus vertretbar. Für die apolaren Bereiche nimmt er an, daß die Sequenz Gly—Pro—Hypro nur in einer der drei Ketten auftritt. Dann bleibt die Deformation erträglich. Für diesen Fall diskutiert er neben der Standardstruktur eine Alternativstruktur mit nur 5 Wasserstoffbrücken, an Stelle von 6, innerhalb benachbarter Dreiergruppen der Tripelhelix (337).

Das Modell von RAMACHANDRAN befriedigt bei der Beurteilung der Röntgenergebnisse und des IR-Dichroismus der Peptidbanden (24) wesentlich besser als die Modelle von RICH und CRICK. Auch stimmen die für die Denaturierung gemessenen thermodynamischen Daten nur dann mit den aus Vergleichsmessungen an Modellsubstanzen berechneten überein, wenn man auf 3 Aminosäuren 2 Wasserstoffbindungen annimmt (154). Die Atome können im Ramachandran-Modell außerdem nur in der für Kollagen spezifischen Form angeordnet sein, während in einem Modell mit nur einer Wasserstoffbindung auf drei Aminosäuren die übergeordnete Schraube verschieden stark verdrillt sein kann (337). Im Ramachandran-Modell ist nur die minus-Struktur mit Wasserstoffbindungen im Uhrzeigersinn möglich (ähnlich Kollagen II von RICH und CRICK). Eine Entscheidung darüber, welches Modell — das von RAMACHANDRAN oder Kollagen II von RICH und CRICK — für Kollagen gültig ist, kann noch nicht endgültig getroffen werden. Möglicherweise sind sogar beide Strukturen gültig; jede für bestimmte Bereiche des Kollagens.

Sicherlich ist ein idealer Kristallzustand nicht über die ganze Länge der Kollagenketten erfüllt. Dazu hat RANDALL (338) einen interessanten Dehnungsversuch ausgeführt: Wenn man Kollagenfasern um ungefähr 8% dehnt und wieder entlastet, dann stellt sich nicht die ursprüngliche Länge wieder her, sondern eine um $1^1/_2$% vergrößerte Länge; bei weiterer Wiederholung des Dehnvorgangs ist aber dann das Verhalten vollkommen elastisch. Die bleibende Dehnung entspricht einer Parallelorientierung von gegeneinander beweglichen Strukturelementen, der elastische Teil der Dehnung ist der Dehnung dieser Strukturelemente selbst zuzuschreiben. Die Weitwinkelinterferenzen zeigen bei schwacher Dehnung überhaupt keine Vergrößerung der Abstände, aber das Schmalerwerden der Interferenzsicheln zeigt die Parallelausrichtung derjenigen kristallinen Strukturbereiche an, die für die Weitwinkelinterferenzen verantwortlich sind. Bei stärkerer Dehnung werden die Identitätsabstände im Weit-

Literaturverzeichnis: SS. 293—314.

winkelbereich gedehnt, aber weit weniger als es der Dehnung der gesamten Faser entspricht. Die Aufweitung der Kleinwinkelperiode ist dagegen in jedem Dehnungsbereich der Dehnung der Gesamtfaser proportional. Dieses Verhalten scheint zu zeigen, daß die den Weitwinkelinterferenzen entsprechenden „kristallin geordneten" Bereiche relativ wenig dehnbar und durch stärker dehnbare Zwischenbereiche getrennt sind. Die aus den Weitwinkelreflexen abgeleiteten Aussagen beziehen sich demnach wahrscheinlich nur auf die erstgenannten Bereiche.

2. Physikalische Eigenschaften des nativen Kollagenmoleküls.

Die mit Hilfe der Röntgenstrukturanalyse ermittelte Struktur des Kollagenmoleküls wurde durch die Bestimmung des Molekulargewichts, der Abmessungen und der Steifheit des nativen Moleküls in Lösung wesentlich ergänzt. Eine weitere Bestätigung des Strukturmodells und einen tieferen Einblick in den molekularen Aufbau brachte die Untersuchung der Denaturierungsprodukte des löslichen Kollagens (Abschnitt V/3, S. 248).

Es war deshalb ein großer Fortschritt, als gefunden wurde, daß ein kleiner Teil des Kollagens mit Neutralsalzlösungen (*132*) und ein meist etwas größerer bei anschließender Extraktion mit sauren Puffern oder sehr verdünnten Säuren (*313, 293*) gelöst werden. Das auf die erstgenannte Art extrahierbare Kollagen muß im ursprünglichen Gewebe schon in löslicher Form vorliegen, denn die Salzlösungen, mit denen man es herauslöst, unterscheiden sich in bezug auf pH-Wert und Ionenstärke nicht wesentlich von der Gewebsflüssigkeit selbst. Die Struktur der zurückbleibenden Kollagenfasern bleibt unzerstört im Gegensatz zur sauren Extraktion, bei der sie weitgehend verändert wird (*134*). Beim neutralsalzlöslichen Kollagen handelt es sich um eine echte Vorstufe des reifen Kollagens (*152*), während sich dieses für das säurelösliche Kollagen nicht bestätigt hat (vgl. Kapitel VIII, S. 285). Diese wurden entsprechend dem wahrscheinlich nur im ersten Falle begründeten Vorstufencharakter als Tropo- bzw. Prokollagen bezeichnet. Heute erscheint die Bezeichnung „lösliche Kollagene" als sinnvoller, da es sich im wesentlichen um Monomere des Kollagens handelt, aus welchen das unlösliche, reife, unter Bildung von intermolekularen Quervernetzungen aufgebaut ist. Die hauptsächlichen Beweise hierfür sind:

a. Die völlige Identität der Aminosäurenzusammensetzung mit der des unlöslichen Kollagens (*134*).

b. Das lösliche Kollagen bildet im Neutralen Fibrillen des nativen Typs (*101*). Diese unterscheiden sich im elektronenmikroskopischen Bild (vgl. SS. 204 und 273) nicht von den im Organismus vorhandenen, nur sind sie im Sauren wieder löslich.

c. Die in Lösungen bestimmte Moleküllänge (s. unten) stimmt mit der aus der periodischen Wiederkehr des elektronenmikroskopischen Querstreifungsmusters in Fibrillen und der Länge von Long-spacing-Segmenten abgeleiteten überein (Kapitel VII, S. 266).

d. Das gleiche gilt für die Moleküldurchmesser, die in Lösungen (s. unten) und röntgenographisch an reifem Kollagen bestimmt wurden.

e. Der gesetzmäßige Zusammenhang zwischen der in Lösung beobachteten Denaturierung und der Schrumpfung reifer Fibrillen zeigt, daß beide Vorgänge durch den Zusammenbruch der Struktur des Kollagenmonomeren verursacht werden (s. Abschnitt VI/2, S. 254 f.).

Das Molekulargewicht des löslichen Kollagens wurde durch methodische Schwierigkeiten, die im Falle eines so asymmetrischen, großen Moleküls auftreten, lange zu hoch gefunden: 7×10^5 *(310)* bis 100×10^5 *(263)*. Die ersten zuverlässigen Bestimmungen gelangen im Arbeitskreis von Doty am löslichen Kollagen der Karpfenschwimmblase (Ichthyocol) *(31)* und der Kalbshaut *(58)*. Für beide wurde sowohl mit der Methode der Lichtstreuung *(41)* als auch nach Scheraga und Mandelkern *(357)* aus Sedimentationskonstante und Viskositätszahl $M = 3,5 \times 10^5$ ermittelt. Dieser Wert wurde in einer Nacharbeit, die $M = 3,3 \times 10^5$ ergab, für Kalbshautkollagen vier verschiedener Tiere bestätigt *(67)*. Rice und Mitarb. fanden für das gleiche Material $M = 2,53 \times 10^5$ bzw. $3,06 \times 10^5$ *(342)*. Young und Lorimer *(427)* fanden $2,8 \times 10^5$ für Kollagen der Kabeljauhaut. Eine gut bewiesene Ausnahme vom Molekulargewicht um 3×10^5 macht das Regenwurmkollagen, welches in doppelt so schweren (und langen) Molekülen in Lösung vorliegt *(276)*. Eine kritische Übersicht über weitere Bestimmungen, teilweise an anderen Kollagenen, gaben Hannig und Engel *(150)*.

Die verläßlichsten Werte für die mit Hilfe der Lichtstreuung *(41)* der Strömungsdoppelbrechung *(389)* und nach Mehl und Mitarb. *(278)* aus der Viskositätszahl bestimmten Länge und Durchmesser sind für Kollagen aus der Karpfenschwimmblase 3000 Å und 13,6 Å *(31)*, aus Karpfenhaut 2800 Å und 15 Å *(427)* aus Kabeljauhaut 2800 Å und 13 Å *(58)* bzw. 2800—3200 Å und 12,7—15 Å *(67)*. Hier, wie beim Molekulargewicht, können die Unterschiede zwischen den Daten für verschiedene Kollagene durchaus auch auf experimentellen Fehlern beruhen.

Die Übereinstimmung dieser Abmessungen mit der elektronenmikroskopisch gefundenen Länge des Moleküls im Fibrillenverband zeigt, daß das Molekül auch in Lösung völlig gestreckt vorliegt. Die große Starrheit des nativen Stäbchenmoleküls wird auch durch die Peterlin-Analyse *(319, 383)* der Winkelabhängigkeit der Lichtstreuung von Kollagenlösungen bestätigt *(67)*. Hierbei liegen die experimentellen Werte auf der theoretischen Kurve für ein völlig starres Stäbchen.

Literaturverzeichnis: SS. 293—314.

Schließlich finden diese Aussagen über die Abmessungen und Steifheit, die mit Hilfe der oft komplizierten und unanschaulichen physikalischen Messungen an Lösungen erhalten wurden, eine direkte und anschauliche Bestätigung. RICE (*341*) sowie HALL und DOTY (*145*) gelangen elektronenmikroskopische Aufnahmen, auf denen die aus verdünnter Essigsäure auf eine Unterlage aufgetrockneten Moleküle als starre Stäbchen von 2800—3000 Å Länge sichtbar sind.

Lösungen des nativen Kollagens zeigen die höchste an Proteinen überhaupt beobachtete spezifische optische Drehung: $[\alpha]_D^{20} = -350°$ für Kollagen aus der Schwimmblase und Haut des Karpfens und Kabeljaus (*31, 427*) und $-408°$ (*58*) bzw. $-415°$ (*67*) für Kalbshautkollagen. Die Drehung ist zum Teil durch die sterische Anordnung der Reste im Molekül bedingt. Sie fällt während der Denaturierung auf $1/_3$ ihres nativen Wertes. Die Wellenlängenabhängigkeit der Drehung folgt der einfachen Drude-Gleichung (*57, 45*), und würde man die für die α-Helix abgeleitete Moffitt-Beziehung (*57*) anwenden, so berechnete sich ein Helixanteil des Kollagens von Null. Dies ist ein weiterer Beweis dafür, daß im Kollagen eine von der α-Helix völlig verschiedene Schraubenanordnung vorliegt.

Eine charakteristische Größe für Kollagen in Lösung ist die Denaturierungs- oder Schmelztemperatur T_m. Die Denaturierung (über deren Verfolgung und Mechanismus s. Kapitel VI, S. 253) erfolgt in einem Temperaturbereich von einigen Graden, und T_m ist definiert als die Temperatur, bei der die optische Drehung oder die Viskosität um die Hälfte der Differenz zwischen den Werten des nativen und denaturierten Zustandes gefallen sind. Sie ist für Kollagene verschiedener Herkunft verschieden und steigt in gesetzmäßiger Weise mit dem Gehalt an Iminosäuren [Summe von Prolin und Hydroxyprolin (*207, 276*) s. Tabelle 1, S. 199]. Die Säugetierkollagene zeigen eine höhere Denaturierungstemperatur (Kalbhautkollagen 37° bei pH 3,7) als die von Fischen [Kabeljauhaut 29° (*40*)]. Die von der Denaturierungstemperatur (und vom Quellungsgrad) abhängige Schrumpfungstemperatur der Kollagenfibrillen scheint allgemein für in warmen Gewässern lebende Fische höher zu sein als für die in kalten Gewässern lebenden (*140*).

Die Stabilisierung durch Iminosäuren hängt sicher mit der in ihnen aufgehobenen Drehbarkeit der NH—Cα-Bindung und der Behinderung der Drehung der Cα—CO-Bindung zusammen. Diese Freiheitsgrade sind auch im denaturierten Zustand aufgehoben, während sie bei nicht cyklischen Aminosäuren nur im nativen Zustand in der Tripelhelix eingefroren sind. Mit zunehmendem Anteil an Aminosäuren wird der denaturierte Zustand energetisch günstiger (Entropieterm!) (*154*) und T_m gesenkt.

Soweit solche Untersuchungen vorliegen, besteht im *nativen* Zustand im Gegensatz zum denaturierten kein Unterschied zwischen dem säurelöslichen und neutralsalzlöslichen Kollagen (*67*).

3. Denaturiertes lösliches Kollagen.

Nach vollständiger Denaturierung findet man die physikalischen Eigenschaften des löslichen Kollagens drastisch verändert. Die negative optische Drehung ist auf ca. $^1/_3$ ihres nativen Wertes gefallen und spiegelt jetzt nur noch die sogenannte Restdrehung der asymmetrischen C-Atome in den freibeweglichen Aminosäuren [z. B. Restdrehung des Prolinrestes $[\alpha]_D^{20} = $ ca. $-250°$ (209)] ohne zusätzliche Festlegung der Liganden wider. Die Viskositätszahl beträgt nur noch etwa $^1/_{20}$ des nativen Wertes und ist charakteristisch für statistisch geknäulte Peptidketten. Auch die Lichtstreuung bestätigt für den Endzustand die geknäulte Form der Denaturierungsprodukte und ein Gewichtsmittel des Molekulargewichts, welches je nach dem Anteil der Komponenten x, β und γ (s. unten) zwischen 1×10^5 und 2×10^5 liegt.

Als erste haben Orekhovich und Shpikiter (310) gefunden, daß das sauer extrahierte denaturierte Kollagen der Rattenhaut sich in der Ultrazentrifuge in zwei Komponenten trennt, von denen sie die langsamer sedimentierende α- und die schnellere β-Komponente nannten. Doty und Nishihara (58) bestätigten dieses Ergebnis für Kalbskollagen und stellten die Hypothese auf, daß die α-Komponente aus einem der drei Stränge der Tripelhelix entstanden ist, die β-Komponente aus den beiden anderen, die jedoch kovalent miteinander verbunden sind. Durch Addition der Molekulargewichte [für α M = ca. 1×10^5 und für β M = ca. 2×10^5 $(58, 67, 150)$] ergibt sich das Molekulargewicht des nativen Moleküls. Später wurde gefunden, daß neutralsalzlösliches Kollagen bei der Denaturierung fast nur in α-Komponenten zerfällt (312). Mit einer verfeinerten Ultrazentrifugentechnik konnten sowohl in diesem wie auch in säurelöslichem denaturiertem Kollagen zusätzlich etwa 10 Gew.-% einer „γ-Komponente" mit M = ca. 3×10^5 gefunden werden $(393, 6, 104)$.

Diese Ergebnisse führten zu der Erkenntnis, daß es drei Typen von Kollagenmonomeren gibt, die nur im denaturierten Zustand unterscheidbar sind: Typ I besteht aus drei Einzelsträngen; Typ II aus einem Einzel- und zwei verbundenen Strängen; in Typ III sind alle drei Peptidstränge miteinander verbunden (104). Letzterer ergibt nach der Denaturierung die γ-Komponente. Das Verhältnis, in dem diese Molekültypen in einer bestimmten Kollagenart anwesend sind, läßt sich aus dem Verhältnis der Komponenten nach der Denaturierung berechnen. Demnach besteht neutralsalzlösliches Kollagen hauptsächlich aus Typ I neben etwa 10% III. In sauren Extrakten aus Kalbshäuten wird das Molverhältnis $\alpha : \beta = $ = $1 : 1$ im allgemeinen nur wenig überschritten (67). Es besteht also hauptsächlich aus Typ II (wieder neben etwa 10% III). Niemals wurde ein kleineres Verhältnis als 1 gefunden (mehr β als α). Dieses wäre auch mit dem Modell der Tripelhelix unvereinbar, da diese ja nicht mehr als eine β-Komponente enthalten kann.

Es liegt die Vorstellung nahe, daß, ausgehend von den Molekülen des Typs I, die, wie erwähnt, eine echte Vorstufe des reifen Kollagens darstellen, im Laufe der Biosynthese mit zunehmender Vernetzung Moleküle des Typs II und schließlich Moleküle des Typs III entstehen. Eine Zunahme des Typs II bei der Alterung von neutralsalzlöslichem Kollagen konnte wirklich in vitro beobachtet werden (_137_). Gegen die Annahme, daß Typ-III-Moleküle den Endzustand dieser Vernetzung darstellen, sprechen bisher zwei Befunde: Ihr Anteil von ca. 10% kommt sowohl im neutralsalzlöslichen wie auch im säurelöslichen Kollagen vor, und sie werden in keiner größeren Menge bei einer denaturierenden Extraktion aus reifem Kollagen herausgelöst (s. unten).

Eventuell könnte die intramolekulare und die intermolekulare Vernetzung (von Molekül zu Molekül) ein einziger kontinuierlicher Vorgang sein. Dafür spricht der Befund von BORNSTEIN und PIEZ (_32, 33_), die menschliches Kollagen mit Neutralsalzlösung, verdünnter Essigsäure und 5 M Guanidiniumchloridlösung extrahierten. Im ersten Falle wurden wie üblich hauptsächlich Typ I-Moleküle und im zweiten Typ II-Moleküle erhalten. (Molverhältnis $\alpha : \beta$ nach der Denaturierung 1:1.) Im letztgenannten Extraktionsmittel liegt das Kollagen geschrumpft, d. h. denaturiert, vor. Der Zusammenhalt der Stränge der Tripelhelix und zwischen den Molekülen ist also mit Ausnahme der kovalenten Bindungen bereits vor der Extraktion zerstört. In dieser Lösung fanden die Autoren eine bedeutend größere Menge an β-Komponente als in den vorherigen Extrakten, die außerdem die Menge an α weit überstieg (Molverhältnis $\alpha : \beta$ etwa 1:1,5). Dies kann nur so erklärt werden, daß ein Teil dieser β-Komponente aus zwei Peptidsträngen _verschiedener_ Tripelhelices, die durch eine intermolekulare Bindung verbunden waren, besteht. Ein Teil der β-Komponente erwies sich außerdem als β_{22}, eine Kombination, die durch intramolekulare Vernetzung gar nicht entstehen kann (s. unten, S. 250).

BORNSTEIN und PIEZ (_32, 33_) fanden in diesem Versuch keine auffallend große Menge an γ-Komponente, welche etwa darauf gedeutet hätte, daß im reifen Kollagen alle Moleküle intramolekular vollständig vernetzt als Typ III vorliegen. Die Bedeutung der γ-Komponente bzw. der Typ III-Moleküle ist demnach noch ungeklärt. Vielleicht handelt es sich um Fehlleitungen von inter- zu intramolekularen Verbrückungen.

In diesem Zusammenhang sei noch erwähnt, daß sich die verschiedenen Molekültypen in einem Punkte auch bereits im nativen Zustand unterscheiden: ihre Fähigkeit, Fibrillen zu bilden, ist verschieden (_419, 420_).

Das Bild über die Komponenten des denaturierten Kollagens ist durch die neuen, hervorragenden Untersuchungen im Arbeitskreis von PIEZ

(*321, 322, 323*) noch komplizierter geworden, hat dabei aber gleichzeitig tiefere Einsichten in den Aufbau des Kollagenmoleküls eröffnet. Es hat sich gezeigt, daß es mindestens (*160a, 320*) zwei chemisch verschiedene α-Komponenten ($\alpha 1$ und $\alpha 2$) und (mindestens) zwei aus ihnen aufgebaute β-Komponenten gibt, die entsprechend den beiden durch Vernetzung möglichen Kombinationen der α-Komponenten β_{11} und β_{12} genannt werden (alte Bezeichnung β_1 und β_2).

Die Aminosäurezusammensetzung der $\alpha 1$- und der $\alpha 2$-Kette ist nach den übereinstimmenden Ergebnissen von Piez (*321*) und aus unserem Arbeitskreis (*359*) etwas verschieden (*Tabelle 7*). Der Gehalt an Ala, Pro, Hypro, Lys ist in der $\alpha 1$-Kette höher, derjenige an Val, Ileu, Leu, His, Hylys niedriger als in der $\alpha 2$-Kette. Die Aminosäurezusammensetzung des Gesamtkollagens entspricht der Annahme, daß das Molekül aus zwei $\alpha 1$- und einer $\alpha 2$-Kette aufgebaut ist. Eine Kombination β_{22} ist demnach durch intramolekulare Vernetzung nicht möglich. Die β_{11}-Komponente hat die gleiche Aminosäurezusammensetzung wie $\alpha 1$, die der β_{12}-Komponente entspricht ihrer Zusammensetzung aus $\alpha 1$ und $\alpha 2$. Die Möglichkeit, daß alle *drei* Ketten chemisch etwas verschieden voneinander sind, ist für Kabeljaukollagen bewiesen (*320*) und beim der-

Tabelle 7. Aminosäureanalysen von Kollagen und seinen Komponenten (Mol/100 Aminosäuren).

Amino-säure ..	Rattenhautkollagen (*321*)*					Kalbshautkollagen (*359*)				
	unge-trennt	$\alpha 1$	$\alpha 2$	β_{11}	β_{12}	unge-trennt	$\alpha 1$	$\alpha 2$	β_{11}	β_{12}
Asp ...	4,6	4,6	4,4	4,5	4,4	4,50	4,25	4,60	4,25	4,45
Glu ...	7,1	7,4	6,6	7,2	7,0	7,60	7,65	7,00	7,60	7,65
Gly ...	33,1	33,0	33,6	33,3	33,8	32,80	32,80	32,85	32,80	32,80
Ala....	10,6	11,2	10,2	11,1	10,7	10,65	11,05	10,30	11,20	10,65
Ser ...	4,3	4,2	4,3	4,1	4,2	3,75	3,80	3,70	3,75	3,70
Thr ...	1,96	1,99	1,98	1,98	2,01	1,75	1,70	1,80	1,75	1,80
Tyr ...	0,24	0,21	0,24	0,18	0,22	0,35	0,30	0,30	0,35	0,30
Val....	2,40	1,96	3,20	1,92	2,56	2,10	1,55	2,75	1,50	2,15
Met ...	0,78	0,80	0,61	0,66	0,70	0,45	0,40	0,40	0,45	0,40
Ileu ...	1,08	0,64	1,61	0,65	1,16	1,25	0,80	1,65	0,90	1,20
Leu ...	2,38	1,81	3,24	2,57	3,24	2,40	1,95	2,70	1,90	2,35
Phe ...	1,13	1,16	1,01	1,21	1,10	1,40	1,30	1,30	1,25	1,30
Pro ...	12,1	12,9	11,3	12,8	12,0	13,15	14,00	12,70	13,55	13,20
Hypro .	9,2	9,6	8,6	9,7	8,8	9,10	9,70	8,90	9,80	9,30
His....	0,49	0,19	0,85	0,20	0,53	0,45	0,25	0,60	0,30	0,45
Lys ...	2,81	3,04	2,24	3,01	2,58	2,60	3,00	2,15	3,00	2,60
Hylys .	0,57	0,43	0,80	0,39	0,62	0,60	0,50	0,70	0,55	0,65
Arg ...	0,51	0,49	0,51	0,50	0,50	5,10	5,00	5,00	5,10	5,05

* In der gleichen Arbeit finden sich Analysen der Komponenten der Kollagene aus Rattenschwanzsehne, Karpfenschwimmblase und der Haut des Katzenhaies.

Literaturverzeichnis: SS. 293—314.

zeitigen experimentellen Stand für andere Kollagene nicht mit voller Sicherheit auszuschließen (*160a*). In diesem Falle würde sich die Zahl der möglichen β-Kombinationen entsprechend erhöhen.

VI. Zerfall und Rückbildung der Sekundär- und Tertiärstruktur (De- und Renaturierung).

1. Denaturierung in Lösung.

Kollagen besitzt wie alle Proteine seine native Struktur nur innerhalb eines bestimmten Bereiches von äußeren Bedingungen. Wenn einer der Parameter, die seine Stabilität beeinflussen — z. B. die Temperatur — einen bestimmten Wert über- bzw. unterschreitet, so tritt ein Zusammenbruch der Tripelhelix ein. Diese Umwandlung ist durch den kooperativen Effekt im Helix-Knäuel-Übergang — die Stabilität jedes Bereiches hängt von der Stabilität der Nachbarbereiche ab — sehr scharf und fast mit dem Schmelzen eines Kristallgitters vergleichbar. Die Temperatur am Mittelpunkt der Umwandlung wird als Denaturierungs- oder Schmelztemperatur T_m bezeichnet (s. Abschnitt V/2, S. 247). Eine Theorie, mit der aus inneren Parametern des Helixaufbaus T_m und die Breite des Übergangsbereiches abgeleitet werden können, besteht bisher nur für die α-Helix (*429, 260, 11*) und die DNA-Doppelhelix (*54*), jedoch noch nicht für die Tripelhelix des Kollagens.

Die Wirkung der Temperatur ist sicherlich komplex. Die Kräfte, die die Struktur stabilisieren (Wasserstoffbrücken und hydrophobe Wechselwirkungen) ändern sich mit der Temperatur. Auch die Konkurrenz von Wasser und eventuell anderen gelösten Molekülen und Ionen (Salze, s. unten) muß zusammen mit vielen anderen temperaturabhängigen Effekten berücksichtigt werden. An einem Punkt, an dem die stabilisierenden Kräfte der allgemeinen natürlichen Tendenz, in einen ungeordneten Zustand überzugehen (Entropiegewinn), nicht mehr standhalten, geht die geordnete native Form in den denaturierten Zustand über.

Sowohl Kationen wie Anionen der meisten Salze vermindern die Stabilität der nativen Tripelhelix in wäßrigen Lösungen (Senkung von T_m); einige wenige steigern ihre Stabilität. Bei jeweils konstant gehaltenem Gegenion lassen sich ihre Wirkungen in folgenden Reihen anordnen (*173, 174*):

Helix ←——————————————→ Knäuel

(Aussalzung ←——————————→ Einsalzung)

SO_4^{--}, Acetat, Cl^-, Br^-, NO_3^-, ClO_4^-, J^-, CNS^-
$(CH_3)_4N^+$, NH_4^+, Rb^+, K^+, Na^+, Cs^+, Li^+, Mg^{++}, Ca^{++}, Ba^{++}.

Diese Reihen entsprechen in auffallender Weise der von Hofmeister (vgl. *61*) bereits um die Jahrhundertwende gefundenen Abstufung der aus- bzw. einsalzenden Wirkung von Anionen und Kationen. Umso größer die aussalzende, desto geringer ist die denaturierende Wirkung. Auf-

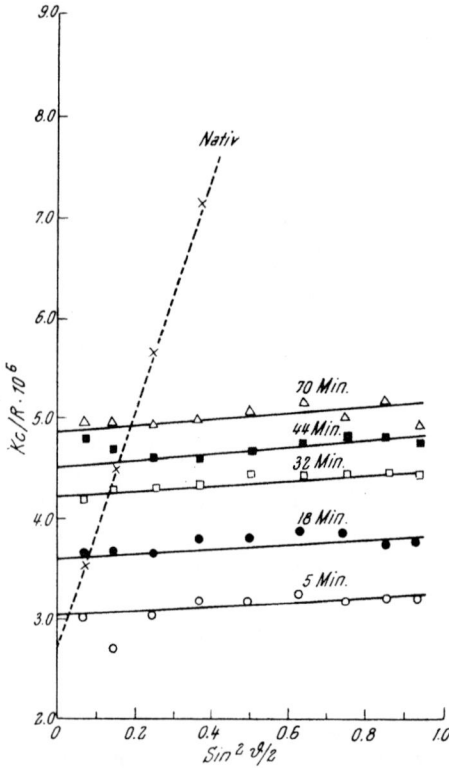

fallend ist weiterhin, daß genau die gleichen Reihen für die Wirkung auf die im Aufbau völlig verschiedenen, durch Wasserstoffbrücken stabilisierten Strukturen von Ribonuclease (*173, 174*) und Desoxyribonucleinsäure (*147*) gefunden wurden. In diesem Zusammenhang ist interessant, daß Guanidiniumthiocyanat auf Polyprolin, welches keine Wasserstoffbrücken enthält, keine Wirkung hat, wohl aber das Kollagenmodellpeptid (Pro·Gly·Pro)$_n$ (s. unten), dessen Tripelhelix durch Wasserstoffbrücken stabilisiert ist, „denaturiert" (*69*).

Von Hippel und Wong (*173*) fanden, daß die Denaturierungstemperatur von Kollagen linear mit der Salzmolarität fällt. Auch für die Geschwindigkeit der Rückkehr der optischen Drehung nach dem Abkühlen (s. Abschnitt 3, S. 255 f.) gilt eine ähnliche Beziehung.

Abb. 17. Verfolgung der isothermen Denaturierung bei 39° mit der Lichtstreuung. (c = 0,0125 g/100 ml; R = reduzierte Streustrahlung; K = Konstante, die von den optischen Daten des Systems abhängt; ϑ = Winkel, unter dem das Streulicht relativ zum eingestrahlten Licht beobachtet wurde.) Nach Engel. [Aus: Arch. Biochem. Biophys. **97**, 150 (1962).]

Trotz dieser Fülle von beobachteten Gesetzmäßigkeiten gibt es bisher noch keine eindeutig richtige Erklärung für den Mechanismus der Wirkung von Salzen bzw. ihrer Ionen (*171, 172, 174*). Die Theorien reichen von einer Änderung der Wasserstruktur bis zur Annahme einer direkten Bindung des Salzes, z. B. an die Peptidbindung (*266, 25*). Letztere Hypothese wird dadurch gestützt, daß Komplexe von LiBr mit Polyprolin und denaturiertem Kollagen aus wäßrigen Lösungen isoliert werden konnten (*156*), in denen jeweils 1 LiBr auf 4 Aminosäurereste kommt. Komplexe von Polyprolin der gleichen stöchiometrischen Zusammensetzung sind mit Perchlorsäure bekannt (*209*).

Literaturverzeichnis: SS. 293—314.

Eine spezifische Wirkung auf die Denaturierungstemperatur von Kollagen übt das Hydroniumion aus. Mit abnehmendem pH wird eine Abnahme von T_m um insgesamt $7°$ bei Kalbhautkollagen beobachtet (*40*), die in ihrem Verlauf genau der Titrationskurve der Carboxylgruppen im Kollagen folgt (*56*). Die durch deren Entladung entstehenden positiven Überschußladungen scheinen die Tripelhelix durch elektrostatische Abstoßung zu schwächen.

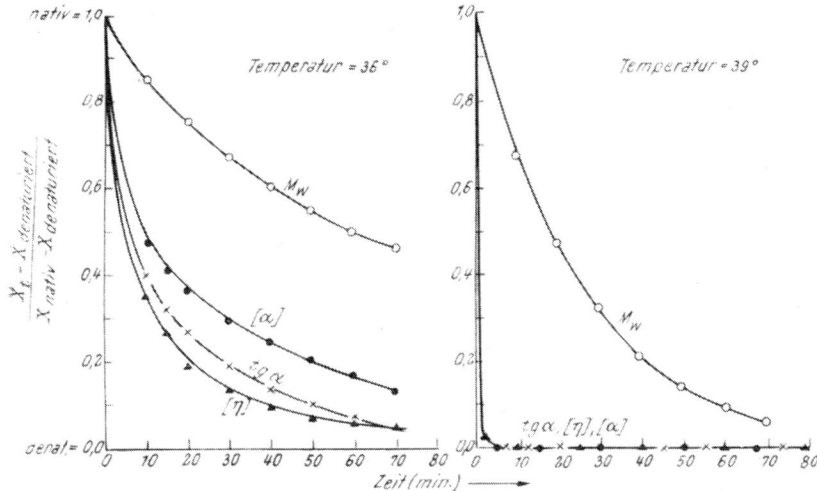

Abb. 18. Abnahme der negativen optischen Drehung, der Viskosität, des Gewichtsmittels des Molekulargewichts und der Anfangsneigung der Winkelabhängigkeit des Streulichtes (tg α) während der Denaturierung bei 36° und 39°. Nach GRASSMANN. [Aus: Das Leder **16**, 32 (1965).]

Der *Mechanismus der Denaturierung* des löslichen Kollagens konnte durch kinetische Messungen weitgehend aufgeklärt werden. DOTY und NISHIHARA (*58*) zeigten, daß der Abfall der optischen Drehung und der Viskosität ungefähr parallel verlaufen. Dies bedeutet, daß die konfigurative innere Ordnung und die hydrodynamische äußere Gestalt gleichzeitig zusammenbrechen. Trotzdem ist die Denaturierung, wie ENGEL (*62, 65*) gezeigt hat, ein mehrstufiger Prozeß. In einer ersten Teilreaktion, die beim pH 3,7 und 37° schon nach 10 Min. abgeschlossen ist, bricht die Stäbchenstruktur zusammen. Aber erst in einer zweiten Stufe, die unter den gleichen Bedingungen etwa 90 Min. erfordert, trennen sich die Einzelkomponenten voneinander, so daß nunmehr die α-, β- und γ-Komponenten frei und in völlig ungeordneter, geknäulter Form vorliegen.

Am anschaulichsten ist dies aus Messungen der Lichtstreuung zu entnehmen, wenn man die erhaltenen Werte im sogenannten Zimm-Diagramm aufträgt, das die Abhängigkeit der Streulichtintensität vom Streuwinkel wiedergibt *(Abb. 17)*. Die Steilheit dieser Kurven (tg α) ist ein Maß für die Asymmetrie der Teilchen (Achsenverhältnis), während der Abschnitt auf der Ordinate den mittlerem Teilchengewicht umgekehrt proportional ist. Die steile Gerade der nativen Lösung ist ein Ausdruck für die stäbchenförmige Gestalt der Tropokollagen-Moleküle. Schon nach 5 Min. ist die Stäbchengestalt verschwunden, die Moleküle sind im wesentlichen kugelförmig geknäuelt. Innerhalb der nächsten 70 Min. sinkt dann das mittlere Teilchengewicht ab (1/M steigt), ohne daß sich in der Gestalt der Moleküle etwas ändert. *Abb. 18* zeigt die zeitliche Änderung von tg α, Molekulargewicht, optischer Drehung und Viskosität bei 36° und 39°. Man sieht, daß das Molekulargewicht sehr viel langsamer absinkt als die übrigen Größen.

Die Existenz einer Zwischenstufe, in der die Ketten noch miteinander zusammenhängen, wird auch dadurch bewiesen, daß ein Material, welches nur bis zum Endwert der ersten Stufe denaturiert wurde, offensichtlich durch den noch vorhandenen *richtigen* Zusammenhang zwischen Ketten viel leichter native Kollagenmoleküle zurückbildet als bis zum Ende der zweiten Stufe (Endwert des Molekulargewichts) denaturiertes Kollagen (s. Abschnitt 3, S. 256).

2. Denaturierung im festen Zustand (Schrumpfung).

Wie erwähnt, ist die treibende Kraft der Denaturierung die Möglichkeit der Peptidketten, in dem umgebenden Lösungsmittel einen ungeordneten, entropiereicheren Zustand anzunehmen, in dem jedes Glied seine maximale Beweglichkeit hat. Damit hat die Konzentration an Lösungsmittel einen Einfluß auf die Denaturierungstemperatur. Im gelösten Zustand ist das Lösungsmittel selbst bei den höchsten erreichbaren Kollagenkonzentrationen immer in einem so großen Überschuß, daß kein Einfluß der Konzentration meßbar ist; anders liegt der Fall bei den mehr oder weniger wasserhaltigen Kollagenfibrillen. Hier beträgt der Wassergehalt nur einen Prozentsatz des Kollagens, und die Denaturierungstemperatur ist wesentlich erhöht. Man spricht von der Schrumpfungstemperatur, entsprechend der Erscheinung, daß sich die Fibrillen (wieder bei einer sehr scharfen Temperatur) stark verkürzen. Auf molekularer Ebene läßt sich dies dadurch erklären, daß die parallel gelagerten, langen Moleküle die kugelsymmetrische Gestalt des denaturierten Knäuels annehmen.

Normalerweise führt man solche Schrumpfungsexperimente so durch, daß man die Fibrillen in Wasser suspendiert, die Einstellung des Quellungsgleichgewichtes abwartet und dann die Temperatur erhöht. Sind die Fibrillen unvernetzt bzw. ungegerbt (vgl. dazu *122a*), so gehen die denaturierten Komponenten während der Schrumpfung in Lösung. Bei reifem, vernetztem unlöslichem Kollagen oder auch bei gegerbten Fibrillen des löslichen Kollagens wird dieses verhindert, und hier kann die schrumpfende Fibrille sehr starke Kräfte entwickeln (ca. 1000 × ihr eigenes Gewicht), welche ein Maß für die große Tendenz sind, den entropiereichen Zustand anzunehmen. Schon früh wurde beobachtet, daß eine Zunahme des Vernetzungsgrades, z. B. durch Gerbstoffe oder durch Alterung, die Schrumpfungstemperatur erhöht (*141*) (vgl. dazu VIII/2, S. 285). Durch die steigende Zahl von Brücken zwischen den Molekülen wird die Quellung und damit das Angebot an Lösungsmittel in der Fibrille verringert. Die Denaturierungstemperatur steigt. FLORY und GARRETT (*73*) fanden, daß die Abhängigkeit der Denaturierungs- bzw. Schrumpfungstemperatur vom Lösungsmittelangebot bekannten thermodynamischen Beziehungen folgt. Sie erhielten durch Extrapolation der Schrumpfungstemperatur von Fibrillen in Gegenwart verschiedener Wassermengen auf unendlichen Überschuß an Wasser genau die Denaturierungstemperatur, die an Kollagenlösungen gemessen wurde.

Obwohl die in Lösung am ausführlichsten untersuchte Renaturierung erst im nächsten Abschnitt behandelt wird, sei die Reversion der Denaturierung im festen Zustand gleich hier beschrieben. Schon früh wurde beobachtet, daß die bei geschrumpften Fibrillen verschwundenen Röntgenreflexe, die für das native Kollagenmolekül charakteristisch sind, beim Abkühlen teilweise wieder zurückkehren (*211*, *36*). Die Reversion der Denaturierung im festen Zustand wird gefördert, wenn man die Fibrillen nach der Schrumpfung unter renaturierenden Bedingungen wieder mechanisch streckt. Dabei ist die Reversibilität umso besser, je höher der Grad der Vernetzung ist. Diese erhält wahrscheinlich das richtige Register zwischen den Peptidketten (man beachte die Analogie zur Renaturierung der γ-Komponente, s. Abschnitt 3, S. 258) und verhindert ein Auseinandergleiten der denaturierten Komponenten unter dem Einfluß der Zugspannung.

Die Möglichkeit, einen Kreisprozeß aufzubauen (z. B. Schrumpfen durch eine hohe Salzkonzentration unter Arbeitsleistung, Überführung in ein Wasserbad, Streckung auf die Ausgangslänge unter Auswaschen des Salzes und Renaturierung sowie erneute Überführung in das Salzbad usw.), macht das Kollagen zu einem interessanten Modell für mechanochemische Vorgänge, in denen chemische Energie direkt in mechanische überführt wird (*426*). In unserem Beispiel wird z. B. eine Salzlösung verdünnt und Arbeit gewonnen.

3. Renaturierung in Lösung.

Schon in frühen Arbeiten wird ein Anstieg der optischen Drehung von abgekühlten Gelatinelösungen beobachtet. HARRINGTON und v. HIPPEL (*155*, *170*) haben diese Untersuchungen an — im Gegensatz zur teilweise hydrolysierten Gelatine — gut definiertem denaturiertem Kollagen wieder aufgenommen. Sie fanden nach Abkühlung auf 4° eine etwa 70%ige Rückkehr der negativen optischen Drehung sowie einen geringen Wiederanstieg der Viskosität. Außerdem wurde beobachtet, daß sich der im denaturierten Zustand erleichterte Angriff von Kollagenase für einen Teil der Bindungen sprunghaft beim Unterschreiten der Denaturierungstemperatur wieder erschwerte (*169*). Die Autoren deuteten dies mit einem 1. Renaturierungsschritt, einer Bildung von Keimen der Polyprolin-II-Struktur innerhalb der Peptidketten. Von diesen ausgehend, soll sich dann in einem 2. Schritt die Polyprolinstruktur der ganzen Kette aufbauen, wodurch der Anstieg der optischen Drehung erklärt wird. Aus diesen Arbeiten ging aber bereits hervor, daß die wirkliche Renaturierung zur nativen Tripelhelix, wenn überhaupt, unter diesen Versuchsbedingungen nur unvollständig ablaufen konnte. Die Viskosität stieg nämlich nicht einmal annähernd auf die für natives Kollagen charakteristischen Werte.

Mit der Zeit ist unsere Kenntnis über die Faktoren, die die Renaturierung beeinflussen, den Grad der Vollständigkeit dieses Vorganges und seinen Mechanismus gewachsen, ohne daß bisher ein völlig abgeschlossenes Bild erhalten worden wäre. Zunächst ist es für das Ergebnis von Renaturierungsversuchen wichtig, ob die Denaturierung bis zur 1. oder 2. Stufe (Abschnitt 1, S. 253) durchgeführt wurde. Die erste Stufe der Denaturierung ist schnell und *vollständig* reversibel, wenn man von der durch den Abfall des Molekulargewichts meßbaren Menge des Materials absieht, welches während der Vollendung der ersten Stufe auch schon die zweite durchlaufen hat. Nach Abkühlen der Lösung auf 4° steigen die vorher auf Endwerte gefallene optische Drehung und Viskosität wieder fast auf ihre nativen Ausgangswerte; das renaturierte Material zeigt in der Ultrazentrifuge die Sedimentationsgeschwindigkeit des nativen; eine erneute Denaturierung erfolgt fast im gleichen Temperaturbereich (gleiche Denaturierungskurve, s. unten) (*64, 227*). Die Beständigkeit gegen proteolytische Enzyme, wie Pepsin und Trypsin (mit Ausnahme von Kollagenase), wird zurückgewonnen (*6, 66*). Auch die Fähigkeit, Quartärstrukturen, wie Fibrillen und Long-spacing-Segmente, zurückzubilden, hat das renaturierte Material voll zurückerlangt (*68*), d. h. die elektronenmikroskopischen Bilder lassen sich nicht von denen nativer Fibrillen und Segmente unterscheiden.

Ganz anders wird das Bild, wenn man die Denaturierung bis zum Endstadium, also bis zum Zerfall in die Einzelkomponenten, durchgeführt hat (*65*). Auch hier steigt zwar die negative optische Drehung auf zirka 70—80% ihres Ausgangswertes an, und es tritt eine Änderung der enzymatischen Angreifbarkeit durch Kollagenase ein (*169*), doch zeigen Viskosität, Lichtstreuung (*65*) und Ultrazentrifugenbild (*63*), daß sich keine nennenswerte Mengen der wirklichen nativen Tripelhelix zurückbilden. Das Teilchengewicht steigt zwar beträchtlich, führt aber nicht zu einem dem Molekulargewicht des lösl. Kollagens entsprechenden Endwert von 330000, sondern steigt bei höheren Konzentrationen weit höher (65), ohne daß eine scharfe Grenze erkennbar wäre. Der tg α des Zimm-Diagramms sowie die Viskosität erreichen nur Bruchteile der für natives Tropokollagen zu fordernden Werte.

Die meisten Polypeptidketten haben sich also spiralisiert, und die Ketten sind zu höheren Aggregaten zusammengetreten, aber diese Aggregate haben weder das Molekulargewicht des lösl. Kollagens noch überhaupt ein definiertes Molekulargewicht, noch sind sie in wesentlichem Umfange stäbchenförmig. Es handelt sich also in der Hauptsache um unspezifische und weitgehend ungeordnete Aggregate, die auch nicht zur Bildung von Fibrillen und Longspacing-Segmenten befähigt sind. Dieser Befund zeigt, daß sich in den Ketten auch ohne Bildung der Tripelhelix die Polyprolin-II-Struktur, welche die hohe negative optische Drehung des Kollagens bedingt, stabilisieren kann.

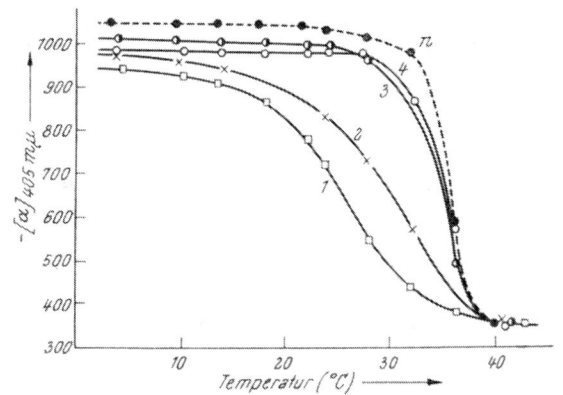

Abb. 19. Denaturierungskurven von nativem und renaturiertem löslichem Kollagen, verfolgt mit der optischen Drehung. Temperaturwechsel von Meßpunkt zu Meßpunkt nach jeweils 15 Minuten. Das native Material (Kurve *n*) wurde vollständig denaturiert (90 Minuten bei 38°) und dann 36 Tage bei 4° isotherm renaturiert (Kurve *1*) bzw. dreimal abwechselnd je 24 Stunden bei 4° und 22° renaturiert (Kurve *2*) und bei letzterem der nicht renaturierte Anteil mit Pepsin (Kurve *3*) bzw. Trypsin (Kurve *4*) verdaut. Konzentration ca. 0,25 g/100 ml in 0,25 m Citratpuffer, pH 3,7. Nach KÜHN, ENGEL, ZIMMERMANN und GRASSMANN. [Aus: Arch. Biochem. Biophys. 105, 387 (1964).]

Daß diese Aggregate in ihrer Temperaturbeständigkeit nicht mit den Molekülen des nativen Kollagens übereinstimmen, zeigt *Abb. 19*, Kurve 1. Während natives Kollagen beim Erwärmen bis zu etwa 30° völlig beständig ist und dann, zwischen etwa 32° und 37°, rasch zusammenbricht, beginnt die thermische Zerstörung dieser unspezifischen Aggregate schon wenig oberhalb 10° und setzt sich über einen weiten Temperaturbereich bis über 30° hinaus fort (64). Die bei 4° gebildeten Aggregate sind also in bezug auf ihre Temperaturbeständigkeit inhomogen, sie enthalten beträchtliche Mengen von Anteilen, deren Wärmestabilität weit geringer ist als diejenige des Kollagenmoleküls.

Das unterschiedliche Renaturierungsverhalten des bis zur ersten und des bis zur zweiten Stufe denaturierten Kollagens wird damit gedeutet (63, 66), daß in den Zwischenstufen nach dem ersten Denaturierungsschritt noch das richtige Register zwischen den Ketten erhalten geblieben ist, welches ein rasches und vollständiges Rückspulen in die native Tripelhelix ermöglicht. Für diese Erklärung spricht auch der Befund von ALTGELT u. Mitarb. (6), daß die γ-Komponente, in der der richtige Zusammenhang der drei Ketten durch kovalente Bindungen erhalten wird, genau so rasch und vollständig renaturiert.

Die geringere Temperaturbeständigkeit der gebildeten unspezifischen Aggregate legt es nahe, die Renaturierung bei höherer Temperatur

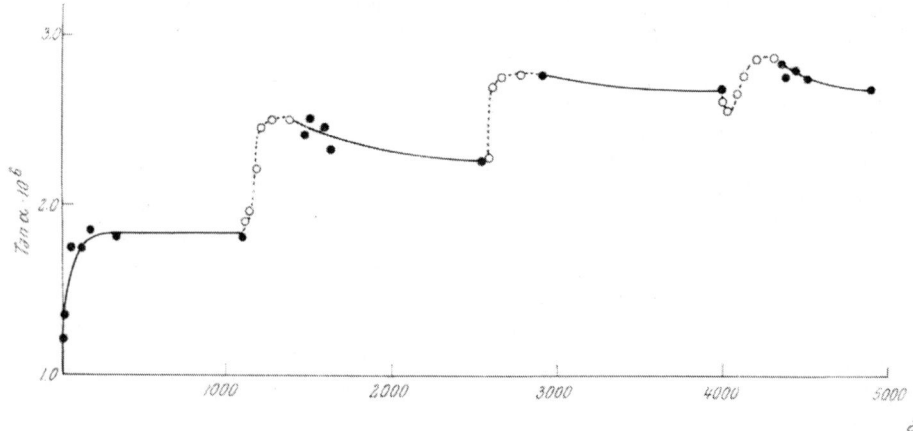

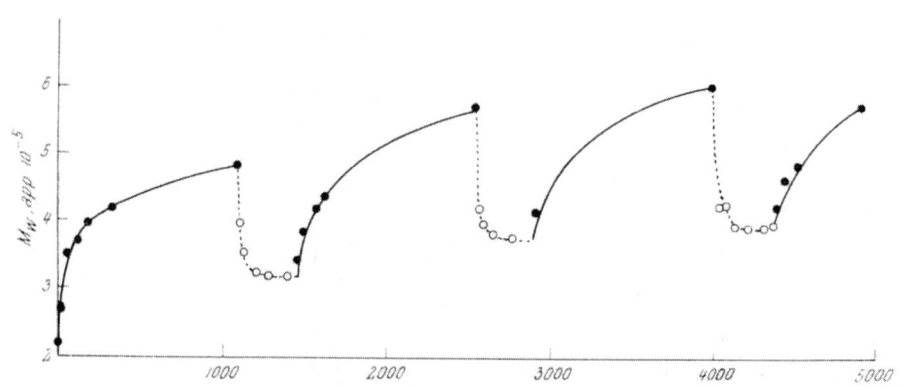

Abb. 20. Renaturierung unter Temperaturwechsel. ●——●Änderungen bei 4°, ○----○Änderungen bei 22°. a Anfangsneigung der Winkelabhängigkeit der Lichtstreuung tg α in Abhängigkeit von der Zeit. b Apparentes d Optische Drehung bei 405 mμ. Nach KÜHN u. Mitarb.

durchzuführen bzw. während der Renaturierung zwischen einer höheren Temperatur und 4° zu wechseln (tempern), um die bei diesen Temperaturen falsch gebildeten Zusammenlagerungen „aufzuschmelzen" und den Ketten Gelegenheit zu geben, die thermodynamisch stabilste Form, die native Tripelhelix, zu bilden (64). Obwohl wir heute von diesen beiden

Literaturverzeichnis: SS. 293—314.

Möglichkeiten die Renaturierung bei einer als optimal bekannten *konstanten mittleren* Temperatur von 25° bevorzugen, sei eine mit den

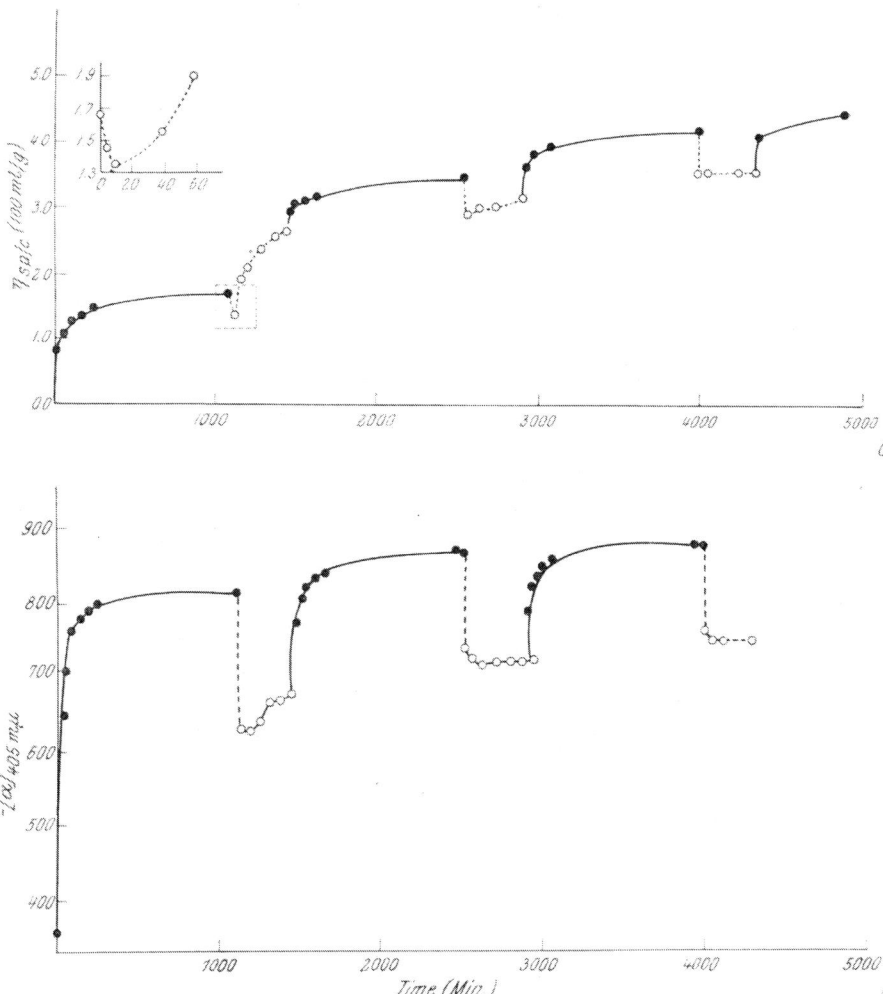

Die Lösung wurde 100 Minuten bei 38° denaturiert und schnell auf 4° abgekühlt. c = 0,055 g/100 ml
Gewichtsmittel des Molekulargewichts Mw, app, gemessen mit der Lichtstreuung. c Reduzierte Viskosität.
[Aus: Arch. Biochem. Biophys. **105**, 387 (1964).]

Methoden der Lichtstreuung, der optischen Drehung und der Viskositätsmessung verfolgte Versuchsreihe bei wechselnder Temperatur
(Schaukelmethode) wiedergegeben *(227)* *(Abb. 20)*, weil sie besonders
anschaulich das Wesen der bei tiefer und bei höherer Temperatur ablaufenden Vorgänge anzeigt. Eine vollständig denaturierte Lösung wurde

17*

zunächst bis zur Konstanz der Meßwerte bei 4° gehalten und dann abwechselnd Temperaturen zwischen 4° und 22° ausgesetzt. Das mittels der Lichtstreuung bestimmte mittlere Teilchengewicht fällt beim Erwärmen auf 22° zuerst rasch, dann langsamer ab und läuft in einen Wert ein, der kleiner ist als das native Molekulargewicht. Dies zeigt, daß ein großer Teil der gebildeten Aggregate wieder zerfällt. Gleichzeitig steigt aber der tg α stark an, was die Ausbildung stäbchenförmiger Aggregate anzeigt. Kühlt man wieder ab, so steigt das Molekulargewicht erneut an, während tg α ein wenig absinkt, aber im ganzen doch auf dem höheren Niveau verbleibt: Offenbar hat neuerdings Assoziation stattgefunden, die nur teilweise zu stäbchenförmigen Gebilden geführt hat.

Auch die Viskosität fällt bei 22° zunächst rasch ab, wie dies bei abnehmendem Molekulargewicht und gleichbleibender Form zu erwarten ist (Abb. 20 c), steigt jedoch nach kurzer Zeit wieder schnell an, obwohl das Molekulargewicht weiter abgenommen hat. Dieses Verhalten kann nur so erklärt werden, daß sich starrere stäbchenförmige Teilchen gebildet haben. Der Verlauf der optischen Drehung zeigt, daß bei 22° zunächst ein sehr großer Teil der bei 4° bereits gebildeten optisch aktiven Helixstrukturen schmelzen. Ein Teil davon bildet sich jedoch bereits bei 22° wieder zurück, offenbar unter Ausbildung stabilerer Konfigurationen. Man kann dieses Spiel wiederholen, wobei die optische Drehung tg α und Viskosität weiter ansteigen, doch wird der Effekt jedes Schrittes von Mal zu Mal kleiner.

Die Denaturierungstemperaturen solcher Lösungen sind, verglichen mit solchen der isotherm renaturierten, stark erhöht (227); sie sind aber, ebenso wie die Viskosität und die Werte von tg α, noch ziemlich weit von den Werten des nativen Kollagens entfernt (Abb. 19, Kurve 2, S. 257). In der Ultrazentrifuge zeigt sich ein Gipfel mit dem Sedimentationsverhalten des nativen Kollagens, dem aber noch ein oder zwei schneller sedimentierende — also entweder schwerere oder nicht stäbchenförmige — Komponenten vorauslaufen. Fibrillen und Segmente konnten in bedeutend besserer Ausbeute erhalten werden als nach isothermer Renaturierung, doch erscheinen die Präparate immer noch durch amorphe Anteile verunreinigt. Offenbar liegen also immer noch Gemische echter Kollagenmoleküle mit unspezifischen Aggregaten vor.

Da native Kollagenmoleküle gegen Pepsin und Trypsin im wesentlichen beständig sind, denaturierte aber spielend angegriffen werden, lag es nahe, den denaturierten Anteil enzymatisch abzubauen und die Spaltstücke etwa durch Dialyse zu entfernen. *Tabelle 8* zeigt die Ausbeuten an enzymbeständigem Kollagen, u. zw. einerseits bestimmt durch direkte

Isolierung und anderseits errechnet auf Grund der Verfolgung der optischen Drehung beim enzymatischen Abbau. Wenn man vollständig denaturiertes Material isotherm bei 4° renaturiert, erreicht demnach die Menge des regenerierten Kollagens nur 14% bzw. 23%, bei Anwendung der Schaukelmethode aber 40—57%. Bei extrem langsamer Abkühlung (Temperaturerniedrigung beispielsweise 5° pro Tag) (227), kann, wie die Werte zeigen, eine relativ gute Ausbildung der stabilen Kollagenmoleküle erreicht werden. Bei Lösungen, die nur bis zur ersten Stufe denaturiert wurden, gelangt man auch bei raschem Abkühlen zu Ausbeuten an nativem Material von rund 60%, und die extrem langsame Abkühlung (oder auch die Schaukelmethode) verbessern das Ergebnis nur unwesentlich. Der fermentresistente Anteil erwies sich in der Ultrazentrifuge, in der Lichtstreuung und in seiner Fähigkeit, Fibrillen und Segmente zu bilden, völlig mit nativem Kollagen identisch. Eine geringfügig kleinere thermische Stabilität (Abb. 19, Kurven 3 und 4, S. 257) kann durch Fehlstellen bzw. Verunreinigungen mit nicht völlig renaturiertem Material erklärt werden.

Damit ist gezeigt, daß aus dem in die völlig geknäuelten Einzelketten zerfallenen Kollagen unter geeigneten Bedingungen die Stäbchenmoleküle

Tabelle 8. Ausbeute an gegen Trypsin bzw. Pepsin beständigen Kollagenmolekülen aus verschieden denaturierten und renaturierten Lösungen.

Denaturierung	Renaturierung	Trypsin-beständig		Pepsin-beständig	
		Ausbeute*	$[\alpha]_{rel.}$**	Ausbeute*	$[\alpha]_{rel.}$**
90 Min. bei 38°	durch Temperaturwechsel zwischen 4 und 22°	0,40	0,52	0,42	0,57
90 Min. bei 38°	durch langsame Abkühlung auf 4°	0,40	0,43		
90 Min. bei 38°	nach rascher Abkühlung auf 4° sieben Tage lang isotherm bei dieser Temperatur	0,14	0,23		
11 Min. bei 38°	durch langsame Abkühlung auf 4°	0,59	0,67		
11 Min. bei 38°	nach rascher Abkühlung auf 4° sieben Tage lang isotherm bei dieser Temperatur	0,58	0,61		

* Als Long-spacing-Segmente isolierter enzymbeständiger Anteil.

** $[\alpha]_{rel.} = \dfrac{[\alpha]_{\text{nach Enzymbehandlung}} - [\alpha]_{\text{denaturiert}}}{[\alpha]_{\text{nativ}} - [\alpha]_{\text{denaturiert}}}$.

der Tripelhelix des Kollagens in guter Ausbeute zurückgebildet werden können. Das Kollagen ist also ein besonders wohluntersuchtes Beispiel dafür, daß in der Aminosäuresequenz der Primärstruktur alle Informationen zur Bildung der Tertiärstruktur des Moleküls enthalten sind [vgl. (101a)]. Ähnliche Resultate wurden an vielen anderen in Struktur und Größe völlig verschiedenen Proteinen gewonnen (66), so daß sich heute die Auffassung durchsetzt, daß in der Biosynthese über die Codierung der Aminosäuresequenz hinaus keine zusätzliche Matritze, die die räumliche Struktur bedingt, notwendig ist.

In den bisher referierten Arbeiten wurde säurelösliches Kollagen, welches im denaturierten Zustand aus $\alpha 1$-, $\alpha 2$-, β_{11}- und β_{12}-Komponenten besteht (s. Abschnitt V/3, S. 248, 250) renaturiert. Kühn und Zimmermann (251) konnten zeigen, daß auch aus denaturiertem neutralsalzlöslichem Kollagen, also aus $\alpha 1$- und $\alpha 2$-Komponenten native Moleküle zurückgebildet werden können. Kürzlich gelang es sogar, isolierte $\alpha 1$- und $\alpha 2$-Komponente allein zu Molekülen zu renaturieren, die normale Long-spacing-Segmente und Fibrillen aufbauen können (246). Versuche von Piez und Carrillo (320a) an isolierten Komponenten führten unter anderen Renaturierungsbedingungen (isotherm bei 15°) nur zu Vorstufen, z. B. $\alpha 1 \alpha 1$- bzw. β_{12}-Doppelhelices.

Läßt man die Renaturierung nur kurze Zeit vor sich gehen und entfernt dann den nicht renaturierten Anteil durch Behandlung mit Pepsin, so kann man Segment-Bruchstücke gewinnen, die das B-Ende enthalten. Daraus kann geschlossen werden, daß die Renaturierung in der Nähe des B-Endes, also des Carboxylendes des Kollagenmoleküls einsetzt (vgl. Abb. 4, S. 205 sowie S. 288, Anmerkung) [Kühn u. Mitarb. (251a)].

Diese Ergebnisse werfen neues Licht auf den *Mechanismus der Renaturierung*, über den bisher noch kein klares Bild besteht. Die Vorstellungen von Harrington und v. Hippel (155, 170) wurden bereits skizziert (S. 256). Im Gegensatz dazu steht die Theorie von Flory und Weaver (74), die annehmen, daß sich als Zwischenstufe gebildete Einzelhelices in *sehr schneller* Reaktion durch Aggregation (zur Tripelhelix) stabilisieren. Letztere Anschauung wird durch die kinetische Verfolgung der Umwandlung von leicht austauschbarem Wasserstoff in schwer austauschbaren, an Wasserstoffbrücken beteiligten, gestützt (26a). Beier und Mitarb. (24a, 24b) konnten zeigen, daß bevorzugt bei tiefen Temperaturen verschiedene Aggregate auftreten (unter diesen wahrscheinlich die von Piez und Carrillo (320a) gefundenen Doppelhelices), die sich durch ihre thermische Stabilität unterscheiden. Durch deren sehr rasche Bildung werden der nativen Tripelhelix die Bausteine entzogen, was erklärt, daß diese bei tiefen Temperaturen fast gar nicht auftritt.

Literaturverzeichnis: SS. 293—314.

Im Hinblick auf das allgemeine Problem der biogenetischen Bildung der Eiweißstrukturen scheint das Beispiel des Kollagens folgendes zu zeigen: Die Spiralisierung der Ketten und die — spezifische oder unspezifische — Assoziation durch zwischenmolekulare Kräfte, z. B. durch Wasserstoffbrücken, sind diejenigen Vorgänge, die bei tiefer Temperatur rasch und spontan erfolgen. Für die Bildung der nativen Moleküle ist aber darüber hinaus die *spezifische* Zuordnung der Ketten notwendig. Im Falle des Kollagens müssen nämlich die Ketten sich so vereinigen, daß sie nach gemeinsamer Verdrillung zur Tripelhelix mit ihren Enden abschneiden. Da das allgemeine Muster, in dem geordnete und ungeordnete Bereiche innerhalb der Ketten abwechseln, für die drei Ketten sehr wahrscheinlich gleich ist [s. S. 262 (*246*)], bedeutet das zugleich, daß die geordneten Bereiche über die gesamte Moleküllänge ineinander einrasten, was zugleich die Ausbildung des thermodynamisch stabilen Zustandes bedeutet. Diese spezifische Zuordnung kann offenbar auf zwei Wegen erreicht werden: Entweder müssen die Ketten durch eine oder wenige vorgebildete kovalente Bindungen bereits „richtig" verknüpft sein (Renaturierung der γ-Komponente), oder sie müssen die Möglichkeit haben, unter Bedingungen, die sowohl für die Schließung wie für die Lösung zwischenmolekularer Verbrückungen günstig sind, den energetisch günstigsten Zustand, den des nativen Moleküls, auszubilden.

Daß vom Tripelhelixmolekül des Kollagens aus unter überschaubaren Bedingungen in vitro die verschiedenen sichtbaren Strukturen des Kollagens gebildet werden können, wird im folgenden Kapitel gezeigt werden.

4. Der De- und Renaturierung des Kollagens analoge Umwandlungen von (Pro · Gly · Pro)$_n$ in Lösung.

Wie bereits erwähnt (S. 242) ist für die Polypeptide (Gly · Pro · Hypro)$_n$ und (Pro · Gly · Pro)$_n$ vom Molekulargewicht 5000 bis 10000 im festen Zustand die Struktur der Kollagentripelhelix nachgewiesen worden (*70, 69, 9*). Das letztgenannte Modellpeptid wurde auch in Lösung untersucht. Es zeigt beim Erwärmen in Wasser eine Umwandlung, deren Schärfe und Schmelztemperatur mit steigendem Molekulargewicht zunimmt (für M = 10000, T_m = ca. 69°) und die von einem Abfall der negativen optischen Drehung und der Viskosität begleitet ist. Die Schmelztemperatur wird wie beim Kollagen durch Salze (Guanidiniumthiocyanat und LiBr) erniedrigt. Nach dem Abkühlen steigen wie bei der Renaturierung von Kollagen in einer verhältnismäßig langsamen Reaktion die optische Drehung und die Viskosität wieder an. Es ist anzunehmen, daß weitere Untersuchungen an derartig einfachen Modellpeptiden einen Einblick in die Elementarvorgänge des Zerfalls und der Rückbildung der Sekundär- und Tertiärstruktur des Kollagens erlauben werden.

VII. Bau und Bildung der Fibrillen und Segmente (Quartärstrukturen).

1. Die nativen Fibrillen.

a. Das Querstreifungsmuster.

Im Jahre 1942 haben unabhängig voneinander Schmitt (*146*) in USA und Wolpers (*412*) in Deutschland eine nur im Elektronenmikroskop sichtbare Querstreifungsperiode von ungefähr 650 Å der Kollagenfibrillen entdeckt. Diese ersten Bilder zeigten lediglich das Abwechseln eines hellen und eines dunklen Streifens. Bald erkannte man (*361, 366, 362, 414, 415*) 6—7 Querstreifen pro Periode, die Schmitt mit a, b_1, b_2, c d und e bezeichnete (*362*). Schließlich konnten Grassmann, Hofmann und Nemetschek (*109, 180, 294*) bis zu 13 Querstreifen pro Periode erkennen. Gute Aufnahmen von Kollagenfibrillen zeigen heute routinemäßig 10—12 Querstreifen (s. Abb. 1, S. 204).

Die hochunterteilte Querstreifung ist nur zu erkennen, wenn die Kollagenfibrillen vorher mit Phosphorwolframsäure, gerbenden Chrom (III)- oder Uranylsalzen behandelt wurden. Ohne diese Behandlung („Anfärbung" oder „Kontrastierung") erkennt man gewöhnlich nur einen sehr verschwommenen Hell- und Dunkelteil. Nur ganz selten konnte auch ohne „Anfärbung" eine Aufteilung der Perioden in mehrere Querstreifen beobachtet werden (*294*).

Die schmalsten, dunklen Querstreifen, die man in Aufnahmen angefärbter Kollagenfibrillen mit hochunterteilter Querstreifung erkennen kann, sind kaum breiter als 15 bis 20 Å. Eine photometrische Auswertung zeigte sogar, daß innerhalb einer einzelnen Fibrille die relativen Lagen der Schwärzungsmaxima der einzelnen Querstreifen längs der Periode mit einer Genauigkeit von 5 Å festgelegt werden können (*294*). Ein Vergleich der an zahlreichen Fibrillen verschiedenster Herkunft und Vorbehandlung durchgeführten Messungen läßt eine weitgehende Übereinstimmung in der Lage der einzelnen Querstreifen erkennen, wobei allerdings deren Intensitäten recht wesentlich schwanken können (*294*). Während die Länge der Perioden durch Kleinwinkelinterferenzen immer ziemlich genau mit 640 Å gemessen wird (*18, 219, 19*), schwankt sie im Elektronenmikroskop gewöhnlich zwischen 500 Å und 800 Å.

b. Die Ursache der Querstreifung.

Über die Ursache der Querstreifung des Kollagens wurden eine Anzahl von Theorien aufgestellt (*121*), von denen nur die von Bear (*19*) erwähnt werden soll. Nach Bear bestehen die dunklen Querstreifen, die „bands", aus Anhäufungen schwerer und polarer Aminosäuren und die hellen Querstreifen, die „interbands", vorwiegend aus leichten

Aminosäuren. Die Interbands stellen die geordneten Bereiche der Periode dar, während die Bands auf Grund der sperrigen Seitenketten der polaren und der schweren Aminosäuren ungeordnet sein sollen. Diese Theorie wurde bald durch experimentelle Befunde über die Primärstruktur des Kollagens von SCHRÖDER et al. (369), KRONER et al. (221, 222) sowie von GRASSMANN und Mitarb. (103, 113) unterstützt.

BEAR erklärte die Verstärkung der Querstreifung durch Behandlung mit ,,Kontrastmitteln'', wie Phosphorwolframsäure (PWS) und Uranylacetat, durch die Annahme, daß die ,,ungeordneten Bands'' im Gegensatz und den ,,geordneten Interbands'' genügend Raum bieten, um solche große Ionen einzulagern. Eine eingehende Untersuchung der Reaktion dieser sogenannten Kontrastmittel, aber auch üblicher gerbender Substanzen, wie z. B. des Chrom(III)-Kations durch KÜHN (231, 232, 84), führt zu einer anderen Vorstellung. An Kollagen mit systematisch veränderten sauren und basischen Seitenketten wurde gezeigt, daß die PWS ausschließlich durch die basischen Seitenketten des Arginins und Lysins gebunden wird (231, 232). Die Bindung an Arginin ist dabei wesentlich fester. Sind die basischen Seitenketten durch Behandlung mit salpetriger Säure zerstört, wird keine PWS mehr durch das Kollagen gebunden. Ein Phosphorwolfram-Ion reagiert stöchiometrisch mit 3 Argininresten.

Die für die Ausbildung einer hochunterteilten Querstreifung verantwortlichen Chrom(III)-Komplexe (229, 230) reagieren dagegen mit den Carboxylgruppen der Seitenketten der Asparagin- und Glutaminsäure. Dabei werden die Carboxylgruppen koordinativ in die Chromkomplexe mit einbezogen. Auch das Uranylkation (84) wird hauptsächlich durch die Carboxylgruppen gebunden. Methylkollagen, in dem die Carboxylgruppen quantitativ durch Methylierung blockiert sind, zeigt, mit Chromkomplexen und Uranylacetat behandelt, im Elektronenmikroskop keine Querstreifung mehr, während sie nach PWS-Anfärbung bei diesem Präparat gut zu sehen ist. Da nach Gerbung mit Chrom(III)-Salzen eine gleich scharfe Querstreifung erhalten wird wie nach Behandlung mit PWS, obwohl nach Chromgerbung zehnmal weniger Masse eingelagert wird (50 mg Chrom/g Kollagen) als mit PWS (500 mg/g Kollagen), liegt die entscheidende Wirkung der sogenannten ,,Anfärbemittel'' offenbar nicht in einer Kontrasterhöhung, sondern mehr in einem Ordnungseffekt. So reagiert z. B. ein dreibasisches Phorphorwolframat-Ion mit drei Argininresten aus drei verschiedenen Kollagenmolekülen innerhalb der Fibrille und fixiert dadurch die basischen Aminosäuren in nebeneinanderliegenden Stellungen. Sie können sich bei der Präparation für das Elektronenmikroskop nicht mehr so leicht gegeneinander verschieben. Die Anhäufungen der schweren und polaren Aminosäuren verschiedener Kollagenmoleküle in den Fibrillen kommen so genauer nebeneinander zu liegen. Auf gleiche Weise wirken die gerbenden Chrom(III)-Komplexe

und das Uranylkation über die Seitenketten der Asparagin- und Glutaminsäure.

Durch einen genaueren photometrischen Vergleich der Querstreifungsmuster von Kollagenfibrillen mit verschiedenem PWS- und Chromgehalt
wurde an Hand der unterschiedlichen Intensitäten einzelner Querstreifen
gezeigt (239, 223, 224), daß die PWS besonders stark in den dunklen
a-Streifen gebunden wird, d. h. daß an diesen Stellen der Periode
basische Aminosäuren angereichert sein müssen. Auf diese Weise wurde
zum ersten Mal die Theorie von Bear (19) über Anreicherungen von
polaren Aminosäuren an den dunklen Querstreifen experimentell im
Elektronenmikroskop nachgewiesen und bestätigt.

Eine weitere Aussage über den Bau der Kollagenfibrillen, vor allem
über die Anordnung der Kollagenmoleküle innerhalb der Fibrillen, gelang
unter Zuhilfenahme der künstlichen Kollagenformen, wie sie aus löslichem
Kollagen hergestellt werden konnten.

2. Der molekulare Bau der verschiedenen Kollagenstrukturen.

Für die verschiedenen Kollagenstrukturen, die auf SS. 203—207 bereits
einführend besprochen worden sind, läßt sich zeigen, daß ihre Entstehung durch verschiedene Anordnung des etwa 2800 Å langen Stäbchenmoleküls des Kollagens zustande kommt.

1956 hat Schmitt (360, 364) folgende Anordnung der Kollagenmoleküle in den einzelnen Kollagenformen vorgeschlagen (Abb. 21):
In den 2800 Å langen Long-spacing-Segmenten (SLS) liegen die Kollagenmoleküle mit den Enden abschneidend parallel, d. h. in gleicher
Richtung nebeneinander. Das symmetrische Querstreifungsmuster der
Long-spacing-Fibrillen (FLS) deutet darauf hin, daß hier die Kollagenmoleküle antiparallel in entgegengesetzter Richtung mit den Enden
abschneidend unter gleichzeitiger Verknüpfung ihrer Enden angeordnet
sind. Die Periode von 700 Å bei den Fibrillen des nativen Typs kommt
dadurch zustande, daß sich die Moleküle um $1/_4$ ihrer Länge gegeneinander
versetzt parallel zusammenlagern. In den folgenden Jahren wurden
diese Anschauungen bis auf einige kleine Korrekturen, die sich im
wesentlichen auf die Überlappung der Enden beziehen, experimentell
bestätigt.

a. Das Querstreifungsmuster der Long-spacing-Segmente (SLS).

Wenn in den SLS die Kollagenmoleküle parallel nebeneinander-
gelegen sind, muß sich in ihrem Querstreifungsmuster die Verteilung der
Aminosäuren längs der Peptidketten direkt widerspiegeln. Behandelt
man die SLS mit PWS (237, 179), welche von den basischen Aminosäuren gebunden wird, so treten diejenigen Querstreifen besonders stark
hervor, in denen vorwiegend Lysin und Arginin angereichert ist. Nach

Literaturverzeichnis: SS. 293—314.

Anfärbung mit Uranylacetat (*179, 238*) dagegen werden besonders die Querstreifen intensiv, die hauptsächlich saure Aminosäuren enthalten. Ein genauer photometrischer Vergleich der Querstreifungsmuster der mit PWS und Uranylacetat behandelten Segmente (*247*) ergab die in *Abb. 22* wiedergegebene Verteilung saurer und basischer Aminosäuren längs des Kollagenmoleküls. So zeigen sich über die Länge des Seg-

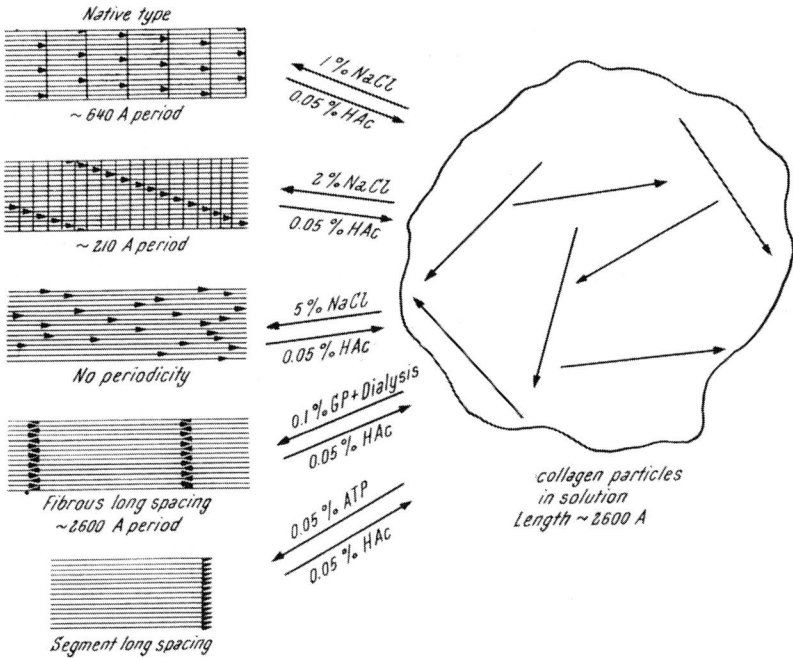

Abb. 21. Schematische Darstellung der Bildung der verschiedenen Kollagenformen aus Kollagenmolekülen (GP = α 1 acid glycoprotein). Nach Schmitt. [Aus: Proc. Amer. Phil. Soc. **100**, 476 (1956).]

mentes getrennte Anreicherungen von basischen und sauren Aminosäuren, die offenbar dem Ladungsmuster des Kollagenmoleküls entsprechen. Die oben besprochenen chemischen Ergebnisse über die Primärstruktur bestätigen die Existenz von Anhäufungen basischer und saurer Aminosäuren entlang der Polypeptidketten des Kollagens. Dieselben photometrischen Untersuchungen von verschieden angefärbten Segmenten wurden später von Hodge und Petruska (*176*) mit gleichen Ergebnissen wiederholt.

Die Entstehung der Long-spacing-Segmente aus sauren Lösungen mit Hilfe von ATP erklärt man folgendermaßen (*247*): Die Kollagenmoleküle besitzen in essigsaurer Lösung vorwiegend positive Ladungen, welche durch Gegen-Ionen (Acetat-Ionen) abgesättigt sind. Gibt man nun zu diesen Lösungen ATP, so werden die Gegen-Ionen durch diese mehrwertigen Anionen ausgetauscht. Die Ladungsdichte der positiven Ladungen auf den Kollagenmolekülen reicht aber nicht aus, die ATP örtlich abzusättigen. Es bleiben negative Ladungen auf den ATP-Ionen frei, die sich mit positiven Ladungen anderer Kollagenmoleküle abzusättigen suchen. Die Folge davon ist die Zusammenlagerung zu den SLS. Die Zusammenlagerung der Kollagenmoleküle gelingt auch mit Hilfe anderer Polyanionen. So entstehen die gleichen Segmente bei Dialyse von Neutralsalzlösungen des Kollagens gegen Essigsäure in Gegenwart von Hyaluronsäure oder Chondroitinsulfat (247). Die besten Segmente entstehen aber mit ATP zwischen pH 2,5 und 3,5. Über pH 4 machen sich die negativen Ladungen der dissoziierenden Carboxylgruppen bemerkbar (247). Es entstehen teilweise wieder Fibrillen des nativen Typs.

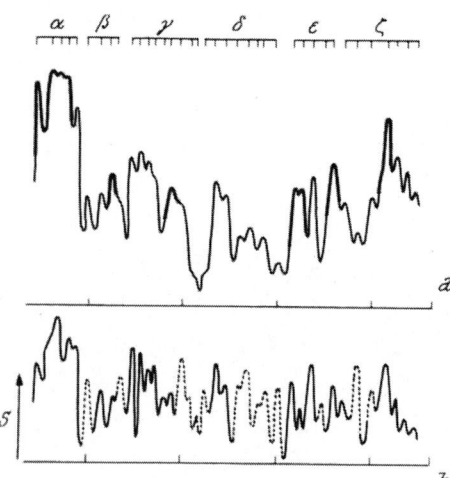

Abb. 22. Photometrische Kurven der Elektronenbilder von Long-spacing-Segmenten. *a* Mit Phosphorwolframsäure und *b* mit Uranylacetat behandelt. Ordinate = $= S = \log I_0/I =$ Schwärzung der Platte durch die Elektronen linear aufgetragen. Stark ausgezogene Linien bedeuten Anhäufungen von basischen Aminosäuren; gestrichelte Linien Anhäufungen von sauren Aminosäuren. Nach Kühn und Zimmer. [Aus: Z. Naturforsch. **16b**, 648 (1961).]

b. Die Anordnung der Kollagenmoleküle in den Fibrillen des nativen Typs.

Der experimentelle Nachweis, daß die Kollagenmoleküle in den Fibrillen des nativen Typs wirklich um $^1/_4$ ihrer Länge gegeneinander versetzt sind, gelang Hodge et al. (*179*) und Kühn et al. (*234, 224*) zur gleichen Zeit auf verschiedenen Wegen. Hodge gelang es, dimorphe Formen zwischen SLS und Fibrillen des nativen Typs herzustellen. Dabei wachsen die Segmente an der Oberfläche der Fibrillen auf, und man kann die Zuordnung des SLS-Querstreifungsmusters zu dem Muster der Fibrillen erkennen. Da sich die native Periode von 700 Å viermal auf die Länge eines Segmentes wiederholt, kommt jeder Streifen des nativen Musters mit vier verschiedenen Querstreifen des Segmentmusters in Berührung. Nachdem durch diese Aufnahmen die Zuordnung bekannt war, wurde eine optische Synthese des Querstreifungsmusters des nativen Typs durchgeführt. Dazu wurden Aufnahmen von SLS-Querstreifungsmustern mehrmals übereinander kopiert, wobei nach jeder Belichtung das Segmentmuster um $^1/_4$ der Länge des Segmentes verschoben wurde.

Literaturverzeichnis: SS. 293—314.

Die auf diese Weise entstandene Abbildung ähnelt dem Querstreifungsmuster des nativen Typs.

KÜHN et al. (234, 224, 247) nahmen an, daß sich die Kollagenmoleküle bei neutralem pH so zusammenlagern, daß die Schwerpunkte der positiven und negativen Ladungen genau nebeneinander zu liegen kommen. Für die Zusammenlagerung der Kollagenmoleküle werden also elektrostatische Kräfte verantwortlich gemacht, die von dem Ladungsmuster der Kollagenmoleküle gesteuert werden. Zeichnet man die Photometerkurven von mit Uranylacetat behandelten Long-spacing-Segmenten um $1/_4$ ihrer Länge so gegeneinander versetzt untereinander, daß sich die Schwerpunkte der positiven und negativen Ladungen möglichst gut decken, dann ergibt die graphische Addition dieser Photometerkurven eine Kurve, wie man sie von einer Fibrille des nativen Typs nach Uranylacetat-Anfärbung beobachtet. Addiert man die Photometerkurven eines mit PWS angefärbten Segmentes, so erhält man für die einzelnen Querstreifen der synthetisierten Kurve die gleichen Intensitäten wie von einer Fibrille nach Phosphorwolframsäure-Anfärbung.

Danach ist auf zwei verschiedenen Wegen gezeigt worden, daß die Kollagenmoleküle innerhalb der Fibrille des nativen Typs um $1/_4$ ihrer Länge gegeneinander versetzt sind und daß das Querstreifungsmuster der 700-Å-Periode durch eine Übereinanderlagerung von vier verschiedenen Vierteln der Kollagenmoleküle zustande kommt. Eine Unklarheit bestand nur noch über die End-an-End-Anlagerung der Moleküle in den Fibrillen.

c. Die Anordnung der Moleküle in den Long-spacing-Fibrillen (FLS).

Die Anordnung der Moleküle in den FLS I und II wurde von KÜHN et al. (248, 249) aufgeklärt. Die Zuordnung des FLS-Querstreifenmusters zu dem SLS-Muster wurde wiederum durch dimorphe Formen zwischen FLS und SLS im Elektronenmikroskop erkannt (247). Die graphische Addition zweier antiparallel angeordneter SLS-Photometerkurven erbrachte den Beweis, daß die Kollagenmoleküle antiparallel, mit den Enden abschneidend in den FLS angeordnet sind. Weiters konnte gezeigt werden, daß sich die Kollagenmoleküle mit ihren Enden um ungefähr 300 Å überlappen. Daraus erklärt sich auch der ungewöhnlich kontrastreiche Dunkelteil in der Querstreifungsperiode der Long-spacing-Fibrillen. Auf dieselbe Weise wurde auch die Anordnung der Moleküle in den FLS II festgestellt (249) (Abb. 23, S. 270).

Die FLS bilden sich am besten aus Neutralsalzlösungen des Kollagens in Gegenwart von Chondroitinsulfat (247). Diese Polyanionen lenken die Zusammenlagerung der Kollagenmoleküle bei dem neutralen pH, bei dem normalerweise Fibrillen des nativen Typs entstehen, in andere Bahnen, die dem Ladungsmuster der Kollagenmoleküle allein nicht entsprechen. Die Kollagenmoleküle haben zwar wie üblich bei neutralem pH das Bestreben, sich um $1/_4$ ihrer Länge gegeneinander zu versetzen und zu den Fibrillen des nativen Typs zusammenzulagern. Da aber das negativ geladene Chondroitinsulfat mit den positiven Ladungen des Kollagens reagiert, wird das Ladungsmuster der Moleküle so verändert, daß es zur antiparallelen Zusammenlagerung kommt.

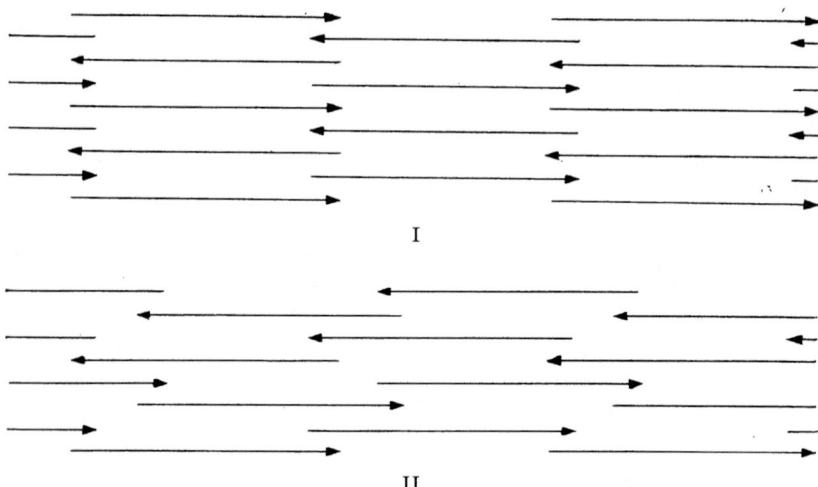

Abb. 23. Anordnung der Kollagenmoleküle in den Long-spacing-Fibrillen. Typ I: Die antiparallel angeordneten Moleküle überlappen sich mit ihren Enden um 10% ihrer Länge. Typ II: Zusätzlich zur antiparallelen Anordnung wie in Typ I sind die Moleküle teilweise gegeneinander versetzt. Nach Kühn und Zimmer. [Aus: Naturwiss. **48**, 220 (1961).]

d. Die End-an-End-Zusammenlagerung der Kollagenmoleküle.

Die Zusammenlagerung der Segmente über ihre Enden wurde als erste von Hodge und Schmitt (*178*) beschrieben. Die Autoren erkannten, vor allen Dingen an mit Ultraschall behandeltem Kollagen, drei Arten von Zusammenlagerung. Einmal können sich die Moleküle über gleiche Enden miteinander verbinden, und zwar über ihre A-Enden oder (zweitens) über ihre B-Enden (A—A, B—B). Drittens erkennt man auch eine A—B-Zusammenlagerung. Für die Endverknüpfung der Moleküle wurden von Hodge et al. (*175*) die sogenannten Telopeptide verantwortlich gemacht. Darunter werden Peptide verstanden, die an den Enden der Moleküle aus der Tripelhelix herausragen und die sich bei der End-an-End-Zusammenlagerung zweier Moleküle zu neuen Peptidschrauben miteinander verdrillen sollen. Bestätigt sahen sich die Autoren durch die Tatsache, daß sie aus mit Pepsin oder Trypsin behandelten Kollagenpräparaten durch Dialyse von essigsauren Lösungen gegen Natriumchlorid bei pH 7 keine Fibrillen des nativen Typs herstellen

Literaturverzeichnis: SS. 293—314.

konnten. Durch die Enzymbehandlung sollen die Telopeptide, welche der Dreikettenschraube des Moleküls nicht mehr angehören, abgespalten werden und so eine Endverknüpfung der Moleküle nicht mehr möglich sein. Die Folge davon wäre auch, daß die Moleküle die Fähigkeit zur Fibrillenbildung verloren haben.

Da Kühn et al. (*242, 236*) bei Dialyse von pepsin- und trypsinbehandelten Kollagenlösungen in Citratpuffer gegen Wasser quantitativ Fibrillen des nativen Typs erhielten, verneinten sie die Rolle der Telopeptide

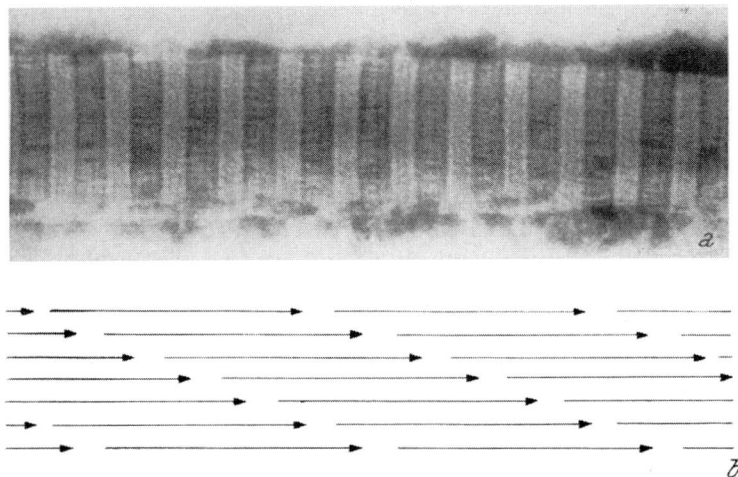

Abb. 24. Fibrillen des nativen Typs mit PWS bei pH 7 negativ angefärbt (Aufnahme: K. Kühn). Die dunklen Teile der Periode stellen die "holes" nach Hodge und Petruska dar (*176*). In den hellen Teilen befinden sich die Überlappungszonen der Moleküle. Dies ist ersichtlich aus der darunter befindlichen schematischen Darstellung der Anordnung der Moleküle in den Fibrillen. [Aus: Aspects of Protein Structure, p. 289. London—New York: Acad. Press (1963).]

für die End-an-End-Aggregation. Sie stellten ein anderes Prinzip der End-an-End-Verknüpfung auf. Es konnte nämlich gezeigt werden, daß bei der A—B-Verknüpfung von Segmenten, die mit ATP aus essigsaurer Lösung gefällt wurden, der letzte Querstreifen eines Segmentes am B-Ende mit dem ersten Querstreifen eines anderen Segmentes am A-Ende zur Überlappung kommt. Auch bei B—B-Zusammenlagerung überlappen sich die beiden letzten Querstreifen zweier verschiedener Segmente an den B-Enden und kommen zur Deckung (*247, 225*). Daraus wurde geschlossen, daß sich bei dieser Art von End-an-End-Anlagerung der Segmente die Kollagenmoleküle um ungefähr 30 bis 50 Å überlappen. Noch eine weitere andersartige A—B-Zusammenlagerung wurde entdeckt (*247, 225*). Bei dieser uberlappen sich die Enden der Molekule ungefähr um 300 Å. Die Überlappung geht immer so vor sich, daß die dunklen Querstreifen der verschiedenen Segmente an den Endregionen zur Deckung kommen. Die End-an-End-Anlagerung geht also nach demselben Prinzip vor sich wie die seitliche Zusammenlagerung der Kollagenmoleküle bei der Fibrillenbildung. KÜHN et al. (247) haben für die End-an-End-Anlagerung der Moleküle in den FLS eine Überlappung der Moleküle um 300 Å und für die Fibrillen des nativen Typs eine Überlappung von 30 bis 40 Å angenommen.

Das Prinzip der Überlappung der Enden der Kollagenmoleküle ist jetzt allgemein anerkannt. Neuerdings konnten aber HODGE und PETRUSKA (176) zeigen, daß die Überlappung von 300 Å nicht nur in den FLS, sondern auch in den Fibrillen des nativen Typs vorliegt.

Mit Hilfe der „negative-staining"-Technik gelang es den Autoren, die Länge der SLS genauer auszumessen, als dies früher möglich war. Nimmt man die Periodenlänge der Fibrillen des nativen Typs zu 1 an, dann beträgt die Länge eines Segmentes und damit des Kollagenmoleküls nicht 4,0, sondern 4,4. Die Moleküle sollten sich also in den Fibrillen um 10% ihrer Länge mit den Enden überlappen. Eine Bestätigung dieser Annahme erhält man durch die Elektronenmikroskop-Bilder von negativ angefärbten Fibrillen des nativen Typs (392) (Abb. 24, S. 271).

Die Periode erscheint in zwei Bereiche geteilt. Der eine, welcher bei einer Periode von 700 Å 400 Å ausmacht, erscheint dunkler und der andere von 300 Å heller. Im hellen Bereich ist bei Behandlung der Fibrillen mit Phosphorwolframsäureüberschuß bei pH 7 weniger PWS eingelagert worden. Dieser ist also dichter gepackt. Der dunkle Bereich ist lockerer aufgebaut und kann mehr PWS aufnehmen. Es wird angenommen, daß der 300-Å-Bereich die Zone der Überlappung der Enden der Kollagenmoleküle in den Fibrillen des nativen Typs ist.

Diese Überlappung der Moleküle in den nativen Fibrillen konnte in letzter Zeit von KÜHN u. a. (228) direkt im Elektronenmikroskop gezeigt werden. Während bei intensiver Behandlung von unlöslichem Kollagen mit Pepsin die intermolekularen Bindungen vollständig gespalten werden und die Kollagenmoleküle monomer in Lösung gehen (s. Abschnitt IV/3, S. 237), kann man bei milder Einwirkung von Pepsin zwischen zwei intermolekularen Bindungen differenzieren. Nach z. B. einstündiger Behandlung mit Pepsin erhält man ein hoch gequollenes, aber nicht lösliches Material; bei Zugabe von ATP bilden sich Long-spacing-Segment-Fibrillen. Diese Fibrillen besitzen eine Querstreifungsperiode von 2500 Å mit dem Querstreifungsmuster der Long-spacing-Segmente. In ihnen sind die Long-spacing-Segmente mit ihren Enden (A—B) miteinander verbunden. Die Enden überlappen sich um 300 Å, wie es für die Fibrillen des nativen Typs angenommen wird. Diese SLS-Fibrillen entstehen, weil durch die milde Enzymbehandlung nur die intermolekularen Bindungen gespalten worden sind, welche die Kollagenmoleküle um $1/4$ ihrer Länge gegeneinander versetzt in den Fibrillen

Literaturverzeichnis: SS. 293—314.

zusammenhalten. Behandelt man dagegen hochgequollenes, nicht mit Pepsin behandeltes, unlösliches Kollagen mit ATP, bilden sich Fibrillen mit dem normalen Querstreifungsmuster aus. Die zweite noch nicht angegriffene intermolekulare Bindung, die die Kollagenmoleküle in den sich überlappenden Endbereichen zusammenhält, kann durch weitere Behandlung mit Pepsin gespalten werden.

3. Die Fibrillenbildung.

Fibrillen des nativen Typs entstehen in vitro einmal bei Dialyse saurer Kollagenlösungen gegen Wasser oder verdünnte, leicht alkalische Puffer. Eine zweite Möglichkeit ist das Erwärmen von Neutralsalzlösungen des Kollagens. Die zweite Methode könnte den Bedingungen in vivo nahekommen.

In sauren Lösungen verfügen die Kollagenmoleküle infolge der undissoziierten Carboxylgruppen nur über positiv geladene Ammonium- und Guanidinium-Ionen. Die elektrostatischen Abstoßungskräfte zwischen den positiven Gruppen halten die Moleküle in Lösung. Die Fibrillen des nativen Typs bilden sich am besten bei Dialyse von Citratpufferlösungen (z. B. pH 3,7; $^1/_{15}$ molar) gegen Leitungswasser (242). Durch die pH-Erhöhung dissoziieren die Carboxylgruppen, die neugebildeten negativen Ladungen reagieren mit den positiven Ladungen anderer Kollagenmoleküle, und es kommt zur Aggregation. Da dieser Vorgang bei einer Dialyse sehr langsam vor sich geht, haben die Moleküle Gelegenheit, sich hochgeordnet zu im Elektronenmikroskop gut quergestreiften Kollagenmolekülen zusammenzulagern.

Auch in Neutralsalzlösungen (z. B. 0,5 m NaCl; pH 7,4) bleibt Kollagen in der Kälte gelöst. Die Ladungen an der Oberfläche der Moleküle werden durch die vorhandenen Ionen abgeschirmt, so daß sie nicht miteinander reagieren können. Erst beim Erwärmen, z. B. auf 30—36°, bilden sich die Fibrillen des nativen Typs aus. Die Fibrillenbildung durch Erwärmen erfolgt in zwei Phasen, einer „lag-period", in der die Kollagenlösungen noch klar bleiben, und einer „growth-phase", in der sich durch Trübung der Lösung eine Fibrillenbildung bemerkbar macht (135, 26, 421). Beide Phasen werden durch steigende Temperatur beschleunigt und durch Erhöhung der Ionen-Stärke verlangsamt. Von besonderem Einfluß ist das pH auf die Fibrillenbildung. Im isoelektrischen Bereich, der bei der angewendeten Ionenstärke ($\mu = 0,25$) bei pH 4,5 gefunden wird, schlägt sich Kollagen sehr leicht und rasch nieder. Die erhaltenen Abscheidungen zeigen lichtmikroskopisch aber nur schlecht ausgebildete Faserstrukturen. Unter stärker alkalischen Bedingungen, im Bereich zwischen pH 6,5 und 9, erfolgt die Fibrillenbildung langsamer. Dafür scheiden sich lange und gut ausgebildete Fasern ab, die ähnlich wie Hautgewebe miteinander verfilzt sind und daher einen gewissen Zusammenhalt zeigen (186, 144, 144a).

Nach Wood (*421*, *417*) kommt es in der „lag-period" zu einer Keimbildung, welche entscheidend für die darauf folgende Fibrillenwachstums-
phase ist. Wird z. B. die Keimbildung immer bei der gleichen Temperatur ausgeführt, während die Temperatur der Wachstumsphase variiert,
so bleibt Anzahl und Durchmesser der gebildeten Fibrillen gleich. Wenn
nach der Keimbildung die entstandenen Keime mit Hilfe einer Ultrazentrifuge entfernt werden, kommt es nur noch sehr zögernd zu einer
Fibrillenbildung. Danach scheint nur ein Teil der Kollagenmoleküle
zu einer Keimbildung befähigt zu sein (*421*, *420*).

Kühlt man nach dem Erwärmen gebildete Kollagenfibrillen auf
4° ab, so geht ein Teil der Fibrillen wieder in Lösung. Der Anteil wird
um so geringer, je länger die Fibrillen vorher erwärmt wurden (*129*).
Genaue Untersuchungen hat Fessler (*71*, *72*) durchgeführt. Es konnten
drei verschiedene Fraktionen bei der Fibrillenbildung durch Erwärmen
und der darauf folgenden Abkühlung beobachtet werden. Fraktion A
enthält die Kollagenmoleküle, die beim Erwärmen ausfallen und sich in
der Kälte wieder auflösen. Fraktion B fällt auch beim Erwärmen nicht
aus, und Fraktion C fällt aus, löst sich aber bei dem darauffolgenden Abkühlen nicht mehr auf. Nach Wood enthält die Fraktion C vor allem
Kollagenmoleküle des Typs II (S. 202) mit einer intramolekularen Quervernetzung (*420*); B dagegen besteht mehr aus Molekülen des Typs I. Es
scheint so, als ob sich die Moleküle des Typs I und II bei der Fibrillenbildung durch Erwärmen verschieden verhalten.

4. Die Beeinflussung der Fibrillenbildung in vitro.

Wenn auch die Zusammenlagerung der Kollagenmoleküle hauptsächlich durch das Ladungsmuster bestimmt wird, so ist es wahrscheinlich,
daß das Bindegewebe über Regulatoren verfügt, die die extrazelluläre
Fibrillenbildung steuern. So ist das Kollagen in den verschiedenen Binde-
gewebsarten auf die verschiedenste Weise angeordnet. Zum Beispiel
bildet es als dicke, parallele Faserbündel die Sehnen. Die Haut besteht
aus einem dreidimensionalen Fasergeflecht, und im lockeren Bindegewebe
liegt ein feinverzweigtes Netzwerk aus dünnen Fasern vor, das eingebettet
ist in einen Überschuß an unstrukturierter Zwischensubstanz. Über den
Mechanismus der Regulation sowie über die Substanzen, die dabei eine
Rolle spielen, ist noch nichts bekannt, wenn auch vermutet wird, daß
die stark negativ geladenen Mucopolysaccharide die Fibrillenbildung in
charaktersitischer Weise beeinflussen.

Wood (*418*) hat den Einfluß dieser Substanzen auf die Fibrillenbildung durch Erwärmen von Neutralsalzlösungen untersucht. Danach
haben Heparin und Desoxyribonucleinsäure einen verzögernden Effekt,
dagegen wird die Fibrillenbildung durch Chondroitinsulfat-A beschleunigt.
Hyaluronsäure und Chondroitinsulfat-B haben keine Wirkung. Gross

Literaturverzeichnis: SS. 293—314.

und KIRK (*135*) haben neben den Mucopolysacchariden auch die Wirkung von niedermolekularen Ionen, unter anderem auch von Aminosäuren, untersucht. Dabei haben die polaren Aminosäuren, wie Asparaginsäure, Glutaminsäure, Arginin und Ornithin, eine verzögernde Wirkung; ebenso wirkt Harnstoff. Glycin, Prolin und Hydroxyprolin üben dagegen keinen oder nur einen sehr geringen Einfluß aus. Arginin und Citrullin verzögern die Fibrillenbildung wesentlich stärker als Guanidin, Kreatin oder Harnstoff in der selben Konzentration (HAFTER und HÖRMANN, unveröffentlicht).

Eine Wirkung auf die Fibrillenbildung in vitro zeigen auch die nichtkollagenen Proteine des Bindegewebes (*233, 244, 186, 144*). So werden bei der Isolierung des neutralsalzlöslichen und säurelöslichen Kollagens aus dem Bindegewebe Substanzen mitextrahiert, die nur schwer vom Kollagen abzutrennen sind. Als Maß für diese Verunreinigungen gilt der Gehalt des Kollagens an Hexosamin und Tyrosin. Hochgereinigte Kollagenpräparate enthalten kein Hexosamin mehr. Der Tyrosingehalt ist bis auf 0,2 bis 0,1 Gew.-% reduzierbar.

Ungereinigte Kollagenpräparate bilden bei der Dialyse saurer Kollagenlösungen gegen Wasser schnell dicke und kontrastreiche Fibrillen mit einem hochunterteilten Querstreifungsmuster. Daneben entstehen auch Long-spacing-Fibrillen (*168, 233*). Mit zunehmender Reinigung durch Umfällen über Fibrillen oder Segmente nimmt das Aggregationsvermögen der Kollagenmoleküle ab. Das heißt, die Fibrillen bilden sich bei Dialyse sowie beim Erwärmen von Neutralsalzlösungen nur noch zögernd. Neben den Fibrillen des nativen Typs erkennt man auch Fibrillen mit einem symmetrischen Querstreifungsmuster, welche ebenfalls eine Querstreifungsperiode von 700 Å aufweisen und Fibrillen mit einer Querstreifungsperiode von 120 Å (*243, 244*) (Abb. 2—3, S. 204 f.). Dieser Reinigungseffekt ist durch Zugabe eines aus der Haut isolierten, serumartigen Proteingemisches wieder rückgängig zu machen. Das Aggregationsvermögen der Kollagenmoleküle wird wieder beschleunigt; es bilden sich ausschließlich wieder Fibrillen des nativen Typs. Neben dem Ladungsmuster der Kollagenmoleküle, das die Aggregation zu den Fibrillen des nativen Typs bewirkt, scheinen auch Begleitkomponenten des Kollagens aus dem Bindegewebe für die richtige Zusammenlagerung eine entscheidende Rolle zu spielen.

Nach Untersuchungen von HÖRMANN und HAFTER (*186*) können sich lichtmikroskopisch gut ausgebildete Fibrillen beim Erwärmen neutraler Kollagenlösungen zwischen pH 6,5 und 9 nur dann bilden, wenn nichtkollagene Begleitsubstanzen des Bindegewebes anwesend sind. Hochgereinigte Kollagenlösungen liefern bei dieser Arbeitsweise lediglich durchscheinende, gelartige Produkte, die im Lichtmikroskop keine Fasern erkennen lassen. Im Elektronenmikroskop erscheinen solche Präparate als sehr dünne, aber quergestreifte Fibrillen (*244*). Durch Zugabe kollagenfreier NaCl-Extrakte aus Haut läßt sich die Fähigkeit zur Ausbildung gut strukturierter Fibrillen wieder herstellen. Für die Förderung der Keimbildung und des Fibrillenwachstums erwiesen sich dabei vor allem die niedrigmolekularen dialysierbaren Anteile wirksam. Die hochmolekularen beschleunigten das Wachstum nicht in dem Maße, verhinderten aber die strukturlose Ausfällung des Kollagens.

Lösliche Kollagenpräparate können auch durch Behandlung mit Proteasen, wie Pepsin, Ficin, Papain und Trypsin, von ihren nichtkollagenen Begleitkomponenten, die dabei proteolytisch zerstört werden, getrennt werden. Die enzymbehandelten Präparate bilden ebenfalls nur noch zögernd Fibrillen, und auch hier kann dieser Effekt durch Zugabe des serumartigen Proteingemisches aus der Haut weitgehend rückgängig gemacht werden (*244*). Über den Reinigungseffekt hinaus verursachen die Proteasen auch am Kollagenmolekül selbst geringe Veränderungen. So beobachtet man eine geringfügige Veränderung der Viskosität, das Auftreten einer geringen Menge freier Aminogruppen (*186*) und einen niedermolekularen dialysierbaren Anteil nach Enzymbehandlung. Hodge et al. (*175, 353*) führen dies auf die Abspaltung kurzer endständiger Peptidbruchstücke (Telopeptide) zurück. Außerdem ist nach Behandlung mit Pepsin eine Spaltung der intramolekularen Bindungen nachgewiesen worden (*353, 228a*).

VIII. Stoffwechsel des Kollagens.

1. Biosynthese*.

Die Kollagen-Biosynthese folgt in ihren prinzipiellen Reaktionen den Mechanismen, die für die Protein-Biosynthese im allgemeinen gelten. Auf zusammenfassende Darstellungen dieses umfangreichen Gebietes sei verwiesen (*159, 77*). Besonderheiten der Problemstellung ergeben sich aber einerseits aus der Frage des Einbaues der beiden Hydroxyaminosäuren Hydroxyprolin und Hydroxylysin, von denen namentlich der ersteren eine entscheidende, wenn auch noch nicht in allen Einzelheiten geklärte Bedeutung für die Struktur des Kollagens zukommt (vgl. S. 222 f.), und anderseits aus der ungewöhnlichen Größe des Kollagenmoleküls. Lange Zeit stand der Einbau der beiden Hydroxysäuren, besonders des Hydroxyprolins, allein im Mittelpunkt aller biosynthetischen Arbeiten auf diesem Gebiet, neuerdings ist aber auch das zweite Problem stark in den Vordergrund gerückt.

Um die Besonderheiten, die im Falle des Kollagens zu erwarten waren und tatsächlich gefunden worden sind, besser von den bekannten

---

* Mitbearbeiter dieses Kapitels: Dr. Josef Hurych, Institut für Arbeitshygiene und Berufskrankheiten, Prag (ČSSR), derzeit Gast unseres Institutes.

Literaturverzeichnis: SS. 293—314.

Schritten der allgemeinen Protein-Biosynthese abgrenzen zu können, soll im folgenden kurz der Dreistufen-Mechanismus der Protein-Biosynthese in Erinnerung gerufen werden:

1. In einem ersten Schritt werden die bei der Synthese zu verknüpfenden Monomeren, die Aminosäuren, aktiviert, d. h. in einen reaktionsbereiten Zustand versetzt. Mit Hilfe von ATP* und einer für jede Aminosäure spezifischen Synthetase werden unter Abspaltung von Pyrophosphat Aminoacyl-Adenylat-Komplexe gebildet:

$$\text{AS} + \text{ATP} + \text{Enzym} \rightleftarrows (\text{AS-AMP-Enzym}) + \text{PP}. \tag{a}$$

Diese Komplexe bleiben mit dem aktivierenden Enzym verbunden. Die Aminoacyl-Adenylatbindung ist außerordentlich energiereich und dementsprechend reaktiv.

2. Es folgt eine Übertragungsreaktion. Der Aminosäurerest wird unter Erhaltung des hohen Energieniveaus esterartig an eine sogenannte „lösliche RNS" oder „Transfer-RNS" gebunden:

$$(\text{AS-AMP-Enzym}) + \text{l-RNS} \rightleftarrows \text{AS-l-RNS} + \text{AMP} + \text{Enzym}. \tag{b}$$

Für jede der bekannten Aminosäuren existiert mindestens eine spezifische Transfer-RNS.

3. Nach diesen vorbereitenden Reaktionen ist das letzte Stadium der Eiweißsynthese die eigentliche Verknüpfung der Einzelbausteine zur Polypeptidkette. Dieser Vorgang spielt sich auf der Oberfläche der Ribosomen ab, kleinen, submikroskopischen Partikeln des Ergastoplasmas der Zelle. Die Polymerisation erfolgt jedoch nur, wenn der Informationsübermittler für die Sequenz der Aminosäuren im Peptid oder Protein, die (im Zellkern gebildete) Matrizen- oder Messenger-RNS, mit den Ribosomen verbunden ist. Notwendig sind außerdem ein polymerisierendes Enzym und als Energiequelle GTP:

$$\text{AS}_1\text{-l-RNS}_1 + \text{AS}_2\text{-l-RNS}_2 + \text{GTP} \rightarrow$$
$$\rightarrow \text{AS}_1\text{-AS}_2\text{-l-RNS}_2 + \text{l-RNS}_1 + \text{GDP} + \text{P}_0. \tag{c}$$

Die Besonderheiten der Biosynthese des Kollagens, also die Abweichungen gegenüber dem angegebenen Schema, sind hauptsächlich bei den Problemen zu suchen, die der Einbau der Hydroxyaminosäuren aufwirft.

---

* *Abkürzungen:*

AS	= Aminosäure,	GDP	= Guanosindiphosphat,
ATP	= Adenosintriphosphat,	P_0	= Orthophosphat,
AMP	= Adenosinmonophosphat,	Pro	= L-Prolin,
PP	= Pyrophosphat,	Hypro	= L-Hydroxyprolin,
RNS	= Ribonukleinsäure,	Lys	= L-Lysin,
l-RNS	= Lösliche Ribonukleinsäure,	Hylys	= L-Hydroxylysin,
GTP	= Guanosintriphosphat,	EDTA	= Äthylendiamintetraacetat.

Ad Reaktion 1. Es gibt keine für Hydroxyprolin oder Hydroxylysin spezifischen Aminoacyl-Adenylat-Synthetasen. Zwar werden radioaktiv markiertes Pro oder Lys in biologisch aktiven Ansätzen normal in Kollagen inkorporiert; bietet man aber Hypro oder Hylys in radioaktiver Form an, so findet man keine Markierung im Kollagen wieder (*386, 411, 285, 375*). Daß für diesen Befund die oben genannte Interpretation — Fehlen der aktivierenden Enzyme — die richtige ist, geht aus den Versuchen von Peterkofsky und Udenfriend (*318*) mit einem Hühner-embryonensystem hervor. Die Autoren verfolgten den Austausch markierten Phosphats zwischen ATP und Pyrophosphat (vgl. Gleichung a). Während die Aktivierung von Pro und anderen Aminosäuren völlig normal verläuft, gelang es nicht, eine Aktivierung von Hydroxyprolin nachzuweisen. Mit einer anderen Versuchsanordnung, die isolierte, aktivierende Enzyme und lösliche RNS enthielt, konnten Manner und Gould (*271*) ebenfalls keine Verwertung von C^{14}-Hypro nachweisen.

Ad Reaktion 2. Die Isolierung des Hypro-l-RNS-Komplexes wurde in mehreren Arbeiten verschiedener Laboratorien beschrieben. Manner und Gould (*271*) wiesen diese Verbindung in Hühnerembryonen und ebenso in zellfreien Systemen der gleichen Herkunft nach Applikation von C^{14}-Pro nach. Später gelang den gleichen Autoren die Gewinnung des entsprechenden Hylys-Komplexes (*272*). Erwähnenswert erscheint in diesem Zusammenhang, daß die Zugabe von l-RNS aus *E. coli* zu den genannten Ansätzen zwar die spezifische Aktivität des inkorporierten Prolins wesentlich ansteigen läßt, aber keinen Einfluß auf die Entstehung des Komplexes C^{14}-Hypro-l-RNS hat. Coronado et al. (*48*) konnten mit einem wirksameren Präparat aus Hühnerembryonen ebenfalls die Bildung von Hypro-l-RNS und die von Hylys-RNS nachweisen. l-RNS aus Hefe stimuliert hier den Einbau von C^{14}-Hypro ebensowenig wie das *E. coli*-Präparat. Später zeigten die gleichen Bearbeiter (*49*), daß unter anaeroben Bedingungen nur sehr wenig C^{14}-Hypro-l-RNS gebildet wird, während die Synthese von C^{14}-Pro-l-RNS-Komplex nicht gehemmt ist. Isoliert man letzteren und reinkubiert ihn aerob, so steigt die Ausbeute an markiertem Hypro-l-RNS auf normale Werte an. Daraus wird geschlossen, daß Prolin in Form seiner l-RNS-Verbindung hydroxyliert wird. Diese Ergebnisse mit ihren Konsequenzen gelten analog für den Übergang Lysin → Hydroxylysin (*49*).

Auch andere Autoren beschrieben das erwähnte Hypro-Zwischenprodukt in Wundgranulationsgewebe (*202*) bzw. in löslichen und ribosomalen Fraktionen von Hühnerembryo-Homogenaten (*392a*). Letztere Autoren lassen allerdings die Möglichkeit offen, daß es sich bei dem Hypro-l-RNS eventuell nur um ein Nebenprodukt handelt.

Andererseits gelang es Peterkofsky und Udenfriend nicht, diese von mehreren Arbeitskreisen beschriebene Schlüsselverbindung zu

Literaturverzeichnis: SS. 293—314.

isolieren (vgl. *318a*). Auch LUKENS (*262a*) fand im Verhältnis zum entsprechenden Pro-Komplex nur sehr geringe Mengen an radioaktivem Hypro-l-RNS und stellt deswegen den Vorläufer-Charakter dieser Verbindung in Frage.

Gegenwärtig ist also noch nicht endgültig entschieden, ob bei der Biosynthese von Hydroxyprolin die Zwischenverbindung Hypro-l-RNS, also der normale Syntheseweg überhaupt existiert. Die alternative Möglichkeit — Hydroxylierung einer höhermolekularen, prolinreichen Zwischenstufe — würde dem Kollagen auch auf dem Gebiet der Biosynthese eine Sonderstellung einräumen. Diese Anschauung hat schon früher manches Für (z. B. *98*, *386*) und Wider (*160*, *325*) gefunden. Bedeutsamer sind einige neue Arbeiten, denen die Erkenntnisse über den genaueren Ablauf der Proteinbiosynthese zugrundeliegen. So gibt es mehrere Hinweise dafür, daß der Hydroxylierungsvorgang erst an der Oberfläche der Ribosomen stattfindet (*318*): Unter anaeroben Bedingungen mit C^{14}-Pro inkubiertes, mikrosomen-gebundenes Polypeptid dient der Prolin-Hydroxylase als Substrat (*328*, *318a*, *216a*). Versuche mit Kollagenase und Trypsin zeigen, daß das Substrat wirklich Peptid- oder Proteincharakter hat; nach Einwirkung der genannten Enzyme ist eine Hydroxylierung nicht mehr möglich (*328a*, *97a*). Ferner wurde gefunden, daß dieses hydroxylierbare Substrat nicht dialysabel und beständig gegen milde Alkalibehandlung sowie gegen Ribonuclease ist, also keine AS-l-RNS sein kann; darüber hinaus liegt sein RNS-Gehalt unter 10% (*328a*). Außerdem findet man nach Hemmung der Proteinbiosynthese durch Puromycin ein Polypeptid, das reich an Prolin, aber arm an Hypro ist und bei weiterer Inkubation hydroxyliert wird (*208*, *328*, *262a*).

Im Widerspruch zu diesen Ergebnissen, die eine höhermolekulare, prolinreiche Vorstufe wahrscheinlich machen, steht eine Arbeit von LOWTHER und Mitarb. (*261*). Diese Autoren fanden praktisch gleiche spezifische Aktivitäten der beiden Iminosäuren im löslichen Kollagen der Mikrosomenfraktion. Ihre Befunde erschweren die Entscheidung darüber, welcher der beiden diskutierten Synthesewege von der Natur tatsächlich beschritten wird. Das Problem erscheint dem Außenstehenden nach wie vor offen.

Daß die OH-Gruppe des Hydroxyprolins atmosphärischem Sauerstoff, nicht Wasser entstammt, war schon länger bekannt. Läßt man Hypro in einer mit O^{18} angereicherten Atmosphäre biosynthetisch entstehen, findet man, daß etwa 60% des Hydroxyl-Sauerstoffs aus der Luft herrühren (*80*, *329*). Für das Hydroxyprolin der pflanzlichen Zellwände liegen vergleichbare Ergebnisse vor (*255*).

Bei Versuchen in tritiiertem Wasser wird Tritium nicht in Hypro inkorporiert (*355*), und bei Inkubation von kollagensynthetisierendem

Material in Stickstoff-Atmosphäre kommt es zu einer bedeutenden Senkung der Hydroxyprolin-Bildung (*318, 355*). Fujimoto und Tamiya (*79*) wollen gefunden haben, daß sich in diesem Punkt die Hydroxylierung des Lysins von der des Prolins unterscheidet. Sie stellten an Hühnerembryonen in vivo fest, daß das Sauerstoffatom der OH-Gruppe einem Wassermolekül entstammt.

Außer Zweifel steht die Beteiligung von Ascorbinsäure als Cofaktor bei der Hydroxylierung von Pro zu Hypro (*98, 347, 349, 43*). Die Fähigkeit zur Bildung von Hydroxyprolin in vivo fällt bei scorbutisch ernährten Meerschweinchen stark ab; Vitamin-C-Gaben normalisieren eine derart verminderte Hydroxyprolin-Synthese nahezu vollständig (*98*). Der Ascorbinsäure-Effekt kann auch mit Hilfe von Gewebsschnitten eindeutig nachgewiesen werden; es besteht ein enger Zusammenhang zwischen der Hydroxylierungsrate im Granulomgewebe von Versuchstieren und dem Ascorbinsäuregehalt der verfütterten Diäten (*43*).

Der Einfluß von Vitamin C auf die Kollagen-Biosynthese ist sehr spezifisch; andere grundlegende Prozesse des biologischen Geschehens, wie Elektronentransport oder Nucleinsäure-Synthese, werden ebensowenig beeinflußt wie die Biosynthese der anderen Proteine (*205*). Der stimulierende Effekt wird allerdings in vivo auch von der *D*-Form (*348*), in vitro sogar von *D*-iso-Ascorbinsäure (*43*) hervorgerufen. Robertson (*348*) weist darauf hin, daß Ascorbinsäure deswegen wohl kaum die Funktion eines Coenzyms haben kann. Es sieht vielmehr so aus, als ob ihre Wirkung in der Bereitstellung von Elektronen oder der Bildung von OH-Radikalen besteht. Die beiden im Moment am stärksten diskutierten Möglichkeiten des Hydroxylierungsmechanismus lassen einer Beteiligung von Ascorbinsäure durchaus Raum. Prockop et al. (*327*) haben Ergebnisse, die für eine elektrophile Substitution des Prolinringes sprechen. Diese Autoren sowie eine japanische Gruppe (*216*) verwendeten 3,4-H³-Prolin für ihre Untersuchungen und stellten eine Tritiumretention von etwa 73% im Verlauf der Hydroxyprolin-Biosynthese fest. Das bedeutet den Verlust von nur einem H (T)-Atom während der Hydroxylierung. Zum gleichen Ergebnis kam auch Lamport (*256*). Man muß also keine ungesättigten Zwischenverbindungen (vgl. *80a*) und keine Konformationsänderungen annehmen; der Luftsauerstoff könnte im Verlauf der Reaktion einen kationischen Komplex mit einem Metallion an der Enzymoberfläche bilden:

$$Me_{enz}^{++} + O_2 \longrightarrow Me_{enz}^{++}\text{-}O_2, \tag{d}$$

$$Me_{enz}^{++}\text{-}O_2 + Pro \longrightarrow Me_{enz}^{++}\text{-}O + Hypro, \tag{e}$$

$$Me_{enz}^{++}\text{-}O + 2 \ominus \longrightarrow Me_{enz}^{++} + O^{--}. \tag{f}$$

Dieser auch für die Hydroxylierung von Steroiden (*367*) postulierte Mechanismus bedarf eines Elektronendonators; Ascorbinsäure könnte diese Wirkung ausüben. Ergebnisse von HURYCH und CHVAPIL (199) zeigen, daß tatsächlich zweiwertige Metallionen bei der Hydroxylierung des Prolins in vivo teilnehmen. Bei der Inkubation der Haut von Hühnerembryonen mit α,α'-Dipyridyl und o-Phenantrolin sinkt die Bildung von kollagengebundenem Hydroxyprolin gegenüber Kontrollen um mehr als 90%. Ähnliche Ergebnisse wurden auch in einer anderen Arbeitsgruppe erhalten [Hemmung des Hydroxylierungsprozesses um etwa 50% in Gegenwart von EDTA (318)]. Andererseits steigt bei Zugabe von Fe^{++}- oder Fe^{+++}-Ionen in zellfreien Systemen die Ausbeute an Hydroxyprolin (328a). Diese Befunde legen eine weitere Erklärungsmöglichkeit für die Ascorbinsäure-Wirkung nahe: Es ist bekannt, daß Ascorbinsäure die Bildung von Eisen-Protein-Komplexen positiv beeinflußt (vgl. 367); sie könnte also auch an der Komplettierung der Holoform des hydroxylierenden Enzyms in ähnlicher Weise beteiligt sein.

Modellversuche lassen ferner die Beteiligung von Vitamin C als Partner bei einer radikalischen Hydroxylierung als möglich erscheinen. HURYCH (42, 197) inkubierte Prolin mit einer Mischung von Ascorbinsäure, komplexgebundenen Fe^{++}-, Cu^+- oder Cr^{+++}-Ionen und H_2O_2, und fand 3—4% des Prolins hydroxyliert. YIP (425a) ersetzte im gleichen System Ascorbinsäure durch Peroxidase und kam zu ähnlichen Ergebnissen. Wahrscheinlich wirken freie Radikale als hydroxylierendes Agens:

$$Fe^{++} + H_2O_2 \longrightarrow Fe^{+++} + OH^- + \cdot OH.$$

Auch Perhydroxylradikale sind denkbar, wie Versuche in Sauerstoffatmosphäre zeigen (198). Interessanterweise läuft die Hydroxylierung auch ab, wenn Metallionen im Ansatz fehlen; Ascorbinsäure kann anscheinend in solchen Fällen die Rolle des Radikalbildners übernehmen, denn bei Anwesenheit von Ascorbinsäureoxydase ist die Hydroxylierungsrate erhöht (198).

Aus solchen Modellsystemen hat man insgesamt vier Hydroxyprolin-Isomere isoliert (die cis- und trans-Formen des 3- und 4-Hydroxyprolins) (256, 198), was auf einen unspezifischen, durch Radikale bewirkten Verlauf der Reaktion hinweist. Für radikalische Mechanismen sprechen auch in vitro-Hydroxylierungen von Lysin (347). Daß in den letzten Jahren mit Hilfe der paramagnetischen Elektronenresonanz im biologischen Geschehen mehrfach Radikale nachgewiesen worden sind (424, 425, 158, 367), ist eine weitere Stütze für diese Annahme.

Allerdings ist auch an einen indirekten Einfluß der Ascorbinsäure auf die Kollagenbiosynthese zu denken. ROSS und BENDITT (352a) konnten auf elektronenmikroskopischem Wege nachweisen, daß die charakteristische Konfiguration der Ribosomen des endoplasmatischen Reticulums zwar in den Fibroblasten normal ernährter Meerschweinchen vorliegt, nicht aber in den Zellen skorbutischer Tiere. Die richtige An-

ordnung wird bereits 8 Stunden nach Verfütterung von ascorbinsäure-
haltigen Diäten zurückgebildet.

Ad Reaktion 3. Wenden wir uns nun dem dritten, dem polymerisieren-
den Schritt der Kollagen-Biosynthese zu. Auch hier folgt in den wesentlichen Zügen unser Spezialfall dem allgemeinen Schema. Schon länger ist
bekannt, daß nach Applikation von C^{14}-Prolin die höchste spezifische
Radioaktivität des Hydroxyprolins in der Mikrosomenfraktion gefunden
wird (*261, 330*). Als weiterer Beweis dafür, daß auch der in Gleichung c
(S. 277) dargestellte Polymerisationsmechanismus im Falle des Kollagens
gilt, darf man die Wirkung des Antibiotikums Puromycin gelten lassen.
Vom Puromycin weiß man, daß es als Analogon der Aminoacyl-l-RNS-Kom-
plexe die Polymerisation der Kette unterbrechen kann. Nach Anwendung
dieses Inhibitors kam es auch im Falle des Kollagens bei in vivo- und in
vitro-Versuchen zu einer Hemmung der Kollagensynthese (*271*). Eine
Beteiligung von Messenger-RNS ist nachgewiesen worden (*24c*). LEVINE
(*258*) hat nach Gaben von C^{14}-Prolin höhermolekulares Material isolieren
können, das C^{14}-Hypro enthielt. Durch Behandlung mit Ribonuklease
konnte das C^{14}-Hypro in eine dialysable Form übergeführt werden.
Es handelte sich hierbei offenbar um eben in Synthese befindliches, also
noch an l-RNS gebundenes Kollagenmaterial.

Die weiteren Fragen, die sich an diese Untersuchungen anschließen,
könnte man etwa so stellen: Werden die α-Ketten des Kollagens, die aus
etwa 1100 Aminosäuren bestehen und ein Molekulargewicht von zirka
100000 aufweisen, als *ein* Peptidfaden synthetisiert? Oder werden irgend-
welche Untereinheiten gebildet [vgl. die Vorstellungen von GALLOP (*82*)
und HODGE (*177, 324*), S. 232 f.] und anschließend durch einen spezifischen
Mechanismus zur Grundeinheit vom Molekulargewicht 100000 zusammen-
gefügt? Nachdem letztere Hypothese schon einmal vertreten (*85*)
und widerlegt (*326*) worden war, sprach sich im vergangenen Jahr SNELL-
MAN (*380*) erneut für eine solche zweistufige Synthese der α-Ketten aus.
Dieser Autor interpretiert seine Versuche so, daß zuerst sogenannte
Nukleotid-Peptide synthetisiert werden, wobei GTP esterartig an das
Carboxyl der endständigen Aminosäure gebunden wird. In der zweiten
Phase werden diese „Unter-Untereinheiten" dann extrazellulär zu
größeren Molekülen verknüpft; als Matrize dient dabei nach SNELLMAN
ein Komplex von Kollagen mit Glykoprotein. Eine aus Leber isolierte
Polymerase wirkt katalysierend auf den Vorgang ein.

Während diese Vorstellung noch zusätzlicher, beweisender Ex-
perimente bedarf, sind inzwischen auch Arbeiten durchgeführt worden,
deren Ergebnisse die zweite Möglichkeit — Synthese der α-Ketten als
Ganzes — unterstützen. In diesem Zusammenhang sei an Untersuchungen
erinnert, die vor allem in der Schule von RICH durchgeführt wurden:
Nicht einzelne Ribosomen sind Ort der Entstehung der Polypeptid-

Literaturverzeichnis: SS. 293—314.

kette, sondern Ribosomenhaufen, deren Größe je nach Länge des sie verknüpfenden messenger-RNS-Fadens schwankt, d. h. je nach der Länge
der zu synthetisierenden Peptid- oder Proteinkette. Diese Zusammenlagerungen nennt man Polyribosomen oder kurz Polysomen; die wichtigste
Literatur zu diesem Thema findet man in (220) zitiert. Einschlägige
Untersuchungen zu der Frage, ob es prinzipiell möglich ist, die Kollagen-
α-Ketten als Ganzes aufzubauen, haben ein positives Ergebnis gehabt.

Zwei Arbeitskreise haben unabhängig voneinander darüber gearbeitet.
KRETSINGER et al. (220) konnten zeigen, daß in einem Hühnerembryonen-
System nach Gaben von C^{14}-Prolin radioaktives Hydroxyprolin in hoch-
molekularer Form mit Polysomen verknüpft ist. Die Sedimentationskonstante für dieses Material wird mit mindestens 400 S angegeben
(zum Vergleich: die entsprechende Größe für die meisten globulären
Proteine beträgt höchstens 250 S). Das bedeutet, daß die kollagen-
synthetisierenden Polysomen wenigstens aus 100 Einzelribosomen bestehen müssen, was für die Synthese selbst so großer Ketten, wie sie hier
in Frage kommen, genügen würde. Die Versuchsergebnisse wurden durch
Behandlung der Ansätze mit Ribonuklease und Natriumdodecylsulfat
untermauert. Erstere zerstört die verbindende Messenger-RNS, und es
entstehen Einzelribosomen; mit ihnen wandert in der Ultrazentrifuge
die Radioaktivität. Das genannte Detergens löst die wachsende Polypeptidkette von den Ribosomen ab; es zeigte sich, daß sie noch die endständige l-RNS enthielt. Die zu gleicher Zeit durchgeführten Arbeiten
von MALT und SPEAKMAN (264) kamen mit Fibroblasten-Präparationen
aus Granulomgewebe (Meerschweinchen) zu den gleichen Ergebnissen.
Die Sedimentationskonstante der Polysomen, auf denen sich die wachsenden Kollagenketten befinden, wird mit 450 S oder mehr angegeben.

Die Markierung applizierter Aminosäuren taucht mit gleicher Geschwindigkeit in den zwei verschiedenen Typen der α-Ketten, α 1 und α 2,
auf (274). Das bedeutet, daß sie gleichzeitig oder jedenfalls sehr kurz
nacheinander entstehen. Wann die α-Einzelketten zur Tripelhelix
spiralisieren, weiß man nicht sicher; doch scheint das nach den Ergebnissen
von KRETSINGER et al. (220) schon während der Synthese auf den Polysomen der Fall zu sein. Die covalente Verknüpfung zweier α-Ketten zu
der β-Einheit dürfte auf jeden Fall erst in einem späteren Stadium erfolgen.
JACKSON (201) konnte zeigen, daß neu entstandenes Kollagen in 0,14 M NaCl
löslich ist und bei Wärmedenaturierung lediglich α-Ketten liefert. Ob die
zuletzt genannten Vorgänge noch intrazellulär oder schon extrazellulär
ablaufen, werden weitere Untersuchungen zeigen.

Diese biochemischen Befunde werden durch elektronenmikroskopische
und autoradiographische Untersuchungen ergänzt. ROSS und BENDITT
(352) verfolgten die Lokalisierbarkeit des radioaktiven Prolins in Fibroblasten. Bereits 30 Minuten nach Verabreichung von H^3-Prolin häuft

sich die Radioaktivität über den Zisternen des endoplasmatischen Retikulums an, wo die Eiweißsynthese ganz allgemein lokalisiert ist. Vier Stunden nach Beginn der Untersuchung wurde autoradiographisch radioaktives Prolin über den peripheren cytoplasmatischen Vakuolen und auch extrazellulär über den Kollagenfasern gefunden. Elektronenmikroskopisch konnten innerhalb der Zelle (im Golgi-Apparat und im endoplasmatischen Retikulum) nur feine Fäserchen gefunden werden, die man als Kollagenvorstufen interpretiert. Nach 24 Stunden befindet sich die Hauptmenge der Radioaktivität im extrazellulären Raum über den Kollagenfasern; nur ein kleiner Teil verbleibt über den peripheren Vakuolen in der Nähe der Zellmembran [vgl. aber (90)]. Nach Ansicht anderer Autoren (374) dokumentiert die Anwesenheit der feinen Fäserchen in diesen Vakuolen den Weg, auf dem Tropokollagenmoleküle, α-Ketten oder eventuell auch kleinere Vorstufen des Kollagens vom Entstehungsort in den extrazellulären Raum gelangen.

2. Reifung und Alterung.

Während auch die Synthese der Tripelhelix des Kollagens noch in der Zelle vor sich zu gehen scheint, geschieht die Zusammenlagerung der Moleküle zu den Fibrillen sowie der Verfestigung zu unlöslichem, reifem Kollagen im extrazellulären Raum. Während dieses Reifungsprozesses durchläuft das Kollagen die schon besprochenen löslichen Vorstufen (vgl. Abschnitt V/2, S. 245 ff.), die so schonend aus dem Gewebe extrahiert werden können, daß sie in vitro die Fähigkeit zur Bildung von Fibrillen des nativen Typs behalten.

Die löslichen Vorstufen des Kollagens treten bei jungen Organismen verstärkt auf. Mit zunehmendem Alter geht ihre Menge im Bindegewebe zugunsten des unlöslichen Kollagens zurück. So nimmt z. B. das neutralsalzlösliche Kollagen in Häuten von Meerschweinchen, das bei jungen Tieren bis zu 10% des Gesamtkollagens beträgt, parallel zur Verringerung der Wachstumsrate der Tiere ab (130). Ebenso nimmt auch das säurelösliche Kollagen mit zunehmendem Alter ab, wenn auch nicht so schnell wie das neutralsalzlösliche Kollagen (311, 39). Ähnliches gilt auch für Rindshäute (101, 181a). Auch der unter denaturierenden Bedingungen lösliche Anteil des säureunlöslichen Kollagens geht mit steigendem Alter der Tiere zurück (182). Achillessehne vom alten Rind ist praktisch vollständig unlöslich. Aus solchen Befunden wurde geschlossen, daß das neutralsalzlösliche und säurelösliche Kollagen Vorstufen des unlöslichen Kollagens darstellen (314).

Durch Isotopenversuche konnte diese Anschauung präzisiert werden. So findet Harkness et al. (152) den schnellsten und stärksten Einbau von C^{14}-Glycin in das neutralsalzlösliche Kollagen. Auch der Abbau ist mit einer Halbwertszeit von 48 Stunden für das neutralsalzlösliche Kollagen sehr rasch. Viel langsamer ist dagegen die Bildung des säurelöslichen und des unlöslichen Kollagens. Die spezifische Aktivität ist bei diesen Fraktionen wesentlich geringer. Der Abbau erfolgt so langsam, daß eine Halbwertzeit nicht angegeben werden kann. Ähnliche Befunde haben auch JACKSON (200) und GREEN et al. (124) an Carrageenin-Tumoren erhalten. Die geringe spezifische Aktivität des säurelöslichen Kollagens wird so gedeutet, daß es kein echter Vorläufer des Kollagens ist (152). Eindeutiger erkennt man dies, wenn die prozentuale Verteilung der gesamten, in das Kollagen eingebauten Aktivität auf die drei Kollagenfraktionen gemessen wird (226, 241). Gegenüber neutralsalzlöslichem und unlöslichem Kollagen werden in das säurelösliche Kollagen nur verschwindend kleine, zu vernachlässigende Mengen von C^{14}-Glycin eingebaut. Das heißt, das neutralsalzlösliche Kollagen geht direkt in das unlösliche Kollagen über. Das säurelösliche Kollagen scheint auf einem für das Unlöslichwerden des Kollagens unbedeutenden Seitenweg zu liegen.

Neben dem löslichen Kollagen verändert sich auch das unlösliche mit zunehmendem Alter. Dies macht sich vor allem im physikalisch-chemischen Verhalten der Fibrillen bei der Schrumpfung bemerkbar. Erwärmt man unlösliche Kollagenfibrillen, z. B. aus Rattenschwanzsehnen, in Wasser oder Ringer-Lösung auf 58—72°, so schrumpfen sie, d. h. sie verkürzen sich auf $^1/_3$ bis $^1/_4$ ihrer ursprünglichen Länge (vgl. Abschnitt VI/2, S. 254).

Die Zugkräfte, die bei dieser Kontraktion auftreten, wurden von VERZÁR (vgl. 397) gemessen. Bei der isotonischen Kontraktion (395) werden die Kollagenfibrillen an dem einen Ende befestigt und an dem anderen Ende mit Gewichten beladen. Je schwerer die Gewichte sind, desto geringer ist die Kontraktion der Fibrillen. Schließlich tritt keine Verkürzung mehr auf („inhibition weight"). Eine andere Methode, die eine genauere Bestimmung der auftretenden Zugkräfte gestattet, ist die isometrische Methode (39a). Hierbei sind die Fibrillen eingespannt, so daß sie sich nicht mehr verkürzen können. Die entstehenden Zugkräfte werden direkt gemessen. Mit steigender Temperatur treten zunehmende Kräfte auf; schließlich wird ein Maximum durchlaufen, oberhalb dessen die Struktur zerstört wird. Die beim Maximum gemessenen Zugkräfte sind wahrscheinlich ein Maß für die Kräfte, die nach Lösung der meisten Wasserstoffbrücken einem Abgleiten der Moleküle gegeneinander entgegenstehen, und damit für die Quervernetzungen.

Mit beiden Methoden hat VERZÁR (394, 399, 39a, 398, 39) gefunden, daß die Zugkräfte mit steigendem Alter der Tiere ansteigen. Die Abhängigkeit vom Alter ist so streng, daß es gelingt, mit dieser Methode Altersbestimmungen von Ratten durchzuführen. Verantwortlich für die ansteigenden Zugkräfte sind die im Alter zunehmenden Quervernetzungen des Kollagens. Führt man in Kollagenfibrillen künstliche Quervernetzungen ein, wie z. B. durch Behandlung mit Formaldehyd, so werden die Altersunterschiede verwischt. Die Zugkräfte steigen dann auch bei jungen Tieren stark an (399). Neben den Zugkräften ist auch die Relaxationszeit der Fasern nach der Schrumpfung ein Maß für das Alter. Belädt man Fibrillen mit einem Gewicht, das noch eine Schrumpfung erlaubt, und kühlt nach der maximalen Verkürzung der Fibrillen das Bad nur ein wenig ab, so erschlafft die Faser spontan (Relaxation). Die Zeit, die sie dazu benötigt, ist die Relaxationszeit. Sie ist bei Fibrillen aus jungen Tieren wesentlich kürzer als bei solchen aus alten Tieren (399). Eine weitere Abhängigkeit vom Alter zeigt nach Verzár (279, 396) die Menge des beim Schrumpfen der Fibrillen in Lösung gehenden Kollagens, das durch Hydroxyprolinbestimmungen gemessen wurde. Noch nicht so stark quervernetztes Kollagen aus jungen Tieren gibt beim Schrumpfen einen größeren Anteil in die Lösungen ab als hochquervernetzte Fibrillen. Narbengewebe zeigt das Altersverhalten der Narbe, nicht des Gesamttiers.

Eine chemische Kontraktion erhält man in konzentrierten Salzlösungen, z. B. Lithiumchlorid, Kaliumjodid oder Natriumperchlorat. Auch hier zeigen die auftretenden Zugkräfte die gleiche Abhängigkeit vom Alter wie bei der Wärmekontraktion (398).

Das gummielastische Verhalten von denaturierten Kollagenfibrillen hat man auch zur Bestimmung der Anzahl der Quervernetzungen benutzt (406). So wurde aus den Ergebnissen ein Abstand der Quervernetzungen in unlöslichem Kollagen von 600—700 Å längs der Kollagenfibrille abgeleitet. Das stimmt gut überein mit der elektronenmikroskopisch sichtbaren Querstreifungsperiode.

Durch die Spaltung der intermolekularen Bindungen mit Hilfe von Pepsin kann auch unlösliches Kollagen wieder nativ in Lösung gebracht werden (vgl. S. 237). Während sich unlösliches Kollagen aus Kalbhaut in ein paar Stunden unter Pepsineinwirkung auflöst, gelingt es nicht, die hochquervernetzte Achillessehne von alten Rindern, die kaum noch eine Quellung zeigen, mit Pepsin in Lösung zu bringen (228).

Abschließend ergibt sich das folgende Bild: In jungen, schnell wachsenden Organismen ist die Kollagen-Biosynthese sehr intensiv. Die löslichen Vorstufen, vor allem das neutralsalzlösliche Kollagen, sind vermehrt. Die sich bildenden säureunlöslichen Kollagenfibrillen können zu einem hohen Prozentsatz mit denaturierenden Agenzien in Lösung gebracht werden. Mit steigendem Alter geht die Synthese des Kollagens zurück. Der Anteil der löslichen Kollagenfraktionen wird geringer. Auch das unlösliche Kollagen macht einen langsamen Reifungsprozeß mit, in dem die Fibrillen mehr und mehr durch kovalente Quervernetzungen verfestigt werden.

Literaturverzeichnis: SS. 293—314.

3. Abbau im Körper.

Untersuchungen von NEUBERGER et al. (*295*), die später von GERBER et al. (*87*) bestätigt wurden, zeigen, daß reifes Kollagen im Körper nur sehr langsam mit Halbwertzeiten in der Größenordnung von 100 Tagen abgebaut wird. Im Gegensatz zum vernetzten und zur Faser strukturierten unlöslichen Kollagen unterliegen die Kollagenvorstufen jedoch einem raschen Stoffwechsel. In Isotopenversuchen an Ratten (*87, 152, 226*) erreicht die Aktivität von neutralsalzlöslichem Kollagen bereits nach einem Tag ihr Maximum und klingt dann verhältnismäßig schnell wieder ab (Halbwertzeit etwa 2—3 Tage). Nur ein kleiner Teil davon wird zur Faserbildung verwendet; die Hauptmenge erleidet Abbau zu niedrigmolekularem Material, das zum Teil im Urin ausgeschieden wird. Offensichtlich besteht ein Zusammenhang zwischen der Ausscheidung von Hydroxyprolin im Harn und dem Wachstum, d. h. der Synthese von Kollagen über leicht abbaubare Zwischenstufen (*204, 226*). Beim jungen Organismus entstammt das reichlich im Harn ausgeschiedene Hydroxyprolin zur Hauptsache den in vergleichsweise großer Menge gebildeten Kollagenvorstufen von niedriger Halbwertzeit. Beim alten Organismus, in dem nur mehr eine geringe Kollagensynthese stattfindet, entstammt es zum größten Teil dem reifen vernetzten Kollagen mit seiner außerordentlich geringen Abbaurate. Für säurelösliches Kollagen von Rattenhaut wurde eine Halbwertzeit von etwa 25 Tagen gefunden (*87*).

Einen besonders ausgeprägten Abbau auch von reifem Kollagen beobachtet man bei der Involution des Uterus post partum (*409, 410a, 356a, 285a*).

Starke Veränderungen des Kollagenstoffwechsels treten auch bei Tieren im Stadium der Metamorphose auf, wenn verschiedene Organe abgestoßen, andere dagegen neu gebildet werden. Besonders gut untersucht sind Schwanz- und Rückenhaut von Kaulquappen (*131*). Schließlich wurde in verschiedenen Entzündungen und in Tumoren, z. B. in den experimentell durch Injektion von Carrageenin erzeugten (*200, 349*), ein gesteigerter Kollagenauf- und -abbau nachgewiesen.

Trotz dieser offensichtlichen Fähigkeit des Organismus, Kollagen abzubauen, haben die Versuche zum Nachweis einer körpereigenen Kollagenase bei Säugetieren nur zu geringem Erfolg geführt. Der Abbau von neutralsalzlöslichem Kollagen, das in einer noch nicht zur Faser strukturierten und somit stabilisierten Form vorliegt und bei Körpertemperatur bereits allmählich denaturiert wird, mag noch kein Problem darstellen, denn denaturiertes Kollagen wird von den verschiedensten Proteasen angegriffen. Für den Abbau der wesentlich stabileren Kollagenfasern, die erst bei 62—64° denaturieren, reicht eine solche Erklärung jedoch nicht mehr aus. Die Unterschiede in der Denaturierungstemperatur von Kollagen verschiedener Entwicklungsstufen dürften für die Unterschiede in der Abbaugeschwindigkeit verantwortlich sein.

Verschiedene Autoren (*409, 356*) haben in Uterusgewebe und anderen Organen cathepsinartige Enzyme nachgewiesen, welche Kollagen bei saurer Reaktion abzubauen vermögen. Diese Fähigkeit läßt jedoch nicht auf das Vorhandensein einer echten Kollagenase schließen, da im sauren Bereich die Denaturierungstemperatur des reifen Kollagens herabgesetzt ist, so daß schon bei Körpertemperatur ein gewisser Abbau möglich erscheint. Unwahrscheinlich dagegen ist es, daß im Organismus tatsächlich so starke pH-Änderungen auftreten könnten, daß sie einen derartigen Effekt erklären würden.

Eine echte Kollagenase wurde von Gross und Mitarb. (*136*) während der Metamorphose von Kaulquappen in der Schwanzflosse nachgewiesen. Das Gewebe, das in Kulturen weiter gezüchtet werden kann, ist in der Lage, ein Fasergel aus löslichem Kollagen bei neutralem pH unter physiologischen Bedingungen aufzulösen. Das Enzym wurde in reiner Form dargestellt (*138*). Es spaltet vom nativen Kollagenmolekül am B-Ende ein Bruchstück ab. Dabei wird keine Änderung der optischen Drehung beobachtet, was für den Erhalt der nativen Konformation spricht. Die beiden Kollagenbruchstücke wurden säulenchromatographisch getrennt, mit Hilfe von ATP als elektronenmikroskopisch sichtbare Segmente gefällt und damit die Stelle der Spaltung des Kollagenmoleküls genau festgelegt. Am langkettigen, das A-Ende enthaltenden Bruchstück findet man carboxylendständig Glycin*, am abgespaltenen kurzen Fragment, das dem B-Ende entstammt, aminoendständig Leucin oder Isoleucin*. Denaturiertes Kollagen wird von der gereinigten Kaulquappen-Kollagenase an zahlreichen Stellen wesentlich weniger spezifisch als natives angegriffen. Das Enzym zeigt eine völlig andere Spezifität als Kollagenase aus *Clostridium histolyticum*.

Was den Abbau von unlöslichem Kollagen im Säugetierorganismus anbelangt, so gewinnt die Vorstellung an Boden, daß dazu zwei Enzyme notwendig sind. Gries und Lindner (*126*) haben 1961 zeigen können, daß Trypsin reifes Kollagen in Gegenwart von Homogenaten von entzündetem cutanen Gewebe wesentlich stärker angreift als in Abwesenheit desselben. Houck und Mitarb. (*193*) berichteten über die Isolierung eines nicht dialysierbaren Faktors aus Haut, der nach kurzer Behandlung mit Trypsin oder Papain in der Lage ist, Kollagen bei pH 5,5 oder 7,0 abzubauen. Schließlich stellte Gross (*138*) fest, daß Mesenchymzellen und Epithelzellen aus Granulationsgewebe für sich allein keine kollagenolytische Aktivität aufweisen, daß sie aber in hinreichende Nähe zueinander gebracht ein Kollagen-Gel aufzulösen vermögen. Während gelöste native

---

* Aus diesem Befund kann man schließen, daß sich die N-terminale Seite des Kollagenmonomeren am A-Ende, die C-terminale am B-Ende befindet.

Literaturverzeichnis: SS. 293—314.

Kollagenmoleküle gegen Pronase, Trypsin und Pepsin weitgehend beständig sind, konnten KÜHN und Mitarb. *(226a)* zeigen, daß durch eine kombinierte Behandlung mit Pronase und Trypsin oder Pronase und Pepsin die Moleküle zerstört werden. Der Angriff beginnt am A-Ende der Moleküle und schreitet von dort aus zum B-Ende fort.

4. Mineralisierung.

Eine interessante Eigenschaft des Kollagens ist die Fähigkeit, Calciumphosphat ein- und anzulagern, eine wichtige Voraussetzung für die Bildung von Knochen, Zähnen und anderen harten Geweben. Die Calcifizierung von Kollagenfasern kann in vitro durch Einbringen in eine metastabile übersättigte Lösung von Calciumphosphat erfolgen. Das niedrigste Ionenprodukt aus Calcium und Phosphat erfordert dabei rachitisches Knorpelkollagen, dann Fibrillen aus neutralsalzlöslichem Kollagen und schließlich solche aus säurelöslichem Kollagen *(381)*.

Die Calcifizierung eines Gewebes setzt die Fähigkeit desselben voraus, Keime für das Kristallwachstum ausbilden zu können, an denen die Apatitkristalle weiterwachsen, bis sie die ganze Fibrille einhüllen. Von den verschiedenen Formen des Kollagens ist dazu nur die native Fibrille mit einer Querstreifungsperiode von 640 Å in der Lage *(96)*. Elektronenmikroskopische Aufnahmen von Kollagen aus embryonalen Knochen durch FITTON JACKSON *(203)* zeigen Apatit-Teilchen, die in regelmäßigen Abständen entsprechend der Identitätsperiode an bestimmten Stellen der Faser eingelagert sind. Ähnliche Aufnahmen kann man auch an Frühstadien der Calcifizierung in vitro erhalten *(93)*. Nach diesen Ergebnissen sind bestimmte Stellen der nativen Kollagenfibrille für die Bildung von Apatit-Keimen befähigt. Nach Vorstellungen von HODGE und PETRUSKA *(176)* handelt es sich dabei um die Hohlräume, welche bei der Zusammenlagerung der Kollagenmonomeren zur nativen Fibrille in der Nähe der überlappenden Endbereiche entstehen (vgl. Abb. 24, S. 271). Die elektronenmikroskopischen Aufnahmen stehen mit dieser Vorstellung in Einklang.

Der Mechanismus der Keimbildung ist noch nicht geklärt. Offensichtlich ist der Primärschritt in einer Bindung von Phosphat an das Kollagen zu suchen. Gereinigtes Kollagen bindet nach Untersuchungen von GLIMCHER *(94, 94a)* in vitro ungefähr 150—170 Mol Phosphatmonomeres. Diese Phosphatreste sind jedoch sehr unterschiedlich gebunden. Die Hauptmenge ist in einer dissoziierbaren Form elektrostatisch angelagert. Eine kleine Menge liegt jedoch in einer kovalent fixierten Form vor. Nach Einlagerung von markiertem Phosphat konnte GLIMCHER *(94, 94a)* in kleiner Menge radioaktive Hydrolysenprodukte erhalten, in denen Phosphat an organische Substanz gebunden war. Die Widerstandsfähigkeit dieser Bruchstücke gegenüber Säure und Alkali sowie ihr elektrophoretisches Verhalten läßt darauf schließen, daß es sich nicht um phosphorylierte Aminosäuren, sondern vielmehr um Zuckerphosphate handelt. Isolierung von phosphathaltigen Bruchstücken aus Kollagenase-hydrolysaten von Kollagen ergab, daß organisch gebundener Phosphor in serin- und glycinreichen Peptiden angereichert ist, die verhältnis-mäßig große Mengen an Zucker enthalten. Auf Grund der stöchio-metrischen Verhältnisse von Phosphat, Serin und Kohlenhydraten in diesen Peptiden sowie der Stabilität des organisch gebundenen Phosphats gegenüber Alkali ist es wahrscheinlich, daß Phosphor hauptsächlich an Kohlenhydrate und nur zu einem geringen Teil an Serin gebunden vor-liegt.

Ältere Vorstellungen gingen davon aus, daß die ε-Aminogruppen des Kollagens für die Phosphorylierung wichtig seien (382). Grund zu dieser Annahme war die Feststellung, daß die Reaktivität dieser Gruppen in calcifiziertem Kollagen sehr niedrig ist und daß sie mit dem Grad der Decalcifizierung zunimmt (382). Vor allem die Entfernung der letzten Apatit-Reste ist mit einer erheblichen Erhöhung der Substituierbarkeit der ε-Aminogruppen durch den Dinitrophenylrest verbunden (423). Die Ergebnisse lassen sich aber auch mit einer räumlichen Blockierung der ε-Aminogruppen durch den Apatit erklären. DNP-Kollagen oder carbobenzoxyliertes Kollagen werden nicht calcifiziert (93).

In der Zwischenzeit ist in Protein aus Zahnschmelz, das ebenfalls zur Calcifizierung fähig ist, sich jedoch von Kollagen in seiner Zusammen-setzung und in seiner Sekundär- und Tertiärstruktur unterscheidet (95), festgestellt worden, daß organisches Phosphat an Serinreste gebunden vorliegt (97). Aus diesem Protein lassen sich viel größere Mengen an O-Phosphorylserin gewinnen, als es möglich ist, organisches Phosphat aus Kollagen zu isolieren. Das Zahnschmelzprotein eignet sich daher besser zur Untersuchung des Calcifizierungsvorganges, und es ist zu er-warten, daß Informationen von dieser Seite auch mehr Licht in den Mechanismus der Apatit-Einlagerung in Kollagen bringen werden.

5. Kollagenosen.

Der Begriff „Kollagenose" wurde 1941 von Klemperer et al. (215) eingeführt. Die Autoren beobachteten bei Lupus erythematodes dis-seminatus einen fibrinoiden Gewebsschaden, der sich lichtmikroskopisch in einem Schwellen der fibrillären Strukturen und in einer steigenden Anfärb-barkeit mit sauren Farbstoffen bemerkbar macht. Das Phänomen der sogenannten fibrinoiden Veränderungen, das bei vielen Bindegewebs-erkrankungen auftritt, zeigt nicht unbedingt eine Veränderung der Kollagenfasern an. Auch Veränderungen der Bindegewebsgrundsubstanz kann als Ursache dafür auftreten (5). So ist es durch elektronenmikro-skopische Untersuchungen nur in den seltensten Fällen gelungen, eine

Literaturverzeichnis: SS. 293—314.

signifikante Veränderung der Struktur der Kollagenfibrillen nachzuweisen. Man ist daher in neuerer Zeit bestrebt, die Bezeichnung „Kollagenosen" durch „Bindegewebserkrankungen" zu ersetzen. Der Begriff
Kollagenose soll nur für solche Erkrankungen gelten, bei denen eine
spezifische Störung oder Veränderung des Kollagen-Stoffwechsels nachgewiesen ist.

Es ist wahrscheinlich, daß bei Bindegewebserkrankungen die ersten
Veränderungen im Stoffwechsel bei den Mucopolysacchariden und den
nichtkollagenen Proteinen einsetzen. Das unlösliche Kollagen mit
seinem sehr trägen Stoffwechsel wird nur sehr selten in Mitleidenschaft
gezogen werden. Dagegen ist ein Eingriff bei den löslichen Vorstufen
des Kollagens, z. B. bei der Fibrillenbildung, durchaus denkbar. Zwei
Ursachen dafür sind denkbar: einmal eine direkte Fehlleistung des
Kollagen-Stoffwechsels; zum zweiten aber auch eine indirekte Beein-
flussung durch Veränderungen der das Kollagen umgebenden Grundsubstanz. Gegenwärtig sind aber die Veränderungen im Kollagen-Stoff-
wechsel bei Bindegewebserkrankungen weitgehend unerforscht.

Es sollen folgende Bindegewebserkrankungen genannt werden: das rheumatische
Fieber, die primärchronische Polyarthritis, der Lupus erythematodes, die Sklerodermie, die Dermatomyositis, das Granuloma anulare und die senile Elastose.
Elektronenmikroskopisch glaubt ALBERTINI (5) beim Granuloma anulare und bei
der senilen Elastose einen Zerfall der Kollagenfibrillen beobachtet zu haben, welcher
zu einer Auflösung führt. Bei der Sklerodermie wurden dagegen keine Veränderungen
beobachtet. Allenfalls ist im Anfangsstadium der Krankheit eine Vermehrung
dünner Fibrillen zu bemerken (212, 217), die mit einer histochemisch erfaßbaren
Grundsubstanzvermehrung einhergeht (36a).

Über die Ursachen dieser Bindegewebserkrankungen weiß man so gut wie
nichts. Die rheumatischen Erkrankungen führt man auf eine Hyperempfindlichkeit
des Antikörper-produzierenden Systems im Organismus zurück. Es sollen sich Auto-
Antikörper bilden, die mit nicht näher identifizierten Strukturen des Bindegewebes
als Antigen reagieren. Diese Antigen-Antikörper-Komplexe vermutet man als eine
Ursache der rheumatischen Veränderungen im Bindegewebe. Es ist aber sicher,
daß diese nicht die alleinige Ursache sein können (287a). Ähnliche Antikörper hat
man auch beim Lupus erythematodes und auch bei der Sklerodermie gefunden.
Andere Bindegewebserkrankungen führt man auf genetische Defekte zurück.
Hierzu gehört das Hurlersche Syndrom (Gargoylismus), das auf einer Überproduktion
von Mucopolysacchariden beruht (281).

Zum Schluß seien zwei wirkliche Kollagenosen genannt: der auf
Vitamin-Mangel beruhende Skorbut und der Osteo-Lathyrismus, der
beim Menschen unbekannt ist und bei Versuchstieren durch β-Aminopropionitril, den Wirkstoff der spanischen Wicke *(Lathyrus odoratus)*,
hervorgerufen wird. Die Ursache von Skorbut ist auf S. 280f. besprochen
worden. Sie beruht darauf, daß zu Hydroxylierungsreaktionen wie die
des Prolins zu Hydroxyprolin die Gegenwart von Ascorbinsäure unbedingt
erforderlich ist. Es ist also die Biosynthese des Kollagens betroffen
(347–349). Bei lathyritischen Tieren findet man einen besonders hohen

Gehalt von neutralsalzlöslichem Kollagen (*257, 240*). Durch Versuche mit C^{14}-markierten Aminosäuren wurde festgestellt, daß der Übergang vom löslichen zum unlöslichen Kollagen stark gehemmt oder blockiert ist (*86, 378, 377, 226*). Es können sich weder intra- noch intermolekulare Bindungen ausbilden. Die Synthese des Kollagenmoleküls ist dagegen nicht gestört; im Gegenteil wird unter Einfluß von β-Aminopropionitril die Produktion der Kollagenmoleküle verstärkt (*226*). Diese Krankheit macht sich vor allem bei jungen Tieren bemerkbar, bei denen es zu verheerenden Deformationen des Knochengerüstes kommt. Die Ursache der verzögerten oder blockierten Ausbildung der intra- und intermolekularen Bindungen ist nicht klar. Eine Veränderung im Kollagenmolekül selbst ist bis jetzt noch nicht festgestellt worden. Der Lathyrismus scheint primär beim Stoffwechsel des Kollagens einzugreifen; die Mucopolysaccharide scheinen nur wenig verändert zu sein (*28a, 208a, 218, 374a*). Der Lathyrismus ist ein wichtiges Hilfsmittel geworden für die experimentelle Untersuchung über den Mechanismus der intermolekularen Quervernetzungen. Kürzlich konnten Martin und Piez (*273a*) und Partridge (*314a*) zeigen, daß auch die Ausbildung der Quervernetzung im Elastin durch β-Aminopropionitril unterbunden wird.

Als Therapeutica gegen Bindegewebserkrankungen werden zumeist antiinflammatorische Substanzen verwendet. Hier sollen nur die Corticosteroide und das Resochin (Chloroquin) genannt werden. Neuerdings spielt auch das Progesteron eine Rolle. Die Wirkung dieser Substanzen auf das Bindegewebe wurde empirisch gefunden, und man ist jetzt bemüht, ihre Wirkung biochemisch näher zu untersuchen. Nach Cortison-Gaben an Versuchtiere wird allgemein eine Verminderung der neutralsalzlöslichen Kollagenfraktion beobachtet (*376, 192, 139, 241, 194*). Bei geringen Cortison-Gaben nimmt das gesamte Kollagen gegenüber Normaltieren zu. Bei höheren Dosen geht der Kollagengehalt zurück. Nach Kühn et al. (*241*), welche den Kollagenstoffwechsel bei mit Prednison behandelten Ratten untersucht hatten, macht sich der allgemeine Eiweiß-antianabole Effekt des Cortisons auch bei der Kollagen-Synthese bemerkbar. Sie ist merklich verringert. Der Übergang vom löslichen zum unlöslichen Kollagen ist dagegen unter Einfluß von Prednison beschleunigt. Die Kollagenmoleküle lagern sich schneller zu geordneten Fibrillen zusammen, und die Ausbildung der intra- und intermolekularen Bindungen ist beschleunigt. Einen umgekehrten Effekt hat das Resochin (*235*). Während die Synthese der Kollagenmoleküle nicht beeinflußt wird, ist der Übergang vom löslichen in den unlöslichen Zustand verzögert. Eine deutliche Wirkung auf den extrazellulären Stoffwechsel des Kollagens zeigt auch das Corpus-luteum-Hormon Progesteron (*245*). Der empirisch gefundene auflockernde Effekt auf das Bindegewebe ist auf eine Verzögerung in der Ausbildung der intra- und intermolekularen Bindungen zurückzuführen. Der Übergang vom löslichen in unlösliches Kollagen ist bei den mit Progesteron behandelten Ratten stark verzögert. Der Effekt ist weitaus größer als bei Resochin.

Adam (*2*) hat das Verhalten von Schwanzsehnen von mit verschiedenen Bindegewebstherapeutica behandelten Ratten bei der Schrumpfung gemessen. Sehnen von mit Cortison-Derivaten behandelten Ratten sind stark verfestigt. Nach hohen Resochin-Gaben erkennt man den umgekehrten Effekt. Die Fasern zeigen im Schrumpfungstest eine geringere Festigkeit als die der normalen Tiere. Diese Befunde sind durch die Stoffwechseluntersuchungen am Kollagen leicht zu erklären. Besonders verfestigt zeigen sich die Sehnen bei Tieren, die mit Thiokomplexen des Goldes behandelt wurden, ein Mittel, das sich besonders gut zur Behandlung von akutem Rheumatismus eignet. ADAM (*1*) konnte zeigen, daß sich das Gold in vivo in die Kollagenfibrillen einlagert. Solche Fibrillen zeigen auch ohne die übliche „Anfärbung" mit Phosphorwolframsäure oder Uranylacetat im Elektronenmikroskop ein hochunterteiltes Querstreifungsmuster mit der üblichen Periode von 600—700 Å. Auch die Arbeiten von VOGEL und THER (*400*) über die Reißfestigkeit der Femurepiphysenfuge bei Ratten stimmt mit den bei den Stoffwechseluntersuchungen gefundenen Veränderungen nach Cortison- und Progesterongaben überein. Die Autoren fanden nach Corticosteroid-Behandlung die Belastungsgrenze der Femurepiphysenfuge erhöht und nach Progesteron-Gaben umgekehrt reduziert.

Aus diesen Stoffwechseluntersuchungen des Kollagens hat KORTING (*217*) praktisch therapeutische Konsequenzen gezogen. So erscheinen z. B. für die Behandlung der Zwischensubstanz-armen späten skleroathrophischen Verlaufsstadien der progressiven Sklerodermie Resochin und Progesteron geeignet, während im Anfangsstadium der Sklerodermie-Krankheit die pathologisch vermehrte Synthese der Zwischensubstanzen sowie des Kollagens durch Cortison reduziert werden sollte.

Literaturverzeichnis.

1. ADAM, M., P. BARTL, Z. DEYL and J. ROSMUS: Reaction of Gold with Collagen in vivo. Experientia **20**, 203 (1964).

2. ADAM, M., Z. DEYL and J. ROSMUS: The Influence of some Anti-Inflammatory Drugs on the Structural Stability of Collagen. Conference on Collagen Problems, Hluboká (ČSSR), 1964; vgl. Collagen Curr. **5**, 166 (1964).

3. ADELMANN, B. und K. KÜHN: Untersuchungen über nicht-kollagene Proteine des Bindegewebes. Intern. Symp. Biochem. Physiol. Connective Tissue. Lyon 1965.

4. AKABORI, S., K. OHNO and K. NARITA: On the Hydrazinolysis of Proteins and Peptides: A Method for the Characterization of Carboxyl-terminal Amino Acids in Proteins. Bull. Chem. Soc. Japan **25**, 214 (1952).

5. ALBERTINI, A. v. and A. VOGEL: Über wirkliche Kollagenosen. Dtsch. med. Wschr. **86**, 1421 (1961).

6. ALTGELT, K., A. J. HODGE and F. O. SCHMITT: Gamma Tropocollagen: A Reversibly Denaturable Collagen Macromolecule. Proc. Nat. Acad. Sci. (USA) **47**, 1914 (1961).

7. AMBROSE, E. J. and A. ELLIOTT: Infra-red Spectra and Structure of Fibrous Proteins. Proc. Roy. Soc. (London) **A 206**, 206 (1951).

8. ANDERER, F. A., H. UHLIG, E. WEBER and G. SCHRAMM: Primary Structure of the Protein of Tobacco Mosaic Virus. Nature **186**, 922 (1960).

9. ANDREEVA, N. S., M. I. MILLIONOVA and Yu. N. CHIRGADZE: Structural Investigations of Polymers Related to Collagen. In: G. N. RAMACHANDRAN, Aspects of Protein Structure, p. 137. Academic Press. London. 1963.

10. Anfinsen, C. B., S. E. G. Åqvist, J. P. Cooke and B. Jönsson: A Comparative Study of the Structures of Bovine and Ovine Pancreatic Ribonuclease. J. Biol. Chem. **234**, 1118 (1959).

11. Applequist, J.: The Helix-Coil Equilibrium in Polypeptides. J. Chem. Phys. **38**, 934 (1963).

12. Astbury, W. T.: X-Ray Studies of Protein Structure. Cold Spring Harbor Sympos. Quant. Biol. **2**, 15 (1934).

13. — The Molecular Structure of the Fibres of the Collagen Group. J. Internat. Soc. Leather Trades' Chem. **24**, 69 (1940).

14. Badger, R. M. und A. D. E. Pullin: The Infrared Spectrum and Structure of Collagen. J. Chem. Phys. **22**, 1142 (1954).

15. Bailey, A. J. and A. J. Hodge: The Subunit Structure of the Tropocollagen Molecule in Relation to the Native-type Fibril and other Ordered Aggregation States. Conference on the Structure and Function of Connective and Skeletal Tissues. St. Andrews (Scotland), 1964.

16—17. Bangle, R., Jr. and W. C. Alford: The Chemical Basis of the Periodic Acid Schiff Reaction of Collagen Fibers with Reference to Periodate Consumption by Collagen and by Insulin. J. Histochem. Cytochem. **2**, 62 (1954).

18. Bear, R. S.: X-Ray Diffraction Studies on Protein Fibers. I. The Large Fiber-Axis Period of Collagen. J. Amer. Chem. Soc. **66**, 1297 (1944).

19. — The Structure of Collagen Fibrils. Adv. Protein Chem. **7**, 69 (1952).

20. — Fibrous Proteins and their Biological Significance. Symp. Soc. Exp. Biol. **9**, 97 (1955).

21. — The Structure of Collagen Molecules and Fibrils. J. Biophys. Biochem. Cytol. **2**, 363 (1956).

22. Bear, R. S., O. E. Bolduan and T. P. Salo: A Model for Collagen Fibril Structure Derived from Small-angle X-Ray Diffraction. J. Amer. Leather Chem. Assoc. **46**, 107 (1951).

23. Beek, J.: Carbohydrate Content of Collagen. J. Res. Nat. Bur. Stand. **27**, 507 (1941).

24. Beer, M., G. B. B. M. Sutherland, K. N. Tanner and D. L. Wood: Infra-red Spectra and Structure of Proteins. Proc. Roy. Soc. (London) A **249**, 147 (1959).

24a. Beier, G. and J. Engel: The Renaturation of Soluble Collagen: Products Formed at Different Temperatures. Biochemistry (in print).

24b. Beier, G., J. Engel und W. Grassmann: Zum Mechanismus der Bildung der Tripelhelix des Kollagens aus den denaturierten Komponenten. Tagungsbericht, Biophysiker Tagung, Wien, 1964; und Collagen Currents **5**, 310 (1965).

24c. Bekhor, I. J. and L. A. Bavetta: Actinomycin D Inhibition of Microsomal-Bound Hydroxyproline Formation in Rabbit Embryo Skin in vitro. Proc. Nat. Acad. Sci. (USA) **53**, 613 (1965).

25. Bello, J.: On the Mechanism of Action of Neutral Salt on the Collagen-Fold. Biochemistry **2**, 276 (1963).

26. Bensusan, H. B. and B. L. Hoyt: The Effect of Various Parameters on the Rate of Formation of Fibers from Collagen Solutions. J. Amer. Chem. Soc. **80**, 719 (1958).

26a. Bensusan, H. B. and S. O. Nielsen: The Deuterium Exchange of Peptide-Group Hydrogen Atoms during the Gelatin→Collagen-Fold Transition. Biochemistry **3**, 1367 (1964).

27. Bergmann, M.: Complex Salts of Amino Acids and Peptides. II. Determination of *l*-Proline with the Aid of Rhodanilic Acid. The Structure of Gelatin. J. Biol. Chem. **110**, 471 (1935).

28. BERGMANN, M. and C. NIEMANN: On Blood Fibrin. A Contribution to the Problem of Protein Structure. J. Biol. Chem. **115**, 77 (1936).

28a. BICKLEY, H. C. and J. L. ORBISON: The Effect of a Lathyrogen on Sulfate Metabolism in Culture of the Strain L Fibroblast. Lab. Invest. **13**, 172 (1964).

29. BLUMENFELD, O. O. and P. M. GALLOP: The Participation of Aspartyl Residues in the Hydroxylamine – or Hydrazine-Sensitive Bonds of Collagen. Biochemistry **1**, 947 (1962).

30. BLUMENFELD, O. O., M. A. PAZ, P. M. GALLOP and S. SEIFTER: The Nature, Quantity, and Mode of Attachment of Hexoses in Ichthyocol. J. Biol. Chem. **238**, 3835 (1963).

31. BOEDTKER, H. and P. DOTY: The Native and Denatured States of Soluble Collagen. J. Amer. Chem. Soc. **78**, 4267 (1956).

32. BORNSTEIN, P., G. R. MARTIN and K. A. PIEZ: Intermolecular Cross-Linking of Collagen and the Identification of a New β-Component. Science **144**, 1220 (1964).

33. BORNSTEIN, P. and K. A. PIEZ: A Biochemical Study of Human Skin Collagen and the Relation between Intra- and Intermolecular Cross-Linking. J. Clin. Investigation **43**, 1813 (1964).

33a. — — Collagen: Structural Studies Based on the Cleavage of Methionyl Bonds. Science **148**, 1353 (1965).

34. BOWES, J. H., R. G. ELLIOTT and J. A. MOSS: The Composition of Collagen and Acid-soluble Collagen of Bovine Skin. Biochem. J. **61**, 143 (1955).

35. BOWES, J. H. and J. A. MOSS: The Reaction of Fluorodinitrobenzene with the α- and ε-Amino Groups of Collagen. Biochem. J. **55**, 735 (1953).

36. BRADBURY, E. M., R. E. BURGE, J. T. RANDALL and G. R. WILKINSON: The Polypeptide Chain Configurations of Native and Denatured Collagen Fibres. Disc. Faraday Soc. No. **25**, 173 (1958).

36a. BRAUN-FALCO, O.: Über das Verhalten der interfibrillären Grundsubstanz bei Sklerodermie. Dermatol. Wschr. **136**, 1085 (1957).

37. BRAUNITZER, G., N. HILSCHMANN und R. MÜLLER: Über den Ort des spezifischen Austausches von Aminosäuren in der β-Kette bei pathologischen menschlichen Hämoglobinen. Z. physiol. Chem. **318**, 284 (1960).

38. BRAUNITZER, G., K. HILSE, V. RUDLOFF and N. HILSCHMANN: The Hemoglobins. Adv. Protein Chem. **19**, 1 (1964).

39. BROCAS, J. and F. VERZÁR: The Ageing of *Xenopus laevis*, a South African Frog. Gerontologia **5**, 228 (1961); Experientia **17**, 421 (1961).

39a. — — Measurement of Isometric Tension during Thermic Contraction as Criterion of the Biological Age of Collagen Fibers. Gerontologia **5**, 223 (1961).

40. BURGE, R. E. and R. D. HYNES: The Thermal Denaturation of Collagen in Solution and its Structural Implications. J. Mol. Biol. **1**, 155 (1959).

41. CANTOW, H.-J.: Streulichtmethode. In: HOUBEN-WEYL, Methoden der organischen Chemie, 4. Aufl., Bd. III, Teil 1, S. 408. Stuttgart: George Thieme. 1955.

42. CHVAPIL, M. and J. HURYCH: Hydroxylation of Proline in vitro. Nature **184**, 1145 (1959).

43. ČMUCHALOVÁ, B. and M. CHVAPIL: The Rôle of Ascorbic Acid in Collagen Biosynthesis. Conference on Collagen Problems, Hluboká (ČSSR), 1964; vgl. Collagen Currents **5**, 165 (1964).

44. COCHRAN, W., F. H. C. CRICK and V. VAND: The Structure of Synthetic Polypeptides. I. The Transform of Atoms on a Helix. Acta Crystallogr. **5**, 581 (1952).

45. COHEN, C.: Optical Rotation and Helical Polypeptide Chain Configuration in Collagen and Gelatin. J. Biophys. Biochem. Cytol. **1**, 203 (1955).

46. COHEN, C. und R. S. BEAR: Helical Polypeptide Chain Configuration in Collagen. J. Amer. Chem. Soc. **75**, 2783 (1953).

47. Corey, R. B. and L. Pauling: Fundamental Dimensions of Polypeptide Chains. Proc. Roy. Soc. (London) B 141, 10 (1953).
48. Coronado, A., E. Mardones and J. E. Allende: Isolation of Hydroxylysyl-sRNA and Hydroxyprolyl-sRNA in a Chick Embryo System. Biochem. Biophys. Res. Comm. 13, 75 (1963).
49. Coronado, A., E. Mardones, J. Celis and J. E. Allende: The Formation of Hydroxyprolyl-sRNA and Hydroxylysyl-sRNA in a Chick Embryo System. VIth Internat. Congr. Biochem., New York 1964, Abstr. I—35.
50. Cowan, P. M., S. McGavin and A. C. T. North: The Polypeptide Chain Configuration of Collagen. Nature 176, 1062 (1955).
51. Cowan, P. M., A. C. T. North and J. T. Randall: High-Angle X-Ray Diffraction of Collagen Fibres. In: J. T. Randall (Edit.), Nature and Structure of Collagen, p. 241. London: Butterworth. 1953.
52. Crick, F. H. C. and A. Rich: Structure of Polyglycine II. Nature 176, 780 (1955).
53. Crosby, N. T., D. G. Higgs, R. Reed, G. Stainsby and A. G. Ward: The Preparation and Properties of Solubilized Collagens. J. Soc. Leather Trades' Chem. 46, 152 (1962).
54. Crothers, D. M. and B. H. Zimm: Theory of the Melting Transition of Synthetic Polynucleotides: Evaluation of the Stacking Free Energy. J. Mol. Biol. 9, 1 (1964).
55. Dasler, W., R. F. Stoner and R. V. Milliser: Effect of Osteolathyrism on Soluble Collagen Fractions of Rat Connective Tissue. Metabolism 10, 883 (1961).
55a. Delaunay, A. and S. Bazin: Mucopolysaccharides, Collagen and Nonfibrillar Proteins in Inflammation. Intern. Rev. Conn. Tiss. 2, 301 (1964).
56. Dick, Y. P., J. Engel and A. Nordwig: The Effect of pH on the Stability of the Collagen Fold. 2nd Meet. Fed. Europ. Biochem. Soc., Vienna, 1965.
57. Djerassi, C.: Optical Rotatory Dispersion. New York: McGraw-Hill. 1960.
58. Doty, P. and T. Nishihara: The Molecular Properties and Thermal Stability of Soluble Collagen. In: G. Stainsby (Edit.), Recent Advances in Gelatin and Glue Research, p. 92. London: Pergamon Press. 1958.
59. Edman, P.: Method for Determination of the Amino Acid Sequence in Peptides. Acta Chem. Scand. 4, 283 (1950).
60. Edmundson, A. B.: Amino-Acid Sequence of Sperm Whale Myoglobin. Nature 205, 883 (1965).
61. Eichler, O.: Die Pharmakologie anorganischer Anionen. Berlin: Springer. 1950.
62. Engel, J.: Untersuchung des Mechanismus der Denaturierung und Renaturierung von löslichem Kollagen mit der Lichtstreuungsmethode. Z. physiol. Chem. 325, 287 (1961).
63. — Physikalisch-chemische Untersuchungen am Tropokollagen. Dissert. Univ. München, 1962.
64. — Über den Einfluß des Denaturierungsgrades und der Renaturierungsbedingungen auf die Rückbildung kollagener Strukturen. Z. physiol. Chem. 328, 94 (1962).
65. — Investigation of the Denaturation and Renaturation of Soluble Collagen by Light Scattering. Arch. Biochem. Biophys. 97, 150 (1962).
66. Engel, J. und G. Beier: Die Rückbildung der Sekundär- und Tertiärstruktur denaturierter Eiweißmoleküle. Kolloid-Z. 197, 7 (1964).
67. — — Vergleich der molekularen Daten von Tropokollagen verschiedener Kalbshäute in nativem und denaturiertem Zustand. Z. physiol. Chem. 334, 201 (1963).

68. ENGEL, J., W. GRASSMANN, K. HANNIG und K. KÜHN: Zur Bildung von Fibrillen und Long-Spacing-Segmenten aus renaturierten Tropokollagenlösungen. Z. physiol. Chem. **329**, 69 (1962).

69. ENGEL, J., J. KURTZ, E. KATCHALSKI und A. BERGER: Polymers of Tripeptides as Collagen Models. II. Conformational changes of Poly(L-prolyl-glycyl-L-prolyl) in Solution. J. Mol. Biol. (in print).

70. ENGEL, J., J. KURTZ, W. TRAUB, A. BERGER and E. KATCHALSKI: On the Mechanism of Collagen De- and Renaturation and on Conformational Changes in Related Polypeptides. Conference on the Structure and Function of Connective and Skeletal Tissues. St. Andrews (Scotland), 1964.

71. FESSLER, J. H.: Some Properties of Neutral-Salt-Soluble Collagen. 1. Biochem. J. **76**, 452 (1960).

72. — Some Properties of Neutral-Salt-Soluble Collagen. 2. Biochem. J. **76**, 463 (1960).

73. FLORY, P. J. and R. R. GARRETT: Phase Transitions in Collagen and Gelatin Systems. J. Amer. Chem. Soc. **80**, 4836 (1958).

74. FLORY, P. J. and E. S. WEAVER: Helix $\rightleftarrows$ Coil Transition in Dilute Aqueous Collagen Solutions. J. Amer. Chem. Soc. **82**, 4518 (1960).

75. FRANZBLAU, C., S. SEIFTER and P. M. GALLOP: The "Nondialyzable" Fraction Obtained from Ichthyocol Digested with Collagenase. Biopolymers **2**, 185 (1964).

76. FRICKE, R. and U. HADDING: Connective Tissue Proteins. Protides of the Biological Fluids. Proc., 10th Coll. Bruges, 1962, p. 52. Amsterdam: Elsevier. 1963.

77. FRUTON, J. S.: Chemical Aspects of Protein Synthesis. In: H. NEURATH (Edit.), The Proteins, Vol. I, p. 190. New York: Acad. Press. 1963.

78. FUJII, T.: Privatmitteilung.

79. FUJIIMOTO, D. and N. TAMIYA: Studies on the Hydroxylation of Lysine in vivo. Biochem. Biophys. Res. Comm. **10**, 498 (1963).

80. — — Studies on Collagen Metabolism with O^{18} as a Tracer. Biochim. Biophys. Acta **69**, 559 (1963).

80a. FUJITA, Y., A. GOTTLIEB, B. PETERKOFSKY, S. UDENFRIEND and B. WITKOP: The Preparation of cis- and trans-4-H^3-L-Prolines and Their Use in Studying the Mechanism of Enzymatic Hydroxylation in Chick Embryos. J. Amer. Chem. Soc. **86**, 4709 (1964).

81. FUNAKOSHI, H. and H. NODA: Estimation of the Average Length of the Interband Regions in Collagen Polypeptide Chain. Biochim. Biophys. Acta **86**, 106 (1964).

82. GALLOP, P. M.: Concerning some Special Structural Features of the Collagen Molecule. Biophys. J. **4**, Suppl. 79 (1964).

83. GALLOP, P. M., S. SEIFTER and E. MEILMANN: Occurrence of "Ester-like" Linkages in Collagen. Nature **183**, 1659 (1959).

84. GEBHARDT, E. und K. KÜHN: Chemische und elektronenoptische Untersuchungen über die Reaktion von Uranylacetat mit Kollagen. Z. anorg. allg. Chem. **320**, 71 (1963).

85. GERBER, G. B. and K. I. ALTMAN: Mechanism of Collagen Synthesis. Nature **189**, 813 (1961).

86. GERBER, G. B., G. GERBER und K. I. ALTMAN: Studies of Collagen Turnover in Lathyritic Rats. Arch. Biochem. Biophys. **96**, 601 (1962).

87. — — — Studies on the Metabolism of Tissue Proteins. I. Turnover of Collagen Labeled with Proline-U-C^{14} in Young Rats. J. Biol. Chem. **235**, 2653 (1960).

88. GERNGROSS, O., K. HERRMANN und W. ABITZ: Über den Feinbau des Gelatinemicells. Biochem. Z. **228**, 409 (1930).

89. Gibian, H.: Mucopolysaccharide und Mucopolysaccharidasen. Wien: Deuticke. 1959.

90. Gieseking, R.: Die Fibrillogenese in verschiedenen Bindegewebsformen des menschlichen Embryos. Beitr. Silikose-Forsch., Sonderband **4**, 231 (1960).

91. — Elektronenoptische Befunde am embryonalen, regenerierenden und proliferierenden Bindegewebe. Beitr. Silikose-Forsch., Sonderband **4**, 249 (1960).

92. Glegg, R. E., D. Eidinger and C. P. Leblond: Some Carbohydrate Components of Reticular Fibers. Science **118**, 614 (1954).

93. Glimcher, M. J.: Specificity of the Molecular Structure of Organic Matrices in Mineralization. In: R. F. Sognnaes, Calcification in Biological Systems, p. 421. Washington, D. C.: Amer. Assoc. Adv. Sci. 1960.

94. — Studies on the Role of Phosphate in the Calcification of Organic Matrices. Conference on the Structure and Function of Connective and Skeletal Tissues. St. Andrew (Scotland), 1964.

94 a. Glimcher, M. J., C. J. Francois, L. Richards and S. M. Krane: The Presence of Organic Phosphorus in Collagens and Gelatins. Biochem. Biophys. Acta **93**, 585 (1964).

95. Glimcher, M. J., U. A. Friberg and P. T. Levine: The Isolation and Amino Acid Composition of the Enamel Proteins of Erupted Bovine Teeth. Biochem. J. **93**, 202 (1964).

96. Glimcher, M. J., A. J. Hodge and F. O. Schmitt: Macromolecular Aggregation States in Relation to Mineralization. The Collagen-Hydroxyapatite System as Studied in vitro. Proc. Nat. Acad. Sci. (USA) **43**, 860 (1957).

97. Glimcher, M. J. and S. M. Krane: The Identification of Serine Phosphate in Enamel Proteins. Biochim. Biophys. Acta **90**, 477 (1964).

97 a. Gottlieb, A. and S. Udenfriend: A Hydroxyproline-Deficient Intermediate in Collagen Synthesis. Federat. Proc. (Amer. Soc. Exp. Biol.) **24**, 358 (1965).

98. Gould, B. S. and J. F. Woessner: Biosynthesis of Collagen. The Influence of Ascorbic Acid on the Proline, Hydroxyproline, Glycine, and Collagen Content of Regenerating Guinea Pig Skin. J. Biol. Chem. **226**, 289 (1957).

99. Grassmann, W.: Untersuchungen über Kollagen. Kolloid-Z. **77**, 205 (1936).

100. — Unsere heutige Kenntnis des Kollagens. Leder **6**, 241 (1955).

101. — Kollagen und Bindegewebe. Svensk Kem. Tidskr. **72**, 4 (1960).

101 a. — Über Eiweißfasern (Vortrag, 24. 10. 1962). Mitt. Max-Planck-Ges. **1963**, 9.

102. Grassmann, W., H. Endres und A. Steber: Esterbindungen in Prokollagen. Z. Naturforsch. **9 b**, 513 (1954).

103. Grassmann, W., K. Hannig, H. Endres und A. Riedel: Zur Bindungsweise des Prolins und Hydroxyprolins. Aminosäuresequenzen des Kollagens, I. Z. physiol. Chem. **306**, 123 (1956).

104. Grassmann, W., K. Hannig und J. Engel: Das quantitative Verhältnis zwischen α- und β-Komponente des denaturierten, löslichen Kollagens in der Ultrazentrifuge sowie Beschreibung einer schneller sedimentierenden γ-Komponente. Z. physiol. Chem. **324**, 284 (1961).

105. Grassmann, W., K. Hannig und A. Nordwig: Über die apolaren Bereiche des Kollagenmoleküls. Aminosäuresequenzen des Kollagens, VI. Z. physiol. Chem. **333**, 154 (1963).

106. Grassmann, W., K. Hannig und M. Plöckl: Eine Methode zur quantitativen Bestimmung der Aminosäurezusammensetzung von Eiweißhydrolysaten durch Kombination von Elektrophorese und Chromatographie. Z. physiol. Chem. **299**, 258 (1955).

107. Grassmann, W., K. Hannig und M. Schleyer: Versuche zur elektrophoretischen Reinigung von Trypsin. Z. physiol. Chem. **316**, 71 (1959).

108. GRASSMANN, W., K. HANNIG und M. SCHLEYER: Zur Aminosäuresequenz des Kollagens, II. Z. physiol. Chem. **322**, 71 (1960).

109. GRASSMANN, W., U. HOFMANN und TH. NEMETSCHEK: Die Querstreifung von Kollagenfibrillen. Naturwiss. **39**, 215 (1952).

110. GRASSMANN, W. und H. HÖRMANN: Endgruppenbestimmung an Kollagen und Gelatine. Z. physiol. Chem. **292**, 24 (1953).

111. GRASSMANN, W., H. HÖRMANN und R. HAFTER: Eine quantitative Bestimmung von Kohlenhydraten als Osazone. Anwendung der Methode auf Kollagen und Prokollagen. Z. physiol. Chem. **307**, 87 (1957).

112. GRASSMANN, W., H. HÖRMANN, A. NORDWIG und E. WÜNSCH: Zur Spezifität der Kollagenase. Z. physiol. Chem. **316**, 287 (1959).

113. GRASSMANN, W., U. HOFMANN, K. KÜHN, H. HÖRMANN, H. ENDRES and K. WOLF: Electron Microscope and Chemical Studies of the Carbohydrate Groups of Collagen. In: R. E. TUNBRIDGE (Edit.), Connective Tissue, p. 157. Oxford: Blackwell Sci. Publ. 1957.

114. GRASSMANN, W., J. JANICKI und F. SCHNEIDER: Über die Einwirkung von Trypsin auf Kollagen. Stiasny-Festschrift, S. 74, Darmstadt: E. Roether Verl. 1937.

115. GRASSMANN, W. und K. KÜHN: Abbau des Kollagens und Prokollagens mit Natriumperjodat und Phenyljodosoacetat. Z. physiol. Chem. **301**, 1 (1955).

116. GRASSMANN, W. und A. NORDWIG: Quantitativer kolorimetrischer Test auf Kollagenase. Z. physiol. Chem. **322**, 267 (1960).

117. GRASSMANN, W., A. NORDWIG und H. HÖRMANN: Über den Bau der apolaren Bereiche der Kollagenfaser: Abbau mit Kollagenase. Aminosäuresequenzen des Kollagens, III. Z. physiol. Chem. **323**, 48 (1961).

118. GRASSMANN, W., A. RIEDEL und TH. ALTENSCHÖPFER: Chemische Veränderungen beim Übergang von gekälkter Haut in Gelatine. Kolloid-Z. **186**, 50 (1962).

119. GRASSMANN, W. und K. RIEDERLE: Über die Konstitution des Glutokyrins. 3. Mitt. zur Kenntnis des Kollagens. Biochem. Z. **284**, 177 (1936).

120. GRASSMANN, W. und H. SCHLEICH: Über den Kohlenhydratgehalt des Kollagens. 2. Mitt. zur Kenntnis des Kollagens. Biochem. Z. **277**, 320 (1935).

121. GRASSMANN, W., F. SCHNEIDER und J. TRUPPKE: Eiweißstoffe und ihre Abbaustufen. In: FLASCHENTRÄGER und LEHNARTZ, Physiologische Chemie, Band I, S. 484, insb. S. 661, Abb. 72. Heidelberg: Springer-Verlag. 1951.

122. GRASSMANN, W., O. v. SCHOENEBECK und G. AUERBACH: Über die enzymatische Spaltbarkeit der Prolinpeptide, II. Z. physiol. Chem. **210**, 1 (1932).

122a. GRASSMANN, W. und P. STADLER: Gerbversuche an Kollagenfibrillen. Leder **12**, 290 (1961).

123. GRASSMANN, W., L. STRAUCH und A. NORDWIG: Reinigung von Kollagenase, II. Z. physiol. Chem. **332**, 325 (1963).

124. GREEN, N. M. and D. A. LOWTHER: Formation of Collagen Hydroxyproline in vitro. Biochem. J. **71**, 55 (1959).

125. GREENBERG, J., L. FISHMAN and M. LEVY: Positions of Amino Acids in Mixed Peptides Produced from Collagen by the Action of Collagenase. Biochemistry **3**, 1826 (1964).

126. GRIES, G. und J. LINDNER: Untersuchungen über den Kollagenabbau bei akuten Entzündungen. Z. Rheumaforschg. **20**, 122 (1961).

127. GRIMM, L. und W. GRASSMANN: Das enzymchemische Verhalten der Lysyl-Prolin-Bindung und der Lysyl-Hydroxyprolin-Bindung. Ein Beitrag zur Spezifität des Trypsins. Z. physiol. Chem. **337**, 161 (1964).

128. Gross, E. and B. Witkop: Nonenzymatic Cleavage of Peptide Bonds: The Methionine Residues in Bovine Pancreatic Ribonuclease. J. Biol. Chem. **237**, 1856 (1952).

129. Gross, J.: Studies on the Formation of Collagen. III. Time-dependent Solubility Changes of Collagen in Vitro. J. Exp. Medicine **108**, 215 (1958).

130. — Studies on the Formation of Collagen. II. The Influence of Growth Rate on Neutral Salt Extracts of Guinea Pig Dermis. J. Exp. Medicine **107**, 265 (1958).

131. — Studies on the Biology of Connective Tissues: Remodelling of Collagen in Metamorphosis. Medicine **43**, 291 (1964).

132. Gross, J., B. Dumsha and N. Glazer: Comparative Biochemistry of Collagen. Some Amino Acids and Carbohydrates. Biochim. Biophys. Acta **30**, 293 (1958).

133. Gross, J., J. H. Highberger and F. O. Schmitt: Some Factors Involved in the Fibrogenesis of Collagen in vitro. Proc. Soc. Exp. Biol. Med. **80**, 462 (1952).

133a. — — — Collagen Structures Considered as States of Aggregation of a Kinetic Unit. The Tropocollagen Particle. Proc. Nat. Acad. Sci. (USA) **40**, 679 (1954).

134. — — — Extraction of Collagen from Connective Tissue by Neutral Salt Solutions. Proc. Nat. Acad. Sci. (USA) **41**, 1 (1955).

135. Gross, J. and D. Kirk: The Heat Precipitation of Collagen from Neutral Salt Solutions: Some Rate-Regulating Factors. J. Biol. Chem. **233**, 355 (1958).

136. Gross, J. and C. M. Lapiere: Collagenolytic Activity in Amphibian Tissues: A Tissue Culture Assay. Proc. Nat. Acad. Sci. (USA) **48**, 1014 (1962).

137. Gross, J. and G. R. Martin: Alternations of Cross Linking in Collagen in Experimental Lathyrism. 141th Meeting Amer. Chem. Soc., March 1962, Abstracts 2C.

138. Gross, J., Y. Nagai, C. M. Lapiere, G. Usuku, H. C. Grillo and A. Z. Eisen: Animal Collagenase and Collagen Metabolism. Conference on the Structure and Function of Connective and Skeletal Tissues. St. Andrews (Scotland), 1964.

139. Günther, Th. und P. M. Carsten: Über die Löslichkeit von Kollagen in Abhängigkeit von Nebennierenrindenhormonen. Naturwiss. **48**, 699 (1961).

140. Gustavson, K. H.: The Chemistry and Reactivity of Collagen, p. 225. New York: Academic Press. 1956.

141. — The Chemistry of Tanning Processes, p. 246. New York: Academic Press. 1956.

142. — Hydrothermal Stability and Intermolecular Organization of Collagens from Mammalian and Teleost Skins. Svensk Kem. Tidskr. **65**, 70 (1953).

143. — The Function of Hydroxyproline in Collagens. Nature **175**, 70 (1955).

144. Hafter, R.: Zur Fibrillenbildung des Kollagens, II. Mitt. Das Verhalten von Tropokollagen und Gelatine unter isoelektrischen Bedingungen. Leder **15**, 237 (1964).

144a. Hafter, R. und H. Hörmann: Der Einfluß von Pepsin auf die Struktur und die faserbildenden Eigenschaften von Kollagen. Z. physiol. Chem. **330**, 169 (1963).

145. Hall, C. E. and P. Doty: A Comparison Between the Dimensions of Some Macromolecules Determined by Electron Microscopy and by Physical Chemical Methods. J. Amer. Chem. Soc. **80**, 1269 (1958).

146. Hall, C. E., M. A. Jakus and F. O. Schmitt: Electron Microscope Observations of Collagen. J. Amer. Chem. Soc. **64**, 1234 (1942).

147. Hamaguchi, K. and E. P. Geiduschek: The Effect of Electrolytes on the Stability of the Deoxyribonucleate Helix. J. Amer. Chem. Soc. **84**, 1329 (1962).

148. HANNIG, K.: Eine Apparatur zur kontinuierlichen Ablenkungselektrophorese auf Filtrierkarton bei gleichzeitiger Anwendung hoher Feldstärken. Z. physiol. Chem. **311**, 63 (1958).

149. — Investigations on the Sequence of Amino Acid Residues in Collagen. Federat. Proc. (Amer. Soc. Exp. Biol.) **19**, 1 (1960).

150. HANNIG, K. und J. ENGEL: Physikalisch-chemische Untersuchungen an Tropokollagenlösungen. Leder **12**, 213 (1961).

151. HANNIG, K., M. R. NAGAMANI und G. BEIER: Unveröffentlicht.

152. HARKNESS, R. D., A. M. MARKO, H. M. MUIR and A. NEUBERGER: The Metabolism of Collagen and other Proteins of the Skin of Rabbits. Biochem. J. **56**, 558 (1954).

153. HARMS, H.: Elastinfärbung mit Orcein. In: Handbuch der Farbstoffe für die Mikroskopie, Teil II, S. 160. Kamp-Lintfort: Staufen Verl. 1957.

154. HARRINGTON, W. F.: On the Arrangement of the Hydrogen Bonds in the Structure of Collagen. J. Mol. Biol. **9**, 613 (1964).

155. HARRINGTON, W. F. and P. H. v. HIPPEL: Formation and Stabilization of the Collagen-Fold. Arch. Biochem. Biophys. **92**, 100 (1961).

156. HARRINGTON, W. F. and J. KURTZ: The Effect of LiBr on the Configuration of Poly-L-proline. Collagen Currents **4**, No. 11, 34 (1964).

157. HARTLEY, B. S.: Amino-Acid Sequence of Bovine Chymotrypsinogen-A. Nature **201**, 1284 (1964).

158. HASHIMOTO, Y., T. YAMANO and H. S. MASON: An Electron Spin Resonance Study of Microsomal Electron Transport. J. Biol. Chem. **237**, 3843 (1962).

159. HAUROWITZ, F.: The Chemistry and Function of Proteins, p. 394. New York: Academic Press. 1963.

160. HAUSMANN, E. and W. F. NEUMAN: Conversion of Proline to Hydroxyproline and its Incorporation into Collagen. J. Biol. Chem. **236**, 149 (1961).

160a. HEIDRICH, H. G. and L. WYNSTON: A Report on the Presence of a Third Alpha Component in Calf and Rat Skin Tropocollagen. Z. physiol. Chem. **341**, 1 (1965).

161. HERINGA, G. C. und N. H. KOLKMEIJER: Verh. Koninkl. Akad. Wetenschap, Amsterdam, Afdeel Natuurk. **29**, 1029 (1926).

162. HERZOG, R. O. und H. W. GONNELL: Weitere Untersuchungen an Naturstoffen und biologischen Strukturen mittels Röntgenstrahlen. Naturwiss. **12**, 1153 (1924).

163. — — Über Kollagen. Ber. dtsch. chem. Ges. **58**, 2228 (1925).

164. — — Über Kollagen. Collegium (Darmstadt) **1926**, 189.

165. HERZOG, R. O. und W. JANCKE: Notiz über Röntgendiagramm des Kollagens (Faserperiode). Z. physik. Chemie B **12**, 228 (1931).

166. HEYNS, K. und G. LEGLER: Über Proteine und deren Abbauprodukte, XVI. Über die Spezifität der Kollagenase aus *Cl. histolyticum* und ihre Anwendung zur Ermittlung der Primärstruktur des Kollagens. Z. physiol. Chem. **315**, 288 (1959).

167. HIGHBERGER, J. H., J. GROSS and F. O. SCHMITT: Electron Microscope Observations of Certain Fibrous Structures Obtained from Connective Tissue Extracts. J. Amer. Chem. Soc. **72**, 3321 (1950).

168. — — — The Interaction of Mucoprotein with Soluble Collagen; an Electron Microscope Study. Proc. Nat. Acad. Sci. (USA) **37**, 286 (1951).

169. HIPPEL, P. H. v., P. M. GALLOP, S. SEIFTER and R. S. CUNNINGHAM: An Enzymatic Examination of the Structure of the Collagen Macromolecule. J. Amer. Chem. Soc. **82**, 2774 (1960).

170. HIPPEL, P. H. v. and W. F. HARRINGTON: The Structure and Stabilization of the Collagen Macromolecule. Proc., Sympos. Protein Structure and Function, Brookhaven Nat. Lab., 1960, No. 13, p. 213.

302   W. Grassmann und Mitarbeiter:

171. Hippel, P. H. v. and K. Y. Wong: The Collagen ⇄ Gelatin Phase Transition. II. Shape of the Melting Curves and Effect of Chain Length. Biochemistry **2**, 1399 (1963).

172. — — The Collagen ⇄ Gelatine Phase Transition. I. Further Studies of the Effects of Solvent Environment and Polypeptide Chain Composition. Biochemistry **2**, 1387 (1963).

173. — — The Effect of Ions on the Kinetics of Formation and the Stability of the Collagen-Fold. Biochemistry **1**, 664 (1962).

174. — — Neutral Salts: The Generality of their Effects on the Stability of Macromolecular Conformations. Science **145**, 577 (1964).

175. Hodge, A. J., J. H. Highberger, G. G. J. Deffner and F. O. Schmitt: The Effects of Proteases on the Tropocollagen Macromolecule and on its Aggregation Properties. Proc. Nat. Acad. Sci. (USA) **46**, 197 (1960).

176. Hodge, A. J. and J. A. Petruska: Recent Studies with the Electron Microscope on Ordered Aggregates of the Tropocollagen Macromolecule. In: G. N. Ramachandran (Edit.), Aspects of Protein Structure, p. 289. London-New York: Academic Press. 1963.

177. Hodge, A. J., J. A. Petruska and A. J. Bailey: The Subunit Structure of the Tropocollagen Molecule and its Relation to Various Ordered Aggregation States. Conference on the Structure and Function of Connective and Skeletal Tissues. St. Andrews (Scotland), 1964.

178. Hodge, A. J. and F. O. Schmitt: Interaction Properties of Sonically Fragmented Collagen Macromolecules. Proc. Nat. Acad. Sci. (USA) **44**, 418 (1958).

179. — — The Charge Profile of the Tropocollagen Macromolecule and the Packing Arrangement in Native-Type Collagen Fibrils. Proc. Nat. Acad. Sci. (USA) **46**, 186 (1960).

180. Hofmann, U., Th. Nemetschek und W. Grassmann: Über die Querstreifung von Kollagenfibrillen und ihre Veränderung im Elektronenmikroskop. Z. Naturforsch. **7 b**, 509 (1952).

181. Hörmann, H.: Chemische Untersuchungen über die Kohlenhydratgruppierung des Kollagens. Leder **11**, 173 (1960).

181 a. — Die Kohlenhydrate des Bindegewebes. Beitr. Silikoseforsch., Sonderband II, 619 (1956).

182. — Zur Frage der Quervernetzung von Kollagen. Leder **13**, 79 (1962).

183. — Chemical Methods for the Conversion of Insoluble Collagen into Tropocollagen. Conference on the Structure and Function of Connective and Skeletal Tissues. St. Andrews (Scotland), 1964.

184. — Untersuchungen zur quantitativen Bestimmung von Carbonsäureestern mit Hydroxylamin. Monatsh. Chemie **96**, 37 (1965).

185. Hörmann, H., K. Dorfner und M. Jellinek: Unveröffentlicht.

186. Hörmann, H. und R. Hafter: Zur Fibrillenbildung des Kollagens. Leder **14**, 293 (1963).

187. Hörmann, H. and K. Th. Joseph: The Presence of Acetyl Groups in Collagen. Biochim. Biophys. Acta **100**, 598 (1965).

188. Hörmann, H., K. Th. Joseph und M. v. Wilm: Acetylierte Aminoendgruppen in den Untereinheiten der Peptidketten des Kollagens. Z. physiol. Chem. **341**, 284 (1965).

189. Hörmann, H., A. Riedel, Th. Altenschöpfer und M. Klenk: Chemische Veränderungen der Haut bei langdauerndem Äscher. Leder **12**, 175 (1961).

190. Hörmann, H. und M. v. Wilm: Kollagensegmente mit symmetrischer Querstreifung. Naturwiss. **51**, 464 (1964).

191. Hörmann, H., K. Th. Joseph und M. v. Wilm: Symp. Lyon 1965.

192. HOUCK, J. C.: Effect of Cortisol and Salicylate upon the Connective Tissue. Amer. J. Pathol. **41**, 365 (1962).

193. HOUCK, J. C., E. R. GOLDSTEIN and Y. M. PATEL: Collagenolytic Activity of the Connective Tissue. Conference on the Structure and Function of Connective and Skeletal Tissues. St. Andrews (Scotland), 1964.

194. HOUCK, J. C. and R. A. JACOB: Connective Tissue. VII. Factors Inhibiting the Dermal Chemical Response to Cortisol. Proc. Soc. Exp. Biol. Med. **113**, 692 (1963).

195. HUGGINS, M. L.: The Structure of Fibrous Proteins. Chem. Rev. **32**, 195 (1943).

196. — The Structure of Collagen. Proc. Nat. Acad. Sci. (USA) **43**, 209 (1957).

197. HURYCH, J.: Biosynthesis of Collagen. Hydroxylation of Proline in vitro. Collagen Currents 4, No. 5, p. 34 (1963).

198. — C. Sc. Dissert., Karlsuniversität Prag, 1965.

199. HURYCH, J. and M. CHVAPIL: Influence of Chelating Agents on the Biosynthesis of Collagen. Biochim. Biophys. Acta **97**, 361 (1965).

200. JACKSON, D. S.: Connective Tissue Growth Stimulated by Carrageenin. 1. The Formation and Removal of Collagen. Biochem. J. **65**, 277 (1957).

201. — Metabolic Studies: Turnover of α- and β-Chains. 141st Meeting Amer. Chem. Soc., Washington D. C. 1962, Abstr. p. 8C.

202. JACKSON, D. S., D. WATKINS and A. WINKLER: Formation of sRNA-Hydroxy-proline in Chick-Embryo and Wound Granulation Tissue. Biochim. Biophys. Acta **87**, 152 (1964).

203. JACKSON, S. F.: Structural Problems Associated with the Formation of Collagen Fibrils in vivo. In: R. E. TUNBRIDGE et al. (Edit.), Connective Tissue, p. 77. Oxford: Blackwell Sci. Publ. 1957.

204. JASIN, H. E., C. W. FINK, W. WISE and M. ZIFF: Relationship between Urinary Hydroxyproline and Growth. J. Clin. Investigation **41**, 1928 (1962).

205. JEFFREY, J. J. and G. R. MARTIN: Ascorbic Acid Dependent Synthesis of Collagen by Embryonic Chick Tibia Grown in Tissue Culture. VIth Intern. Congr. Biochem., New York 1964, Abstr. I—90.

206. JOSEPH, K. TH. and S. M. BOSE: N-Terminal Amino Acids of the Constituent Proteins of Hides and Skins. Bull. Central Leather Res. Inst. 4, No. 9 (1958).

207. JOSSE, J. and W. F. HARRINGTON: Role of Pyrrolidine Residues in the Structure and Stabilisation of Collagen. J. Mol. Biol. **9**, 269 (1964).

208. JUVA, K. and D. J. PROCKOP: Puromycin Inhibition of Collagen Synthesis as Evidence for Ribosomal or Post-ribosomal Site for the Hydroxylation of Proline. Biochim. Biophys. Acta **91**, 174 (1964).

208 a. KARNOVSKY, M. J. and M. L. KARNOVSKY: Metabolic Effects of Lathyrogenic Agents on Cartilage in vivo and in vitro. J. Exp. Medicine **113**, 381 (1961).

209. KATCHALSKI, E., A. BERGER and J. KURTZ: Behaviour in Solution of Polypeptides Related to Collagen. In: G. N. RAMACHANDRAN (Edit.), Aspects of Protein Structure, p. 205. London-New York: Academic Press. 1963.

210. KATCHALSKY, A. und A. OPLATKA: Persönliche Mitteilung.

211. KATZ, J. R.: Zur Polymorphie der hochmolekularen Substanzen. III. Die Mutarotation der Gelatine im Zusammenhang mit der Änderung des Röntgenspektrums beim Gelatinieren. Rec. trav. chim. Pays-Bas **51**, 835 (1932).

212. KEECH, M. K.: The Effect of Collagenase on Human Skin Collagen. Comparison of Different Age Groups and of Cases with and without Collagen Disease. Yale J. Biol. Med. **26**, 27 (1954).

213. KEECH, M. K. and R. REED: Enzymic Elucidation of the Relation between Collagen and Elastin: An Electron-microscopic Study. Ann. Rheum. Diseases **16**, 35 (1957).

214. Keil, B. and F. Šorm: On the Structure of Chymotrypsinogen A. 1st Meeting Fed. Europ. Biochem. Soc., London 1964, Abstr. D-5.

215. Klemperer, P., A. D. Pollack and G. Baehr: Pathology of Disseminated Lupus erythematodes. Amer. Med. Assoc. Arch. Pathol. **32**, 596 (1941).

216. Konno, K. and T. Tetsuka: The Isotopic Studies on the Conversion of Proline to Hydroxyproline. J. Biochemisty (Tokyo) **52**, 466 (1962).

216a. — — Studies on the Connective Tissue. III. Biosynthesis of Collagen by the Granuloma Tissue in vitro. J. Biochem. (Tokyo) **56**, 581 (1964).

217. Korting, G. W., H. Holzmann und K. Kühn: Biochemische Bindegewebsuntersuchungen in Analogie zum Sklerodermie-Problem. Med. Welt Nr. **34**, 1751 (1964).

218. Kowalewski, K. and M. A. Emery: Effect of Lathyrus Factor and of an Anabolic Steroid on Healing of Fractures in Rats, Studied by S^{35} Uptake Method. Acta Endocrinol. **34**, 317 (1960).

219. Kratky, O. und A. Sekora: Die Auffindung von großen Netzebenenabständen bei Känguruh-Schwanzsehne. Makromolek. Chem. [3] **1**, 113 (1944).

220. Kretsinger, R. H., G. Manner, B. S. Gould and A. Rich: Synthesis of Collagen on Polyribosomes. Nature **202**, 438 (1964).

221. Kroner, T. D., W. Tabroff and J. J. MacGarr: Peptides Isolated from a Partial Hydrolysate of Steer Hide Collagen. J. Amer. Chem. Soc. **75**, 4084 (1953).

222. — — — Peptides Isolated from a Partial Hydrolysate of Steer Hide Collagen. II. Evidence for the Prolyl-Hydroxyproline Linkage in Collagen. J. Amer. Chem. Soc. **77**, 3356 (1955).

223. Kühn, K.: Über die Ausbildung einer hochunterteilten Querstreifung des Kollagens nach Gerben mit Phosphorwolframsäure und mit basischen Chromsalzlösungen. Leder **9**, 217 (1958).

224. — Über den Ursprung des Querstreifungsmusters bei Kollagen. Leder **11**, 110 (1960).

225. — Die End-an-End-Verknüpfung der Tropokollagenmoleküle. Leder **13**, 86 (1962).

225a. — Investigations of the Renaturation of Tropocollagen. In: G. N. Ramachandran (Edit.), Aspects of Protein Structure, p. 279. London: Academic Press. 1963.

226. Kühn, K., M. Durruti, P. Iwangoff, F. Hammerstein, K. Stecher, H. Holzmann, und G. W. Korting: Untersuchungen über den Stoffwechsel des Kollagens, I. Der Einbau von [^{14}C]-Glycin in Kollagen bei lathyritischen Ratten. Z. physiol. Chem. **336**, 4 (1964).

226a. Kühn, K. und M. Eggl: Ein kollagenolytischer Effekt bei kombinierter Einwirkung von Pronase, Pepsin und Trypsin auf Kollagen (in Vorbereitung).

227. Kühn, K., J. Engel, B. Zimmermann and W. Grassmann: Renaturation of Soluble Collagen. III. Reorganization of Native Collagen Molecules from Completely Separated Units. Arch. Biochem. Biophys. **105**, 387 (1964).

228. Kühn, K. und P. Fietzek: Die Einwirkung von Proteasen auf unlösliches Kollagen (in Vorbereitung).

228a. Kühn, K., P. Fietzek und J. Kühn: Die Einwirkung von proteolytischen Enzymen auf unlösliches Kollagen. Naturwiss. **50**, 444 (1963).

229. Kühn, K. und E. Gebhardt: Chemische und elektronenoptische Untersuchungen über die Reaktion von Chrom(III)-komplexen mit Kollagen. Z. Naturforsch. **15b**, 23 (1960).

230. — — Chemische und elektronenoptische Untersuchungen über die Reaktion von basischen Chrom(III)-komplexen mit Kollagen. Leder **11**, 147 (1960).

231. KÜHN, K., W. GRASSMANN und U. HOFMANN: Über die Bildung der Phosphorwolframsäure in Kollagen. Naturwiss. **44**, 538 (1957).

232. — — — Die elektronenmikroskopische „Anfärbung" des Kollagens und die Ausbildung einer hochunterteilten Querstreifung. Z. Naturforsch. **13 b**, 154 (1958).

233. — — — Über die Bildung der Kollagenfibrillen aus gelöstem Kollagen und die Funktion der kohlenhydrathaltigen Begleitkomponenten. Z. Naturforsch. **14 b**, 436 (1959).

234. — — — Über den Aufbau der Kollagenfibrillen aus Tropokollagenmolekeln. Naturwiss. **47**, 258 (1960).

235. KÜHN, K., F. HAMMERSTEIN, K. STECHER, M. DURRUTI, H. HOLZMANN und G. W. KORTING: Untersuchungen über den Stoffwechsel des Kollagens. IV. Der Einbau von C^{14}-Glycin in Kollagen bei mit Resochin behandelten Ratten. (in Vorbereitung).

236. KÜHN, K., K. HANNIG und H. HÖRMANN: Die Einwirkung von Trypsin und Pepsin auf natives Kollagen. Leder **12**, 237 (1961).

237. KÜHN, K., U. HOFMANN und W. GRASSMANN: Über die Verteilung der basischen Aminosäuren in der Tropokollagenmolekel. Naturwiss. **46**, 512 (1959).

238. — — — Über die Verteilung der sauren Aminosäuren in der Tropokollagenmolekel. Naturwiss. **47**, 15 (1960).

239. KÜHN, K., U. HOFMANN, W. GRASSMANN und E. GEBHARDT: Veränderung der Intensität der Querstreifen des Kollagens durch Einlagerung von Phosphorwolframsäure. Naturwiss. **45**, 521 (1958).

240. KÜHN, K., H. HOLZMANN und G. W. KORTING: Quantitative Bestimmungen der löslichen Kollagenvorstufen nach Anwendung einiger Bindegewebstherapeutica. Naturwiss. **49**, 134 (1962).

241. KÜHN, K., P. IWANGOFF, F. HAMMERSTEIN, K. STECHER, M. DURRUTI, H. HOLZMANN und G. W. KORTING: Untersuchungen über den Stoffwechsel des Kollagens, II. Der Einbau von [^{14}C]-Glycin in Kollagen bei mit Prednison behandelten Ratten. Z. physiol. Chem. **337**, 249 (1964).

242. KÜHN, K., J. KÜHN und K. HANNIG: Die Einwirkung von Trypsin auf gelöstes Kollagen. Z. physiol. Chem. **326**, 50 (1961).

243. KÜHN, K., J. KÜHN und G. SCHUPPLER: Kollagenfibrillen mit anormalem Querstreifungsmuster. Naturwiss. **51**, 337 (1964).

244. KÜHN, K., G. SCHUPPLER und J. KÜHN: Veränderungen im Verhalten des Kollagens mit zunehmender Reinigung und nach Behandlung mit proteolytischen Enzymen. Z. physiol. Chem. **338**, 10 (1964).

245. KÜHN, K., K. STECHER, P. IWANGOFF, F. HAMMERSTEIN, M. DURRUTI, H. HOLZMANN and G. W. KORTING: Studies on the Metabolism of Collagen, III. The Incorporation of [^{14}C]-Glycine into the Collagen of Rats Treated with Progesterone. Biochem. Z. (in print).

246. KÜHN, K., CH. TKOCZ, B. ZIMMERMANN and G. BEIER: Long-Spacing-Segments from Renatured $\alpha 1$-Subunits of Collagen. Nature (in print).

247. KÜHN, K. und E. ZIMMER: Eigenschaften des Tropokollagenmoleküls und deren Bedeutung für die Fibrillenbildung. Z. Naturforsch. **16 b**, 648 (1961).

248. — — Über die Anordnung der Tropokollagenmolekeln in den Long-Spacing-Kollagenfibrillen. Naturwiss. **48**, 219 (1961).

249. — — Über eine neue Form der Long-Spacing-Fibrillen des Kollagens. Naturwiss. **48**, 220 (1961).

250. KÜHN, K., E. ZIMMER, P. WAYKOLE und P. FIETZEK: Die Wirkung von Alkali auf Kollagen. Die Änderung des Ladungsmusters der Tropokollagenmoleküle sowie die Spaltung der intra- und intermolekularen Bindungen. Z. physiol. Chem. **333**, 209 (1963).

251. Kühn, K. and B. Zimmermann: Renaturation of Soluble Collagen. IV. Regeneration of Native Collagen Molecules from α-Subunits. Arch. Biochem. Biophys. **109,** 421 (1965).

251a. Kühn, K., B. Zimmermann und Ch. Tkocz: Elektronenmikroskopische Beiträge zur chemischen Struktur des Kollagens. Sitz. Ber. Bayr. Akad. Wissensch. (4. Juli 1965).

252. Kühne, W.: Unters. physiol. chem. Inst. Heidelberg **1,** 219 (1877).

253. Küntzel, A. und F. Prakke: Morphologie und Feinbau der kollagenen Faser. Biochem. Z. **267,** 243 (1933).

254. Lakshmanan, B. R., C. Ramakrishnan, V. Sasisekharan and Y. T. Thathachari: X-Ray Diffraction Pattern of Collagen and the Fourier Transform of the Collagen Structure. In: N. Ramanathan (Edit.), Collagen, p. 117. London: Interscience Publ. 1962.

255. Lamport, D. T. A.: Oxygen Fixation into Hydroxyproline of Plant Cell Wall Protein. J. Biol. Chem. **238,** 1438 (1963).

256. — Hydroxyproline Biosynthesis: Loss of Hydrogen during the Hydroxylation of Proline. Nature **202,** 293 (1964).

257. Levene, C. I. and J. Gross: Alterations in State of Molecular Aggregation of Collagen Induced in Chick Embryos by β-Aminopropionitrile (Lathyrus-Factor). J. Exp. Medicine **110,** 771 (1959).

258. Levine, M.: Conversion of Proline to Hydroxyproline in Rat Skin, in vitro. 46th Annu. Meeting, Federat. Amer. Soc. Exp. Biol., Atlantic City 1962, Abstr. No. 120.

259. Levy, M., L. Fishman and G. Cabrera: ε-Amino Group of Lysine, a Branch Point in Collagen Structure. Federat. Proc. (Amer. Soc. Exp. Biol.) **19,** 343 (1960).

260. Lifson, S. and A. Roig: On the Theory of Helix-Coil Transition in Polypeptides. J. Chem. Phys. **34,** 1963 (1961).

261. Lowther, D. A., N. M. Green and J. A. Chapman: Morphological and Chemical Studies of Collagen Formation. II. Metabolic Activity of Collagen Associated with Subcellular Fractions of Guinea Pig Granulomata. J. Biophys. Biochem. Cytol. **10,** 373 (1961).

262. Lucas, F., J. T. B. Shaw and S. G. Smith: The Amino Acid Sequence in a Fraction of the Fibroin of *Bombyx mori.* Biochem. J. **66,** 468 (1957).

262a. Lukens, L. N.: Evidence for the Nature of the Precursor That is Hydroxylated during the Biosynthesis of Collagen Hydroxyproline. J. Biol. Chem. **240,** 1661 (1965).

263. McEwen, M. B., M. I. Pratt and J. Randall: Scattering of Light by Collagen Solutions. In: Nature and Structure of Collagen, p. 158. London: Butterworth Sci. Publ. 1953.

264. Malt, R. A. and P. T. Speakman: Ribosomal Aggregates Associated with the Production of Collagen. Life Science **3,** 81 (1964).

265. Manahan, J. and I. Mandl: Peptides Isolated from Collagenase-Collagen Digests. Biochem. Biophys. Res. Comm. **4,** 368 (1961).

266. Mandelkern, L. and W. E. Stewart: The Effect of Neutral Salts on the Melting Temperature and Regeneration Kinetics of the Ordered Collagen Structure. Biochemistry **3,** 1135 (1964).

267. Mandl, I.: Privatmitteilung.

268. — Collagenases and Elastases. Adv. Enzymology **23,** 163 (1961).

269. Mandl, I., L. T. Ferguson and S. F. Zaffuto: Exopeptidases of *Clostridium histolyticum.* Arch. Biochem. Biophys. **69,** 565 (1957).

270. Mandl, I., S. Keller and B. Cohen: Microbial Elastases. A Comparative Study. Proc. Soc. Exp. Biol. Med. **109,** 923 (1962).

271. MANNER, G. and B. S. GOULD: Collagen Biosynthesis. The Formation of Hydroxyproline and Soluble-Ribonucleic Acid Complexes of Proline and Hydroxyproline by Chick Embryo in vivo and by a Subcellular Chick Embryo System. Biochim. Biophys. Acta 72, 243 (1963).

272. MANNER, G., B. S. GOULD, A. LEBI and A. SCHANTZ: Collagen Biosynthesis. Formation of an sRNA-Hydroxylysine Complex. 47th Annu. Meeting Federat. Amer. Soc. Exp. Biol., Atlantic City 1963, Abstr. 411.

273. MARGOLIASH, E.: Primary Structure and Evolution of Cytochrome c. Proc. Nat. Acad. Sci. (USA) 50, 672 (1963).

273a. MARTIN, G. R. and K. A. PIEZ: Biosynthesis and Maturation of Elastin. Collagen Symp. Somerville, N. J.: Ethicon, Inc., 1965.

274. MARTIN, G. R., K. A. PIEZ and M. S. LEWIS: The Incorporation of [^{14}C]-Glycine into the Subunits of Collagens from Normal and Lathyritic Animals. Biochim. Biophys. Acta 69, 472 (1963).

275. MASER, M. D. and R. V. RICE: Biophysical and Biochemical Properties of Earthworm-Cuticle Collagen. Biochim. Biophys. Acta 63, 255 (1962).

276. — — Soluble Earthworm Cuticle Collagen: A Possible Dimer of Tropocollagen. J. Cell Biol. 18, 569 (1963).

277. MECHANIC, G. L. and M. LEVY: An ε-Lysine Tripeptide Obtained from Collagen. J. Amer. Chem. Soc. 81, 1889 (1959).

278. MEHL, J. W., J. L. ONCLEY and R. SIMHA: Viscosity and the Shape of Protein Molecules. Science 92, 132 (1940).

279. MEYER, A. und F. VERZÁR: Altersveränderung der Hydroxyprolin-Abgabe bei der thermischen Kontraktion von Kollagenfasern. Gerontologia 3, 184 (1959).

280. MEYER, K., E. DAVIDSON, A. LINKER and P. HOFFMAN: The Acid Mucopolysaccharides of Connective Tissue. Biochim. Biophys. Acta 21, 506 (1956).

281. MEYER, K., M. M. GRUMBACH, A. LINKER and P. HOFFMAN: Excretion of Sulfated Mucopolysaccharides in Gargoylism (Hurler's Syndrom). Proc. Soc. Exp. Biol. Med. 97, 275 (1958).

282. MICHAELS, S., P. M. GALLOP, S. SEIFTER and E. MEILMAN: Studies on the Specificity of Collagenase. Biochim. Biophys. Acta 29, 451 (1958).

283. MILLIONOVA, M. I. and N. S. ANDREEVA: A Model of the Chain Configuration in Poly-(glycyl-L-proline). Biophysics (Biofizika) 3, 250 (1958).

284. — — On the Structure of Molecular Chain of Collagen. — The Structure of a Glycyl-L-proline Polymer. Biophysics (Biofizika) 2, 294 (1957); 4, No 3, 138 (1959).

285. MITOMA, C., T. E. SMITH, F. FRIEDBERG and C. R. RAYFORD: Incorporation of Hydroxyproline into Tissue Proteins by Chick-Embryos. J. Biol. Chem. 234, 78 (1959).

285a. MORRIONE, T. G. and S. SEIFTER: Alteration in the Collagen Content of the Human Uterus during Pregnancy and Post Partum Involution. J. Exp. Medicine 115, 357 (1962).

286. Moss, J. A.: The Carbohydrate of Collagen. Biochem. J. 61, 151 (1955).

287. MUIR, H.: Chemistry and Metabolism of Connective Tissue Glycosaminglycans (MPS). Intern. Rev. Conn. Tiss. Res. 2, 101 (1964).

287a. MÜLLER, W.: Die Serologie der chronischen Polyarthritis. Berlin: Springer-Verlag. 1962.

288. NAGAI, Y.: Collagenase Digestion of Collagen. J. Biochemistry (Tokyo) 50, 486 (1961).

289. NAGAI, Y. and H. NODA: The Specificity of Collagenase. Biochim. Biophys. Acta 34, 298 (1959).

290. Nagai, Y., S. Sakakibara, H. Noda and S. Akabori: Hydrolysis of Synthetic Peptides by Collagenase. Biochim. Biophys. Acta **37**, 567 (1960).

291. Nagai, Y., S. Yagisawa and H. Noda: Hydrolysis of Peptides Containing Sarcosine and Alanine by Collagenase. J. Biochemistry (Tokyo) **51**, 382 (1962).

292. Nagamani, M. R.: Darstellung und Isolierung von Untereinheiten des Kollagens nach milder Hitzedenaturierung und alkalischer Behandlung. Diplomarbeit, Univ. München, 1963; Diss. 1965.

293. Nageotte, J.: Coagulation fibrillaire in vitro du collagène dissous dans un acide dilué. C. R. hebd. séances Acad. Sci. **184**, 115 (1927).

294. Nemetschek, Th., W. Grassmann und U. Hofmann: Über die hochunterteilte Querstreifung des Kollagens. Z. Naturforsch. **10 b**, 61 (1955).

295. Neuberger, A., J. C. Perrone and H. G. B. Slack: The Relative Metabolic Inertia of Tendon Collagen in the Rat. Biochem. J. **49**, 199 (1951).

296. Neuman, R. E. and M. A. Logan: The Determination of Collagen and Elastin in Tissues. J. Biol. Chem. **186**, 549 (1950).

297. — — The Determination of Hydroxyproline. J. Biol. Chem. **184**, 299 (1950).

298. Nishigai, M., Y. Nagai and H. Noda: Partial Collagenase Digestion of the Fiber Structure of Collagen. J. Biochemistry (Tokyo) **48**, 152 (1960).

299. Nishihara, T.: Solubilization of Insoluble Collagen Fibers and Reconstitution Thereof. U. S.-Patent 3,034,852 (1962).

300. Nishihara, T. and T. Miyata: The Effects of Proteases on the Soluble and Insoluble Collagen and the Structure of the Insoluble Collagen Fiber. Collagen Sympos. **3**, 66 (1962).

301. Noda, H. and R. W. G Wykoff: The Electron Microscopy of Reprecipitated Collagen. Biochim. Biophys. Acta **7**, 494 (1951).

302. Nordwig, A.: Kollagenase und ihre Wirkung. Leder **13**, 10 (1962).

303. Nordwig, A., G. Beier und Y. P. Dick: Unveröffentlicht.

304. Nordwig, A. and Y. P. Dick: Cleavage with Cyanogen Bromide of Methionyl Bonds in Collagen. Biochim. Biophys. Acta **97**, 179 (1965).

305. Nordwig, A., H. Hörmann, K. Kühn und W. Grassmann: Weitere Versuche zum Abbau des Kollagens durch Kollagenase. Aminosäuresequenzen des Kollagens, IV. Z. physiol. Chem. **325**, 242 (1961).

306. Oakley, C. L. and N. G. Banerjee: Bacterial Elastases. J. Pathol. Bacteriol. **85**, 489 (1963).

308. Ogle, J. D., R. B. Arlinghaus and M. A. Logan: Studies on Peptides Obtained from Enzymic Digests of Collagen with Evidence for the Presence of an Unidentified Compound in this Protein. Arch. Biochem. Biophys. **94**, 85 (1961).

309. — — — 3-Hydroxyproline, a New Amino Acid of Collagen. J. Biol. Chem. **237**, 3667 (1962).

310. Orekhovich, V. N. and V. O. Shpikiter: Physicochemical Nature of Procollagen. In: G. Stainsby (Edit.), Recent Advances in Gelatin and Glue Research, p. 87. London: Pergamon Press. 1958.

311. — — Procollagens. Science **127**, 137 (1958).

312. Orekhovich, V. N., V. O. Shpikiter, V. I. Mazourov et O. V. Kounina: Procollagènes. Classification, métabolisme, action des protéinases. Bull. soc. chim. biol. (Paris) **42**, 505 (1960).

313. Orekhovich, V. N., A. A. Tustanovski, K. D. Orekhovich and N. E. Plotnikova: Procollagen of Hide. Biokhimiya (USSR) **13**, 55 (1948).

314. Orekhovich, V. N., A. A. Tustanovski and N. E. Plotnikova: C. R. (Doklady) Acad. Sci. (USSR) **60**, 837 (1954).

314a. Partridge, S. M.: The Constitution and Biosynthesis of the Cross Linkages in Elastin. Symp. Internat. Biochim. Physiol. Tissu Conjonctif. Lyon, 1965.

315. PAULING, L. and R. B. COREY: The Pleated Sheet, a new Layer Configuration of Polypeptide Chains. Proc. Nat. Acad. Sci. (USA) **37**, 251 (1951).

316. — — The Structure of Fibrous Proteins of the Collagen-Gelatin Group. Proc. Nat. Acad. Sci. (USA) **37**, 272 (1951).

317. PAULING, L., R. B. COREY and H. R. BRANSON: The Structure of Proteins: Hydrogen-bonded Helical Configurations of the Polypeptide Chain. Proc. Nat. Acad. Sci. (USA) **37**, 205 (1951).

318. PETERKOFSKY, B. and S. UDENFRIEND: Conversion of Proline to Collagen Hydroxyproline in a Cell-free System from Chick Embryo. J. Biol. Chem. **238**, 3966 (1963).

318a. — — Enzymatic Hydroxylation of Proline in Microsomal Polypeptide Leading to Formation of Collagen. Proc. Nat. Acad. Sci. (USA) **53**, 335 (1965).

319. PETERLIN, A.: Light Scattering and Small Angle X-Ray Scattering by Macromolecular Coils with Finite Persistence Lenght. J. Polymer Sci. **47**, 403 (1960).

320. PIEZ, K. A.: Nonidentity of the Three α-Chains in Codfish Skin Collagen. J. Biol. Chem. **239**, PC 4315 (1964).

320a. PIEZ, K. A. and A. L. CARRILLO: Helix Formation by Single- and Double-Chain Gelatins from Rat Skin Collagen. Biochemistry **3**, 908 (1964).

321. PIEZ, K. A., E. A. EIGNER and M. S. LEWIS: The Chromatographic Separation and Amino Acid Composition of the Subunits of Several Collagens. Biochemistry **2**, 58 (1963).

322. PIEZ, K. A. and J. GROSS: The Amino Acid Composition and Morphology of Some Invertebrate and Vertebrate Collagens. Biochim. Biophys. Acta **34**, 24 (1959).

323. PIEZ, K. A., M. S. LEWIS, G. R. MARTIN and J. GROSS: Subunits of the Collagen Molecule. Biochim. Biophys. Acta **53**, 596 (1961).

324. PETRUSKA, J. A. and A. J. HODGE: A Subunit Model for the Tropocollagen Macromolecule. Proc. Nat. Acad. Sci. (USA) **51**, 871 (1964).

325. POPENOE, E. A. and D. D. VAN SLYKE: The Formation of Collagen Hydroxylysine. J. Biol. Chem. **237**, 3491 (1962).

326. PROCKOP, D. J.: Synthesis of Collagen. Nature **194**, 477 (1962).

327. PROCKOP, D. J., P. S. EBERT and B. M. SHAPIRO: Studies with Proline-3,4-H³ on the Hydroxylation of Proline during Collagen Synthesis in Chick Embryos. Arch. Biochem. Biophys. **106**, 112 (1964).

328. PROCKOP, D. J. and K. JUVA: Hydroxylation of Proline in Particulate Fractions from Cartilage. Biochem. Biophys. Res. Comm. **18**, 54 (1965).

328a. — — Synthesis of Hydroxyproline in vitro by the Hydroxylation of Proline in a Precursor of Collagen. Proc. Nat. Acad. Sci. (USA) **53**, 661 (1965).

329. PROCKOP, D. (J.), A. KAPLAN and S. UDENFRIEND: Oxygen-18 Studies on the Conversion of Proline to Collagen Hydroxyproline. Arch. Biochem. Biophys. **101**, 499 (1963).

330. PROCKOP, D. J., B. PETERKOFSKY and S. UDENFRIEND: Studies on the Intracellular Localization of Collagen Synthesis in the Intact Chick Embryo. J. Biol. Chem. **237**, 1581 (1962).

331. PROCKOP, D. J., O. PETTENGILL and H. HOLTZER: Incorporation of Sulfate and the Synthesis of Collagen by Cultures of Embryonic Chondrocytes. Biochim. Biophys. Acta **83**, 189 (1964).

332. PROCKOP, D. J. and S. UDENFRIEND: A Specific Method for the Analysis of Hydroxyproline in Tissues and Urine. Analyt. Biochem. **1**, 228 (1960).

333. RAMACHANDRAN, G. N.: Structure of Collagen. Nature **177**, 710 (1956).

334. RAMACHANDRAN, G. N. and G. KARTHA: Structure of Collagen. Nature **174**, 269 (1954).

335. — — Structure of Collagen. Nature **176**, 593 (1955).

336. Ramachandran, G. N. and V. Sasisekharan: Structure of Collagen. Nature **190,** 1004 (1961).

337. Ramachandran, G. N., V. Sasisekharan and Y. T. Thathachari: Structure of Collagen at the Molecular Level. In: N. Ramanathan (Edit.), Collagen, p. 81. London: Interscience Publ. 1962.

338. Randall, J. T.: Observations on the Collagen System. J. Soc. Leather Trades' Chem. **38,** 362 (1954).

339. Randall, J. T., R. D. B. Fraser, S. F. Jackson, A. V. M. Martin and A. C. T. North: Aspects of Collagen Structure. Nature **169,** 1029 (1952).

340. Randall, J. T., R. D. B. Fraser and A. C. T. North: The Structure of Collagen. Proc. Roy. Soc. (London) **B 141,** 62 (1953).

341. Rice, R. V.: Reappearance of Certain Structural Features of Native Collagen after Thermal Transformation. Proc. Nat. Acad. Sci. (USA) **46,** 1186 (1960).

342. Rice, R. V., E. F. Casassa, R. E. Kerwin and M. D. Maser: On the Length and Molecular Weight of Tropocollagen from Calf Skin. Arch. Biochem. Biophys. **105,** 409 (1964).

343. Rich, A. and F. H. C. Crick: The Structure of Collagen. Nature **176,** 915 (1955).

344. — — The Structure of Collagen. In: G. Stainsby (Edit.), Recent Advances in Gelatin and Glue Research, p. 20. London: Pergamon Press. 1958.

345. — — The Molecular Structure of Collagen. J. Mol. Biol. **3,** 483 (1961).

346. Rigby, B. J.: Thermal Transitions in Normal and Deuterated Rat Tail Tendon, Human Skin, and Tuna-fish Skin. Biochim. Biophys. Acta **62,** 183 (1962).

347. Robertson, W. van B.: The Biochemical Rôle of Ascorbic Acid in Connective Tissue. Ann. New York Acad. Sci. **92,** 159 (1961).

348. — D-Ascorbic Acid and Collagen Synthesis. Biochim. Biophys. Acta **74,** 137 (1963).

349. Robertson, W. van B. and B. Schwartz: Ascorbic Acid and the Formation of Collagen. J. Biol. Chem. **201,** 689 (1953).

350. Rogulenkova, V. N., M. I. Millionova and N. S. Andreeva: On the Close Structural Similarity between poly-Gly-*L*-Pro-*L*-Hypro and Collagen. J. Mol. Biol. **9,** 253 (1964).

351. Romeis, B.: Mikroskopische Technik, S. 167. München: Leibnitz Verl. 1948.

352. Ross, R. and E. P. Benditt: A Comparison of the Utilization of Proline-H[3] in Healing Wounds as Seen by Electron Microscope-Autoradiography. 47th Annu. Meeting Federat. Amer. Soc. Exp. Biol., Atlantic City, 1963, Abstr. No. 178.

352a. — — Altered Configuration of Ribosomes in Scorbutic Fibroblasts. Federat. Proc. (Amer. Soc. Exp. Biol.) **23,** 441 (1964).

353. Rubin, A. L., D. Pfahl, P. T. Speakman, P. F. Davison and F. O. Schmitt: Tropocollagen: Significance of Protease-Induced Alterations. Science **139,** 37 (1963).

354. Sanger, F.: The Terminal Peptides of Insulin. Biochem. J. **45,** 563 (1949).

355. Scharpensel, H. W. und G. Wolf: In vitro-Kollagen-Synthese in foetalen Hautpräparaten. Z. Tierphysiol. Tierernähr. Futtermittelk. **14,** 347 (1959).

356. Schaub, M. C.: Degradation of Young and Old Collagen by Extracts of Various Organs. Gerontologia **9,** 52 (1964).

356a. — A Collagen Degrading Mechanism in the Rat Uterus During Post-Partum Involution. Helv. Physiol. Pharm. Acta **21,** 227 (1963).

357. Scheraga, H. A. and L. Mandelkern: Consideration of the Hydrodynamic Properties of Proteins. J. Amer. Chem. Soc. **75,** 179 (1953).

358. SCHERRER, P.: Zsigmondys Lehrbuch der Kolloidchemie, 3. Auflage, S. 408. Leipzig: O. Spanner. 1920.

359. SCHLEYER, M.: Trennung der α- und β-Komponenten aus Tropokollagen. Aminosäuresequenzen des Kollagens, V. Z. physiol. Chem. **329**, 97 (1962).

360. SCHMITT, F. O.: Macromolecular Interaction Patterns in Biological Systems. Proc. Amer. Philos. Soc. **100**, 476 (1956).

361. — Ultrastructure and the Problem of Cellular Organization. Harvey Lect. **40**, 249 (1944—45).

362. SCHMITT, F. O. and J. GROSS: Further Progress in the Electron Microscopy of Collagen. J. Amer. Leather Chem. Assoc. **43**, 658 (1948).

363. SCHMITT, F. O., J. GROSS and J. H. HIGHBERGER: A New Particle Type in Certain Connective Tissue Extracts. Proc. Nat. Acad. Sci. (USA) **39**, 459 (1953).

364. — — — States of Aggregation of Collagen. Sympos. Soc. Exp. Biol. **9**, 148 (1955).

365. SCHMITT, F. O., C. E. HALL and M. A. JAKUS: Electron Microscope Investigations of the Structure of Collagen. J. Cell. Comp. Physiol. **20**, 11 (1942).

366. — — — Fine Structure in the Fiber-Axis Macroperiod of Collagen Fibrils. J. Appl. Phys. **16**, 263 (1945).

367. SCHNEIDER, W. und HJ. STAUDINGER: Zum Wirkungsmechanismus von Vitamin C. Klin. Wschr. **42**, 879 (1964).

368. SCHROEDER, W. A., L. HONNEN and F. C. GREEN: Chromatographic Separation and Identification of Some Peptides in Partial Hydrolysates of Gelatin. Proc. Nat. Acad. Sci. (USA) **39**, 23 (1953).

369. SCHROEDER, W. A., L. M. KAY, J. LeGETTE, L. HONNEN and F. C. GREEN: The Constitution of Gelatin. Separation and Estimation of Peptides in Partial Hydrolysates. J. Amer. Chem. Soc. **76**, 3556 (1954).

370. SCHROHENLOHER, R. E., J. D. OGLE and M. A. LOGAN: Two Tripeptides from an Enzymatic Digest of Collagen. J. Biol. Chem. **234**, 58 (1959).

371. SCHWARZ, W. und H. J. MERKER: Die Fibrillogenese in verschiedenen Bindegewebsformen des menschlichen Embryos. Beitr. Silikose-Forsch., Sonderband **4**, 231 (1960).

372. SEIFTER, S., P. M. GALLOP and C. FRANZBLAU: Some Aspects of Collagenolytic Action. Trans. New York Acad. Sci. **23**, 540 (1961).

373. SEIFTER, S., P. M. GALLOP, L. KLEIN and E. MEILMAN: Studies on Collagen. II. Properties of Purified Collagenase and its Inhibition. J. Biol. Chem. **234**, 285 (1959).

374. SHELDON, H. and F. B. KIMBALL: Studies on Cartilage. III. The Occurrence of Collagen within Vacuoles of the Golgi Apparatus. J. Cell Biol. **12**, 599 (1962).

374a. SHINTANI, Y. K. and H. E. TAYLOR: Radioautographic Study of $S^{35}O_4$ Uptake by Epiphyseal Cartilage in Experimental Lathyrism. Canad. J. Biochem. Physiol. **40**, 565 (1962).

375. SINEX, F. M., D. D. VAN SLYKE and D. R. CHRISTMAN: The Source and State of the Hydroxylysine of Collagen. J. Biol. Chem. **234**, 918 (1959).

376. SIUKO, H., J. SÄVELÄ and E. KULONEN: Effect of the Hydrocortisone on the Formation of Collagen in Guinea Pig Skin. Acta Endocrinol. **31**, 113 (1959).

377. SMILEY, J. D., H. YEAGER and M. ZIFF: Collagen Metabolism in Osteolathyrism in Chick Embryos: Site of Action of β-Aminopropionitrile. J. Exp. Medicine **116**, 45 (1962).

378. SMITH, D. J. and R. C. SHUSTER: Biochemistry of Lathyrism. I. Collagen Biosynthesis in Normal and Lathyritic Chick Embryos. Arch. Biochem. Biophys. **98**, 498 (1962).

379. SMITH, E. L.: Dipeptidases. In: S. P. COLOWICK and N. O. KAPLAN (Edit.). Methods in Enzymology, Vol. II, p. 93. New York: Academic Press. 1955.

380. SNELLMAN, O.: A Glycoprotein from Reticulin Tissue. — Studies on a Matrix for the Biosynthesis of Collagen. — Biosynthesis of Collagen Performed in vitro with the Aid of a Matrix Occurring in Reticulin. Acta Chem. Scand. 17, 1049, 1057, 1062 (1963).

381. SOBEL, A. E.: Multiple Mechanism of Nucleation. Proc., 2nd Europ. Sympos. on Calcified Tissues, Lüttich 1964.

382. SOLOMONS, C. C. and J. T. IRVING: The Reaction of some Hard- and Soft-Tissue Collagens with 1-Fluoro-2,4-Dinitrobenzene. Biochem. J. 68, 499 (1958).

383. STACY, K. A.: Light Scattering in Physical Chemistry. London: Butterworth Sci. Publ. 1956.

384. STADLER, P. und W. TRÄSCH: Fluoreszenzfärbungen an tierischem Hautmaterial. Leder 10, 213 (1959).

385. STEGEMANN, H.: Mikrobestimmung von Hydroxyprolin mit Chloramin T und p-Dimethylaminobenzaldehyd. Z. physiol. Chem. 311, 41 (1958).

386. STETTEN, M. R.: Some Aspects of the Metabolism of Hydroxyproline, Studied with the Aid of Isotopic Nitrogen. J. Biol. Chem. 181, 31 (1949).

387. STEVEN, F. S. and G. R. TRISTRAM: The Origin of N-Terminal Residues in Acetic Acid-Soluble Calf-Skin Collagen. Biochem. J. 83, 245 (1962).

387a. STRAUCH, L. und W. GRASSMANN: Reinigung von Kollagenase, III. Gewinnung enzymatisch einheitlicher Kollagenase. Z. physiol. Chem. (im Druck).

388. TAKAHUSHI, T. and W. YOKOYAMA: Physico-Chemical Studies on the Skin and Leather of Marine Animals. 12. The Content of Hydroxyproline in the Collagen of Different Fish Skins. Bull. Jap. Soc. Scient. Fisheries 20, 525 (1954).

389. TAYLOR, E. W. and W. CRAMER: Birefringence of Protein Solutions, and Biological Systems. I.; II. Studies on TMV, Tropocollagen, and Paramyosin. Biophys. J. 3, 128, 143 (1963).

390. THOMAS, J., D. F. ELSDEN and S. M. PARTRIDGE: Partial Structure of Two Major Degradation Products from the Cross-linkages in Elastin. Nature 200, 651 (1963).

391. TRISTRAM, G. R.: The Amino Acid Composition of Proteins. In: H. NEURATH and K. BAILEY (Edit.), The Proteins, Vol. I, Part A, p. 181. New York: Academic Press. 1953.

392. TROMANS, W. J., R. W. HORNE, G. A. GRESHAM and A. J. BAILEY: Electron Microscope Studies on the Structure of Collagen Fibrils by Negative Staining. Z. Zellforsch. 58, 798 (1963).

392a. URIVETZKY, M., J. M. FREI and E. MEILMAN: Cell-Free Collagen Biosynthesis and the Hydroxylation of sRNA-Proline. Arch. Biochem. Biophys. 109, 480 (1965).

393. VEIS, A., J. ANESEY and J. COHEN: The Long Range Reorganization of Gelatin to the Collagen Structure. Arch. Biochem. Biophys. 94, 20 (1961).

394. VERZÁR, F.: Veränderungen der thermoelastischen Kontraktion von Sehnenfasern im Alter. Helv. Physiol. Pharmacol. Acta 13, C 64 (1955).

395. — Das Altern des Kollagens. Helv. Physiol. Pharmacol. Acta 14, 207 (1956).

396. — Nachweis der Zunahme der Bindung von Hydroxyprolinen im Kollagen der Haut mit dem Alter. Gerontologia 4, 105 (1960).

397. — Aging of the Collagen Fiber. In: D. A. HALL (Edit.), International Revue of Conn. Tiss. Res. Vol. II, p. 243. New York: Academic Press. 1964. Vgl. dazu Übersicht von VERZÁR.

398. — Differenzierung verschiedener Vernetzungen (Crosslinks) des Kollagens durch Spannungsmessungen. Z. physiol. Chem. 335, 38 (1963).

399. VERZÁR, F. and K. HUBER: Thermic-Contraction of Single Tendon Fibres from Animals of Different Age after Treatment with Formaldehyde, Urethane, Glycerol, Acetic Acid and Other Substances. Gerontologia **2**, 81 (1958).

400. VOGEL, G. und L. THER: Untersuchungen über den Einfluß endokriner Faktoren auf die Reißfestigkeit bindegewebiger Strukturen. Naunyn Schmiedeberg's Arch. exp. Pathol. Pharmakol. **246**, 72 (1963).

400a. WALDSCHMIDT-LEITZ, E. und O. ZEISS: Über Protofibroin, die kristalline Hauptkomponente der Seidenfaser. Z. physiol. Chem. **300**, 49 (1955).

401. WALSH, K. A., D. L. KAUFFMAN, K. S. V. SAMPATH KUMAR and H. NEURATH: On the Structure and Function of Bovine Trypsinogen and Trypsin. Proc. Nat. Acad. Sci. (USA) **51**, 301 (1964).

402. WATSON, J. D.: Die Beteiligung der Ribonucleinsäure an der Proteinsynthese. Angew. Chem. **75**, 439 (1963).

403. WATSON, M. R.: The Chemical Composition of Earthworm Cuticle. Biochem. J. **68**, 416 (1958).

404. WATSON, M. R. and N. R. SILVESTER: Studies of Invertebrate Collagen Preparations. Biochem. J. **71**, 578 (1959).

405. WEIGERT, C.: Über eine Methode zur Färbung elastischer Fasern. Zbl. Pathol. **9**, 289 (1898).

406. WIEDERHORN, N. M. and G. V. REARDON: Studies Concerned with the Structure of Collagen. II. Stress-Strain Behavior of Thermally Contracted Collagen. J. Polymer Sci. **9**, 315 (1952).

407. WILLIAMS, A. P.: The Chemical Composition of Snail Gelatin. Biochem. J. **74**, 304 (1960).

408. WILM, M. v.: Chemische Veränderungen an den funktionellen Gruppen des Kollagens. Dissert., Univ. München, 1965.

409. WOESSNER, J. F., Jr.: Catabolism of Collagen and Non-collagen Protein in the Rat Uterus during Post-partum Involution. Biochem. J. **83**, 304 (1962).

410. — The Determination of Hydroxyproline in Tissue and Protein Samples Containing Small Proportions of this Imino Acid. Arch. Biochem. Biophys. **93**, 440 (1961).

410a. WOESSNER, J. F., Jr. and T. H. BREWER: Formation and Breakdown of Collagen and Elastin in the Human Uterus during Pregnancy and Post-Partum Involution. Biochem. J. **89**, 75 (1963).

411. WOLF, G., and C. R. A. BERGER: The Metabolism of Hydroxyproline in the Intact Rat. Incorporation of Hydroxyproline into Protein and Urinary Metabolites. J. Biol. Chem. **230**, 231 (1958).

412. WOLPERS, C.: Kollagenquerstreifung und Grundsubstanz. Klin. Wschr. **22**, 624 (1943).

413. — Zur elektronenmikroskopischen Darstellung elastischer Gewebselemente. Klin. Wschr. **23**, 169 (1944).

414. — Das Scheiben- und das Lamellenstadium der Kollagenquerstreifung. Makromolek. Chem. **2**, 37 (1948).

415. — Kollagenquerstreifung und Hitzeschrumpfung. Biochem. Z. **318**, 373 (1948).

416. WOOD, G. C.: 4th Internat. Congr. Biochem., Vienna, 1958, Abstr. No. 2—92, p. 26.

417. — The Formation of Fibrils from Collagen Solutions. 2. A Mechanism of Collagen-Fibril Formation. Biochem. J. **75**, 598 (1960).

418. — The Formation of Fibrils from Collagen Solutions. 3. Effect of Chondroitin Sulphate and Some other Naturally Occuring Polyanions on the Rate of Formation. Biochem. J. **75**, 605 (1960).

419. Wood, G. C.: The Heterogeneity of Collagen Solutions. Biochem. J. **82**, 2 P (1962).

420. — The Heterogeneity of Collagen Solutions and its Effect on Fibril Formation. Biochem. J. **84**, 429 (1962).

421. Wood, G. C. and M. K. Keech: The Formation of Fibrils from Collagen Solutions. 1. The Effect of Experimental Conditions: Kinetic and Electron-Microscope Studies. Biochem. J. **75**, 588 (1960).

422. Wünsch, E. und H. G. Heidrich: Darstellung von Prolinpeptiden, III. Ein neues Substrat zur Bestimmung der Kollagenase. Z. physiol. Chem. **332**, 300 (1963).

423. Wuthier, R. E., P. Grøn and J. T. Irving: The Reaction of 1-Fluoro-2,4-dinitrobenzene with Bone. Studies on the Relationship between Bone Collagen and Apatite. Biochem. J. **92**, 205 (1964).

424. Yamazaki, I., H. S. Mason and L. (H.) Piette: Identification by Electron Paramagnetic Resonance Spectroscopy, of Free Radicals Generated from Substrates by Peroxidase. J. Biol. Chem. **235**, 2444 (1960).

425. Yamazaki, I. and L. H. Piette: Mechanism of Free Radical Formation and Disappearance during the Ascorbic Acid Oxidase and Peroxidase Reactions. Biochim. Biophys. Acta **50**, 62 (1961).

425a. Yip, C. C.: The Hydroxylation of Proline by Horseradish Peroxidase. Biochim. Biophys. Acta **92**, 395 (1964).

426. Yonath, J., A. Oplatka and A. Katchalsky: Equilibrium Mechanochemistry of Collagen Fibers. Conference on the Structure and Function of Connective and Skeletal Tissues. St. Andrews (Scotland), 1964.

427. Young, E. G. and J. W. Lorimer: A Comparison of the Acid-Soluble Collagens from the Skin and Swim Bladder of the Cod. Arch. Biochem. Biophys. **92**, 183 (1961).

428. Yphantis, D. A.: Rapid Determination of Molecular Weights of Peptides and Proteins. Ann. New York Acad. Sci. **88**, 586 (1960).

429. Zimm, B. H. and J. K. Bragg: Theory of Phase Transition between Helix and Random Coil in Polypeptide Chains. J. Chem. Phys. **31**, 526 (1959).

(Eingelaufen am 6. April 1965.)

Some Applications
of Nuclear Magnetic Resonance Spectroscopy
in Natural Product Chemistry.

By L. M. JACKMAN, Melbourne.

With 17 Figures.

Contents.

I. Introduction.

In the last decade, the organic chemist's approach to Natural Product Chemistry has undergone a considerable change. It is now apparent that the structures of the majority of natural products can be determined absolutely and in a surprisingly short time by the method of X-ray crystallography. This method is however expensive, and at present it is not feasible to determine the structures of many natural products, particularly simple ones, by this method. Even so, the organic chemist is now conscious that the expenditure of several years on the determination of a structure of a natural product is no longer justified. Fortunately, recent developments in various spectroscopic methods now enable the organic chemist to solve structures much more efficiently than in the past.

For a number of years organic chemists have relied to an increasing extent on infrared, ultraviolet and visible spectroscopy in structural work. More recently, three new methods have been widely adopted. These are optical rotary dispersion, mass spectrometry and nuclear magnetic resonance (n. m. r.) spectroscopy.

It is the purpose of the present article to outline the principles of the application of nuclear magnetic resonance spectroscopy to structural and stereochemical problems in natural product chemistry and to illustrate such applications. While it is invariably true that an investigator who has a thorough understanding of the principles of a given physical method will extract more information than a colleague whose knowledge is more superficial, the demands on the average organic chemist today are such that he cannot hope to master the physical and mathematical background of each of the physical methods he employs. Accordingly, the following presentation of the subject will be purely descriptive.

Nuclei of certain isotopes possess a mechanical spin. Since nuclei are also charged particles this spin is associated with a magnetic moment so that such nuclei resemble small bar magnets. Such nuclei when placed in a uniform magnetic field (H_0) will interact with this field but, because of quantum restrictions, only certain orientations of the nuclear magnets are possible. The maximum number of orientations is given by the spin number I of the particular isotope. I can assume the values $0, \frac{1}{2}, 1, \frac{3}{2} \ldots$ The value of zero characterises isotopes which do not have a mechanical spin. Nuclei for which $I = \frac{1}{2}$ have two possible orientations in the magnetic field; while those for which $I = 1$ have three possible orientations, and so on. Each orientation corresponds to a spin energy state of the nucleus. Transitions between the spin energy states can be stimulated by electromagnetic radiation of the appropriate frequency. The difference in energy between spin states, i. e. the transition energy, depends on the spin number and magnetic moment (μ) of the nucleus and on the strength of the uniform magnetic field, the relationship being given by the equation

$$h\,\nu = \mu\,H_0/I.$$

The value of the uniform magnetic field H_0 generally employed is in the region 9000 to 24000 gauss and the resulting transition frequencies then lie in the radio-frequency region of the electromagnetic spectrum. For instance, the resonance frequency of protons lies in the range 40 to 100 Mc. per second. Thus, in nuclear magnetic resonance spectroscopy the nuclei are placed in a uniform magnetic field and are irradiated with radiofrequency radiation. Absorption of energy is detected either by holding the value of the magnetic field constant and scanning the frequency or, conversely, by holding the frequency constant and varying the magnetic field. Spectrometers in current use employ the latter method. The nuclei of the abundant isotopes of carbon and oxygen have $I = 0$; i. e. they do not possess a magnetic moment and are therefore not available for study by this method. The proton, on the other hand, has a spin number of $^1/_2$ and its nuclear magnetic resonance absorption can readily be detected. Practically all applications of n. m. r. spectroscopy in natural product chemistry involve the study of proton magnetic resonance. Information relating to the structure of molecules becomes available from studying the way in which the value of the applied magnetic field H_0 experienced by a particular proton is modified by its electronic environment and by neighbouring magnetic nuclei. These topics are discussed below.

One or two practical considerations of considerable importance will be mentioned at this stage. The modification of H_0 by the environment of a given proton is an extremely small effect, often of the order of one part in a hundred million. Accordingly, it is necessary to work with extremely high resolution in order to observe such effects. Thus, the uniformity or homogeneity of the magnetic field must be of the order of one part in a hundred million and this is the principle reason for the high cost of n. m. r. spectrometers.

The contribution of neighbouring nuclear magnets to the field experienced by a given proton is made up of two terms. One of these, and by far in a way the larger, arises from direct interaction through space. However, this effect is averaged to zero if measurements are carried out in the liquid phase in which the molecules containing the magnetic nuclei are undergoing random rotation. The second term does not average to zero and this will be discussed in detail later. In the solid phase the first term produces extensive broadening of the resonance lines and thus obliterates the fine differences which are of use in structural organic chemistry. For this reason all high resolution n. m. r. spectra must be determined in the liquid phase. Thus, solid compounds must be studied in solution in solvents which do not themselves contain protons. Generally the solvents of choice are carbon tetrachloride and deutero-chloroform. Sometimes other solvents such as trifluoracetic acid, pyridine, dimethyl-sulphoxide and dimethylformamide have to be used for reasons of solubility. These solvents, of course, give rise to intense absorption in some regions of the proton magnetic resonance spectrum but may be transparent in regions of interest. With the most refined techniques currently available useful spectra can be obtained with as little as one or two mg. of material dissolved in o.1 ml. of a solvent.

The determination of the structure of a natural product usually involves four steps. The first stage is the determination of the empirical and molecular formulae. This is usually followed by the demonstration of the presence of various classes of functional groups. The next stage is to determine the number of functional groups in each particular class. Finally, it is necessary to determine the sequence in which the functional groups and intervening atoms occur within the molecule. Proton magnetic resonance spectroscopy provides useful information at each of these stages. Thus, the method can be used for carrying out hydrogen analyses frequently with an accuracy greater than currently available from conventional combustion data. Proton magnetic resonance spectra frequently indicate the various classes of protons present in the molecule and from this information the various functional groups present in the molecule can often be inferred. The numbers of protons in each particular class can be counted. Finally, from a consideration of fine detail within the spectrum it is often possible to establish the sequence in which groups of protons exist in the molecule.

Once the structure has been defined, the question of relative stereochemistry of the molecule has to be established and here again n. m. r. spectroscopy can be most useful.

The following Chapter dealing with the principles of application of n. m. r. spectroscopy in structural and stereochemical analysis is designed to indicate the way in which the above information is obtained.

II. Principles of Application.

1. Determination of Empirical Formulae.

The transition of a proton from one spin state to the other is fundamentally the same irrespective of the chemical environment of the proton. For this reason the transition probability and hence intensity of absorption is the same for every proton in the molecule, provided certain aspects of experimental technique are observed. The intensity of absorption of a given proton is determined by measuring the area under the absorption band which it gives rise to in the n. m. r. spectrum. Thus, if the total area enclosed by the nuclear magnetic resonance spectrum of a solution, containing a known concentration of a compound of unknown structure, is compared with the area of a solution containing known concentration of a compound of known structure, it is possible to derive the percentage of hydrogen in the former compound. To do this, the two determinations must be carried out under precisely the same conditions and it must be established that the result is independent of the amount of radio-frequency power used in both determinations*.

Sometimes it is possible to choose a reference compound the absorption of which occurs in a region where the sample is transparent. In this case it is possible to determine the percentage of hydrogen in the sample compound by examining a solution of a mixture of the unknown and reference substances and comparing the intensity of its absorption with that of the reference.

---

* The use of too great a radio-frequency power can lead to a phenomenon known as saturation. When saturation occurs the intensities of absorptions from various protons are no longer necessarily proportional to the numbers of each type present.

References, pp. 360—362.

The area under an absorption band can be determined by the usual methods of graphical integration, for example by use of a planimeter or by counting squares. However, all commercial spectrometers now have provision for electronic integration so that the integral of the absorption spectrum can be directly recorded. An example is provided by Fig. 14 (p. 342). The relative intensity of a given absorption band is proportional to the corresponding step in the integrated spectrum.

2. Determination of the Classes of Protons in a Molecule.
The Chemical Shift.

Fig. 1 shows the spectrum of ethanol determined at low resolution. It is seen to consist of three bands which clearly arise from the three

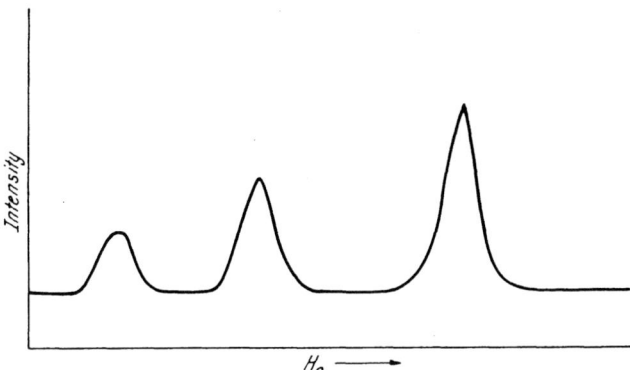

Fig. 1. Low resolution spectrum of ethanol.

classes of protons in the molecule namely the hydroxylic proton, the two protons of the methylene group and the three protons of the methyl group. The separation of the proton resonances or absorptions into bands characteristic of different chemical environments is known as the chemical shift. Before discussing the origin of this effect and the way in which it can be used to provide structural information, it is necessary to establish a quantitative form in which the chemical shift is expressed.

Frequently, chemical shifts of the order of a few cycles per second (c./sec.) in a total proton resonance frequency of one hundred million c./sec. must be measured. Consequently, an absolute determination of the band position is not feasible since this would require a knowledge of the value of H_0 to one part in hundred million. To overcome this difficulty it is convenient to introduce an arbitrary scale based on the line position of a suitable reference compound which is added to each sample investigated. The reference compound which is now almost invariably employed for spectra determined on non-aqueous solutions is tetramethylsilane,

$Si(CH_3)_4$. This substance gives rise to a single sharp line in the proton magnetic resonance spectrum and its position lies to higher values of the applied field than almost all protons present in organic molecules. The separation between this line and the various lines in the spectrum of a sample compound can readily be determined with an accuracy of the order of 0.1 to 0.2 c./sec. It will be seen later that the chemical shift or separation between the lines arising from different protons in the spectrum is proportional to the strength of the applied magnetic field H_0. Since various commercial spectrometers operate at different field strengths, it is not convenient to express the chemical shift in c./sec. Instead, these separations are converted to field independent units in the following way. The separation between the tetramethylsilane absorption line and that of a proton in the reference sample is defined as a positive quantity Δ c./sec. if the latter occurs at lower fields. (In the rare cases in which a sample absorption occurs at higher fields Δ is taken as negative.) One convenient field independent expression of the chemical shift is then given as

$$\delta = \frac{\Delta \cdot 10^6}{\text{spectrometer frequency}}.$$

This is known as the δ scale and on this scale tetramethylsilane is at zero parts per million. An alternative scale which is commonly used is the tau scale and is simply defined as $\tau = 10 - \delta$. The tau scale will be used throughout the remainder of this article to express the chemical shifts of protons in various environments.

The chemical shift is the result of secondary magnetic fields associated with the circulation of extra-nuclear electrons which are induced by the applied magnetic field. These secondary fields either augment or diminish the applied field at a given proton and their strength is proportion to H_0. The circulation of electrons may take place either about the nucleus under observation or about neighbouring nuclei. The magnitude of the former depends on electron density at the nucleus concerned and may be correlated with electronegativity and inductive effects. The circulation of the electrons around a particular proton shield that proton from the applied magnetic field. Thus, the proton of a methyl group attached to an electronegative substituent such as oxygen will be less shielded than the protons of a C-methyl group because in the former the electronegative substituent reduces the electron density in the vicinity of the protons and hence reduces the shielding effect. The contribution of shielding to a given proton from circulations of electrons around neighbouring nuclei can be either positive or negative, depending on the nature of the electronic system and on its orientation with respect to the proton. This aspect will be dealt with in the Section dealing with stereochemistry (p. 333). Since the intensity of electron circulations, and hence their associated

References, pp. 360—362.

shielding effect, depends on the magnitude of the applied field the chemical shift measured as a frequency will, as pointed out earlier, also depend on the magnitude of the applied field.

While the theory of proton chemical shifts is fairly well understood in a qualitative fashion, the use of chemical shift data in structural work is based largely on correlation tables and on the use of carefully chosen model compounds. Several compilations of chemical shift data are now available (20, 22, 31, 50) and there are many uncorrelated data available in the literature. Thus it is possible to recognise a large number of common groups in organic molecules from the position of the absorption of their protons in the nuclear magnetic resonance spectrum and this procedure will be illustrated in the various Sections dealing with specific natural products (p. 337).

3. The Number of Protons in a Given Class. Intensities.

Having established the various classes present in a molecule, the next problem is to count the number of protons in each class. This is done by determining the intensities of the absorptions arising from various classes of protons by integrating the absorption bands. As discussed above, the area under the absorption bands will be proportional to the number of protons which give rise to the absorption, and for a single molecular species this means that the intensities of the various absorption bands arise from different classes of protons in each class. Of course, the accuracy with which this can be done is often limited by the fact that the absorptions from various classes of protons overlap. Nevertheless, this process is an extremly valuable one which yields much important information.

4. The Sequence of Groups in Molecules. Electron Coupled Spin-Spin Interaction.

The absorption bands arising from single protons or from groups of equivalent protons (i. e. protons with exactly the same chemical shift) frequently exhibit well defined fine structure when examined under high resolution. *Fig. 2* shows the spectrum of acetaldehyde in which it can be seen that the aldehyde proton gives rise to a quartet and that the protons of the methyl group give rise to a doublet. It should be further noticed that the quartet is symmetrical, the spacing of the quartet components is the same as that of the doublet, and that the intensities of the components of the quartet are approximately in the ratio $1 : 3 : 3 : 1$, while the two components of the doublet are approximately equal. This multiplicity arises from the magnetic interaction between the two classes of protons which is not averaged to zero by the random motion of molecules in the liquid state. The phenomenon is known as electron coupled

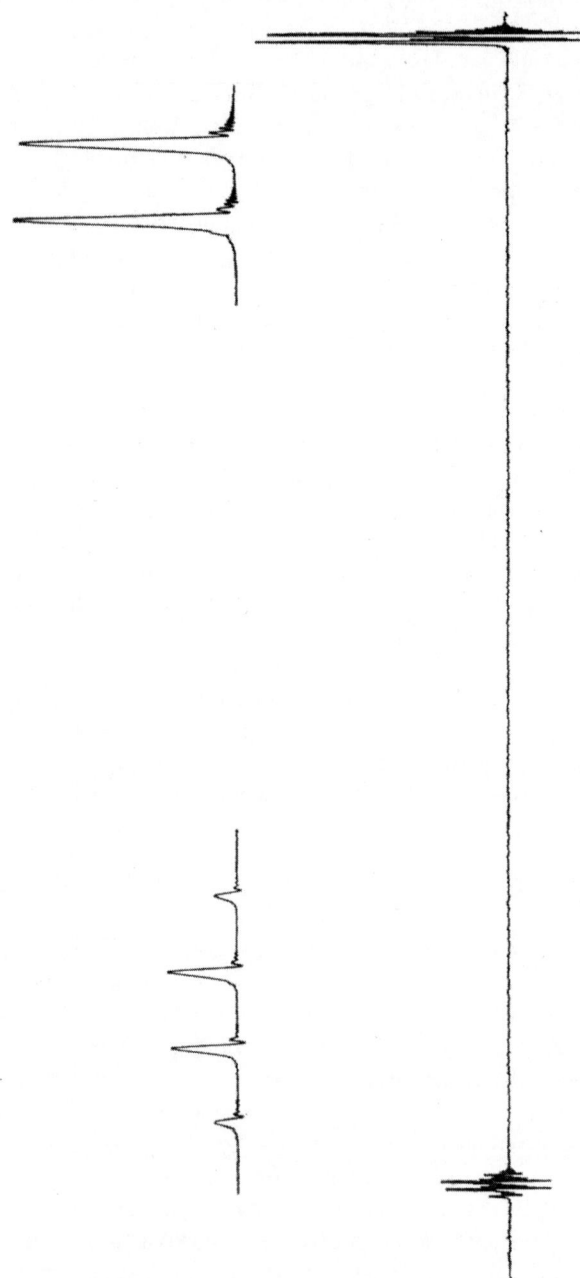

Fig. 2. High resolution spectrum of acetaldehyde (upper spectra are expanded traces of the multiplets).

spin-spin interaction since the interaction takes place via the electrons which constitute the intervening bonds between the interacting nuclei. With the knowledge that nuclear magnets can contribute, via the intervening electrons, to the magnetic field experienced by a given proton, it is possible to account for the multiplicities of the absorption bands in acetaldehyde and in the absorption bands associated with many arrangements of protons in organic molecules.

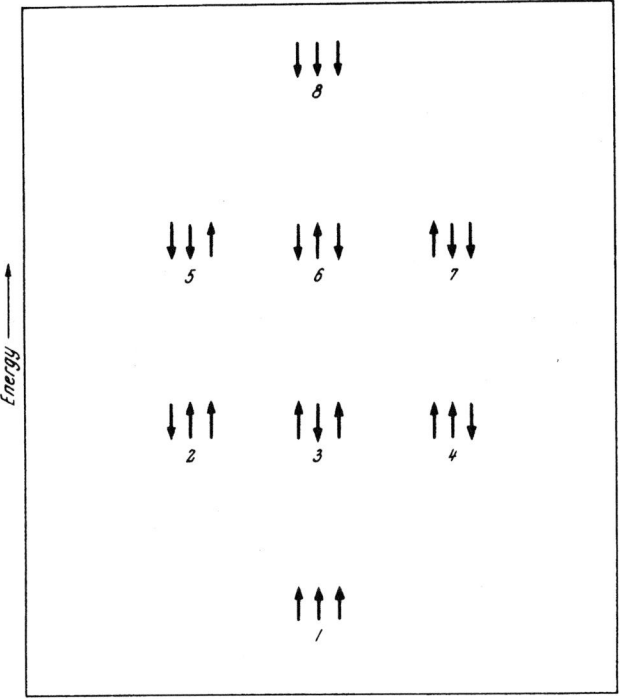

Fig. 3. Diagrammatic representation of spin-spin coupling.

Consider the contribution made by the methyl protons to the magnetic field experienced by the aldehyde proton in acetaldehyde. This contribution will depend on the orientations of the three methyl protons. As explained above a proton can be in one of two spin energy states. Each of these two states is associated with a particular orientation of the proton magnet with respect to the applied magnetic field. One orientation will have a component parallel to the applied field while the other will have a component anti-parallel to the applied magnetic field. For a group of three nuclei, each of which can assume one of two orientations, there will be eight arrangements of the nuclear magnets. These are depicted in *Fig. 3.* To a high degree of approximation each of these eight arrange-

ments is equally populated. However, since the three protons are identical
it is seen that the arrangements labelled 2, 3 and 4 in Fig. 3 are equivalent
and the same is true for the arrangements labelled 5, 6 and 7. Thus,
there will be a total of four different spin configurations of the three
methyl protons and those corresponding to the two groups of three will
be three times as probable as the remaining two (1 and 8). The first group
of three (2, 3, 4) differs from arrangement 1 by one change in spin. Similarly, the second group of three (5, 6, 7) differs from the first group of
three by one change in spin, and finally the arrangement 8 differs from
the preceding group of three also by one change in spin. Thus, if a collec-
tion of acetaldehyde molecules is considered, it is predicted that the
aldehyde protons will experience four fields which differ by constant
increments so that they will give rise to four equally spaced absorption
lines, the intensities of which will be in the ratio of 1 : 3 : 3 : 1. Since the
aldehyde proton can adopt one of two orientations it follows that the
three protons of the methyl group will give rise to just two lines of equal
intensity and, as the interaction between the two classes of protons is
mutual, the spacing of these two lines will be the same as the spacings in
the quartet arising from the aldehyde proton.

This analysis of the interactions between groups of protons predicts
that a proton or group of protons interacting with one neighbouring
proton will absorb as a doublet; those interacting with two magnetically
equivalent protons will give rise to a triplet with an intensity ratio of
1 : 2 : 1; those interacting with three magnetically equivalent protons
will give rise to a quartet with an intensity ratio 1 : 3 : 3 : 1, etc. The
spacing of the components of the multiplet, measured in c./sec., is known
as the spin-spin coupling constant, J. The spin-spin coupling constant
is independent of the strength of the applied magnetic field and its value
depends on structural and stereochemical factors (see below).

The above analysis is adequate provided the chemical shift between
the interacting nuclei is large compared with the spin-spin coupling
constant, and, if this condition is obeyed, the extension to cases in which
a given proton interacts *unequally* with two or more neighbouring protons
is straightforward. *Fig. 4* shows diagrammatically how the fine structure
of an absorption band arising from a proton coupled unequally to two
other protons of very different chemical shifts is developed. When the
chemical shifts and coupling constants are of the same order, much more
complicated spectra arise and these frequently are unrecognisable in
terms of the simple patterns described above. In the limit at which all
the chemical shifts are exactly equal (i. e. the protons are equivalent)
no spin-spin splitting is observed. Thus, methane gives rise to a singlet.
In cases where the chemical shifts and coupling constants are of the
sam eorder, useful information can still be extracted from the spectrum

References, pp. 360–362.

by using quantum mechanical methods to analyse the multiplets. This procedure often requires extensive calculations which can only be carried out with a computer. Organic chemists who wish to extract the maximum information from nuclear magnetic resonance spectra need to master these mathematical methods and are referred to the excellent monograph by ROBERTS (46) which provides a simple introduction to the subject.

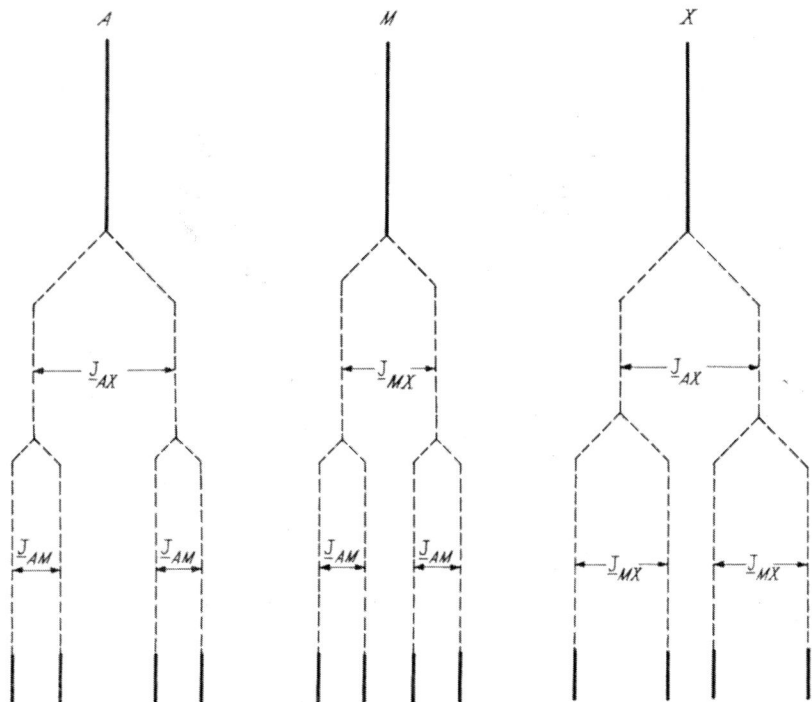

Fig. 4. First order derivation of the AMX spectrum (definition of AMX see below).

A convenient symbolism for describing more complicated spin systems has been introduced by POPLE, SCHNEIDER and BERNSTEIN (45). A, B, C, ... represent coupled nuclei for which the coupling constants are of the same order as the chemical shifts. The series is arranged in order of increasing τ-values. M, N, ...; and X, Y, ... are similar groups. Coupling between the groups A, B, ...; M, N, ...; and X, Y, ... occurs but the relevant coupling constants are small compared with the analogous chemical shifts. The equivalence of nuclei is indicated by subscript integers. For example, the spin system in acetaldehyde is designated as AX_3 since $J_{AX} \ll [\tau_X - \tau_A]$ and the three protons of the methyl group are equivalent. The proton considered in connection with Fig. 4 is part of an AMX system. As a final example, the protons of crotonaldehyde constitute an $AMNX_3$ system, A being the aldehyde proton, M and N the olefinic protons which have similar chemical shifts, and X_3 the three equivalent protons of the methyl group.

As spin-spin interaction is electron coupled, the spin-spin coupling constant is a property of bonds. It depends, in fact, on the delocalisation of electrons in the bonds which separate the interacting nuclei. Accordingly, the effect is rapidly attenuated as the number of intervening bonds is increased. Significant spin-spin splitting is thus observed between two protons attached to the same carbon atom (provided they are non-equivalent) and also between protons separated by three bonds, for example in acetaldehyde. In special circumstances, particularly in unsaturated systems, small but significant spin-spin splitting is observed between protons separated by four or five bonds. Although a number of useful theoretical treatments of coupling constants have been published, the utility of spin-spin coupling constants in structural investigations depends largely on empirical correlations, a number of which are now available (6, 31, 54).

The observation and interpretation of spin-spin splitting is the means by which the sequence of groups in molecules is established by n. m. r. spectroscopy. To take a simple example, the observation of a doublet equivalent in area to three protons near $\tau = 9.0$, the region characteristic of methyl groups on saturated carbon atoms, conclusively shows the presence in the molecule of one C-methyl group attached to a saturated carbon which is bonded to one proton, i. e. $\diagdown CH$—CH_3. If elsewhere in the spectrum an absorption band, consisting of a quartet or a multiplet of quartets, the spacings of which correspond exactly to those of the doublet near 9.0 is observed, this can, with fair certainty, be assigned to the methine proton which is coupled to the methyl group, and the chemical environment of this methine proton can be deduced from its chemical shift. In this way it is often possible to establish quite extensive sequences of groups in molecules. In a number of cases the interrelation of two absorption bands can only be established by mathematical analysis of the type referred to above, thus emphasising how necessary it is for the organic chemist to master the mathematical technique for the analysis of spectra.

The process of establishing sequences of groups in molecules described above frequently fails because, while it may be possible to observe a discrete multiplet from one group of protons it may be impossible to recognise the absorptions of the protons to which this group is coupled, since they may be obscured by absorptions of other protons in the molecule. An ancillary technique known as spin-decoupling or multiple resonance often helps to overcome this difficulty. This technique will now be described in some detail as it is of considerable importance to the organic chemist.

Spin-spin splitting of the absorption bands between two mutually coupled protons of different chemical shift is only observable if the half-life of the protons in any one of their two spin states is sufficiently long. More precisely, the observation of spin-spin splitting requires that the half-life in any one spin state is appreciably greater than the reciprocal of the coupling constant between the protons. If the opposite condition, namely the half-life being much less than the reciprocal of the coupling

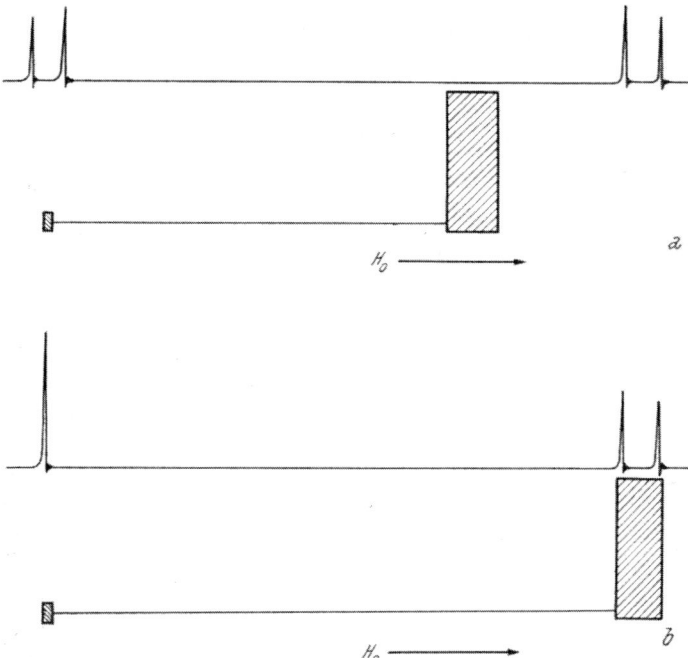

Fig. 5. Diagrammatic representation of field sweep spin decoupling.

constant, applies the interacting protons absorb as if they were not coupled. Unless one or more of the interacting protons is undergoing rapid chemical exchange, the half-lives of protons in any one spin state are much greater than the reciprocal of coupling constants between protons so that spin-spin multiplicity is generally observable. However, it is possible to reduce the half-lives of the spin states artifically by irradiating the nuclei at their resonance frequencies. This is the basis of the spin decoupling method.

Consider the case of two mutually interacting protons, labelled A and X, of different chemical shifts; under the normal conditions of observation both A and X will give rise to doublets. If, however, the absorption of X is observed while A is being strongly irradiated at its resonance frequency,

X will give rise to a singlet since the half-life of A in any one of its two spin states will be reduced to a value considerably less than the reciprocal of the coupling constant between A and X. Of course, signals from both A and X are detected by the spectrometer but, by the use of various electronic techniques, it is possible to reject the signal from the intensely irradiated nucleus A and record only the absorption spectrum of X. There are basically two methods of carrying out spin decoupling experiments. In one method the strong radiofrequency field used for limiting the half-life of A in its spin states and the weak radio-frequency field used for serving the absorption spectrum of X are generated with an accurately known but variable frequency relation. The entire spectrum is then scanned by varying the applied magnetic field. If the difference in frequency between the strong and the weak radiation does not correspond to the difference in chemical shift between A and X, the observed spectrum will be normal *(Fig. 5 a)*. If, however, this frequency difference is adjusted to correspond exactly to the difference in chemical shift between A and X then, as the absorption of X is being observed, A is being strongly irradiated and decoupling occurs *(Fig. 5 b)*. Thus, the determination of the difference in frequencies between the strong and weak radiation necessary to produce decoupling provides an indirect measurement of the chemical shift between A and X. This method is called the field-sweep method.

In the second method, known as the frequency-sweep method, the strong radiofrequency field is held at a constant frequency and the frequency of the weak field is continually varied for the purpose of scanning the spectrum, the applied magnetic field kept constant throughout the experiment. Decoupling is then achieved by first adjusting the applied magnetic field so that the frequency of the strong radiofrequency field corresponds to the absorption frequency of a multiplet in the spectrum and the remainder of the spectrum is then scanned by varying the frequency of the weak radiofrequency field *(Fig. 6)*. The precise frequency difference between the strong and weak radiation is known throughout the spectrum, so again an indirect measurement of chemical shift is possible.

The field-sweep method is experimentally the easier technique but is somewhat difficult to apply if the multiplet under observation is very broad since the strong field is only effective if its frequency lies fairly close to the resonance frequency of the nucleus to be decoupled. This difficulty is obviated in the frequency-sweep method since throughout the experiment the frequency of the strong field corresponds to the resonance frequency of the nucleus to be decoupled. The major difficulty in the frequency-sweep method is to maintain this exact equality between the frequency of the strong radiation and of the resonance frequency of the decoupled nucleus. This problem can be overcome and some commercial spectrometers now offer this facility. In principle, it is possible to irradiate strongly at more than one frequency, so triple or, in general, multiple resonance experiments can be carried out. In this way it is possible to simplify a complex multiplet arising from a proton which is coupled to several other protons.

Spin-decoupling has three major applications in structural organic chemistry. Firstly, it can provide an unambiguous proof that two multiplets in a spectrum arise from mutually coupled protons. Not infrequently equality in splittings or considerations based on mathematical analysis of multiplets suggest that two multiplets are related in this way and a spin-decoupling experiment can then be performed to confirm this conclusion with complete certainty.

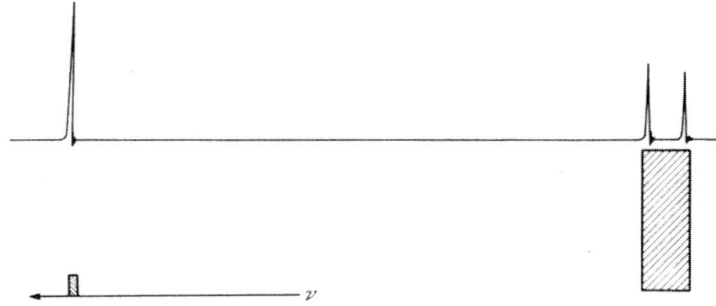

Fig. 6. Diagrammatic representation of frequency sweep spin decoupling.

The second and probably the most important application is to the location of hidden absorption bands. In dealing with the spectra of complex organic molecules, the situation frequently exists in which a well resolved multiplet is observed in one region of the spectrum but the absorption of the protons which cause this multiplicity are obscured by overlying absorptions from a number of other protons. The chemical shifts of these other protons can be measured by determining the frequency differences between the strong and weak radiation necessary to cause simplification of the well resolved multiplet. Thus, spin-decoupling permits the determination of the chemical shifts of some protons which do not give rise to discreetly resolved absorptions in the spectrum.

Finally, the spin-decoupling method can be used to simplify the complex multiplets which frequently arise from a number of mutually interacting protons. Thus, the need for detailed mathematical analysis may sometimes be reduced or circumvented in this way. The frequency-sweep method is ideally suited for this type of application. It should be pointed out, however, that the method is only of limited value in dealing with systems for which all the chemical shifts are of the same order as the coupling constants.

5. An Example of Application to Structure Determination.

In order to clarify the various principles of application dealt with above an example of the determination of structure by n. m. r. spectroscopy will now be given. This particular example embodies most aspects of the method.

(1) Columbin.   (2)

+ -O-CHO
+ =O
+ -O-

The compound under investigation is palmarin, $C_{20}H_{22}O_7$, a congener of the bitter principle, columbin (1) (5, 11). Degradation by a logical series of steps converts palmarin to a compound which can be assigned the partial structure (2) (12). The structural problem is therefore reduced to the determination of the positions of attachment of the ether oxygen and of the formate ester, and the position of the keto group. The n. m. r. spectrum of (2) in deuterochloroform is reproduced in *Fig. 7*. The numbers of protons responsible for the various absorption bands were determined by integration. Certain assignments can be made on the basis of the known features of the structure. Thus, the absorption between 8.5 and 9.1 arising from ten protons can be assigned to the two angular methyl groups and the four methylene protons in one of the carbocyclic rings. The absorptions of the two methyl groups are clearly singlets which is consistent with their attachment to fully substituted carbon atoms. The absorption near 6.1 which arises from two protons exhibits the typical features of the AB region of an ABX spectrum. This band may be assigned to the two protons of the methylene group adjacent to the ether oxygen of the lactone ring. It is anticipated that these two protons will have rather similar chemical shifts, that they will be mutually coupled and that they will be separately coupled to the proton at $C_{(8)}$. If so desired, this multiplet could be mathematically analysed and the chemical shift between the two protons as well as the various coupling constants determined. However, in the present case this procedure is not necessary as the absorption arises from a known feature of the molecule. The protons of the $C_{(11)}$ methylene group give rise to a pair of doublets at 6.8 and 7.6 with a coupling constant of 16.8 c./sec., a value characteristic of coupling between two protons attached to the same saturated carbon atom. Recognition of these two doublets and of their mutual relation is not immediately obvious but can be clearly demonstrated by a spin-decoupling experiment using the field-sweep method. Thus, the doublet at 6.83 is partially collapsed if observed while simultaneously irradiating the doublet centred at 7.86 (Fig. 7). The doublet at 1.76 can be assigned on the basis of its chemical shift to the "aldehydic" proton of the formate ester. The absorption at 3.89 (one proton), 6.40 (one proton), 6.66 (one proton), and between 7.6 and 8.2 (two protons excluding one of the $C_{(11)}$ protons) have now to be

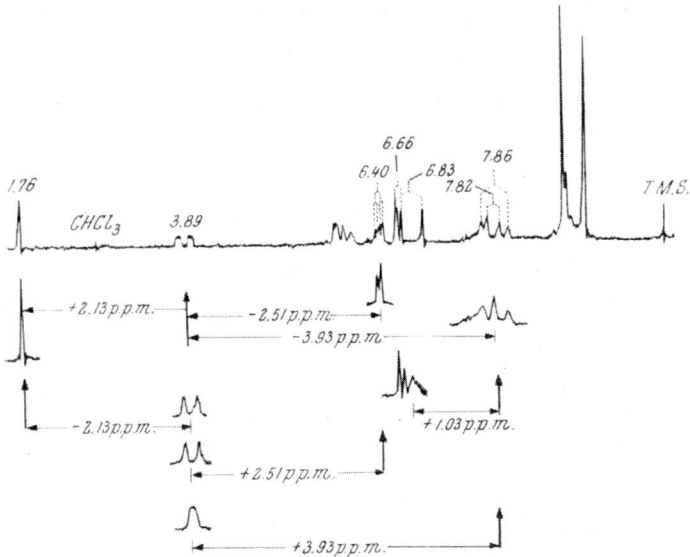

Fig. 7. Single and double resonance spectra of a degradation product (2) of palmarin, at 60 Mc./sec.

assigned. It is noticed that the absorption of the formate ester proton at 1.76 is a doublet ($J \sim$ 1 c./sec.). This splitting must be due to a small coupling of the formate proton across four bonds to a proton on the same carbon atom as the ester group. This shows that the formyloxy group is attached to a secondary carbon atom. It is now a simple matter to locate the position of absorption of the proton on the secondary carbon atom. It is found that the doublet at 1.76 collapses to a singlet if observed with a simultaneous irradiation of the proton giving rise to the multiplet at 3.89. Inspection of the multiplet at 3.89 shows that it consists of 8 lines i. e. it is a double-double doublet. Following the line of argument relating to Fig. 4 (p. 325) it can be seen that this proton is coupled to three other protons with coupling constants of 4.7, 2.3 and ca. 1 c./sec., respectively. The last of these coupling constants refers, of course, to the interaction with the formyl proton which has already been established by spin-decoupling. The protons responsible for the 8.9 and 2.3 c./sec. coupling constants are located by spin decoupling at 7.82 and 6.40, respectively and, in view of this difference in chemical shift, they cannot be attached to the same carbon atom. Thus, the partial structure (3) may be written. If the proton absorbing at 3.89 is irradiated during the observation of the absorption at 7.82, the latter appears as a singlet. Thus, the proton absorbing at 7.82 is coupled only to the proton of the secondary ester. Its τ-value indicates that it is a simple methine proton and in view of the lack of further splitting the remaining two substitutents on this methine grouping must be fully substituted carbon atoms. The partial structure (3) can therefore be extended to (4).

(3)　　　　　　　(4)　　　　　　　(5)

The absorption at 6.40 is a double doublet, one spacing of which is due to the 2.3 c./sec. coupling with the proton attached to the same carbon atom as the formyloxy group. From its τ-value this absorption must be assigned to a proton α to a singly bonded oxygen atom, leading to the partial structure (5). The second splitting (J equals 4.7 c./sec.) can only arise from coupling with a proton on an adjacent carbon atom. Attempts to remove this splitting by spin-decoupling failed. However, experimental difficulties precluded the use of frequency differences between the strong and weak radiation of less than 30 c./sec. and so it may be concluded that the two protons involved in this interaction differ by less than 30 c./sec. in their chemical shift. Examination of the spectrum reveals a doublet ($J = 4.7$ c./sec.) at 6.66. This band overlaps one component of the doublet centred at 6.83 but not when the latter is collapsed by spin-decoupling (see Fig. 7, p. 331). Like the proton absorbing at 6.40, that absorbing at 6.66 can also be reasonably assigned to a proton α to singly bonded oxygen and the partial structure (6) may be written. Since there is only one oxygen atom available for the two indicated at carbon atoms 2 and 3 in structure (6), the ether must be an oxirane. Since the proton absorbing at 6.66 is coupled only to that absorbing at 6.40 the carbon atom 4 does not carry a proton. The chemical shifts of the two protons of the oxirane rings are atypical since protons of the oxirane system usually absorb approximately one part per million to higher fields (i. e. to higher τ-values) than protons adjacent to oxygen in other ethers. The observed values are, however, in agreement with

References, pp. 360—362.

those expected for an $\alpha\beta$-epoxyketone so that the partial structure (7) can be written. Finally, the coupling constant of 4.7 c./sec. between the protons at $C_{(2)}$ and $C_{(3)}$ in (7) lies within the range found for a *cis* coupling constants in oxiranes. Consideration of partial structures (2) and (7) leads to the structure (8) for palmarin. Because of the peculiar geometry of ring A in palmarin, detection of the oxirane ring by conventional chemical methods was not possible (5).

(6)   (7)   (8) Palmarin.

This example indicates the way in which sequences of protons in different environments can be established provided a reasonable analysis of the n. m. r. spectrum can be made. The spin-decoupling technique greatly facilitated the analysis and allowed considerable reliance to be placed on the interpretation of the spectrum.

6. Applications to the Elucidation of Relative Stereochemistry and Conformation.

Both the chemical shift and the magnitude of spin-spin coupling are critically dependent on geometrical factors. In cases in which this dependence is understood it is possible to use chemical shift and coupling constant data to derive the relative orientations of protons in molecules and hence determine the relative stereochemistry of groups. It is important to realize that many organic molecules are dynamic equilibrium mixtures of various conformations and that in almost all of such systems the observed nuclear magnetic resonance parameters have values which correspond to time averages over these conformations. Thus, the chemical shift of the protons of the methyl group in α-bromopropionic acid and also the coupling constant of these protons with the adjacent α-proton are the average of the three conformations (9), (10) and (11). In this particular example the three conformations are, of course, equally populated. In other systems, however, the conformations are not necessarily equally populated and the observed parameters have values which correspond to an appropriately weighted average of their values in each conformation.

It was pointed out earlier that the contribution to the shielding of a given proton from circulations of the electrons about neighbouring nuclei depends both in magnitude and sign on the orientation of the proton to

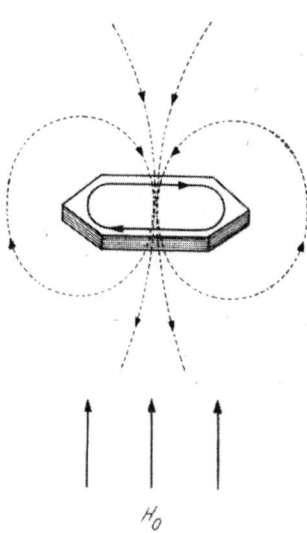

$$(9) \qquad\qquad (10) \qquad\qquad (11)$$

the group of electrons which cause the shielding effect. The best example of this effect is provided by the π-electrons of the benzene nucleus. *Fig. 8* depicts the induced circulations of the π-electrons and the associated secondary magnetic field which occur when the applied magnetic field is perpendicular to the plane of the aromatic ring. If the applied magnetic field lies in the plane of the aromatic ring no such circulations can occur. In liquid benzene, in which the molecules are undergoing random rotation, the net secondary magnetic field will be the average over all possible orientations of the ring with respect to the applied magnetic field, so that the actual shielding field will take on the general characteristics of that depicted in Fig. 8 but will, of course, be smaller in magnitude. It is seen that the protons attached to the aromatic ring lie in a region in which the secondary magnetic field is parallel to the applied field and, accordingly, they are deshielded. In a complicated molecule containing a benzenoid nucleus it is possible that a given proton can occupy a region above the plane of the ring and it will therefore be shielded by the secondary magnetic field associated with the circulation of the π-electrons. A shielding effect of the type just discussed falls off rapidly with distance. Nevertheless, it can be predicted that the benzene π-electron system could make significant contributions to the shielding of a proton at distances as great as 5 or 6 Å (35). Clearly, the situation can arise in which the position of a proton relative to an aromatic nucleus in a molecule will vary considerably from one stereoisomer to another. In such cases

Fig. 8. Shielding field of benzene.

comparison of the chemical shift of this proton in a pair of diastereoisomers will disclose its disposition to the aromatic ring and hence its relative configuration in the two isomers.

The nature of the shielding associated with some other groups of electrons has been established either theoretically or empirically. One particularly useful result concerns the carbonyl group. The shielding field associated with the carbonyl group is roughly depicted in *Fig. 9 (31)*. Protons lying in the plane of the trigonal carbon atom and in close proximity to the carbonyl group are invariably deshielded. Thus, it is a relatively

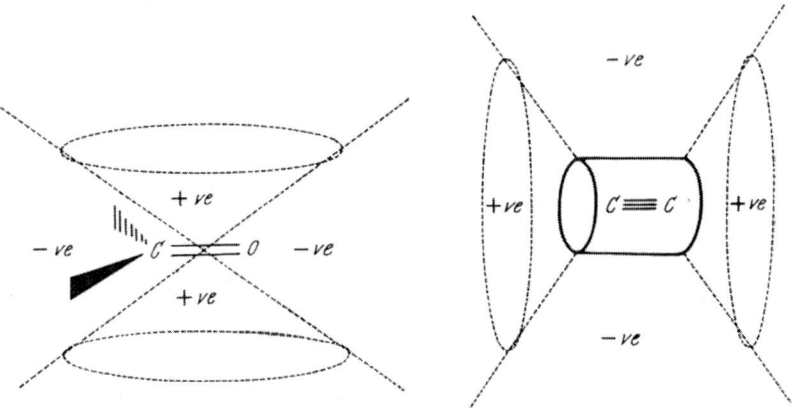

Fig. 9. Shielding field of the carbonyl group.   Fig. 10. Shielding field of the triple bond.

simple matter to establish the configurations of α, β-unsaturated acids, esters, ketones and aldehydes by comparing the chemical shifts of β-protons or the protons of β-alkyl substituents in the *cis* and *trans* isomers. In general protons which are *cis* to the carbonyl group will be deshielded; the effect being of the order of 0.8 p. p. m. for a β-proton (*33, 44*) and of about 0.3 p. p. m. for the protons of an alkyl substituent in α, β-unsaturated esters (*33*). Similar effects have been noted for the carbon-carbon double bond although here the magnitude of the shielding contribution seems to be rather less (*4, 28, 31*). In the case of the carbonyl group the observed shifts almost certainly contain a significant contribution from the direct electrostatic perturbation of the electron density around the proton under observation. No contribution of this type is possible with the non-polar carbon-carbon double bond. Predictions can also be made as to the nature of the long-range shielding associated with the carbon carbon triple bond and the carbon-nitrogen triple bond. Both these groups are axially symmetric and *Fig. 10* shows the signs of the shielding regions in the vicinity of these groups.

A number of other long-range shielding effects have been observed, although the precise nature of their origin is not understood. In general, almost all bonds and lone pairs of electrons may be expected to give rise to long-range shielding, although not necessarily of the magnitude of those groups of electrons discussed in the preceding paragraph. One particularly useful correlation concerns the chemical shift of axial and equatorial protons in alicyclic systems. It has been found that in simple alicyclic rings axial protons are more shielded than the corresponding equatorial protons (40, 49). There is not doubt that many other correlations relating to long-range shielding will be developed in the future and that chemical shift data will become increasingly useful for solving stereochemical problems.

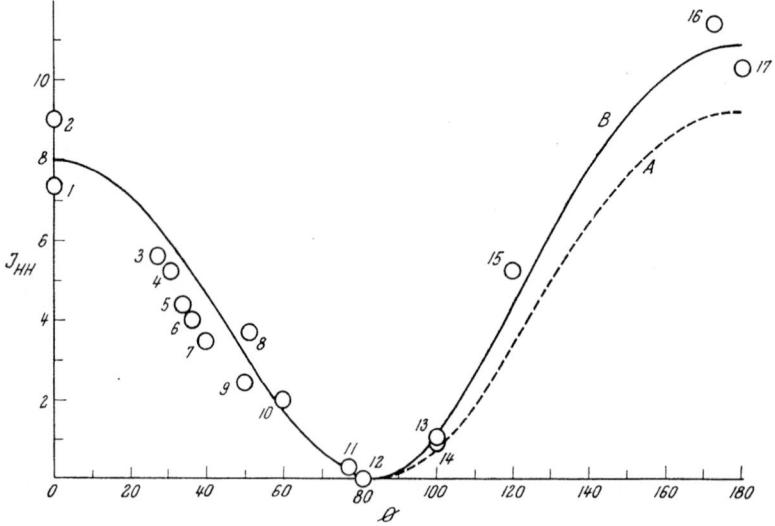

Fig. 11. The relation between dihedral angle, ∅, and the vicinal coupling constant; according to CONROY. [From: Adv. Organ. Chem. 2, 265 (1960).]

The second means by which stereochemistry can be elucidated relates to the angular dependence of the spin-spin coupling constant. The principle of this method is based on theoretical calculations by KARPLUS (36). He predicted that the coupling constant between a pair of vicinal protons in an "ethane type" system would vary with the dihedral angle between them, the dihedral angle being the angle observed between the two carbon-hydrogen bonds when the molecule is viewed along the axis of the carbon-carbon bond. *Fig. 11* shows the angular dependents of this spin-spin interaction. KARPLUS' predictions have been justified empirically (22) and the values in Fig. 11 are those which agree best with experiment. Some caution is necessary in using the KARPLUS relation, since the spin-spin coupling constant is also dependent an normal bond angles (e. g., the HCC bond angle) and apparently to some extent on the nature of the substituents. Thus, Fig. 11 applies to systems in which the normal bond angles are close to tetrahedral. Modified KARPLUS relations are obviously required for other systems; for instance in 3-, 4- and 5-membered rings in which the valence angles are no longer tetrahedral. Many data are now available for such systems and also for the vicinal interactions in double bonds. It should be emphasized that predictions based on the magnitudes of coupling constants are usually highly reliable and that they provide a most valuable way of establishing relative stereochemistry. Similar correlations have been established for long range coupling constants in unsaturated systems (54). In particular, the interactions between an olefinic proton and β-allylic protons and those between α- and β-allylic protons have been considered, although as yet little use of these correlations has been made in practice.

III. Examples in Natural Product Chemistry.

A complete coverage of investigations in which nuclear magnetic resonance spectroscopy has been used in the determination of structure and stereochemistry of natural products is beyond the scope of this article since such examples now number many hundreds. Instead, consideration will be given to applications to some specific classes of natural products. The n. m. r. spectroscopy of certain classes of compounds, for instance steroids, carotenoids and flavonoids, have been the subject of extensive experimental surveys. As a result, accurate correlations are available which are of considerable use in structural and stereochemical work. The findings of these investigations will be surveyed briefly. Other classes of natural products are made up of such diverse structures as to make a systematic survey of their nuclear magnetic resonance spectral properties impossible. Thus, the alkaloids in general cannot be treated in this way, although some valuable correlations have been made for specific types of alkaloids. In dealing with classes of this type, specific examples will be considered and these have been chosen so as to illustrate the general usefulness of the nuclear magnetic resonance method or to draw attention to certain difficulties which may be encountered in the interpretation of results.

1. Steroids.

With most physical methods employed by the organic chemist, the steroid nucleus has permitted extensive and accurate correlations of the observed parameters with structure and stereochemistry, and proton magnetic resonance spectroscopy is no exception. A number of publications dealing solely with proton magnetic resonance spectroscopy of steroids have appeared and, in addition, there are numerous other publications in which chemical shift and coupling constant data are reported.

Fig. 12 is a typical example of a steroid spectrum. The region between 7.0–9.0 invariably contains a broad band envelope arising from the protons of the methylene and methine groups of the basic steroid nucleus. In addition to this absorption, the spectra of steroids always exhibit certain well resolved resonances, the chemical shifts of which are of considerable use in structural work. In particular, chemical shifts of the protons at $C_{(18)}$ and $C_{(19)}$, that is of the angular methyl groups, have been related to the presence of structural features in rings *A, B, C* and *D*.

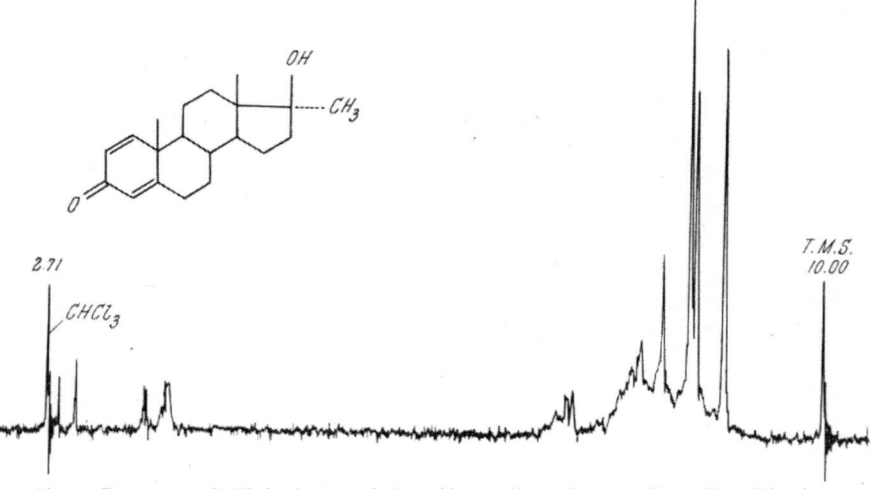

Fig. 12. The spectrum (60 Mc./sec.) of a typical steroid; according to ZÜRCHER. [From: Helv. Chim. Acta **44,** 1380 (1961).]

The first systematic study of steroid spectra was made by SHOOLERY and ROGERS (49) in 1958. Their correlations for the protons of the 19-angular methyl group have been considerably extended by ZÜRCHER (59) who has shown that functional groups in various positions make constant contributions to the chemical shift of these protons. ZÜRCHER's findings are reproduced in *Table 1*. Since long range shielding effects depend on the orientation of the proton with respect to the shielding group, it follows that the effect of substituents in ring *A* will depend on the stereochemistry of the *A/B* ring junction. Similarly, the

(12)

3 β,17 β-Diacetoxy-11-oxo-5 α-Δ⁸⁽⁹⁾-androstene.

effects of substituents in ring *D* depend on the stereochemistry of the *C/D* ring junction. The additivity of these effects is remarkable. For most steroids having the substituents referred to in Table 1 it is possible to calculate the chemical shift of the 19-protons with an accuracy of $\pm$ 0.03 p. p. m. For example,

References, pp. 360—362.

the predicted and observed values for 3 β,17 β-diacetoxy-11-oxo-5 α-$\Delta^{8(9)}$-androstene (12) are 8.84 and 8.87, respectively. ZÜRCHER (59) has made an interesting comparison *(Table 2)* of the shifts produced by structural variations in ring *A* with those produced by the same variations at the symmetry related positions in ring *B*.

Table 1. Contributions (p. p. m.) of Functional Groups to the τ-values of the Protons of the 19-Methyl Group in the Steroid Nucleus (59).

Position*	C=O	α—OH	β—OH	α—OAc	β—OAc	Δ**	α—Ep***	β—Ep***
1 (5α)	— 0.38							
(5β)	— 0.21							
2 (5α)	0.01					0.01		
3 (5α)	— 0.23	— 0.01	— 0.04	— 0.02	— 0.05			0.04
(5β)	— 0.10	0.02	— 0.04	0.00	— 0.04			
4 (5α)	0.02							
(5β)	— 0.19							
5						— 0.19		
6	0.04		— 0.22		— 0.20			
7	— 0.28	— 0.04	— 0.02	— 0.10		— 0.01		
8						— 0.09†	0.10	
9						— 0.15	— 0.20	— 0.12
11	— 0.23	— 0.14	— 0.26	— 0.11		0.03	— 0.07	— 0.19
12	— 0.09	0.00	— 0.01	0.00				
14 (14α)						— 0.01	— 0.02	— 0.10
15 (14α)	— 0.02		— 0.02					
(14β)	0.03							
16 (14α)					— 0.02		— 0.02	
17 (14α)	— 0.02	— 0.01	— 0.01		— 0.01			

* Symbols in parenthesis indicate the stereochemistry of the *A/B* or *C/D* ring junctions.

** Endocyclic double bond. *** Epoxide. † $\Delta^{8(9)}$.

Table 2. Comparison of the Contributions to the τ-value of the 19-Methyl Group by Substituents in Ring *A* with those of the Same Substituents in the Symmetry Related Positions in Ring *B*, in 5α-Steroids (59).

Substituent Ring A			Substituent Ring B
3-Oxo	— 0.23	— 0.28	7-Oxo
4-Oxo	0.03	0.04	6-Oxo
Δ^4-3-Oxo	— 0.41	— 0.40	Δ^5-7-Oxo
Δ^2	— 0.01	— 0.01	Δ^7
3 α-Hydroxy	— 0.01	— 0.04	7 α-Hydroxy
3 β-Hydroxy	— 0.04	— 0.02	7 β-Hydroxy
3 α-Acetoxy	— 0.03	— 0.10	7 α-Acetoxy

Correlations between the chemical shift of the 18-angular methyl group and structure have been given by Shoolery and Rogers (49), and Cox, Bishop and Richards (25) have established correlations for the chemical shifts of both angular methyl groups in a series of 11-keto-steroids. The chemical shift of a methyl group in the 6-position has

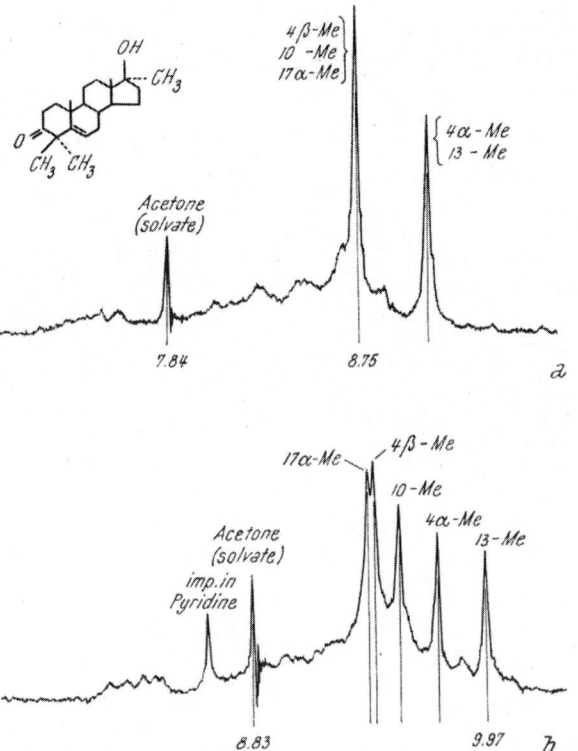

Fig. 13. The spectra of 4,4,17 α-trimethyl-17 β-hydroxy-5-androsten-3-one in (a) deuterochloroform and (b) pyridine; according to Slomp and MacKellar. [From: J. Amer. Chem. Soc. 82, 999 (1960).]

been shown to depend on its configuration (51). The difference in the chemical shifts of the 18-methyl group in normal and D-homo-steroids have been correlated with the nature and configuration of substituents at positions 16, 17 and 17 a (55).

The presence of alkyl and alkenyl side-chains complicates the "methyl" region of the spectra of steroids. However, Slomp and MacKellar (53) have shown that the absorptions arising from methyl groups in the side-chains of cholestane, ergost-22-one, stigmast-22-ene, and sitostane can be assigned and hence recognised in the spectra of steroids. The same authors (52) have also shown that pyridine as a solvent causes

References, pp. 360—362.

márked changes (0.3–1.0 p. p. m.) in chemical shifts of methyl groups in steroids, and that this effect can be useful in analysing the spectra of steroids which contain a number of methyl groups. *Fig. 13* provides an excellent example of this technique.

Since the absorptions of angular methyl groups can be so readily assigned, proton magnetic resonance spectroscopy has proved useful for demonstrating substitution at $C_{(18)}$. For example, JEGER and his colleagues (*16*) were able to show very simply by nuclear magnetic resoance that the photolysis of 20-ketopregnanes involved cyclisation between the 18- and 20-positions.

For a comprehensive treatment of the proton spectra of steroids, the reader is referred to the recent book by BHACCA and WILLIAMS (*15*).

2. Carbohydrates.

The most important use of proton magnetic resonance spectroscopy in carbohydrate chemistry is in establishing the stereochemistry and conformation of pyranose and furanose rings. The foundations of these applications were provided, in 1957, by LEMIEUX, KULLNIG, BERNSTEIN and SCHNEIDER (*40*), who established that the chemical shifts of protons attached directly to the pyranose ring, and also of protons of O-acetyl groups, depended on conformation. They further showed that the coupling constants between protons attached to contiguous carbon atoms of the pyranose ring could be used to establish their geometric relation, a result which was subsequently rationalised by the KARPLUS equation (p. 336).

Most of the early correlations concerned chemical shifts and coupling constants for anomeric protons since, being attached to the same carbon atom as two oxygen atoms, these absorb in a characteristic region (4.0– –5.5). However, by working at 100 Mc./sec., by using spin-decoupling, and by choosing suitable derivatives, it has since proved possible to assign most absorptions in the spectra of a number of monosaccharides. It is now known that the original correlations of LEMIEUX et. al. (*40*) apply quite well to all protons of the pyranose ring. These correlations may be stated as follows.

(i) An axial proton absorbs at higher τ-values than the corresponding equatorial proton, the shift being 0.1–0.70 p. p. m.

(ii) The protons of an axial acetoxyl group absorb at lower τ-values than the proton of the corresponding equatorial group, the shift being 0.1–0.2 p. p. m.

(iii) The coupling constant between two vicinal and axial protons is 6–11 cps. The coupling constants between two vicinal protons which are either axial-equatorial or diequatorial is ca. 3 cps.

Of these correlations the third is undoubtedly the most reliable and has the advantage that it can be used even if only one isomer is available, whereas the application of the first two correlations really requires a comparison of data for the axial and equatorial protons. Of course, all three correlations are unreliable in those molecules in which the two chair conformations of the pyranose ring are similarly populated. In

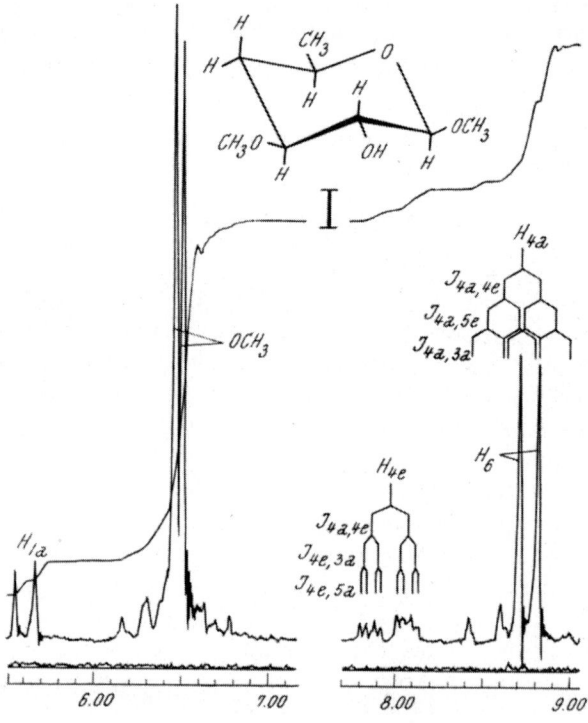

Fig. 14. The spectrum of methyl chalcoside in pyridine at 60 Mc./sec.; according to Woo, Dion and JOHNSON. [From: J. Amer. Chem. Soc. **84**, 1066 (1962).]

these cases, the observed spectrum is a time averaged one in which the observed shifts and coupling constants are the weighted mean of those of the two conformations, the weighting being proportional to the relative population of the two conformations.

An excellent example of the application of coupling constant data in the elucidation of conformation and stereochemistry is provided by the proton magnetic resonance study of chalcose (**13**) and one of the methyl-chalcosides (**14**) (**56**).

The spectrum of the glycoside is shown in *Fig. 14*. The region 9–7.5 contains the absorptions of the C-methyl group (doublet at 8.8) and of the protons of the $C_{(4)}$ methylene group which consists of two chemically

References, pp. 360—362.

shifted multiplets at 8.7 and 8.0, the former being partly obscured by the absorption of the C-methyl group. The multiplet at 8.0 exhibits one large (12.5 c./sec.) and two small splittings. The large splitting is clearly due to the geminal coupling constant between the two methylene protons. The multiplet at 8.7 has a spacing of 35 c./sec. between its outermost lines, which is equal to the sum of the geminal coupling constant and the two vicinal coupling constants. The sum of the two vicinal coupling constants is therefore 22.5 c./sec. so that both must be due to diaxial interactions. Hence, the substituents at both the 3- and 5-positions are equatorial. The configurations of the remaining two asymmetric centres are established by consideration of the splitting of the absorption at 5.60 due to the anomeric proton. This absorption is a simple doublet the separation (ca. 7 cps.) of which indicates that this proton is involved in a diaxial coupling with the adjacent 2-proton. Thus all substituents in (14) are equatorial. Chalcose itself is the other anomer since the absorption of the anomeric proton in the free sugar is a triplet with a sparing of 2–3 cps. due to coupling with the hydroxylic proton and to an equatorial-axial coupling with the 2-proton.

(13)
Chalcose.

(14)
Methyl-chalcoside.

(15)

The configuration of a number of other pyranose derivatives have been established in a similar way (29, 30, 43).

The extension of the above arguments to problems of conformation and configuration of furanose sugars has been made by several investigators. JARDETZKY (34) has analysed the spectra of a number of ribonucleosides, and desoxyribonucleosides in terms of the KARPLUS equation (p. 336) and has made predictions regarding the conformation of the furanose rings in these systems. A particularly detailed study of the coupling constants between the protons of the furanose ring has been made by ABRAHAM, HALL, HOUGH and McLAUCHLAN (1, 2). These authors have made a careful study of a number of derivatives of 1, 2-O-isopropylidene-α-D-xylohexofuranose derivatives and have shown that the furanose ring adopts the skew conformation (15). They have also developed for these systems the empirical "KARPLUS equation" which can be used for establishing relative configurations:

$$J = J_0 \cos^2 \Theta - 0.28 \text{ cps.},$$

where
$$J_0 = 9.27 \text{ for } 0 \leqslant \Theta \leqslant 90°,$$
$$= 10.36 \text{ for } 90 \leqslant \Theta \leqslant 180°.$$

3. Carotenoids and Acyclic Terpenes.

The combination of visible light absorption and proton magnetic resonance spectroscopy has proved remarkably effective for determining the structures of carotenoids. The former method establishes the nature of the conjugated polyene system (37), while the latter technique is particularly valuable for deriving the structures of the end groups since these are usually characterised by two or three methyl groups which give rise to well defined proton absorption signals. In this way the structures of a number of carotenoids have been elucidated using only a few milligrams of material.

The characteristic chemical shifts of methyl groups in carotenoids of known structures have been established by BARBER, DAVIS, JACKMAN and WEEDON (7). A feature common to the spectra of all carotenoids is the absorption at 7.95–8.40 associated with the methyl groups attached directly to the polyene system. These methyl groups are of two types, those situated within the polyene chain (in-chain) and those attached to the terminal carbon atoms of the conjugated system (end-of-chain). The former absorb consistently at 8.03 (in $CDCl_3$) whereas the latter absorb at slightly higher fields (8.15–8.35), the precise position depending on the nature of the adjacent end-group. Thus, in lycopene and related acyclic carotenoids the end-of-chain methyl groups give rise to absorption near 8.18. If, however, the terminal double bonds are part of cyclic systems, as in β-carotene, the end-of-chain methyl groups absorb near 8.30.

The 1, 3, 3-trimethylcyclohexenyl and related end groups in carotenoids are readily characterised by the chemical shifts of the protons of the *gem*-dimethyl groups. In the examples (16), (17) and (18) these two methyl groups are rendered equivalent by the rapid interchange between the two identical half-chain conformations of the cyclohexene ring. The methyl groups of the end group (19) are also equivalent although in this case the two conformations are not equivalent. In contrast, each of the methyl groups in the end groups of methyl azafrin (20), β-carotene di-epoxide (21) and aurochrome (22) give rise to distinct absorptions.

The most common acyclic end group is that of lycopene (23). In this system the two methyl groups attached to the non-conjugated double bond are in different environments and hence their protons have different chemical shifts, as **shown**. The absorption at 8.40 is assigned to the methyl group which is *cis* to the adjacent methylene group (see below)

References, pp. 360—362.

and, in general, methyl groups of the type (24) also absorb at 8.40. Thus, the methyl groups in lycopene (in CCl_4) give rise to absorption bands at 8.06 (in-chain), 8.20 (end-of-chain), 8.35 and 8.40 which have the relative intensities 2 : 1 : 1 : 1.

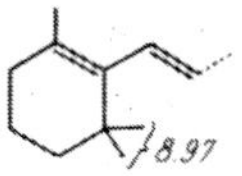

(16) 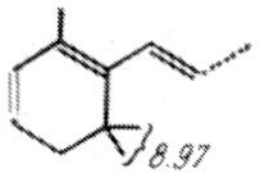   (17) 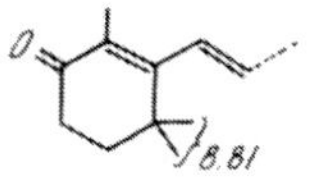   (18)

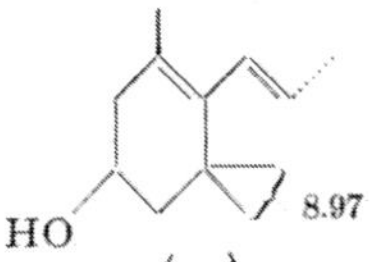

(19)

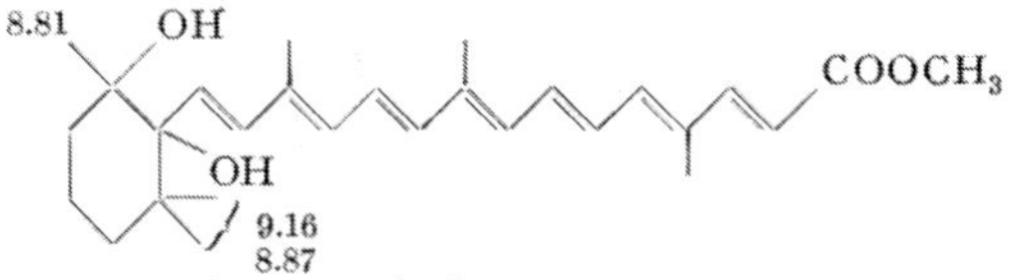

(20)
Methyl azafrin.

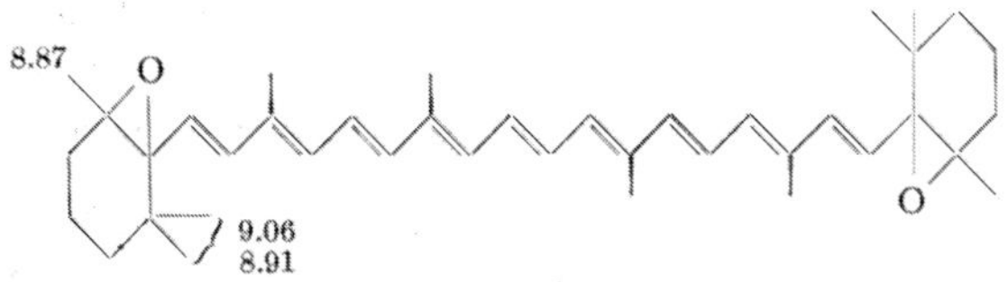

(21)
β-Carotene diepoxide.

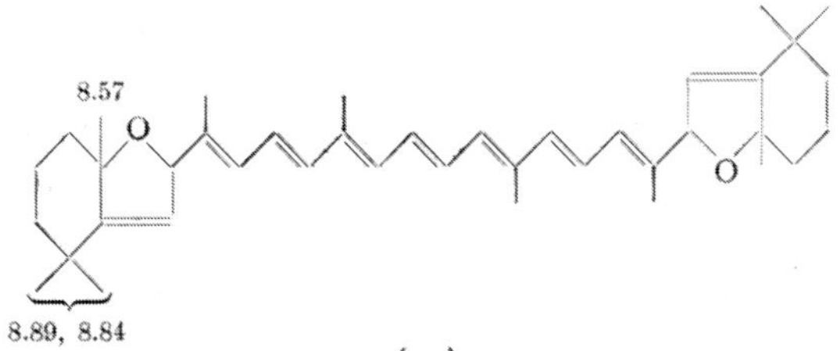

(22)
Aurochrome.

(23)
Lycopene.

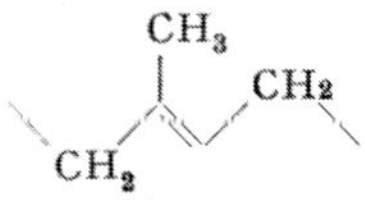

(24)

Postulated biosynthetic precursors of lycopene which differ only in the degree of unsaturation have been isolated from various natural sources and their structures have been either established or confirmed by n. m. r. spectroscopy (*27*). For instance, neurosporene which has one less con-jugated double bond than lycopene is assigned the structure (**25**) since the relative intensities of the absorption bands at 8.06, 8.20, 8.35, and 8.40 are 3 : 2 : 2 : 3. The alternative formulation (**26**) would require the relative intensities to be 3 : 3 : 1 : 3.

(**25**)
Neurosporene.

(**26**)

A number of carotenoids isolated from bacteria have oxygenated acyclic end-groups of the types (**27**), (**28**) and (**29**), and these structural features can be readily identified by proton resonance spectroscopy.

$R = H$ or CH_3 $R = H$ or CH_3 $R = H$ or CH_3

(**27**)   (**28**)   (**29**)

In this way it has been possible to establish the structures of spirilloxan-thin (**30**) (*10*), 3,4-dehydrorhodopin (**31**) (*32*), chloroxanthin (**32**), sphero-idene (**33**) and spheroidenone (**34**) (*27*).

(**30**)
Spirilloxanthin.

(**31**)
3,4-Dehydrorhodopin.

(**32**)
Chloroxanthin.

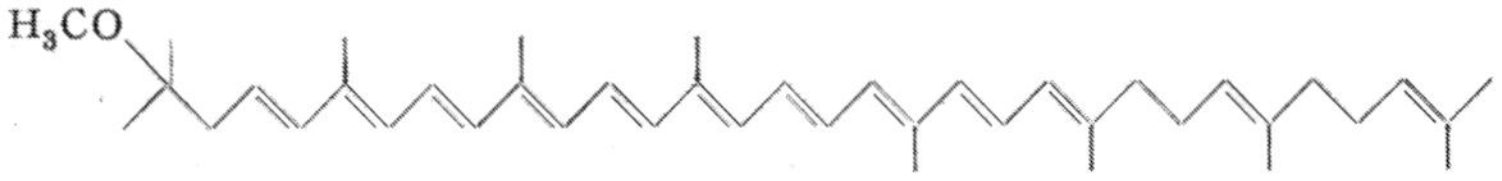

(33)

Spheroidene.

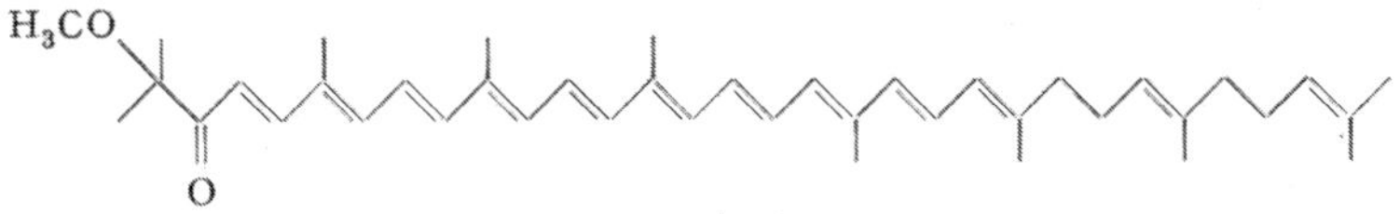

(34)

Spheroidenone.

Proton magnetic resonance spectroscopy has also played a major role in the elucidation of the paprika pigments, capsorubin and capsanthin (9). The former was known to contain a nonadienone system and hydroxylated end-groups. The hydroxyl groups could be oxidised to keto groups, the carbonyl stretching frequencies of which were 1740 cm.$^{-1}$, suggestive of a cyclopentanone. The nuclear magnetic resonance spectrum exhibits sharp singlets at 9.16, 8.80 and 8.63 each of which is equivalent to six protons. Clearly, these bands must arise from pairs of equivalent methyl groups which are attached to fully substituted saturated carbon atoms. It follows that the two end groups of capsorubin are identical. On biogenetic grounds, the most likely structure of the end-group is that of a 1,2,2-trimethylcyclopentanol. In agreement with this formulation it was found that the three singlets arising from aliphatic methyl groups of synthetic 4,8,13,17-tetramethyl-1,20-di-(1,2,2-trimethylcyclopentyl)-eicosa-2,4,6,8,10,12,14,16,18-nonaene-1,20-dione (35; R=H) were uniformly displaced to higher fields from the analogous absorptions in the spectrum of the tetraketone derived from oxidation of the two secondary alcoholic functions in capsorubin. This result not only confirms the presence of the 1,2,2-trimethylcyclopentane ring system but indicates that the hydroxyl group is in the 4-position so that the corresponding carbonyl group deshields the three methyl groups to the same extent. Capsorubin is therefore assigned the structure (35; R=OH) which has recently been confirmed by synthesis (23).

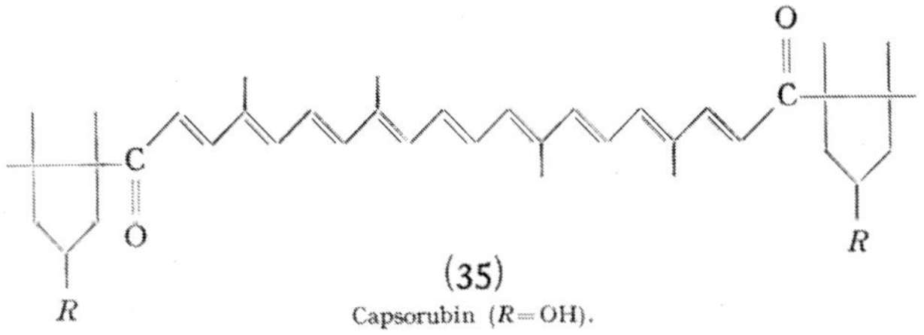

(35)

Capsorubin (R=OH).

Capsanthin was shown by proton magnetic resonance spectroscopy to have one end group the same as capsorubin. It had already been established (57) that the other end group was identical with those (19, p. 345) of β-citraurin, so that capsanthin has the structure (36).

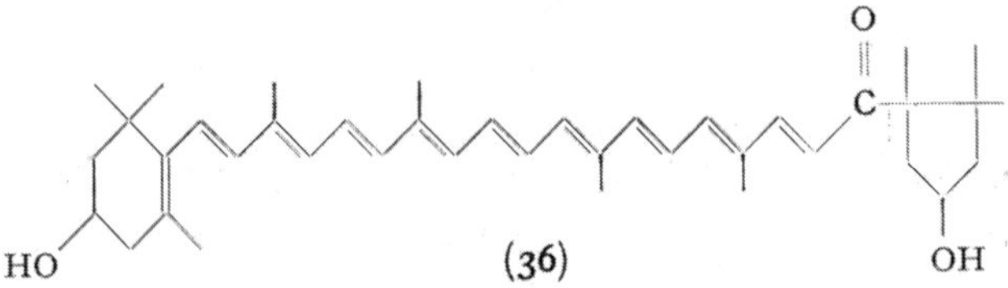

(36)

Capsanthin.

Nuclear magnetic resonance spectroscopy has proved to be most valuable for determining the stereochemistry of trisubstituted double bonds, of the type —C(CH₃)=CH—, which are frequently present in *acyclic terpenes*. An example is provided by the use of chemical shift data to confirm the assignment (58) of geometric configurations to nerol (37) and geraniol (38). The allylic alcohols were interrelated with the corresponding α,β-unsaturated acid either by synthesis of the acids and subsequent reduction with lithium aluminium hydride or by a two-stage oxidation of the alcohols with manganese oxide followed by silver oxide to the acids (17, 18). The configurations of the two acids was then established by the τ-values of the β-methyl groups in the spectra of the corresponding esters using rules enumerated in Section II/6 (p. 333). These methods have also been used to establish the configuration of the double bond in natural phytol and of the 2-double bond in natural farnesol (17).

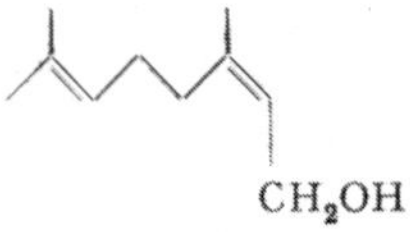

(37)

Nerol.

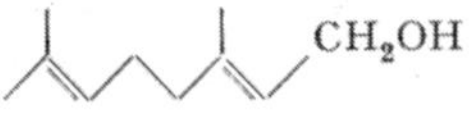

(38)

Geraniol.

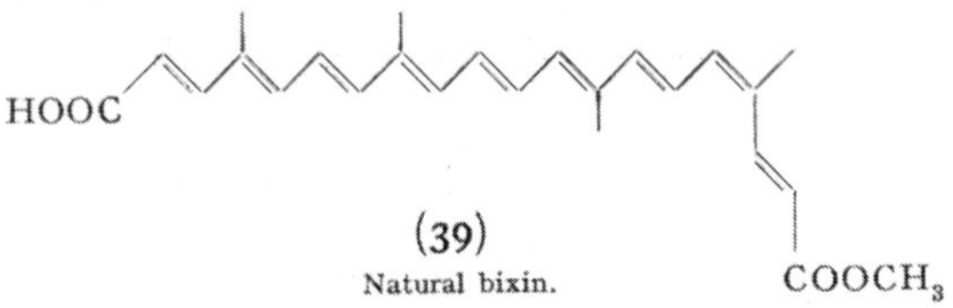

(39)

Natural bixin.

A useful correlation between the stereochemistry of double bonds of the type —C(CH₃)=CH—CH₂— and the chemical shift of the methyl group has been developed by BATES and GALE (13). They have shown that the methyl group, in the isomer in which it is *cis* to the methylene group, is usually at higher fields by 0.05–0.1 p. p. m.

N. m. r. spectroscopy has also been used to show that natural bixin has the configuration (39) (8).

4. Flavonoids and Related Compounds.

A number of structural and stereochemical problems in the chemistry of this class of compounds have been solved by MASSICOT and MARTHE (42), BATTERHAM and HIGHET (14), and CLARK-LEWIS, JACKMAN and SPOTSWOOD (21) by n. m. r. spectroscopy, and three extensive surveys of chemical shift and coupling constant data have been reported. The compounds investigated include flavones (40), flavanones (41), flavans (42), isoflavones (43) and aurones (44).

(40)
Flavones.

(41)
Flavanones.

(42)
Flavans.

(43)
Isoflavones.

(44)
Aurones.

The structure of the heterocyclic ring can generally be established from its proton spectrum. Thus it is possible to distinguish between flavones and isoflavones since the 3-proton of the former is found at 3.3 compared with 2.2 (1.7 in dimethylsulphoxide) in the latter. The olefinic protons in aurones are reported in the range 3.52–3.35 (in dimethyl-sulphoxide). 3-Hydroxyflavones can be recognised by the absence of absorption due to protons attached directly to the heterocyclic rings. The assignment of absorptions to olefinic protons of the heterocyclic rings must be made with care, as in flavonoids which are heavily substituted with oxygen atoms and isoprenoid residues, the distinction between an olefinic proton and the proton of a penta- or 1,2,4,5-tetrasubstituted benzene ring may not be easy (47).

Saturated heterocyclic rings also give rise to characteristic proton absorptions. The spectra of flavanones contain typical ABX multiplets arising from the 2-proton and the two 3-protons (*42*). The 2-proton is generally a double-doublet near 4.5, the precise position depending on the substitution of ring *B*, while the protons at the 3-position give rise to a multiplet of eight lines near 7.0. 3-Hydroxyflavones which occur in nature have the *trans* configuration and give rise to an AB quartet in the region 4.4–5.6 with a characteristic coupling constant of ca. 12 c./sec. The 3-proton is usually at higher field (5.2–5.6) but is moved by about 1 p. p. m. to lower field in the spectrum of the corresponding 3-acetoxy compound.

3-Hydroxyflavones have absorptions at 5.0–5.4 (2-proton), 5.8–6.0 (3-proton; 4.3–4.8 in the corresponding acetates) and 6.7–7.4 (4-protons). The multiplicities of these absorptions depend on the stereochemistry. 3,4-Dihydroxyflavans also give characteristic absorptions, particularly in the case of the diacetates in which the 4-proton absorbs in the range 3.6–4.2. In all cases of flavans containing hydroxyl substituents in the heterocyclic ring, assignments to protons attached to the same carbon

Table 3. Chemical Shifts for Protons of Ring *A* in Flavonoids.

Compound Class	Position of Proton			
	5	6	7	8
6-*Hydroxy*				
Flavanone	2.83	—	2.92	3.07
7-*Hydroxy*				
Flavone	1.90	2.88	—	2.92
3-Hydroxyflavone	2.92—2.93	3.05—3.07	—	3.08—3.12
Flavanone	2.35	2.46	—	2.68
Isoflavanone	1.98	3.03	—	3.12
5,7-*Dihydroxy*				
Flavone	—	3.73	—	3.45
3-Hydroxyflavone	—	3.70—3.75	—	3.45—3.54
Flavanone	—	4.05—4.07	—	4.05—4.07
3-Hydroxyflavanone	—	4.12—4.15	—	4.07—4.08
3-Hydroxyflavan	—	4.25—4.27	—	4.05—4.08
Isoflavone	—	3.76—3.77	—	3.58—3.62
5,7-*Dimethoxy*				
Flavone	—	3.64	—	3.17
3-Hydroxyflavanone	—	3.83—3.90	—	3.83—3.90
Flavan	—	3.73—3.90	—	3.90—3.93
7,8-*Dimethoxy*				
Flavan	2.83—3.25	3.26—3.48	—	—

* Measured in hexadeutero-dimethylsulphoxide (*14*).

** Measured in deuterochloroform (*21, 42*).

atoms as the hydroxyl groups can be confirmed by examining the spectra of the corresponding acetates, since acetylation is associated with a down field shift of 1.0–1.20 p. p. m.

Most naturally occurring flavonoids have one or more hydroxyl substituents in each of the aromatic rings, and nuclear magnetic resonance can be used to indicate the substitution patterns. The free phenols are usually insoluble in chloroform or carbon tetrachloride and spectra have to be obtained with solutions in hexadeuterodimethylsulphoxide (14) or, alternatively, the corresponding fully methylated derivatives can be examined (21, 42).

The most common hydroxylation pattern for ring A has oxygen atoms at the 5- and 7-position so that the aromatic ring is of the phloroglucinol type. In these compounds, the 6- and 8-protons absorb in the range 3.7–4.1, often as AB systems exhibiting the *meta*-coupling constant, 2–3 c./sec. The 7,8-oxygen pattern is also quite common and the 5- and 6-protons give rise to AB multiplets in the range 2.8–3.5, in this case with the *ortho*-coupling constant of ca. 8 c./sec. *Table 3* summarises the

Table 4. Chemical Shifts for Protons of Ring B in Flavonoids.

Compound Class	Position of Proton				
	2'	6'	3'	5'	4'
*Unsubstituted**					
Flavone............	ca. 1.95	ca. 1.95	ca. 2.37	ca. 2.37	ca. 2.37
3-Hydroxyflavone ..	ca. 1.75	ca. 1.75	ca. 2.37	ca. 2.37	ca. 2.37
Flavanone	2.54–2.58	2.54–2.58	2.54–2.58	2.54–2.58	2.54–2.58
*4'-Hydroxy**					
Flavone............	2.01	2.01	3.10	3.10	—
3-Hydroxyflavone ..	1.90	1.90	3.02	3.02	—
Flavanone	2.66	2.66	3.18	3.18	—
Isoflavone.........	2.62	2.62	3.17	3.17	—
*4'-Methoxy***					
3-Hydroxyflavone ..	2.52	2.52	3.03	3.03	—
Flavan	2.54–2.67	2.57–2.67	3.05–3.12	3.05–3.12	—
*3',4'-Dihydroxy**					
Flavone............	2.40	2.50	—	3.00	—
3-Hydroxyflavone ..	2.22–2.26	2.32–2.35	—	2.99–3.02	—
Flavanone	3.09–3.10	3.19–3.20	—	3.19–3.20	—
Flavan	3.08–3.23	3.30	—	3.30	—
*3',4',5'-Trihydroxy**					
3-Hydroxyflavanone	2.68–2.70	2.68–2.70	—	—	—
*3',4',5'-Trimethoxy***					
3-Hydroxyflavanone	2.67	2.67			

* Measured in hexadeutero-dimethylsulphoxide (14).

** Measured in deuterochloroform (21, 42).

available data for the chemical shifts of the protons of ring *A*. It is also possible to recognise certain hydroxylation patterns in ring *B*. Thus, 3',4',5'-trimethoxyflavans give singlets at 3.3, equivalent in intensity to two protons. In the 4'-hydroxy and methoxy compounds, the four protons of ring *B* constitute an A_2B_2 (or, more rigorously, AA'BB') system, the spectrum of which can often be resolved. *Table 4* summarises the chemical shift data for the protons of ring *B*. Of course, in some compounds the absorptions of the protons of ring *A* and *B* overlap and the analysis of the spectrum is either impossible or leads only to partial assignments.

The relative stereochemistry of 3-substituted flavanones and of 3-, 4-, and 3,4-di-substituted flavans can usually be established from a consideration of vicinal coupling constants and the KARPLUS equation. In all cases, the heterocyclic ring appears to adopt the chair or half-chair conformation in which the 2-aryl group is quasi-equatorial.

MASSICOT and MARTHE (*42*) have analysed the ABX spectrum of the heterocyclic ring protons of 6,7-dimethoxyflavanone, and have shown the two vicinal coupling constants to be 13.5 and 3.2 c./sec. The former is clearly a diaxial interaction, thus establishing the equatorial character of the 2-aryl group in flavanones. All 3-hydroxy- and 3-acetoxy-flavanones which have been examined exhibit vicinal coupling constants of ca. 12 c./sec. and are therefore assigned the *trans* (di-equatorial)-configuration, although in the case of the naturally occurring compounds the possibility of epimerisation during isolation cannot be excluded. Both *cis*- and *trans*-3-bromoflavanones have been prepared and exhibit vicinal coupling constants of 1.7 and 8.5 c./sec., respectively. Apparently, the bulky bromine atom caused some distortion of the normal conformation of the heterocyclic ring (*21*).

The conformation of the flavan ring has also been established as having the 2-aryl group essentially equatorial. The 2,3-coupling constants in 3-hydroxy- and 3-acetoxy-flavans are ca. 1.0 and 8.0 c./sec. in the *cis* and *trans* isomers, respectively. In the *cis* isomer the two 3,4-coupling constants are 2–3 and 4 c./sec. compared with 4–5 and 7–10 c./sec. for the analogous interactions in the *trans* isomer. These data indicate the quasi-equatorial character of the 2-aryl group in both isomers. The

Table 5. Coupling Constant Data for 3,4-Disubstituted Flavans (*21, 24*).

Epimer	$J_{2,3}$ (c./sec.)	$J_{3,4}$ (c./sec.)
cis-cis	1.0	3.9–4.6
2,3-*cis*-3,4-*trans*	1.0	2.4–3.2
2,3-*trans*-3,4-*cis*	9.5–10.9	3.1–3.5
trans-trans	8.5–10.0	6.8–7.8

References, pp. 360—362.

2,3-coupling constants provide a very easy means of establishing the configuration of a 3-hydroxy-flavan. Examination of the four epimeric 3,4-dihydroxyflavans has shown that coupling constant data again provides a simple proof of relative stereochemistry (*21, 24*). The coupling constant data for these systems is assembled in *Table 5*.

Closely related to the flavonoids are the rotenoids which have the basic skeleton (**45**) and a survey of the nuclear magnetic resonance spectra of this group has been reported by CROMBIE and LOWN (*26*). Most rotenoids have methoxyl groups at the 2- and 3-positions so that the 1- and 4-protons give rise to singlets in the aromatic region of the spectrum. The 4-protons, being well removed from the regions of structural and stereochemical variation, absorb within a very narrow range, viz. 3.5–3.62, throughout the series investigated. The chemical shift of the 1-proton, on the other hand, is very sensitive to its steric relation to the 12-keto group. In normal rotenoids, which have a *cis-B/C* ring junction, the 1-proton is found in the range 3.15–3.43. This is in marked contrast to the *trans*-series and the dehydrorotenoids (**46**) in which the 1-proton absorbs at 1.9–2.0 and 1.6–1.7, respectively. In both cases the 1-proton lies in close proximity to, and in the trigonal plane of the carbonyl group (cf. Fig. 9, p. 335). Significant variations in the chemical shift of the 1-proton are found in the spiro-derivatives (**47**), the observed value of 3.7 indicating shielding by the carbonyl group. Conversion of the carbonyl group to an enol acetate shifts the position of the 1-proton to 2.6–2.7.

(45)
Rotenoids.

(46)
Dehydrorotenoids.

(47)
Spiro-derivatives.

The protons of rings *B* and *C* give rise to complex multiplets in the region 5.0–6.5 which overlaps with the absorptions of methoxyl groups. It is probable that this region of the spectrum measured at 100 Mc./sec. could be sufficiently well resolved to permit analysis, and the 12a, 6a-coupling constant should give the stereochemistry of the *B/C* junction.

Many naturally occurring rotenoids have an isoprenoid substituent in ring *D*, the methyl groups of which have highly characteristic proton spectral properties making the identification of the nature of the isoprenoid residue a very simple matter. Since isoprenoid groups are also common in flavonoids, rotenoids, coumarins, and xanthones as well as, of course, terpenes and steroids, a survey of the n. m. r. characteristics of the more important types will now be given.

The isopropenyldihydrofuran structure (48) is found in rotenone and a number of other rotenoids. It is most readily recognised by a single methyl group absorption near 8.3 and olefinic absorption, equivalent to two protons, in the range 4.8–5.4. Its presence is readily confirmed by hydrogenation of the double bond since the product possesses an isopropyl group which gives a doublet ($J = 6$–7 c./sec.) equivalent to six protons, or two doublets, each equivalent to two protons, at 8.6–8.8. The closely related isopropylfuran structure (49) gives a six proton doublet at 8.6–8.8. It can be distinguished from the isopropyldihydrofuran structure on the basis of the position of absorption of the methine proton of the isopropyl group, if necessary using spin decoupling methods to locate it. The aromatic proton of the structure (49) is found at 3.3.

(48)
Isopropenyldihydrofuran.

(49)
Isopropylfuran.

(50)
Dimethylchromene.

The dimethylchromene structure (50) is found in a variety of natural products, and nuclear magnetic resonance data for this group in rotenoids, flavones (*19*), isoflavones (*47*), and coumarins (*3*) have been recorded. The *gem*-dimethyl group gives a sharp singlet (six protons) near 8.5 and the olefinic protons absorb as a pair of doublets ($J = 10$ c./sec.) at 3.2–3.3 and 4.3. The olefinic absorption is replaced by a pair of triplets ($J = 6.5$–7 c./sec.) at 7.1–7.3 and 8.1–8.2 in the spectrum of the chroman obtained by hydrogenation of the double bond.

(51)
Morellin.

The isoprenoid residue is often simply a 3,3-dimethylallyl group. This gives rise to broad singlets at 8.2 (three protons) and 8.3 (three protons), a broad doublet ($J = 7$ c./sec.) near 6.6 (two protons) and a broad triplet at 4.9 (one proton). In the coumarin, avicennin, one isoprenoid residue is present as a 3-methylbutadien-1-yl group (3). The 1- and 2-protons give an AB spectrum (2.58, 3.30; $J = 16$ c./sec.) indicating the presence of a *trans* disubstituted double bond. The methylene and methyl protons are at 4.85 and 7.98, respectively.

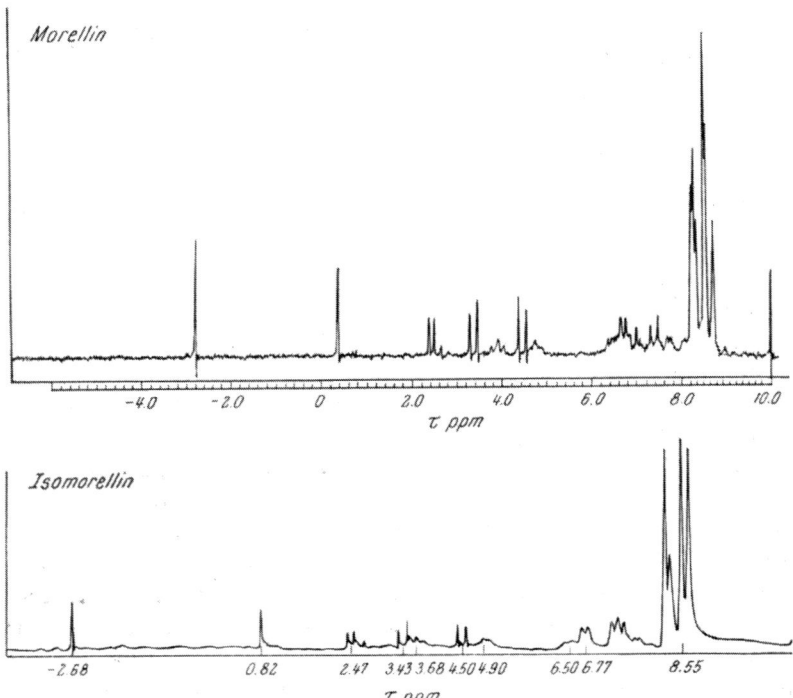

Fig. 15. The spectra of (a) morellin, and (b) isomorellin in deuterochloroform at 60 Mc./sec.; according to VENKATARAMAN et. al. (From: Tetrahedron Letters 1963, 459.)

An excellent example of the usefulness of n. m. r. spectroscopy in detecting isoprenoid resonances is provided by the study of the xanthone derivative morellin (51) which contains four such residues (38). The spectra of morellin and isomorellin, which differ simply in the configuration of the double bond of the α,β-unsaturated aldehyde, are reproduced in *Fig. 15*. The characteristic absorptions of the 3,3-dimethylallyl group and the 1,1-dimethylchromene are clearly observed. In addition, the presence of the residue, $-CH_2CH=C(CH_3)CHO$ can be established. The remaining isoprene unit is less well defined although the *gem*-dimethyl group gives a recognisable signal in the spectrum of isomorellin.

5. Alkaloids.

The great diversity of structures exhibited by the alkaloids as a class prevents a systematic survey of characteristic n. m. r. features. The two examples given below will, however, serve to indicate the value of the technique in alkaloid chemistry.

Table 6. N. M. R. Data for Ochotensimine.

	Intensity	Multiplicity
2.88	1	AB system; $J = 8$ c./sec.
3.20	1	AB system; $J = 8$ c./sec.
3.47	1	singlet
3.70	1	singlet
4.03	2	singlet
4.37	1	singlet
5.10	1	singlet
6.17	3	singlet
6.38	3	singlet
6.55	1	AB system; $J = 18$ c./sec.
7.05	1	AB system; $J = 18$ c./sec.
7.10	4	complex multiplet
7.87	3	singlet

The first example concerns the structure of ochotensimine, $C_{22}H_{23}NO_4$, (41) and is typical of the way in which n. m. r. spectra of a compound and some of its degradation products readily yield extensive structural information. The spectral data for ochotensimine is presented in *Table 6*. These data show the presence of two aromatic rings. The absorptions at 6.17 and 6.38 must be attributed to two aromatic methoxyl groups (assuming —COOCH₃ is eliminated by the absence of carbonyl absorption in the infrared spectrum), and the band at 4.03 is characteristic of an aromatic at methylenedioxy group. However, the AB system centered at 3.0 clearly arises from a pair of *o*-oriented aromatic protons so that there must be two aromatic rings (the molecular formula excludes more than two). The absorptions at 3.47 and 3.70 which exhibit no mutual splitting evidently arise from two *p*-oriented aromatic protons. This information indicates the presence of the structural features (52) and (53) or (54) and (55).

(52)   (53)   (54)   (55)

The presence of the grouping $\diagup$C$=$CH$_2$ is conclusively demonstrated by the observation that the two singlets at 4.37 and 5.10 are absent in the spectrum of dihydro-ochotensimine (obtained by catalytic hydrogenation) which instead contains a doublet ($J = 7.2$ c./sec.), equivalent to three protons, at 9.05 which is characteristic of the grouping $\diagup$CH$-$CH$_3$. The very large coupling constant for the AB system near 6.8 can be associated with a geminal spin-spin interaction and more specifically, both the coupling constant and the chemical shifts suggest an α-methylene group of an indane ring (39). The lack of further splitting of the AB spectrum shows the absence of β-protons. The alkaloid also contains one N-methyl group (7.87).

Hofmann degradation of the methiodide of dihydro-ochotensimine and Emde reduction of ochotensimine yield the same base ($C_{23}H_{27}NO_4$). This compound has a C-methyl group which absorbs as a closely spaced triplet at 8.07 but contains no absorption characteristic of olefinic protons. This information places the methyl group, and hence the $\diagup$C$=$CH$_2$ grouping in the parent alkaloid, at the second α-position of the indane ring. The partial structure (56) can therefore be written. A second Hofmann degradation eliminates the nitrogen atom and the spectrum of the neutral product contains a complex ABC absorption characteristic of styrene. The presence of a vinyl group in this compound is confirmed by the spectrum of its dihydro derivative which contains the familiar quartet (7.48) and triplet (8.87) of an aromatic ethyl group. This partial structure may therefore be expanded to (57) and ochotensimine must have one of the structures which accord with the oxygenation patterns given above.

(56)   (57)

Many alkaloids give rather diffuse spectra in the region 9.0–6.0, arising from methylene and methine protons, which are variously affected by nitrogen atoms in these molecules. This is well illustrated by the spectrum of methoxy-N-despropionyl-aspidoalbine (48) (Fig. 16 a) even when determined at 100 Mc./sec. This particular example is interesting since it has proved possible to extract extra information from the complex region, 7.0–9.0, by the use of multiple spin-decoupling. An important feature of the structure (58) of the alkaloid is the grouping —O—CH₂— —CH₂— and its attachment to the molecule at two angular positions.

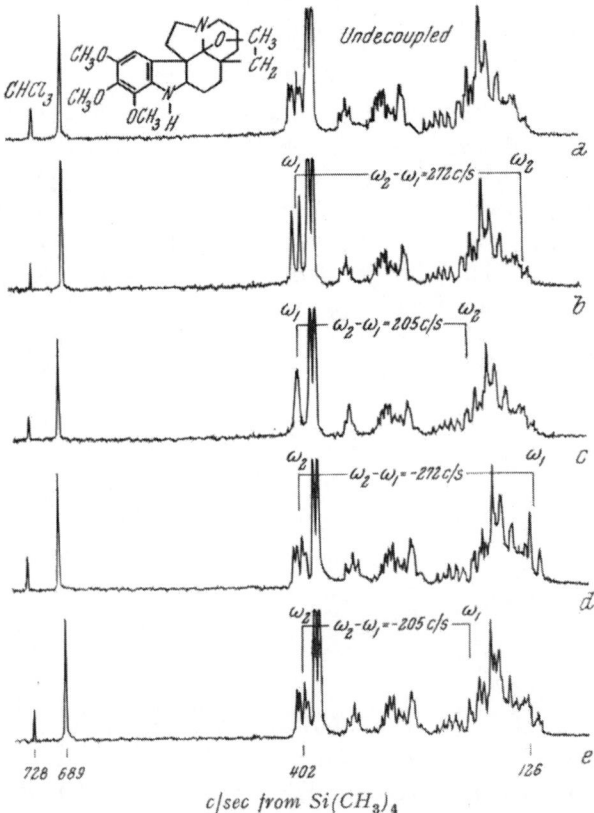

(58)

Methoxy-N-despropionyl-aspidoalbin.

(The Aspidosperm alkaloids have two less O-functions.)

Fig. 16. The single and double resonance spectra of methoxy-N-despropionyl-aspido-albine in deuterochloroform at 100 Mc./sec. In these spectra ω_2 refers to the strong radiofrequency field; according to SHOOLERY. [From: Discuss. Faraday Soc. **34**, 104 (1962).]

References, pp. 360—362.

The spectrum exhibits a double doublet at 5.98. This absorption is equivalent to two protons and its position allows its assignment to a methylene group adjacent to oxygen. The spectra in Fig. 16 (b–e) show the results of simple spin-decoupling experiments. These conclusively locate the two protons, A and B, responsible for the splitting of the methylene group signal, at 8.03 and 8.70, respectively. Finally, *Fig. 17 a* shows that the signal from proton A collapses to a sharp singlet if both the

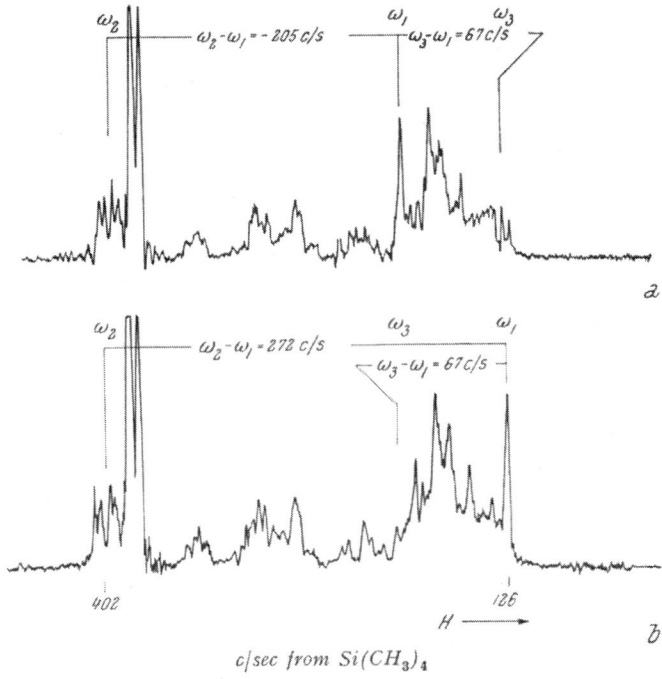

Fig. 17. The triple resonance spectra of methoxy-N-despropionyl-aspidoalbine in deuterochloroform at 100 Mc./sec. In these spectra ω_2 and ω_3 refer to the two strong radiofrequency fields; according to Shoolery. [From: Discuss. Faraday Soc. **34**, 104 (1962).]

methylene group and proton B are simultaneously irradiated. Similarly *(Fig. 17 b)*, proton B gives a singlet on simultaneous irradiation of proton A and the methylene group. It follows that A and B are coupled to each other and to the methylene group which establishes the presence of the grouping —O—CH_2—CH_2—. Since the triple resonance experiments collapsed the signals from protons A and B to singlets, neither of these protons are coupled to other protons in the molecule and the attachment of the second methylene group must be via an angular position. As, apart from the signals from the three methoxyl groups, there is no other absorption attributable to protons adjacent to oxygen, the oxygen atom must also be attached to an angular position.

References.

1. Abraham, R. J., L. D. Hall, L. Hough and K. A. McLauchlan: A Proton Resonance Study of the Conformations of Carbohydrates in Solution. Part I. Derivatives of 1,2-O-Isopropylidene-α-D-xylohexofuranose. J. Chem. Soc. (London) **1962**, 3699.

2. Abraham, R. J., K. A. McLauchlan, L. D. Hall and L. Hough: The Conformations of some Furanose Derivatives in Solution by Proton Magnetic Resonance. Chem. and Ind. **1962**, 213.

3. Arthur, H. R. and W. D. Ollis: A Revised Structure for Avicennin. J. Chem. Soc. (London) **1963**, 3910.

4. Ayer, W. A., C. E. McDonald and J. B. Stothers: The Stereochemistry of Maleopimaric Acid and the Long Range Shielding Effect of the Olefinic Bond. Canad. J. Chem. **41**, 1113 (1963).

5. Balasubramanian, S. K., D. H. R. Barton and L. M. Jackman: Diterpenoid Bitter Principles. Part V. The Constitution of Palmarin and its Congeners. J. Chem. Soc. (London) **1962**, 4816.

6. Banwell, C. N. and N. Sheppard: (H—H) Coupling Constants in the Nuclear Magnetic Resonance Spectra of Hydrocarbon Groupings. Discuss. Faraday Soc. **No. 34**, 115 (1962).

7. Barber, M. S., J. B. Davis, L. M. Jackman and B. C. L. Weedon: Studies in Nuclear Magnetic Resonance. Part I. Methyl Groups of Carotenoids and Related Compounds. J. Chem. Soc. (London) **1960**, 2870.

8. Barber, M. S., A. Hardisson, L. M. Jackman and B. C. L. Weedon: Studies in Nuclear Magnetic Resonance. Part IV. Stereochemistry of the Bixins. J. Chem. Soc. (London) **1961**, 1625.

9. Barber, M. S., L. M. Jackman, C. K. Warren and B. C. L. Weedon: Carotenoids and Related Compounds. Part IX. The Structures of Capsanthin and Capsorubin. J. Chem. Soc. (London) **1961**, 4019.

10. Barber, M. S., L. M. Jackman and B. C. L. Weedon: The Structures of Spirilloxanthin and Related Carotenoids. Proc. Chem. Soc. (London) **1959**, 96.

11. Barton, D. H. R. and D. Elad: Colombo Root Bitter Principles. Part II. The Constitution of Columbin. J. Chem. Soc. (London) **1956**, 2090.

12. Barton, D. H. R., K. H. Overton and A. Wylie: Diterpenoid Bitter Principles. Part IV. Investigations on the Constitution of Palmarin. J. Chem. Soc. (London) **1962**, 4809.

13. Bates, R. B. and D. M. Gale: Stereochemistry of Trisubstituted Double Bonds in Terpenoids. J. Amer. Chem. Soc. **82**, 5749 (1960).

14. Batterham, T. J. and R. J. Highet: Nuclear Magnetic Resonance Spectra of Flavonoids. Austral. J. Chem. **17**, 428 (1964).

15. Bhacca, N. S. and D. H. Williams: Applications of N. M. R. Spectroscopy in Organic Chemistry. Illustrations from the Steroid Field. San Francisco: Holden-Day, Inc. 1964.

16. Buchschacher, P., M. Cereghetti, H. Wehrli, K. Schaffner und O. Jeger: Über Steroide und Sexualhormone. 212. Mitt. Photochemische Umwandlungen von 20-Keto-pregnan-Verbindungen. Helv. Chim. Acta **42**, 2122 (1959).

17. Burrell, J. W. K., R. F. Garwood, L. M. Jackman, E. Oskay and B. C. L. Weedon: Carotenoids and Related Compounds. Part XIII. Stereochemistry and Synthesis of Geraniol, Nerol, Farnesol and Phytol. J. Chem. Soc. (London) (in press, 1965).

18. Burrell, J. W. K., L. M. Jackman and B. C. L. Weedon: Stereochemistry and Synthesis of Phytol, Geraniol and Nerol. Proc. Chem. Soc. (London) **1959**, 263.

19. BURROWS, B. F., W. D. OLLIS and L. M. JACKMAN: Sericetin. Proc. Chem. Soc. (London) **1960**, 177.

20. CHAMBERLAIN, N. F.: Determining Molecular Structure by Nuclear Magnetic Resonance of Hydrogen. Analyt. Chemistry **31**, 56 (1959).

21. CLARK-LEWIS, J. W., L. M. JACKMAN and T. M. SPOTSWOOD: Nuclear Magnetic Resonance Spectra, Stereochemistry and Conformation of Flavan Derivatives. Austral. J. Chem. **17**, 632 (1964).

22. CONROY, H.: Nuclear Magnetic Resonance in Organic Structural Elucidation. Adv. Organ. Chem. **2**, 265 (1960).

23. COOPER, R. D. G., L. M. JACKMAN and B. C. L. WEEDON: Stereochemistry of Capsorubin and Synthesis of its Optically Inactive Epimers. Proc. Chem. Soc. (London) **1962**, 215.

24. COREY, E. J., E. M. PHILBIN and T. S. WHEELER: Stereochemistry of Flavan-3,4-diols. Tetrahedron Letters **1961**, 429.

25. COX, J. S. G., E. O. BISHOP and R. E. RICHARDS: Proton Resonance Spectra of Some 11-Keto-steroids. J. Chem. Soc. (London) **1960**, 5118.

26. CROMBIE, L. and J. W. LOWN: Proton Magnetic Studies of Rotenone and Related Compounds. J. Chem. Soc. (London) **1962**, 775.

27. DAVIS, J. B., L. M. JACKMAN, P. T. SIDDONS and B. C. L. WEEDON: The Structures of Phytoene, Phytofluene, ζ-Carotene and Neurosporene. Proc. Chem. Soc. (London) **1961**, 261.

28. FRASER, R. R.: The Establishment of Configuration of Diels-Alder Adducts by N. M. R. Spectroscopy. Canad. J. Chem. **40**, 78 (1962).

29. HALL, L. D., L. HOUGH, K. A. McLAUCHLAN and K. PACHLER: A Proton Resonance Study of the Conformation of Carbohydrates in Solution. Some Pyranose Derivatives. Chem. and Ind. **1962**, 1465.

30. HOFHEINZ, W. und H. GRISEBACH: Zur Biogenese der Makrolide. VII. Die Stereochemie der Mycaminose. Z. Naturforsch. **17 B**, 355 (1962).

31. JACKMAN, L. M.: Applications of Nuclear Magnetic Resonance Spectroscopy in Organic Chemistry. London: Pergamon. 1959.

32. JACKMAN, L. M. and S. L. JENSEN: Bacterial Carotenoids. IX. The Constitution of the Third Member of the P 481-Group (3,4-Dehydro-rhodopin). Acta. Chem. Scand. **15**, 2058 (1961).

33. JACKMAN, L. M. and R. H. WILEY: Studies in Nuclear Magnetic Resonance. Part III. Assignment of Configurations of $\alpha\beta$-Unsaturated Esters and the Isolation of Pure *trans-β*-Methyl-glutaconic Acid. J. Chem. Soc. (London) **1960**, 2886.

34. JARDETZKY, C. D.: Proton Magnetic Resonance Studies on Purines, Pyrimidines, Ribose Nucleosides and Nucleotides. III. Ribose Conformation. J. Amer. Chem. Soc. **82**, 229 (1960).

35. JOHNSON, C. E., Jr. and F. A. BOVEY: Calculation of Nuclear Magnetic Resonance Spectra of Aromatic Hydrocarbons. J. Chem. Physics **29**, 1012 (1958).

36. KARPLUS, M.: The Analysis of Molecular Wave Functions by Nuclear Magnetic Resonance Spectroscopy. J. Physic. Chem. **64**, 1793 (1960).

37. KARRER, P. and E. JUCKER: Carotenoids. New York: Elsevier. 1950.

38. KARTHA, G., G. N. RAMACHANDRAN, H. B. BHAT, P. M. NAIR, V. K. V. RAGHAVAN and K. VENKATARAMAN: The Constitution of Morellin. Tetrahedron Letters **1963**, 459.

39. KEVILL, D. N., G. A. COPPENS, M. COPPENS and N. H. CROMWELL: Reactions of 2-Bromo-2-(α-halogenobenzyl)-1-indanones. J. Organ. Chem. (USA) **29**, 382 (1964).

40. LEMIEUX, R. U., R. K. KULLNIG, H. J. BERNSTEIN and W. G. SCHNEIDER: Configurational Effects on the Proton Magnetic Resonance Spectra of Six-membered Ring Compounds. J. Amer. Chem. Soc. 80, 6098 (1958).
41. McLEAN, S. and MEI-SIE LIN: Ochotensimine: A Novel Benzyl-isoquinoline Alkaloid. Tetrahedron Letters 1964, 3819.
42. MASSICOT, J. et J.-P. MARTHE: Résonance magnétique nucléaire de produits naturels. III. Étude de quelques dérivés flavoniques et substances apparentées. Bull. soc. chim. France 1962, 1962.
43. MIYAMOTO, M., Y. KAWAMATSU, M. SHINOHARA, Y. ASAHI, Y. NAKADAIRA, H. KAKISAWA, K. NAKANISHI and N. S. BHACCA: Chromose A. Tetrahedron Letters 1963, 693.
44. NAIR, M. D. and R. ADAMS: The Structure of Ridellic Acid and the Stereochemistry of Necic Acids. J. Amer. Chem. Soc. 83, 922 (1961).
45. POPLE, J. A., W. G. SCHNEIDER and H. J. BERNSTEIN: High Resolution Nuclear Magnetic Resonance, p. 98. New York: McGraw-Hill. 1959.
46. ROBERTS, J. D.: An Introduction to the Analysis of Spin-Spin Splitting in High-resolution Nuclear Magnetic Resonance Spectra. New York: W. A. Benjamin. 1961. Nuclear Magnetic Resonance. Applications to Organic Chemistry. New York, Toronto, London: McGraw-Hill. 1959.
47. SCHWARZ, J. S. P., A. I. COHEN, W. D. OLLIS, E. A. KACZKA and L. M. JACKMAN: The Extractives of Piscidia erythrina L. I. The Constitution of Ichthynone. Tetrahedron 20, 1317 (1964).
48. SHOOLERY, J. N.: Recent Applications of High Resolution N. M. R. to the Determination of Molecular Structure. Discuss. Faraday Soc. No. 34, 104 (1962).
49. SHOOLERY, J. N. and M. T. ROGERS: Nuclear Magnetic Resonance Spectra of Steroids. J. Amer. Chem. Soc. 80, 5121 (1958).
50. SILVERSTEIN, R. M. and G. C. BASSLER: Spectrometric Identification of Organic Compounds. New York: Wiley and Sons. 1963.
51. SLOMP, G., Jr. and B. R. McGARVEY: Nuclear Magnetic Resonance Studies on 6-Methyl Steroids. J. Amer. Chem. Soc. 81, 2200 (1959).
52. SLOMP, G. and F. A. MacKELLAR: Nuclear Magnetic Resonance Studies using Pyridine Solutions. J. Amer. Chem. Soc. 82, 999 (1960).
53. — — Nuclear Magnetic Resonance Studies on Some Hydrocarbon Side Chains of Steroids. J. Amer. Chem. Soc. 84, 204 (1962).
54. STERNHELL, S.: Long-range H'—H' Spin-Spin Coupling in Nuclear Magnetic Resonance Spectroscopy. Rev. Pure and Appl. Chem. 14, 15 (1964).
55. TRENNER, N. R., B. H. ARISON, D. TAUB and N. L. WENDLER: Proton Magnetic Resonance Shifts in Steroid D-Homoannulation. Proc. Chem. Soc. (London) 1961, 214.
56. WOO, P. W. K., H. W. DION and L. F. JOHNSON: The Stereochemistry of Chalcose, a Degradation Product of Chalcomycin. J. Amer. Chem. Soc. 84, 1066 (1962).
57. ZECHMEISTER, L. und L. v. CHOLNOKY: Untersuchungen über den Paprika-Farbstoff. X. Citraurin aus Capsanthin. Liebigs Ann. Chem. 530, 291 (1937).
58. ZEITSCHEL, O.: Über das Nerol und seine Darstellung aus Linalool. Ber. dtsch. chem. Ges. 39, 1780 (1906).
59. ZÜRCHER, R. F.: Protonenresonanzspektroskopie und Steroidstruktur. I. Das C-19-Methylsignal in Funktion der Substituenten. Helv. Chim. Acta 44, 1380 (1961).

(Received, March 31, 1965)

Namenverzeichnis. Index of Names. Index des Auteurs.

Kursiv gedruckte Seitenzahlen beziehen sich auf Literaturverzeichnisse.
Page numbers printed in *italic* refer to References.
Les chiffres en *italique* indiquent les pages de bibliographie.

Sachverzeichnis. Index of Subjects. Index des Matières.

Fortschritte der Chemie organischer Naturstoffe. Progress in the Chemistry of Organic Natural Products. Progrès dans la chimie des substances organiques naturelles. Herausgegeben von **L. Zechmeister,** California Institute of Technology, Pasadena, California, U. S. A.

Springer-Verlag / Wien · New York

Bisher erschienen:

Erster Band: Mit 41 Abbildungen im Text. VI, 371 Seiten. Gr.-8⁰. 1938.
Ganzleinen S 348.—, DM 72.25, $ 17.20

Zweiter Band: Mit 24 Abbildungen im Text. VII, 366 Seiten. Gr.-8⁰. 1939.
Ganzleinen S 348.—, DM 72.25, $ 17.20

Dritter Band: Mit 10 Abbildungen im Text. VI, 252 Seiten. Gr.-8⁰. 1939.
Ganzleinen S 264.—, DM 55.45, $ 13.20

Vierter Band: Mit 47 Abbildungen im Text. VIII, 499 Seiten. Gr.-8⁰. 1945.
Ganzleinen S 474.—, DM 99.10, $ 23.60

Fünfter Band: Mit 34 Abbildungen. VIII, 417 Seiten. Gr.-8⁰. 1948.
Ganzleinen S 305.—, DM 50.40, $ 12.—

Sechster Band: Mit 32 Abbildungen. VIII, 392 Seiten. Gr.-8⁰. 1950.
Ganzleinen S 338.—, DM 55.80, $ 13.30

Siebenter Band: Mit 12 Abbildungen. VII, 330 Seiten. Gr.-8⁰. 1950.
Ganzleinen S 325.—, DM 53.70, $ 12.80

Achter Band: Mit 47 Abbildungen. XI, 400 Seiten. Gr.-8⁰. 1951.
Ganzleinen S 427.—, DM 70.50, $ 16.80

Neunter Band: Mit 20 Abbildungen. XI, 535 Seiten. Gr.-8⁰. 1952.
Ganzleinen S 498.—, DM 82.50, $ 19.60

Zehnter Band: Mit 19 Abbildungen. IX, 529 Seiten. Gr.-8⁰. 1953.
Ganzleinen S 498.—, DM 83.—, $ 19.80

Elfter Band: Mit 67 Abbildungen. VIII, 457 Seiten. Gr.-8⁰. 1954.
Ganzleinen S 448.—, DM 74.80, $ 18.—

Zwölfter Band: Mit 15 Abbildungen. X, 550 Seiten. Gr.-8⁰. 1955.
Ganzleinen S 497.—, DM 82.80, $ 19.80

Dreizehnter Band: Mit 48 Abbildungen. XII, 624 Seiten. Gr.-8⁰. 1956.
Ganzleinen S 645.—, DM 107.50, $ 25.60

Vierzehnter Band: Mit 38 Abbildungen. VIII, 377 Seiten. Gr.-8⁰. 1957.
Ganzleinen S 450.—, DM 75.—, $ 17.85

Fünfzehnter Band: Mit 81 Abbildungen. VI, 244 Seiten. Gr.-8⁰. 1958.
Ganzleinen S 246.—, DM 41.—, $ 9.75

Weitere Bände siehe nächste Seite!

Zu beziehen durch Ihre Buchhandlung

SPRINGER-VERLAG / WIEN · NEW YORK

Fortsetzung von vorhergehender Seite

Sechzehnter Band: Mit 27 Abbildungen. VI, 226 Seiten. Gr.-8⁰. 1958.
Ganzleinen S 240.—, DM 40.—, $ 9.50

Siebzehnter Band: Mit 57 Abbildungen. X, 515 Seiten. Gr.-8⁰. 1959.
Ganzleinen S 498.60, DM 83.10, $ 19.80

Achtzehnter Band: Mit 65 Abbildungen. X, 600 Seiten. Gr.-8⁰. 1960.
Ganzleinen S 618.—, DM 103.—, $ 24.50

Neunzehnter Band: Mit 16 Abbildungen. VIII, 420 Seiten. Gr.-8⁰. 1961.
Ganzleinen S 490.—, DM 78.—, $ 19.50

Über den Inhalt der Bände gibt der Verlag bereitwilligst Auskunft.

Zwanzigster Band: Mit 33 Abbildungen. XIII, 509 Seiten. Gr.-8⁰. 1962.
Ganzleinen S 604.—, DM 96.—, $ 24.—

Inhalt: **Birkinshaw, J. H.,** and **C. E. Stickings.** Nitrogen-containing Metabolites of Fungi. — **Freudenberg, K.** Forschungen am Lignin. — **Schindler, O.** Die Ubichinone (Coenzyme Q). — **Mors, W. B., M. Taveira Magalhães** and **O. R. Gottlieb.** Naturally Occurring Aromatic Derivatives of Monocyclic α-Pyrones. — **Harborne, J. B.** Anthocyanins and their Sugar Components. — **Baschang, G.** Aminozucker, Synthesen und Vorkommen in Naturstoffen. — **Wiesner, K.** Structure and Stereochemistry of the Lycopodium Alkaloids. — **Narayanan, C. R.** Newer Developments in the Field of Veratrum Alkaloids. — **Vinograd, J.,** and **J. E. Hearst.** Equilibrium Sedimentation of Macromolecules and Viruses in a Density Gradient. — **Horowitz, N. H.,** and **S. L. Miller.** Current Theories on the Origin of Life.

Einundzwanzigster Band: Mit 14 Abbildungen. VII, 362 Seiten. Gr.-8⁰. 1963.
Ganzleinen S 479.—, DM 76.—, $ 19.—

Inhalt: **Bonner, J.** The Biosynthesis of Rubber. — **Oroshnik, W.,** and **A. D. Mebane.** The Polyene Antifungal Antibiotics. — **Muxfeldt, H.,** und **R. Bangert.** Die Chemie der Tetracycline. — **Brockmann, H.** Anthracyclinone und Anthracycline (Rhodomycinone, Pyrromycinone und ihre Glykoside). — **Jaenicke, L.,** und **C. Kutzbach.** Folsäure und Folat-Enzyme. — **Crombie, L.** Chemistry of the Natural Rotenoids.

Generalregister / Cumulative Index / Index Général. Bände I—XX. 1938—1962.
XVI, 369 Seiten. Gr.-8⁰. 1964. Ganzleinen S 378.—, DM 60.—, $ 15.—

Zweiundzwanzigster Band: Mit 8 Abbildungen. VII, 370 Seiten. Gr.-8°. 1964.
Ganzleinen S 554.—, DM 88.—, $ 22.—

Inhalt: **Schaffner, K.** Photochemische Umwandlungen ausgewählter Naturstoffe. — **Billek, G.** Stilbene im Pflanzenreich. — **Halsall, T. G.,** and **R. T. Aplin.** A Pattern of Development in the Chemistry of Pentacyclic Triterpenes. — **Grove, J. F.** Griseofulvin and Some Analogues. — **Scheuer, P. J.** The Chemistry of Toxins Isolated from Some Marine Organisms. — **Keller-Schierlein, W., V. Prelog** und **H. Zähner.** Siderochrome.

Subskribenten auf die „Fortschritte der Chemie organischer Naturstoffe" erhalten die Bände zu einem um 10% ermäßigten Vorzugspreis.

Zu beziehen durch Ihre Buchhandlung